OFFSHORE ENGINEERING

ELECTRICAL

VOLUME 1

Michael J. Dennis

Published by New Generation Publishing in 2013

Copyright © Michael J Dennis 2013

First Edition

www.newgeneration-publishing.com

New Generation Publishing

About This Book

This book, in two volumes, has been produced for engineers and technicians in the electrical discipline who work, or wish to work, in the offshore oil and gas industry.

The book aims to provide all of the knowledge elements required by electrical engineers and technicians who will be assessed as authorised electrical persons, to work on offshore rigs and production platforms. Naturally the specific systems and company procedures for such assessment can only be learnt on the job.

Knowledge is provided from elementary electrical theory, power generation and distribution, electrical systems and units, safety, protection and commissioning; covering all aspects of electrical engineering offshore. Where it is considered helpful, reference is made to the differences or similarities to onshore systems. All parts of the book are fully illustrated.

Self assessment questions and answers are provided at the end of each section, to ensure knowledge gain and retention.

A comprehensive contents list is provided at the start of the book as well as an index at the end. Thereby each volume becomes an ongoing reference tool.

The author spent 24 years in the offshore oil and gas industry. He has worked offshore as an electrical engineer on production platforms, and has been responsible for the electrical training of engineers and technicians from many drilling and production companies operating in the UK and Norwegian sectors of the North Sea, the Middle East, Africa and the Far East. He therefore has extensive knowledge of electrical engineering gained on the job, as well as from writing training programmes, lecturing and training and assessment of engineers and technicians throughout the offshore production and electrical disciplines.

Contents
Volume 1

PART 1
ELECTRICAL THEORY

CHAPTER 1 ELECTRON THEORY

1.1   STRUCTURE OF THE ATOM

It used to be a common thing for people to say 'We all use electricity, but no one knows what it really is'. This is no longer true, ever since, at about the turn of the century, Lord Rutherford gave to the world his theory of the structure of the atom, shown in Figure 1.1.

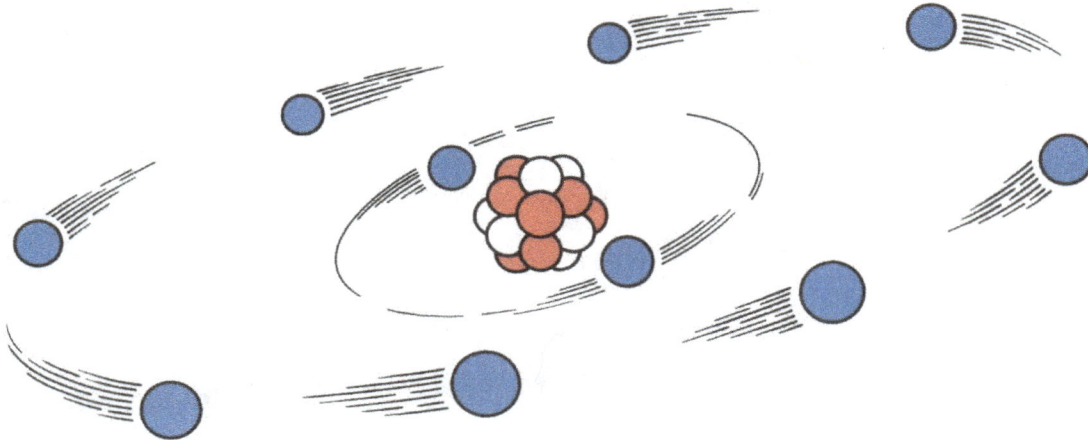

FIGURE 1.1
NEON ATOM

He said that every atom had a nucleus carrying positive electric charges, and around it, circulating in orbit, were one or more electrons each carrying a single negative electric charge. Moreover the number of negative electrons was such as exactly to neutralise the positive charge on the nucleus. The attraction between the negative electrons and the positive nucleus keeps the fast-moving electrons in orbit exactly like the gravitational attraction between the planets and the sun. And, like in the solar system, the electrons (planets), coloured blue in the figure, are much smaller than the nucleus, being about one-thousandth of its size.

The nucleus itself is not a single element but consists of many particles called 'protons', each one carrying a single positive electric charge. The protons are shown coloured red in the figure, so that normally in all matter there are equal numbers of protons in the nucleus and of the orbiting electrons.

Later it was discovered there was a third type of nuclear element - the 'neutron'. This is a particle, about the same size as a proton, but without any electrical charge, so that it has no effect on the number of orbiting electrons but merely makes the atom heavier. These neutrons appear white in the nucleus of the figure.

The chemical behaviour of all the elements which constitute matter depends on the number of orbiting electrons. This number varies from 1 in the lightest element (hydrogen) to 92 in the heaviest (uranium). Until recently it was thought that the list stopped there, but advances in nuclear physics have identified elements with up to 103 electrons in man-made matter; plutonium has 94.

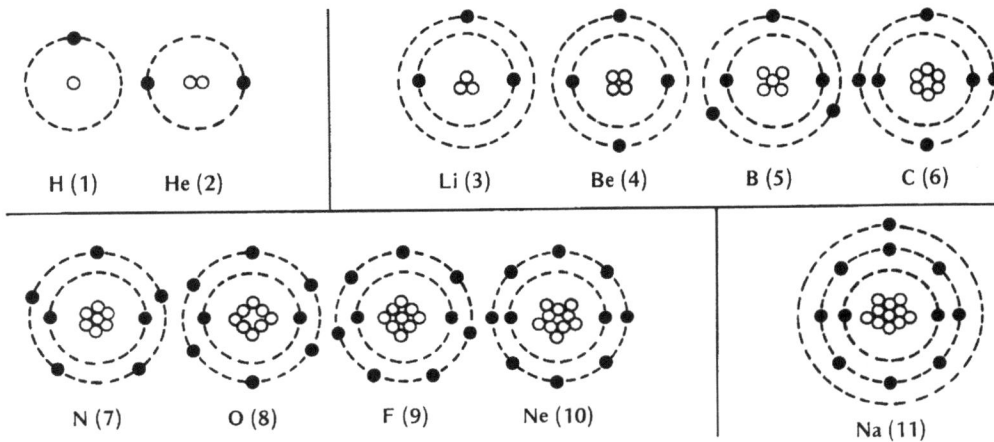

FIGURE 1.2
ELEVEN LIGHTEST ELEMENTS

Figure 1.2 shows the first eleven elements of the list. The lightest, hydrogen, has one proton and one electron. The next, helium, has two of each, then comes lithium with three, beryllium with four, and so on to the metal sodium (Na) with eleven. Some of these elements are well known (hydrogen (1), carbon (6), nitrogen (7), oxygen (8)); others are less common. The figure shows only the protons in the nucleus. There are in most cases also neutrons, but as they do not affect the orbiting electrons they are not shown.

One peculiar thing should be noted about the electron orbits: they form themselves into rings (or more strictly 'shells'). The innermost ring cannot contain more than two electrons. Any additional ones go into a second ring (like lithium) until that ring is full, It cannot contain more than eight (neon), after which a third ring begins to fill (sodium). That too cannot contain more than eight, after which a fourth ring begins to fill. This has a most important effect on the behaviour of electricity.

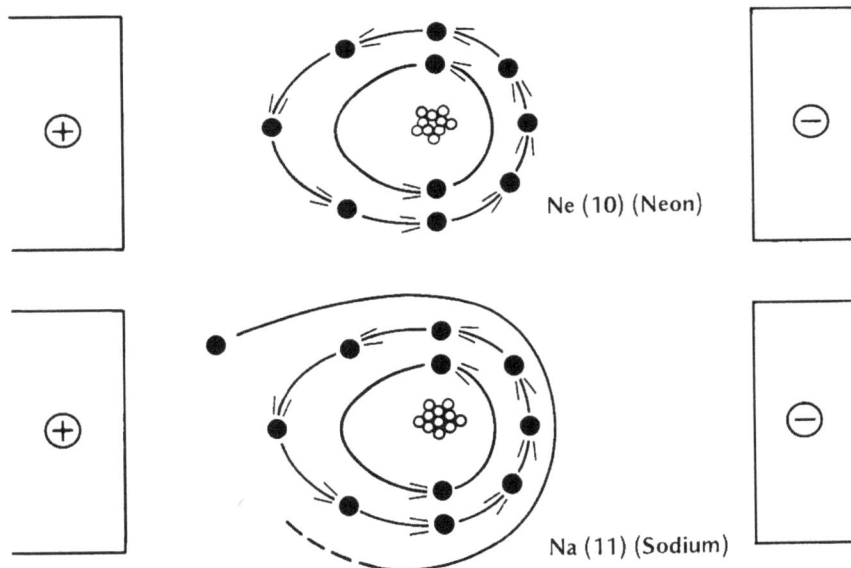

FIGURE 1.3
ATOMS UNDER ELECTRIC FIELD

When a ring is full, such as with helium (2) or neon (10), the electrons are tightly bound together and are difficult to displace. In the upper part of Figure 1.3 a neon atom is shown under the influence of a strong electric field. The negative electrons are attracted towards the positive end, but, though their orbits are distorted, they are not broken up. Such elements form the 'inert gases' such as helium, neon, argon, which will not combine with anything.

In the lower half of the figure is a sodium atom in the same electric field. Here there is a lone electron in the third ring, and it is quite loosely bound to the nucleus. Quite a small electric field is sufficient to break it out of its orbit and cause it to seek the positive pole. Such elements, where electrons can move easily, form many of the 'metals'; they are used for most electric conductors.

It is easy to see now why metals conduct electrons easily whereas many other elements do not - the latter are called 'insulators'.

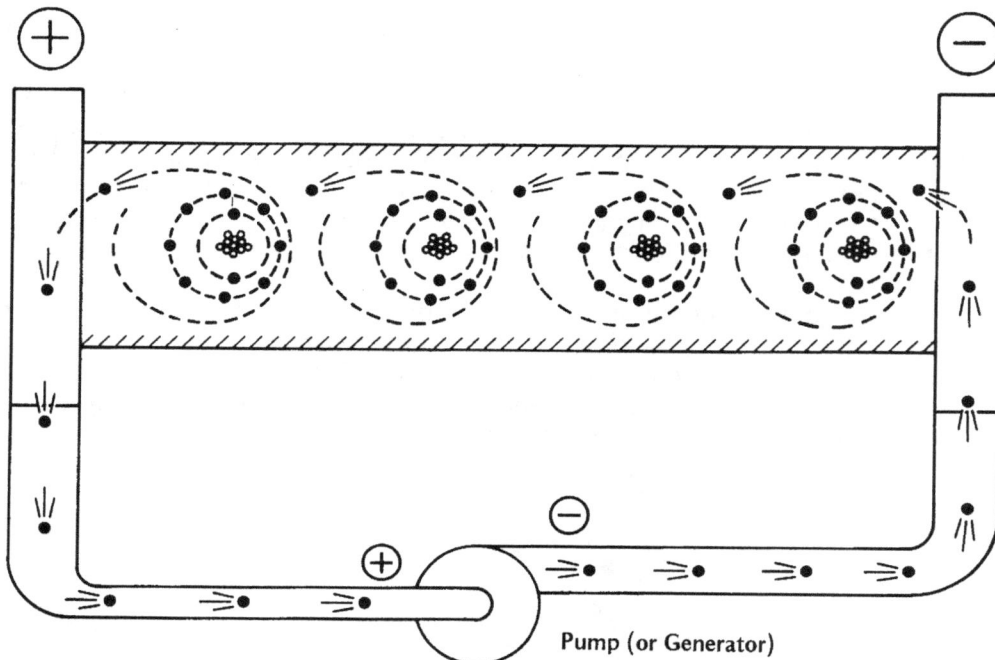

FIGURE 1.4
ELECTRON CIRCULATION

Figure 1.4 shows a conductor between two poles of an electric field. This field is maintained by an electron 'pump', or generator, which is driving them round from left to right in the pump and from right to left in the conductor. If the conductor is a metal, the electric field continually breaks off electrons which pass from one atom to the next and so on back to the pump. There is a continuous flow of electrons through the conductor and round the loop - we say there is an 'electric current' in the circuit.

For reasons going back into history the charge on an electron was found to be negative, so it is always attracted towards the positive pole - that is to say, in Figure 1.4 the electron flow in the conductor is from right to left, and the left-hand pole must be the positive. But convention has ordained that currents flow from positive to negative, and therefore the conventional electric current is said to be from left to right - that is, against the electron flow. The original decision was a disaster, but it is too late to change now, and this is a cross we just have to bear. In this book current will always be considered as flowing from positive to negative.

CHAPTER 2 PRIMARY AND SECONDARY CELLS

2.1   ORIGINAL VOLTAIC CELL

About the year 1799 Alessandro Volta, an Italian scientist (after whom the unit 'volt' is named), discovered that, if two dissimilar metals were separated by a certain liquid which was in contact with both, an 'electromotive force' (emf), or 'voltage', appeared across the metals and, when they were connected together, an electric current flowed. Volta used zinc and copper discs separated by a piece of cloth moistened with a solution of common salt. Later the voltage was found to be 1.0V no matter what the size of the metal discs. Of course at that time no such unit as the 'volt' was known. This was the original 'Voltaic Cell' and was used, with modifications, for 150 years as almost the sole source of electric currents for the experimenters of the day.

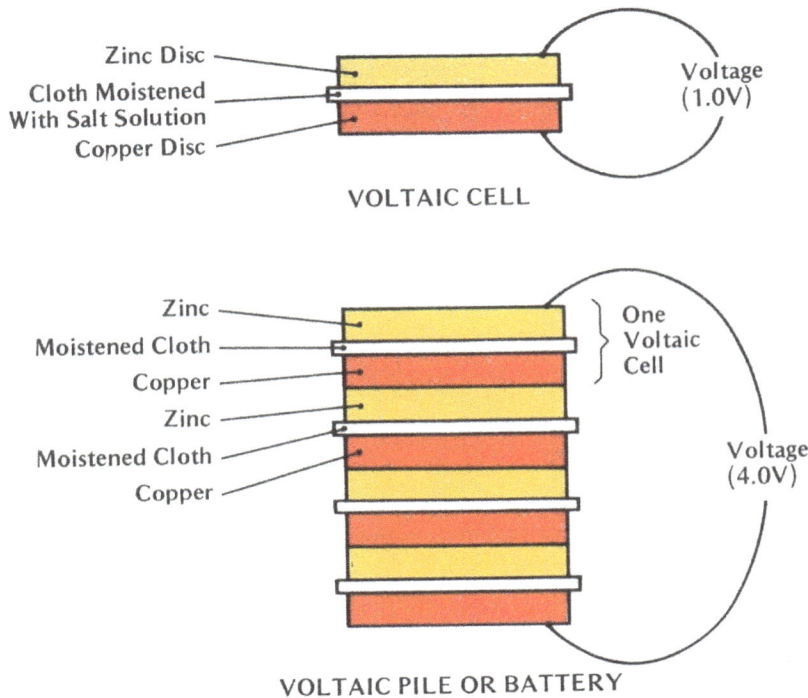

VOLTAIC CELL

VOLTAIC PILE OR BATTERY

FIGURE 2.1
PRIMARY CELLS

If several such cells are arranged in a pile as in the lower part of Figure 2.1 (as Volta did), each cell is in fact in series with its neighbour, and the voltages of each individual cell add together to give a much higher voltage than could be achieved with only one. Such a series collection is referred to as a 'battery of cells' (by analogy with a battery of guns) - or nowadays simply as a 'battery'.

2.2   PRIMARY CELLS

Although Volta used zinc and copper as his main elements, almost any two dissimilar metals will produce a like result. Only the voltages would be different, depending on the metals actually used.

Such voltaic cells, whatever the materials used, are termed 'primary cells.' They do not have to be charged, as in a modern battery, but derive their electrical energy from chemical action between the metals and the surrounding conducting liquid, called the 'electrolyte'. As power is extracted, one of the two metals gradually erodes away as the chemical action proceeds, and the process is not reversible - that is to say, the cell cannot be recharged. It will continue to give power until the erosion is complete.

An example of the application of primary cells was in the domestic bell systems of some old houses, which used a battery of 'Leclanche' cells. Each consisted of a glass jar filled with sal ammoniac and containing a zinc rod and a carbon plate inside a porous pot. After prolonged but intermittent use the zinc rod eventually eroded away and had to be replaced.

Certain types of primary cell are still used in laboratories as a voltage reference; they are called 'standard cells', and their voltage is very accurate and constant. One such is the 'Weston Standard Cell'. Its electrodes are mercury (positive) and mercury-cadmium amalgam (negative) with an electrolyte of a solution of cadmium sulphate. The emf of this cell is 1.0183V at 20°C, varying by only one part in 25 000 per °C change. Such a cell is never used as a source of current, it is purely a very accurate voltage reference.

A common use for the primary cell today is the so-called 'dry battery'. This uses zinc and carbon for its two elements, the zinc forming the outer case and the carbon taking the form of a rod down the centre. The electrolyte is a stiff moist paste packed into the space between the zinc case and carbon rod. It is not strictly dry, but it cannot spill and can be sealed in the case.

Except in such special applications primary cells are little used in power plants today, being replaced by the 'secondary cell', which is rechargeable.

2.3 SECONDARY CELLS

Like the primary cell, the secondary cell, also sometimes called an 'accumulator' or 'storage battery', consists of two metal plates immersed in a conducting liquid electrolyte. They fall into two types: 'lead-acid' and 'alkaline'.

Lead-acid Cells

In the original lead-acid cell both plates were of pure lead, immersed in dilute sulphuric acid. In this state the metals are not dissimilar, and no emf (i.e. voltage) is developed between them. But after an external 'charging' current has been passed through the cell from one plate to the other, a chemical action takes place; one plate becomes covered with spongy lead and the other with lead peroxide. In this state the cell behaves like a primary one, the spongy lead plate being the negative and the lead peroxide the positive. When these plates are connected together externally, a current will flow, driven by the cell's emf or internal voltage.

This construction was soon found to be rather unsatisfactory, as the plates' coatings rapidly fell off. Now each plate consists of a grid of lead-antimony alloy, into which the active materials are inserted in the form of a hard paste.

As shown in Figure 2.2, positive and negative groups of plates are interleaved. The capacity of the cell is determined by the number and total surface area of the plates, which are kept from touching each other by 'separators'. These were originally of wood, but nowadays porous plastic is often used. If, due to too heavy a rate of discharge, some of the active material becomes dislodged from its grid, it will gradually accumulate in the bottom of the cell until its level rises to touch the bottoms of the plates. This will short-circuit them and rapidly discharge the battery. If the cell is sealed, its life is finished, but if the plates can be removed, the cells can be flushed out, refilled and used again.

Negative Terminal

Vent/Filler Cap

Positive Terminal

Negative Plate

Positive Plate

Separator

Case

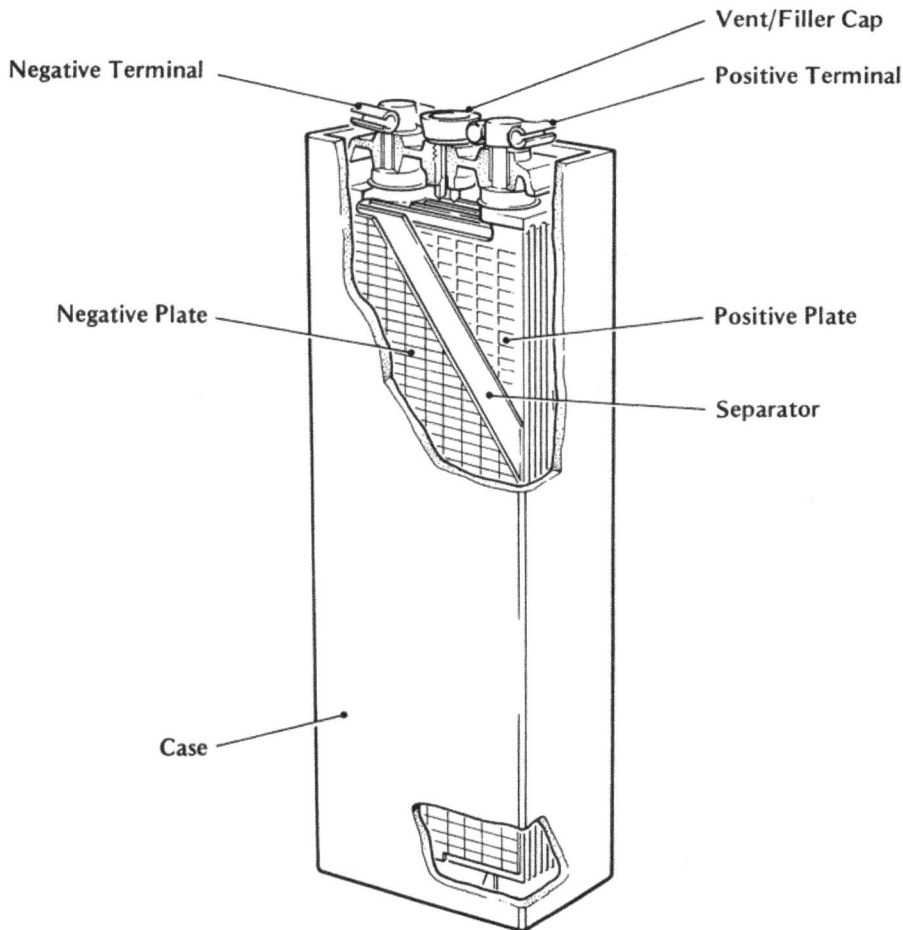

FIGURE 2.2
TYPICAL LEAD ACID CELL

The sulphuric acid electrolyte is diluted to a relative density between 1.200 and 1.300, depending on the make and duty of the battery. When fully charged, an open-circuit voltage of about 2.1V appears across the cell's terminals.

As power is taken out, derived from the chemical action, the sulphuric acid begins to form lead sulphate at both plates and in doing so loses density. The cell voltage also falls steadily. Once the acid density has fallen to 1.150 (the figure differs somewhat with different makes of cell), the cell is regarded as 'discharged', and further extraction of power could damage it.

It must then be recharged (not possible with primary cells). The current is reversed by applying an external voltage to the cell higher than the cell's own voltage. The chemical action is thereby reversed, the lead sulphate returning to the electrolyte and raising its density again to its original level, at the same time restoring the cell voltage.

If a cell is left in a discharged state for long periods, the lead sulphate on both plates will harden; the cell is then said to be 'sulphated'. In this condition attempts to recharge it may not remove the sulphate coatings, in which case the cell has become unfit for further use. **Lead-acid cells must never be left for long periods in a discharged state.**

The ability to discharge and recharge is a property only of the secondary cell. Its efficiency is such that it takes about 1.4 times the power to recharge it as can be obtained out of it on discharge. The lost power appears as heat.

Another property of the secondary cell is that, if discharged too quickly, not only is active material liable to be dislodged from the plates but also the normal chemical processes are upset and limit the output obtainable. For this reason the rated discharge current is always specified as a certain maximum rate - for example 50A at a 5-hour rate, or 70A at a 10-hour rate. If these rates are exceeded, the cells will not give their full output for the stated times.

The capacity of a cell is expressed in 'ampere-hours' (Ah). For example, if a battery is rated 300Ah, it should sustain a discharge rate of 1A for 300 hours, or 5A for 60 hours, or 10A for 30 hours, so long as the discharge rate does not exceed the stated rating. If it does, the rated ampere-hour capacity will not be achieved.

The easiest way to test a lead-acid cell's state of charge is to measure the electrolyte density with a hydrometer. The full-charge density varies somewhat between makes of battery, but it is always stated by the maker. A typical figure is 1.250.

Since a cell can only lose liquid by evaporation or by 'gassing', it is only water that is lost, not the acid. Therefore liquid should only be replaced by topping up with pure (distilled) water, never by acid (unless the acid has actually been spilt).

When a secondary cell is recharged, the power going into it is at first absorbed by causing chemical changes. But once these changes are complete the power, if continued, starts to electrolyse the water, forming hydrogen and oxygen at the two plates; the cell is said to be 'gassing'. Gas bubbling is a sign of a completed charge, but if allowed to continue it will release a considerable quantity of hydrogen and oxygen mixed in its most explosive proportions.

Ventilation of battery rooms is therefore very important, and charging should on no account continue, at least at full rate, unless ventilation is available. Most platforms now have ventilation monitors which stop, or at least reduce, charging on failure of air flow.

Alkaline Cells

A different class of secondary cell is known as 'alkaline'. It was invented by Edison and originally consisted of plates of nickel hydrate and iron oxide immersed in an electrolyte of a solution of potassium hydroxide (with some lithium hydroxide added) in distilled water. These were the original nickel-iron (or 'NiFe') cells; they had a cell voltage of about 1.2V.

The chief advantages of the alkaline call over the lead-acid are:

- longer life

- greater reliability

- less maintenance

- lighter weight per Ah

- greater robustness against vibration and shock

- good high-rate discharge performance

- ability to accept high rates of charge

- better charge retention during long periods of rest

- rapid voltage recovery after heavy discharge

- immunity from harm if over-discharged.

Alkaline cells differ from lead-acid not only in their voltage but also because their electrolyte does not lose density as the cell discharges. The electrolyte is necessary to sustain the chemical actions but is not itself affected - it is in fact a 'catalyst'. The only way to know the state of charge is to measure the cell voltage. When it has fallen to about 0.8V, the cell is considered to be discharged.

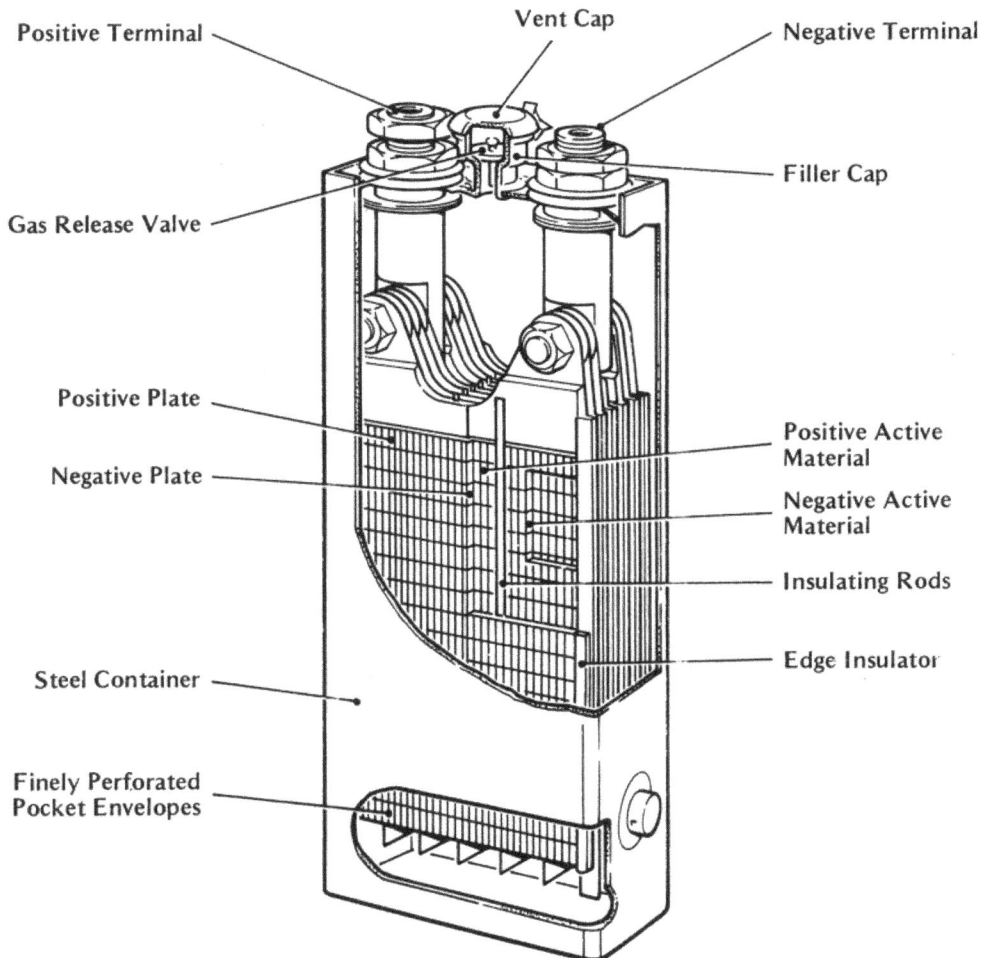

FIGURE 2.3
TYPICAL ALKALINE (NICKEL-IRON) CELL

A typical alkaline cell is shown in Figure 2.3. During charging, the positive nickel hydrate plates become heavily oxidised, whereas the negative iron oxide plates are reduced to pure iron, and the cell voltage rises to about 1.2V. This amounts simply to the transfer of oxygen from one plate (the positive) to the other and does not call for any chemical changes in the electrolyte. It is for this reason that in an alkaline cell the electrolyte density does not change with the state of charge. On discharge the chemical process is reversed, oxygen returning to the iron to re-form it into iron oxide, and in so doing the cell voltage is reduced.

Like the lead-acid cells, alkaline cells are rated in ampere-hours at a specified maximum discharge rate. However, they are very robust and can stand heavy discharges without damage, though in that case they will not give their full rated ampere-hour output.

More recently the iron in alkaline cells was replaced by cadmium to give the nickel-cadmium ('Nicad') cell. Most platforms and shore installations use this type exclusively rather than the lead-acid or nickel-iron. It has about 20% higher voltage per cell.

The foregoing description of the nickel-iron cell applies in large part also to the nickel-cadmium, except that the cell's open-circuit voltage lies between 1.4 and 1.5V, and at 1.1V the cell is regarded as discharged.

The nickel-cadmium cells are manufactured in either plastic or steel containers. The latter have greater strength against severe vibration and shock and also have advantages when operating in extreme climates. Plastic containers on the other hand are completely free from corrosion, especially in salt-laden atmospheres, and, being translucent, allow the electrolyte level to be checked at a glance and the electrolyte topped up if necessary.

2.4   CHARGING OF SECONDARY CELLS

The apparatus for charging secondary cells and the voltage variations during charge are described in Part 3 Electrical Power Distribution.

CHAPTER 3 MAGNETISM AND ELECTROMAGNETISM

3.1   MAGNETISM

Long ago, in the middle ages, it was found that a mineral called 'lodestone', which is in fact an iron ore, attracted small iron objects. So 'magnetism', which is a natural phenomenon, was discovered. It got its name from a district in Asia Minor called 'Magnesia' where lodestone was found.

It was found too that an iron bar or needle, if rubbed with lodestone, could also be made to attract small pieces of iron - that is, the magnetism could be imparted from the lodestone to the iron. Such a needle, if placed on a wooden raft and left to float in a bowl of water, always tended to lie in a rough North to South direction. So we had the first primitive compass.

At this stage it never occurred to anybody that there was any connection whatever between magnetism and electricity, and it was left to Faraday to bring these two together in 'electromagnetism'.

3.2   ELECTROMAGNETISM

In 1820 Oersted discovered that an electric current flowing in a wire caused a magnetic field around it. This can easily be detected by placing a small compass near the wire and observing the movement of the needle when current is switched on. This is shown in Figure 3.1.

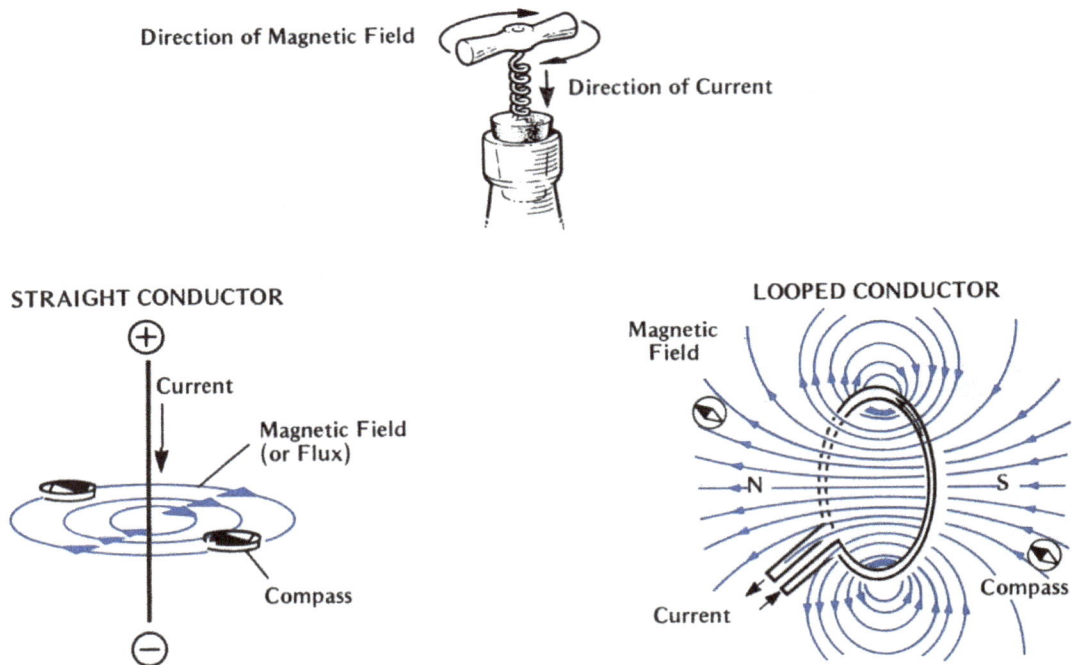

FIGURE 3.1
MAGNETIC FIELD AROUND A CONDUCTOR

The effect can be intensified by bending the wire into a loop. The magnetic fields from each bit of the wire are brought together inside the loop, where the magnetic field is concentrated and intensified.

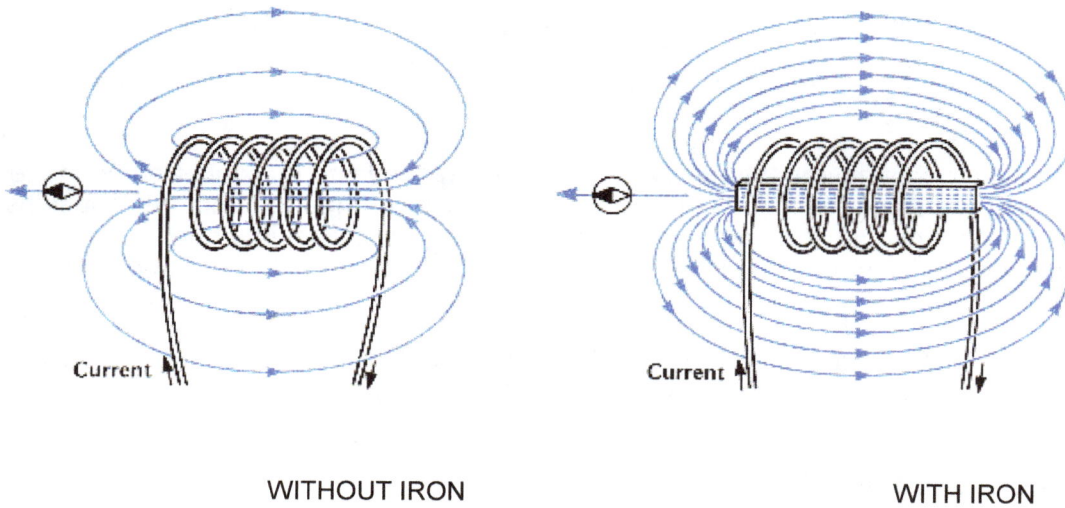

WITHOUT IRON   WITH IRON

FIGURE 3.2
MAGNETIC FIELD AROUND A COILED CONDUCTOR

If now the wire is bent into several loops, or a 'helix', as shown in Figure 3.2, the magnetic fields of each 'turn' are superimposed, and the field down the middle is still further intensified. The result is a 'coil' which, when current flows in it, produces an artificial magnet, called an 'electromagnet'. Unlike a natural magnet, whose magnetism is always present, an electromagnet can be switched on or off at will.

If iron is introduced inside the coil, the magnetic strength is still further increased, and 'permanent' magnets can be made this way. Very powerful electromagnets can be built, which are widely used: they can actuate solenoids or valves directly; they can drive any device needing a fore-and-aft motion; and they are used with cranes in scrap-yards for picking up large weights of scrap-iron. On a smaller scale they are used to operate relays and switching devices.

Although it may not at first seem so, solenoids and other electromagnets operate just as well with alternating as with direct current.

CHAPTER 4 SIMPLE D.C. CIRCUIT - OHM'S LAW

4.1   VOLTS AND AMPS

In Figure 4.1 the electron flow was maintained by a 'pump' or generator in an anti-clockwise direction, but, because of a historical mistake, current was deemed to flow in a clockwise direction.

In an equivalent water circuit, the pump delivers a pressure (measured in psi, kgf/cm^2 or bars), and the water flow is measured in gal/min, or m^3/s or other unit.

So for an electric circuit pressure is measured in VOLTS (after Volta, the early Italian experimenter) and current in AMPERES (after an early French pioneer). Instruments are made which indicate pressures in volts and currents in amperes - all switchboards have voltmeters and ammeters.

On platforms and large installations pressures tend to be very high, involving thousands or tens of thousands of volts. In those cases the 'kilovolt' (equals one thousand volts) is usually taken as the unit of pressure. On most platforms the main generation pressure is 6.6kV, or 6.6 thousand volts. For domestic appliances and small services 440 or 250 volts is usual on platforms and 415 or 240 volts ashore.

4.2   CURRENT FLOW - OHM'S LAW

Once the units of pressure and current flow were established, a German experimenter named Georg Simon Ohm discovered a very important relationship between them.

It has already been seen that some materials (mainly metals) allow electrons to move freely (but not all as freely as each other), whereas others do not do so and tend to resist such movement - again some more so than others.

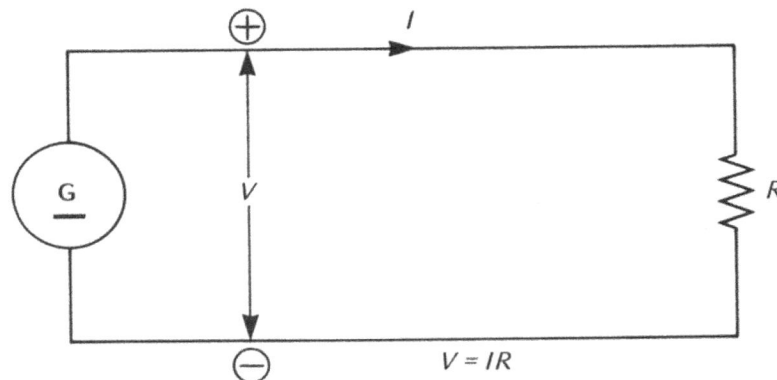

FIGURE 4.1
OHM'S LAW (D.C.).

Ohm discovered that, for a given sample of material, the current flowing I (in amperes) was directly proportional to the pressure applied V (in volts). In other words, for that given sample, the ratio of voltage to current was constant:

$$V/I = \text{const}$$

This was true for any one sample, but the constant itself differed from sample to sample. The ratio is called the 'resistance' of that sample, symbol R. It can be considered as opposition to the flow of electrons - like friction.

Ohm's Law can then be stated:

$$V/I = R$$

or

$$V = IR$$

where R is the resistance of the sample and differs from sample to sample. If V is measured in volts and I in amperes, R is measured in 'ohms'.

4.3   HEATING

An important result stems from this. Since the resistance R of a conductor is akin to friction in the mechanical equivalent, it might be expected that loss of energy by heating might result from a current flow.

This indeed is so. Whenever current is forced by pressure of voltage to flow through a conductor which has resistance (and all conductors do, even metals), heat is generated in that conductor. The rate of heat generation is proportional to the resistance (in ohms) and to the **square** of the current (in amperes squared). That is to say, the heat generated is I^2R, and, since it represents a continuing loss of energy, it is expressed in the energy-rate unit 'watts' (W).

It is important to remember that current flowing in **any** conductor, be it cable, generator, motor or transformer, gives rise to heat, which must be conducted away if the temperature is not to rise to a level which can damage the insulation and possibly lead to flashover or breakdown and severe damage, or even danger to life.

To reduce heat generation either the current (I) or the resistance (R) must be reduced (for example, by increasing the cross-section of the conductor).

CHAPTER 5 RESISTANCES IN SERIES AND PARALLEL

5.1 SERIES CIRCUITS

The term 'in series' means that two or more circuits are supplied one after the other in any single circuit, as shown diagrammatically in Figure 5.1.

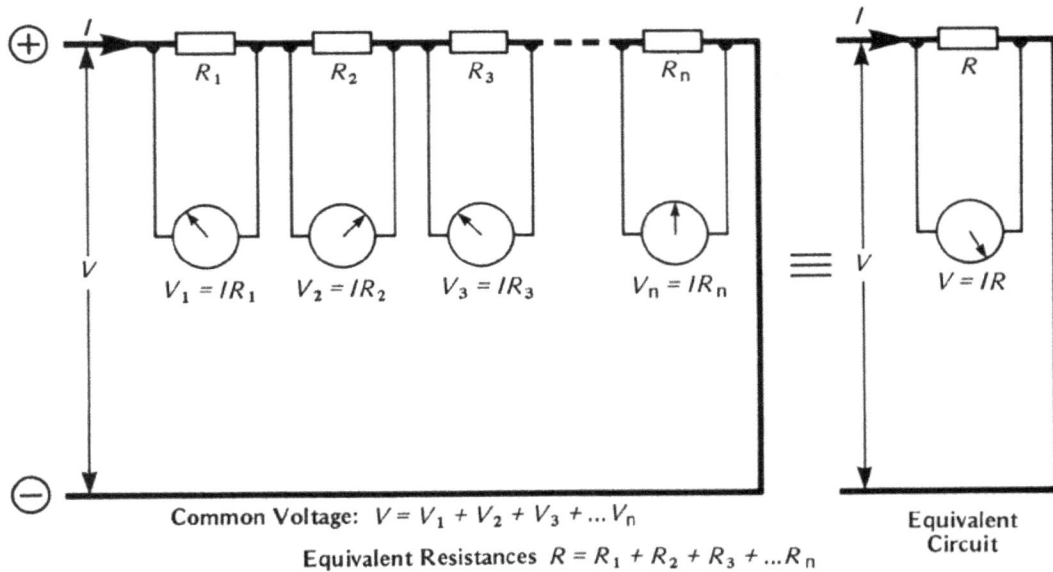

Common Voltage: $V = V_1 + V_2 + V_3 + ... V_n$

Equivalent Resistances $R = R_1 + R_2 + R_3 + ... R_n$

Equivalent Circuit

FIGURE 5.1
SERIES RESISTANCES

Since there is only one single path from the power source through the circuits and back again, the same current flows through all. The voltage, or 'pressure', is reduced by resistance according to Ohm's Law: each circuit element causes a 'voltage drop' across it, very similar to the 'loss of head' due to fluid flow in a hydraulic system. Also the sum of the individual volt-drops is equal to the applied voltage.

By Ohm's Law the volt-drop V_1 across the load R_1 is

$$V_1 = I \times R_1$$

and similarly $\qquad V_2 = I \times R_2,$ $\;I$ being the same for each of the cases.

Therefore: $\qquad V_1 + V_2 + ... = IR_1 + IR_2 + ...$

$$= I(R_1 + R_2 + ...)$$

But the sum of the individual volt-drops $V_1 + V_2 + ... = V$ (the applied voltage).

By Ohm's Law for the equivalent circuit $V = IR.$

Hence: $\qquad IR = I(R_1 + R_2 + ...)$

or $\qquad R = R_1 + R_2 + ...$

15

That is to say, the total resistance of a series circuit is equal to the sum of all the individual resistances.

It is evident that the failure of any single component in a series circuit interrupts the supply to all; also that each element of load must work at a reduced voltage. For these reasons the series arrangement of loads is seldom used in power circuits.

5.2 PARALLEL CIRCUITS

The term 'in parallel' means that the circuits are so arranged that there is a separate path through each, as shown in Figure 5.2.

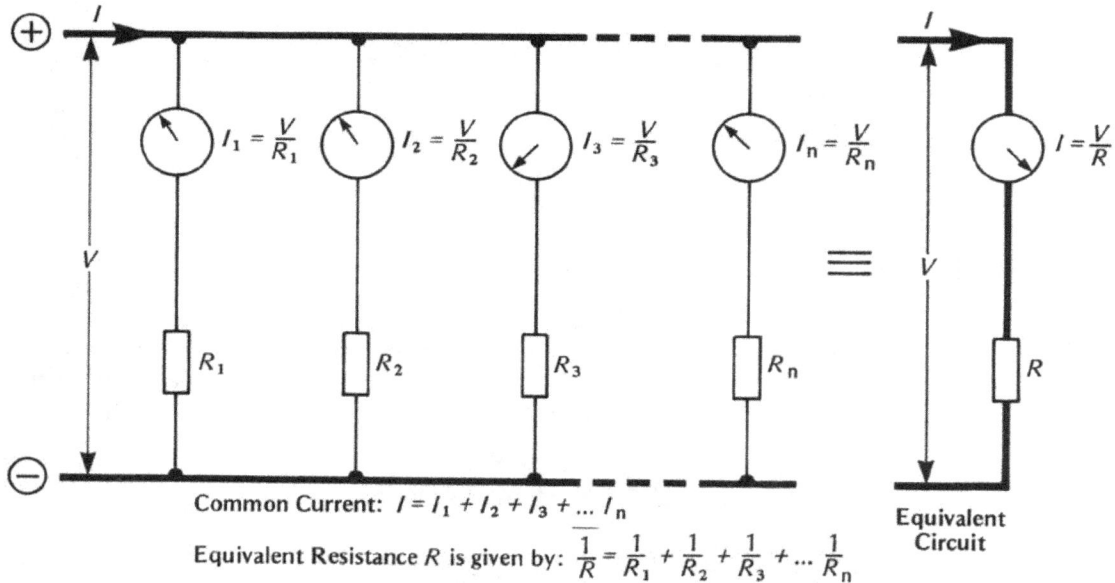

Common Current: $I = I_1 + I_2 + I_3 + \dots I_n$

Equivalent Resistance R is given by: $\dfrac{1}{R} = \dfrac{1}{R_1} + \dfrac{1}{R_2} + \dfrac{1}{R_3} + \dots \dfrac{1}{R_n}$

FIGURE 5.2
PARALLEL RESISTANCES

The voltage applied to every circuit element is the same throughout. The total current divides between the circuits according to the resistance of each element, so that the current flowing through each individual circuit is less than the total, and the sum of the currents flowing through the individual elements is equal to the total available current.

By Ohm's Law the current I_1 flowing through the load R_1 is:

$$I_1 = V/R_1$$

and similarly V being the same for each of the loads

$$I_2 = V/R_2$$

Therefore:

$$I_1 + I_2 + \dots = V/R_1 + V/R_2 + \cdots$$

which can be written

$$V(1/R_1) + V(1/R_2) + \dots$$

or

$$V(1/R_1 + 1/R_2 + \dots)$$

But the sum of the currents flowing in each individual circuit

$$I_1 + I_2 + \ldots = I \text{ (the total current)}$$

Hence:
$$I = V(1/R_1 + 1/R_2 + \ldots)$$

or
$$1/V = 1/R_1 + 1/R_2 + \ldots$$

But by Ohm's Law for the equivalent circuit
$$V = IR$$
or
$$I/V = 1/R$$

Hence:
$$1/R = 1/R_1 + 1/R_2 + \ldots$$

That is to say, the inverse of the equivalent resistance of a set of parallel circuits is equal to the sum of the inverses of each individual resistance.

It is evident that for the power engineer the parallel circuit has two important practical advantages. First, the failure of any element of load has no effect on the rest; they continue to receive a supply at the correct voltage and to draw the current which each individually requires. Second, all apparatus is supplied at the same voltage. Consequently, the parallel circuit is used almost exclusively for power supply in industrial plant.

It should be observed in passing that the characteristic of the series circuit, in which resistances in series have the effect of reducing the voltage at different points of the circuit, finds wide practical application in electronic apparatus such as radio, control, and 'solid-state' measuring equipment.

5.3   SUMMARY

If R is the single resistance equivalent to a number of individual resistances $R_1, R_2, R_3, \ldots$, then:

(a)   if $R_1, R_2, R_3, \ldots$, are in series, $R = R_1 + R_2 + R_3 + \cdots$

(b)   if $R_1, R_2, R_3, \ldots$, are in parallel, $1/R = 1/R_1 + 1/R_2 + 1/R_3 + \ldots$

CHAPTER 6 ELECTROMAGNETIC INDUCTION

6.1   FARADAY'S LAW

One day in 1837 Michael Faraday was working in his laboratory when by accident he dropped a magnet into a coil of wire which happened to be connected to a galvanometer. He noticed, to his surprise, that the galvanometer needle gave a kick when this happened. He was even more surprised to see, when he took the magnet out, that the needle kicked the other way.

This started a train of thought which finally led to a monumental discovery which was to become the whole basis of modern electrical engineering: it was the theory of 'Electromagnetic Induction'.

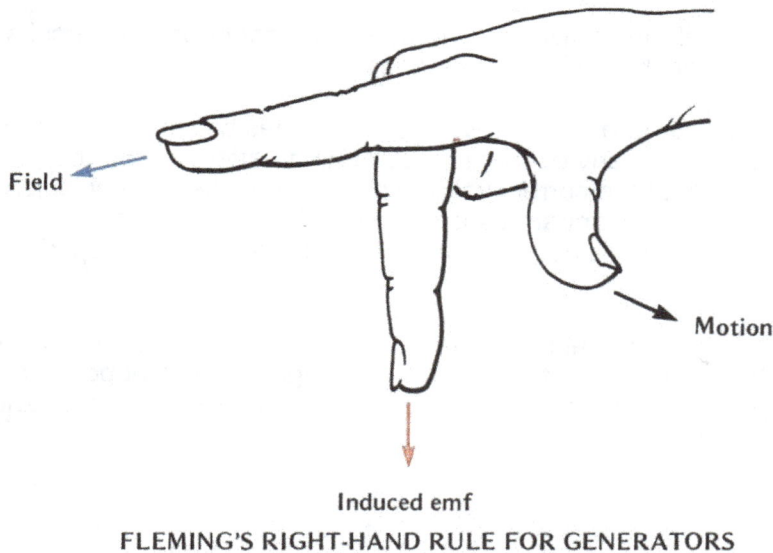

FLEMING'S RIGHT-HAND RULE FOR GENERATORS

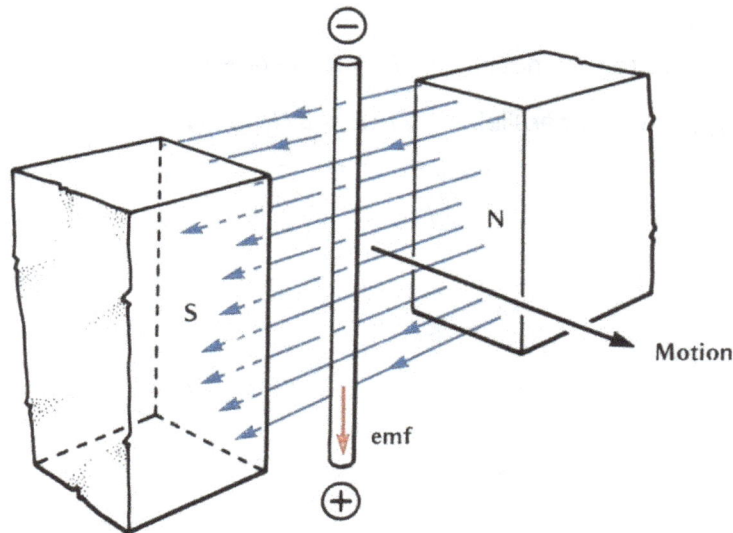

ELECTROMAGNETIC INDUCTION

FIGURE 6.1
FARADAY'S LAW OF ELECTROMAGNETIC INDUCTION

These are heavy words, but in short they mean that, if a conductor is moved in a magnetic field, then an 'electromotive force' (emf) - that is, a voltage - is induced in that conductor. This is shown in Figure 6.1. It follows that, if the ends of the conductor are connected to a load, then an electric current, driven by that voltage, will flow from the conductor, through the load and back again.

Whereas Oersted had shown that an electric current moving in a wire gives rise to an artificial magnetic field, Faraday showed the opposite - that if a wire moves in a magnetic field an artificial charge, or voltage, will be created in that wire. Electricity and magnetism were now firmly tied together by these two great discoveries.

Here then is the basis of electrical power generation. We start with a magnetic field, either a natural magnet or an artificial electromagnet of Oersted's type, and cause a conductor or a number of conductors to move past it, from which the current can be extracted as they are moving. How this is done in practice is discussed later, but first look at one other rule which determines how the directions of field, voltage and movement are related.

Figure 6.1 shows 'Fleming's Right-hand Rule for Generators'. If the right hand is held with the thumb, forefinger and centre finger extended mutually at right angles as shown in the figure, then, with the magnetic field in the direction (North to South) pointed by the forefinger and the motion of the conductor in the direction indicated by the thumb, the centre finger will point in the direction in which the emf (i.e. voltage) is induced in that conductor (and in which current will flow when connected to a load).

The magnitude of the voltage induced in the moving conductor depends on the strength of the magnetic field and the speed of movement, and on nothing else.

Use is made of these laws and rules when considering the Principles of Generation described in Chapter 14'.

CHAPTER 7 FORCES ON A CONDUCTOR

7.1   CURRENT-CARRYING CONDUCTOR IN A MAGNETIC FIELD

Figure 7.1 is similar to Figure 6.1 in that it shows a conductor in a magnetic field, but in this case there is a current from an external source being passed through that conductor.

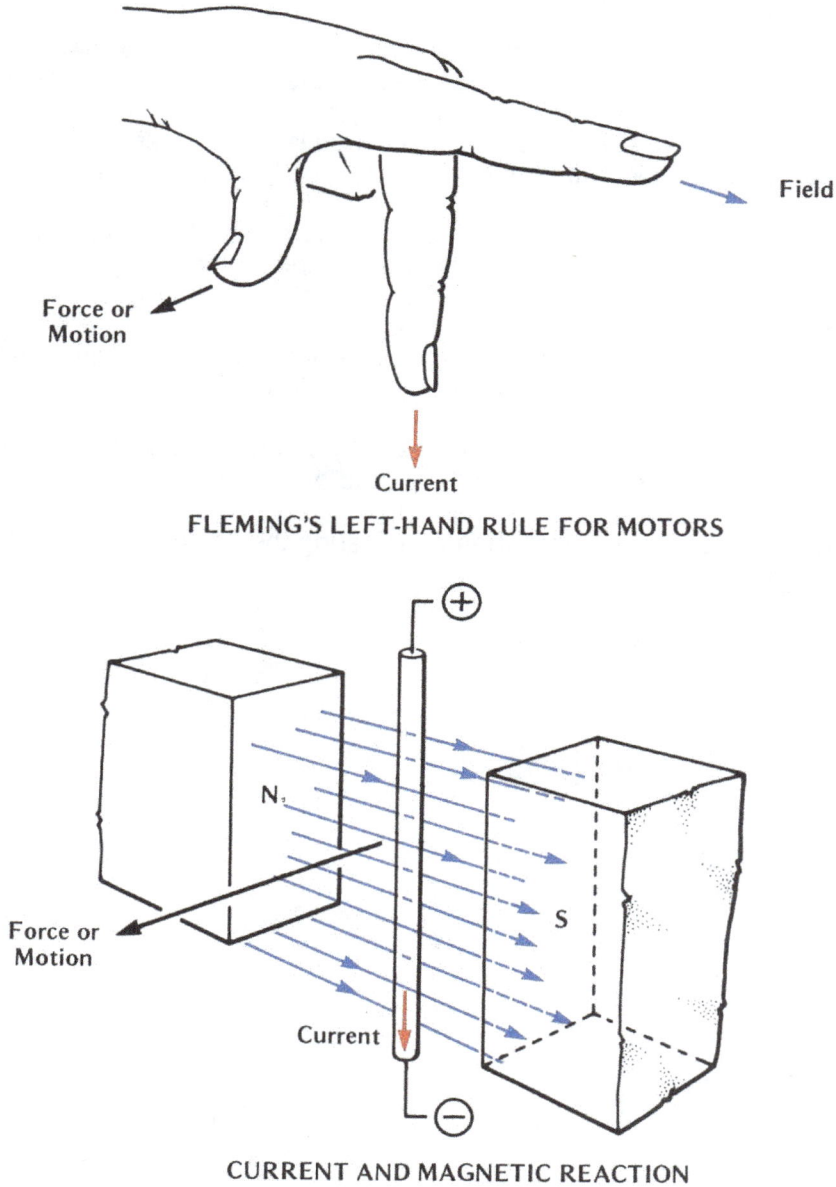

Field

Force or
Motion

Current

FLEMING'S LEFT-HAND RULE FOR MOTORS

N

S

Force or
Motion

Current

CURRENT AND MAGNETIC REACTION

FIGURE 7.1
FORCES ON A CONDUCTOR

The reaction between the current and the magnetic field through which it is passing causes a mechanical sideways force on the conductor. If the conductor is free to move, it will move sideways in the direction of the force. This is the basis of operation of all electric motors.

There is for motors a 'Fleming's Left-hand Rule' corresponding to his Right-hand Rule for generators. If the left hand is held with the thumb, forefinger and centre finger extended mutually at right angles, then, with the magnetic field in the direction (North to South) pointed by the forefinger and the direction of current in the direction indicated by the centre finger, the thumb will point in the direction of the mechanical force on the conductor (or of its motion if it is free to move).

The magnitude of the force on the conductor depends on the strength of the magnetic field and the strength of the current, and on nothing else.

Use is made of these laws and rules when considering the Principle of Operation of Motors described in Part 5 Electric Motors.

CHAPTER 8 INDUCTANCE

8.1   WHAT IS INDUCTANCE?

Wherever a magnetic field is produced by an electric current passing through a circuit, that circuit displays the phenomenon of 'inductance'.

Before looking at the effects of inductance on a d.c. circuit, it will be useful to see what is its nature by looking at a mechanical analogy.

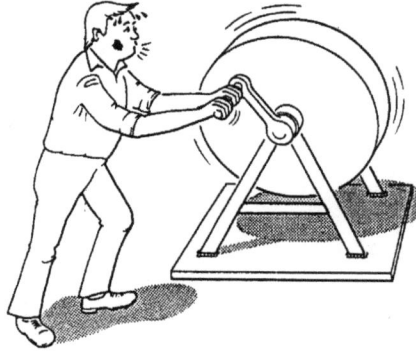

FIGURE 8.1
GRINDSTONE ANALOGY

Suppose there is a large grindstone with a turning handle (Figure 8.1). It is old, and its bearings are stiff and rusty, giving a lot of friction. If we try to turn the handle, even slowly, we must overcome this friction, causing heat and loss of energy at the bearings and making ourselves hot with the effort expended.

But there is another type of opposition to our attempts to turn the wheel - its inertia. It is heavy, and in order to accelerate it we must not only overcome friction but also provide it with an accelerating force in order that it shall gather speed. The greater the weight or inertia, the greater the force needed to accelerate. Also, the greater the acceleration desired, the greater the force we must apply. *(This is Newton's Second Law of Motion.)*

An electric circuit exhibits the same effects. It has resistance, and, in order for a current to flow, a pressure in the form of a voltage is needed to overcome it.

But an electrical circuit has inertia too. It opposes, like the grindstone, any attempt to speed up the current or to cause it to grow. And the faster it has to grow, the greater the voltage needed to be applied, quite apart from that needed to overcome resistance. This inertia in an electrical circuit is called 'inductance' and is due to the fact that any electric current causes magnetisation, and that effect is greatly increased by the presence of iron (which magnetises easily).

Some circuits, especially those without coils and without iron, have resistance but very little inductance. They are referred to as 'resistive circuits'. Others which have coils, and especially those with iron such as generators, motors and transformers, have both resistance and considerable inductance. They are referred to as 'inductive circuits'. In the fairly rare cases where the resistance is so small that it can be neglected compared with the inductance (say the grindstone with ball bearings) the circuit is called 'purely inductive'.

How inductance arises in a circuit due to its magnetisation and causes it to display electrical inertia, or 'sluggishness', is explained in the following paragraphs.

8.2   THE INDUCTOR (OR 'CHOKE')

Faraday's Law of Electromagnetic Induction, as explained in Chapter 6, states that, if a conductor moves in a magnetic field, an emf (or voltage) is induced in it. Such movement need only be relative; it is equally true if the magnetic field moves past a stationary conductor.

Movement implies change - that is to say, Faraday's Law applies also to any conductor around which the field is **changing**, that is, growing or decreasing.

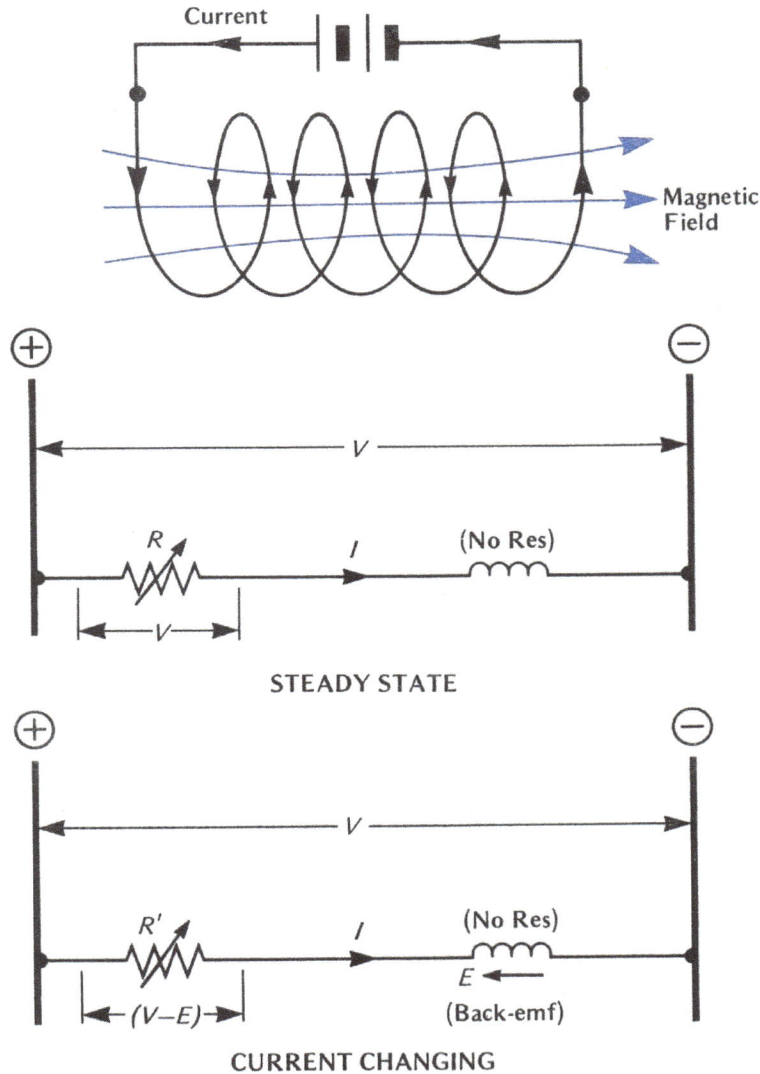

STEADY STATE

CURRENT CHANGING

FIGURE 8.2
INDUCTANCE AND BACK-EMF

Suppose there were a coil of wire through which a current is flowing, as in Figure 8.2. Then, by Oersted's principle, there is a magnetic field concentrated along its axis. If now the current started to **change** - say to increase - the magnetic field through all the turns of the coil would also be increasing. This is then a changing field which, by Faraday's Law, induces in each turn an emf (or voltage), and its direction would be such as to oppose the change - that is, to try to prevent the current in this case increasing.

What happens is shown diagrammatically in Figure 8.2. A voltage V is applied through a variable resistance R to the coil. For any given setting of R the current I through the coil (assumed to have no resistance of its own) is given by Ohm's Law:

$$I = V/R$$

If now R is decreased to R' with a view to increasing the current in the coil, the increasing current gives rise to an induced voltage E in the coil in a direction opposed to V. This induced voltage E is called the 'back-emf' of the coil. Consequently the net voltage appearing across R' is no longer V but is now $(V - E)$, and, by Ohm's Law:

$$I = (V - E)/R$$

Although R has been reduced to R', I is not proportionately higher because E reduces the effective voltage. In other words, Ohm's Law does not seem to apply in this case.

The back-emf E depends on the **rate of change** of current $\left(\frac{di}{dt}\right)$ through the coil and on the physical construction, including the number of turns, of the coil. It is written:

$$E = -L\frac{di}{dt}$$

As stated $\frac{di}{dt}$ is the rate of change of current (positive if increasing), and L is a property of the coil. The minus sign indicates that its direction opposes the increasing current, so that E is then negative.

L is called the 'inductance' of the coil; for any given coil it is a fixed quantity, but it differs from coil to coil. The presence of iron in the core increases L considerably. Inductance is measured in the unit 'henry' (H).

A coil carrying electric current, especially one with an iron core, becomes thereby magnetised, and an electromagnet is a store of energy. The energy stored in a coil of inductance L (in henrys) and carrying a current I (in amperes) is:

$$\tfrac{1}{2}LI^2 \ \text{(joules)}$$

(Compare the kinetic energy of a mass m moving with velocity v - it is $\tfrac{1}{2}mv^2$.) If ever the current in the coil is stopped, this energy has to be given up, in one form or another.

8.3 SWITCHING AN INDUCTIVE CIRCUIT WITH RESISTANCE

A special case arises when a voltage is suddenly switched on to a circuit containing resistance R and an inductance L (assumed to have no resistance of its own). Before the switching no current at all was flowing. When the switch is closed the current starts to flow and tries to build up, but this change is opposed by a back-emf proportional to the rate of build-up and which reduces the effective voltage to $(V - E)$.

As the current increases, its rate of rise slows down; so therefore does the back-emf E, and the net voltage $(V - E)$ approaches nearer and nearer to V. Eventually the current levels off, and, since there is now no change, there is no back-emf, and the full voltage V appears across the resistance R, giving the steady current by Ohm's Law

$$I = V/R$$

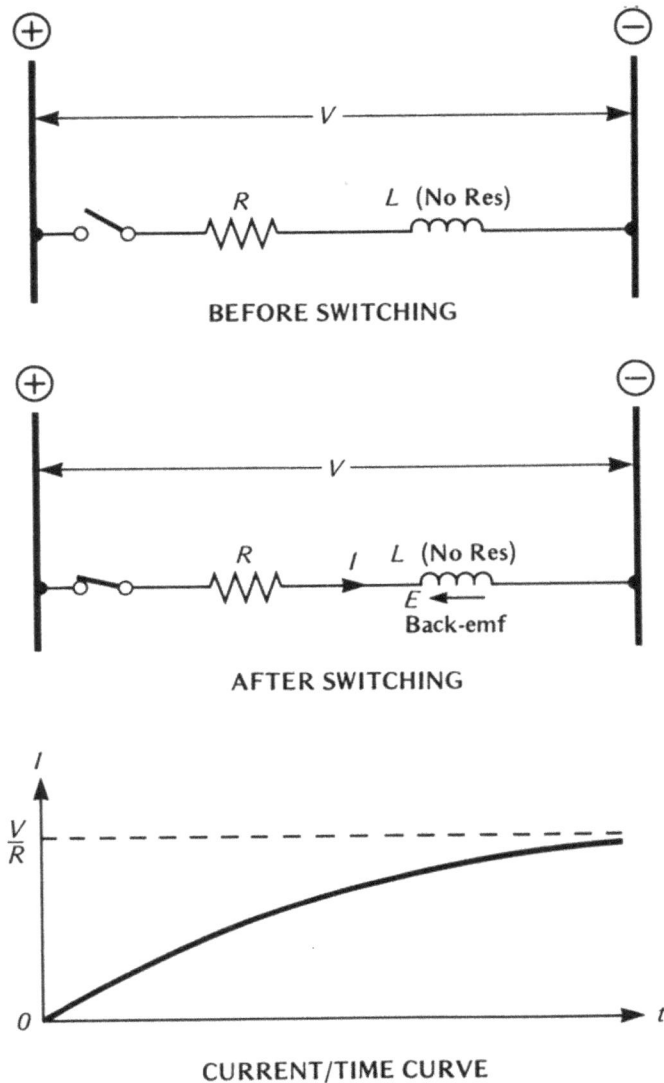

**FIGURE 8.3
SWITCHING ON AN INDUCTIVE CIRCUIT**

This type of current rise, shown in Figure 8.3, is known as 'exponential' and is found in all branches of physics where the rate of change depends on the amount already present. In this case, where a voltage is suddenly applied to a circuit containing inductance and resistance, the current rises, not suddenly, but at a reduced rate, or 'sluggishly', the rate falling off 'exponentially' until it finally settles down at a value given by Ohm's Law, namely

$$I = V/R$$

In the discussion so far the coil has been assumed to be inductive but to have no resistance of its own $(L$ but no $R)$. In practice of course, all coils must have some resistance, but it is convenient to regard that resistance as separate from the purely inductive coil.

One aspect of this treatment should be realised. Since the back-emf depends on the rate of change of current, $\frac{di}{dt}$ any attempt to **stop** the current suddenly by opening the switch causes the rate $\frac{di}{dt}$ to rise steeply towards infinity, and therefore a very large back-emf would be induced to oppose the change - it would be many times greater than the applied voltage V. This greatly increased voltage would appear across the open switch contacts (which could be regarded as a resistance of very high ohmic value) and would cause severe sparking or arcing at the switch contacts and possibly voltages dangerous to personnel.

Therefore a d.c. inductive circuit of any size must never be simply broken by a switch. Special precautions must be taken, one of which is shown in Figure 8.4.

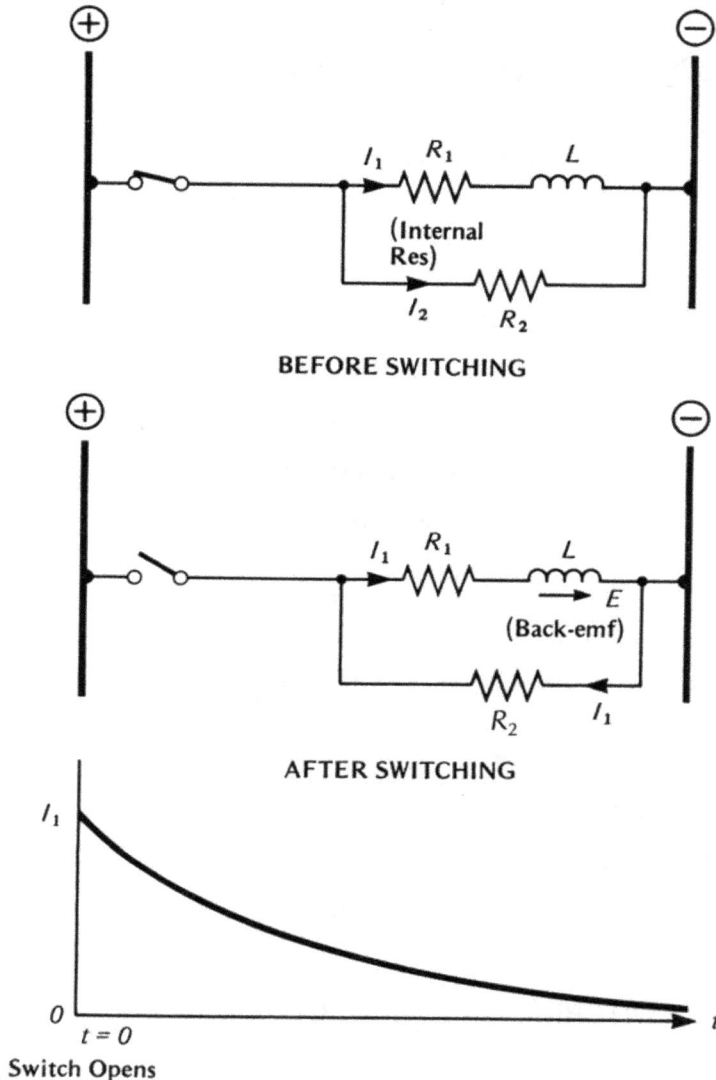

BEFORE SWITCHING

AFTER SWITCHING

CURRENT/TIME CURVE

FIGURE 8.4
DISCHARGING INDUCTANCE

The inductive coil, which in practice has some resistance of its own (R_1), is shunted by another resistance (R_2). In normal use the switch is closed and current flows in parallel through both the coil and the shunt resistance, the I^2R_2 energy in the latter being wasted as heat.

When the switch is opened, the current already flowing in the coil, instead of being stopped, finds a backward path through R_2 and continues to circulate round the coil and the shunt resistance. Eventually the stored energy in the coil will be dissipated in heat loss in both R_2 and R_1 $(= I^2R_2 + I^2R_1)$, and the current will fall exponentially to zero. The rate of change, even at the beginning, is therefore quite slow, so the back-emf is also low, and the voltage appearing across the switch contacts is quite small and causes little sparking - it is in fact only equal to the volt-drop IR_2 across the shunt resistance at the start. The slow decay of current in the coil may however delay the release of whatever mechanism the coil is driving, such as the opening of a solenoid-operated valve.

A shunt resistance used in this way is often called a 'discharge resistance' because it discharges and dissipates the energy stored in the coil and reduces contact sparking. The greater its ohmic value the quicker the discharge of energy (I^2R_2), but the greater the 'spark voltage' (IR_2) appearing across the switch contacts. It is always necessary to make a compromise, taking into account also the time delay for the coil's discharge.

8.4 TIME CONSTANT

It has been shown that, when an inductive circuit is switched on, the rate of build-up of current is relatively slow, and it slows down still more in an exponential curve as the steady-state condition, given by Ohm's Law, is approached. It is necessary to quantify such exponential curves so as to distinguish between fast and slow changes. For this purpose a 'time constant' is used.

There is a general, but wrong, idea that the time constant is a measure of the time taken for the current (or other quantity) to settle down, but a moment's thought will show that, in theory, it never quite reaches its steady state, and it is therefore not possible to pin-point any instant in time at which it has done so.

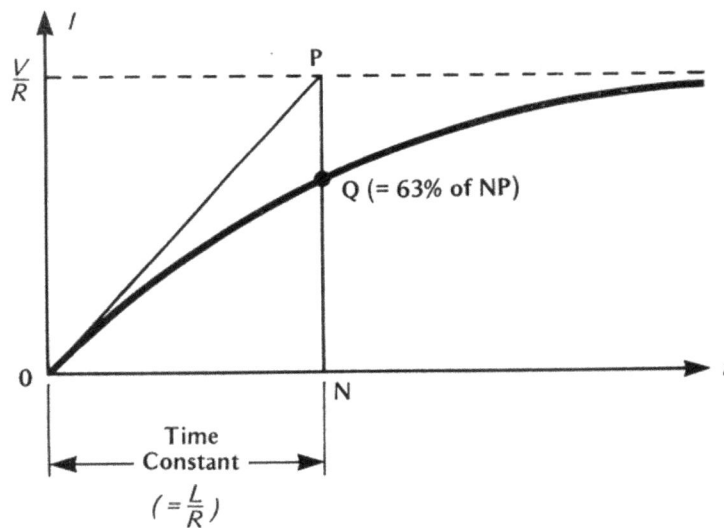

FIGURE 8.5
TIME CONSTANT OF AN INDUCTIVE CIRCUIT

Figure 8.5 shows a typical exponential rising curve of, for example, a current after switch-on. If the circuit consists of an inductance L (henrys) and a total resistance R (ohms), it can be shown mathematically that the initial rate of rise (i.e. the slope at time $t=0$) when a voltage V is applied is equal to $\frac{V}{L}$ amperes per second.

If the tangent is drawn at this point and extended to cut the final steady-state line at a point P, and if a perpendicular PN is dropped from P to the time axis, then the time represented by ON is called the 'time constant' of that curve. It is expressed in seconds (or milliseconds).

It can also be shown mathematically that for any exponential curve at all the value of current at point Q - that is, after a time equal to the time constant - is $(1 - 1/e)$ of its ultimate value. 'e' is the 'exponential number' equal to 2.718, so that at Q the current has risen to 63% of its ultimate steady value (**not**, it should be noted, 100%), no matter how fast or slow it is rising.

The time constant (in seconds) of a circuit such as that shown in Figure 8.3, and whose current/time curve is repeated in Figure 8.5, is equal to $\frac{L}{R}$, where L is in henrys and R in ohms. From this it can readily be seen that a highly inductive circuit (large L) is likely to have a long time constant, and also that any inductive circuit with low resistance (i.e. a purely inductive' circuit) will also have a long time constant. To reduce the constant and so to speed up the response it is necessary either to reduce L, or to increase R, or both. The time constant depends on both.

Increasing V, and so the slope at the start, would also reduce the response time (but not the time constant), but this is not always possible, since V is usually the unalterable system voltage. This is however done in the 'field forcing' of generators - see Part 2 Electrical Power Generation.

Similarly the current of an inductance discharging through a shunt resistance as shown in Figure 8.4 will decay exponentially as shown there, and its time constant will again be $\frac{L}{R}$. In this case however R will be the total resistance $(R_1 + R_2)$ of the discharge loop and will include both the shunt discharge resistance and the internal resistance of the inductance itself. Since $(R_1 + R_2)$ on discharging will be far greater than R_1 alone on closing, the decay time constant of a discharging inductance fitted with a discharge resistor, such as a machine's field, is in general much shorter than the build-up time - that is to say, it discharges more rapidly than it builds.

The correct symbol for time constant is 'τ' the Greek letter 'tau'.

CHAPTER 9 CAPACITANCE

9.1   THE CAPACITOR

A capacitor, or 'condenser' as it used to be called, is a device for storing electrostatic energy. In the early days, when the only sources of electricity were the electrostatic machines such as the Van der Graaf and Wimshurst, electrical energy so created was stored in 'Leyden Jars' for future use (Figure 9.1). For many years the property of capacitance was measured in the unit 'jar'.

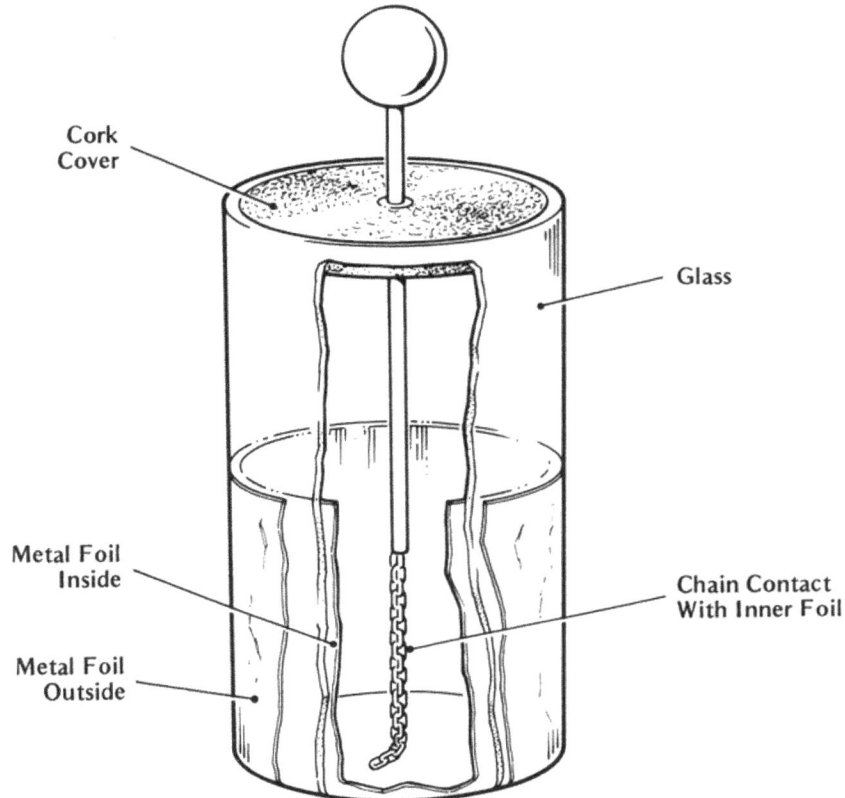

FIGURE 9.1
LEYDEN JAR

The modern condenser or capacitor is the direct descendent of the Leyden jar, and the modern unit of measurement is the 'farad' (F). The farad is however an extremely large unit much too large for practical use, so the unit one-millionth of a farad, or one microfarad (μF), is in general use. One jar is equal to about one-thousandth of a microfarad (0.001μF or 10^{-9} F).

Care is needed to distinguish between the following:

Capacitor: the actual device for storing the energy
Capacitance: the ability to store the energy, measured in μF

The word 'capacity' should not be used in this connection.

The ability to store energy is best explained by looking at the construction of a capacitor. It consists of several parallel metal plates, in flat or cylindrical form, separated from a similar set of metal plates by a thin insulating substance such as glass, mica or paraffin-waxed paper. These insulating layers are called the 'dielectric'. When a potential difference (or voltage) is applied across each set of plates, a strong electric field is set up between them through the thin dielectric. The closer the plates are together, the stronger the electric field. This field causes electric strain in the material of the dielectric, causing it to behave like a spring which has been squeezed in a vice. When an external circuit is provided between the two sets of plates - say by a wire connecting them - this electric spring is released and gives up its energy.

Although the property of capacitance is the principal reason for providing a capacitor, capacitance is found in many other places, often where it is not wanted. When it exists in this way it is called 'self-capacitance'. It is particularly noticeable in cables and overhead power lines, but it is also to be found in machine windings and transformers - in fact anywhere where conductors are arranged close to one another with a thin layer of insulation between. Self-capacitance exists not only between adjacent conductors but also between conductors and earth or cable-sheath.

9.2 CHARGE AND DISCHARGE OF CAPACITOR

The mechanism of charge can best be understood by considering the mechanical analogy of Figure 9.2.

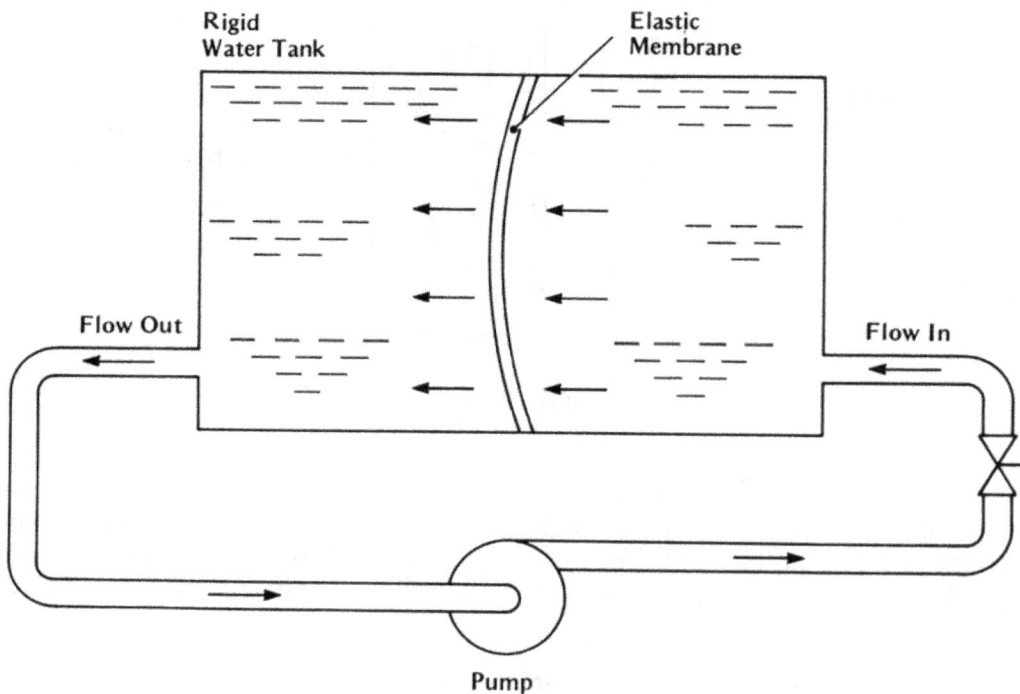

FIGURE 9.2
CAPACITANCE - AN ANALOGY

A large, rigid tank is completely full of water which is regarded as incompressible. Down the middle is a flexible elastic membrane. One side of the tank is connected through a valve to a water supply under pressure, and the other to suction.

Initially the valve is closed. Both sides of the membrane are at equal pressure and the membrane is undistorted. If now the valve is opened and water admitted under pressure, it will flow into the right side of the tank and out from the left side. The water movement through the tank itself, being over a wide cross-section, will be small compared with the movement of water in the pipes. As the water in the tank is displaced from right to left, so the membrane becomes distorted to the left and stretches, imposing increasing pressure on the right-hand side. Eventually, when the distortion is such as to produce a pressure equal to that of the incoming water, the water flow will cease. A definite volume of water will have entered the tank on the right-hand side, and an equal amount will have departed from the left. The stretched membrane will be in a state of elastic strain.

The valve can now be closed, leaving the membrane in the position shown. The right side of the tank is under pressure and the static energy is stored in the stretched elastic membrane. Although the water can move in either direction through the external piping, in considerable quantity in the case of a large tank, there is no **transfer** of water within the tank across the membrane.
Dielectric

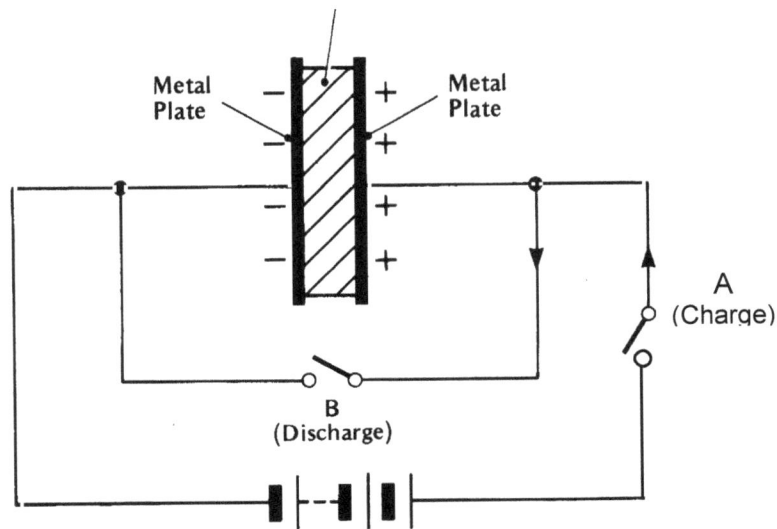

FIGURE 9.3
CAPACITOR

An electric capacitor (Figure 9.3) behaves in much the same way. A d.c. voltage is applied across the two plates of a capacitor by closing battery switch 'A', so that one plate is at a higher potential than the other. The dielectric, which can be regarded as an 'electrically compressible' substance, is subject to a strong electric field which puts it into a state of electric strain, just as the stretched membrane was in a state of mechanical strain.

If the battery switch 'A' is now opened, the capacitor will be left in that state of strain - it is said to be 'charged' - and it will remain so until discharged or until it discharges itself by internal leakage. Some large oil-filled capacitors have been known to hold their full charge for many months.

The current entering one side and leaving the other side of the capacitor is the 'charging current', exactly akin to the water entering one side and leaving the other side of the tank.

If switch 'B' is now closed, the two plates are short-circuited together, and the charge on the positive plate is conveyed back to the negative, driven by the dielectric 'spring' unwinding. The stored energy has been released, and the dielectric has relaxed.

The amount of energy that could be stored in the water tank analogy depended on the volume of water and the elastic properties of the membrane. In the case of a capacitor the amount of energy that can be stored depends on the total plate area, on their distance apart and on the electrical properties of the dielectric used. Taking these into account, the ability of any given capacitor to store electric energy is called its 'capacitance', symbol C.

If the given capacitor has a capacitance of C farads, and a voltage V is applied across it, the amount of electrostatic energy stored is

$$\tfrac{1}{2}CV^2 \quad \text{joules}$$

(This should be compared with the magnetic energy $\tfrac{1}{2}LI^2$ stored in an inductance (see Chapter 8), or with the kinetic energy $\tfrac{1}{2}mv^2$ stored in a moving mass.)

The water analogy showed that passing the water in and out through pipes and under pressure caused the tank to store energy, and on reversal to allow the stretched membrane to relax. Similarly passing a current into a capacitor 'charges' it and causes it to store electric energy; reversing that current discharges it and recovers the energy (Figure 9.4).

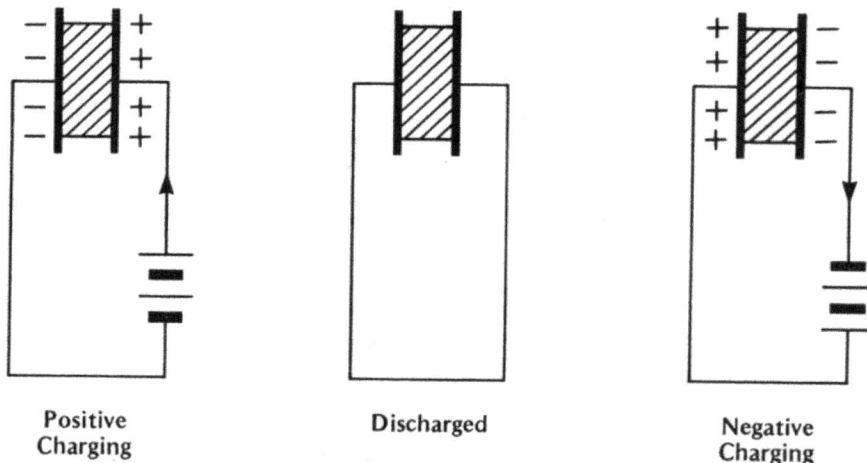

FIGURE 9.4
POSITIVE AND NEGATIVE CHARGING

The analogy can be taken a little further. If the water pump were reversed so that water enters the tank from the left and leaves it from the right, the membrane would simply stretch to the right until its pressure balanced the incoming pressure from the left. Similarly, if the charging current is reversed (Figure 9.4), the left plate of the capacitor would become positive and the right negative. The electric field across the dielectric would still be present but reversed in direction and would still be in a state of electric strain.

So it is possible, in general, to charge a capacitor in either direction, and it will store the same amount of energy (depending on the voltage) in either case: this energy is $\tfrac{1}{2}CV^2$.

An exception to this statement applies to electrolytic capacitors which, for chemical reasons, may not be reverse-charged. These capacitors consist of a single spiral aluminium foil coated wth a very thin film of aluminium oxide which acts as the dielectric. The electrolyte itself (ammonium borate) acts as the second 'plate' and makes contact with the metal case. The very thin dielectric film allows the 'plates' to come very close to each other and so to increase the capacitance greatly. In fact the electrolytic capacitor has a far greater capacitance, size for size, than the conventional type and is now widely used, especially in electronic circuits.

However, any attempt to reverse the polarity will destroy the oxide film, and their application is therefore limited. The polarity of such capacitors is clearly marked on them to prevent their reverse connection. One suitable use for them is for smoothing a rectified a.c. circuit, where the d.c. polarity is always maintained.

9.3   TIME CONSTANT

Figure 9.5 shows a capacitor, capacitance C, charging from a d.c. source V through a resistance R

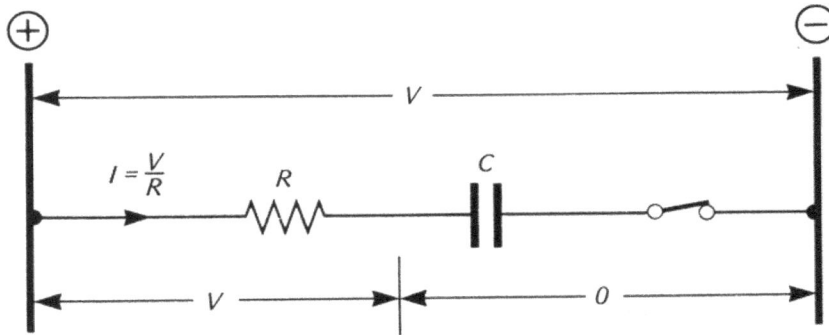

(a) SWITCH JUST CLOSED

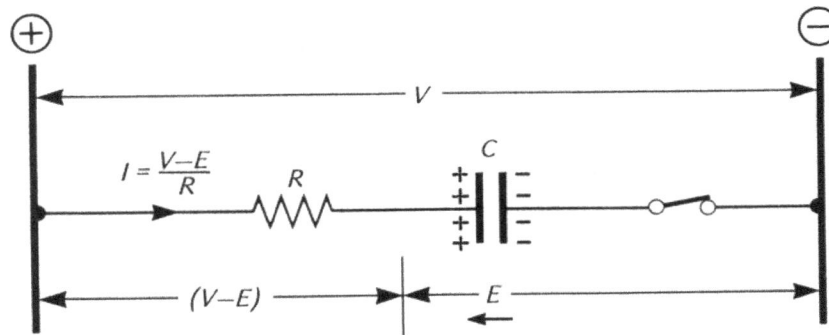

(b) TIME PARTLY ELAPSED

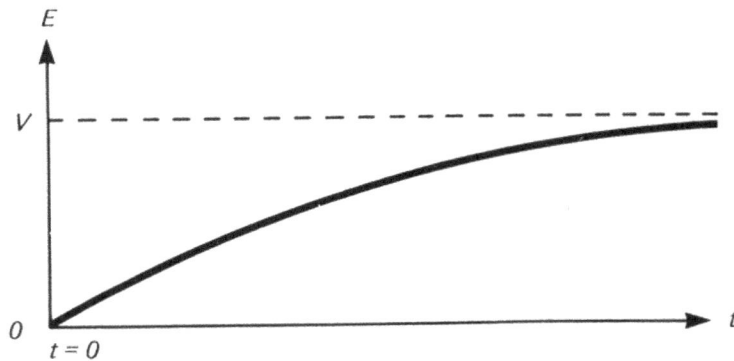

(c) CAPACITOR VOLTAGE (E) v. TIME

FIGURE 9.5
VOLTAGE / TIME CURVE

When the switch is closed (Figure 9.5(a)) the capacitor is at first without charge, which means that there is no potential difference, or voltage, across its plates. At the instant of closing therefore, the full applied voltage V appears across the resistance only, and, by Ohm's Law, the current I through it is $\frac{V}{R}$; this flows round the loop and into the capacitor and is therefore the initial charging current of the capacitor.

After a short time (Figure 9.5(b)) the current has produced some charge in the capacitor - suppose it has acquired a voltage E. This must be in a direction to oppose the applied voltage V and to reduce its effectiveness; it is very similar to the back-emf in an inductance when the current is rising (see Chapter 8).

The effective voltage trying to charge the capacitor is now reduced to $(V - E)$, with E growing all the time. This is the voltage appearing across the resistor, so the charging current I is falling steadily. This is once again the classic case of the rate of charge depending on the amount present, which, as in the inductive case, gives an exponential voltage/time curve, as shown in Figure 9.5(c).

The voltage-charge (E) on the capacitor rises exponentially until it eventually equals the applied voltage, at which point $E = V$, the charging current stops and the charge voltage remains steady at the value V.

The rate of rise of this curve is determined by its 'time constant', the meaning of which was explained for an inductive curve in Chapter 8. In the case of a capacitive curve such as that of Figure 9.5, the time constant has the value CR.

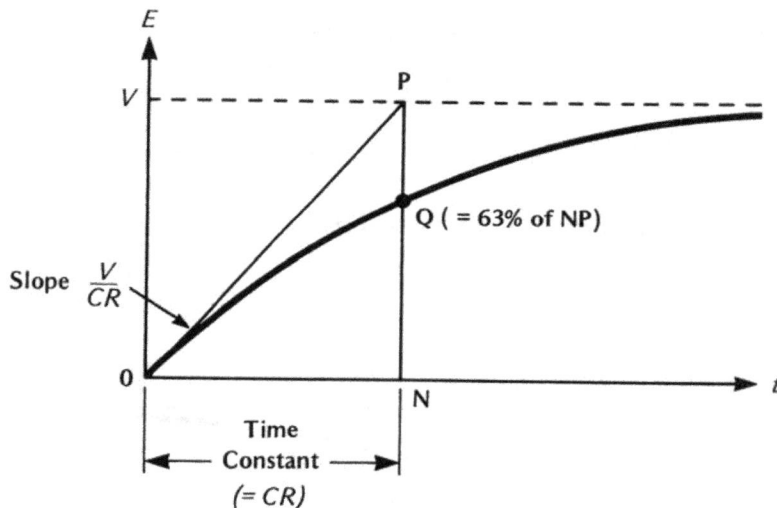

FIGURE 9.6
TIME CONSTANT OF A CAPACITIVE CIRCUIT

Figure 9.6 is given the same treatment as Figure 8.4 in the preceding chapter. A tangent is drawn at the origin, cutting the final steady-voltage line at P. A perpendicular PN is drawn; ON then represents the time constant of this particular capacitor, and the rate of rise (gradient) of the voltage at the start is $\frac{V}{CR}$ volts per second.

It should be noted that, the higher the resistance R or the capacitance C, the longer the time constant, and this is extensively used to provide time-delay circuits.

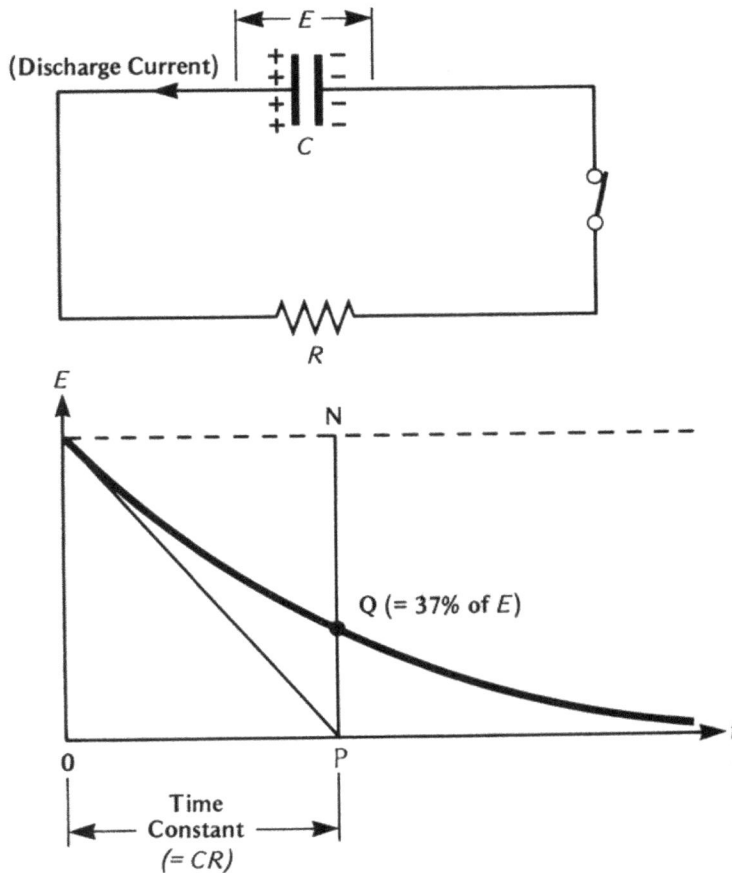

FIGURE 9.7
DISCHARGING CAPACITOR

If a capacitor is charged, and is then switched to discharge through a resistance, the voltage would **decay** exponentially towards zero, and the discharge current would follow the same pattern (Figure 9.7). The time constant would still be CR and could be arrived at as before by drawing a tangent at $t = 0$. The amount of decay at the time equal to the time constant (time P) is again $(1 - 1/e)$ or about 63%, leaving 37% still to decay.

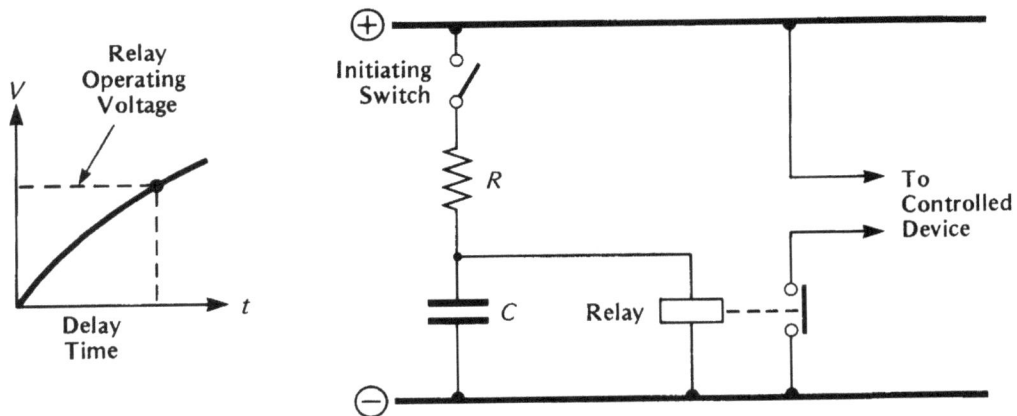

FIGURE 9.8
R-C TIME-DELAY CIRCUIT

Figure 9.8 shows a simple delay circuit for operating some device. When the switch is closed, the R-C circuit begins to charge at a rate determined by its time constant, the capacitor's voltage rising all the time. When it reaches a certain predetermined level, it becomes sufficient to operate a relay which only then switches on the device to be controlled.

If R is made variable, the time constant can be altered at will and so the time delay made adjustable. This type of delay device is widely used because it consists of simple and cheap elements (a capacitor and a resistor) and no moving parts.

9.4 CAPACITORS IN PARALLEL AND SERIES

The behaviour of capacitors when placed in parallel or series is best explained by considering how the energy is disposed between them.

Parallel

If a voltage V is applied to, say, three parallel capacitors each of capacitance C, then the full voltage is applied to each, and the energy stored by each is $\frac{1}{2}CV^2$ (Figure 9.9). The total energy stored by the three is therefore three times this, namely $\frac{3}{2}CV^2$

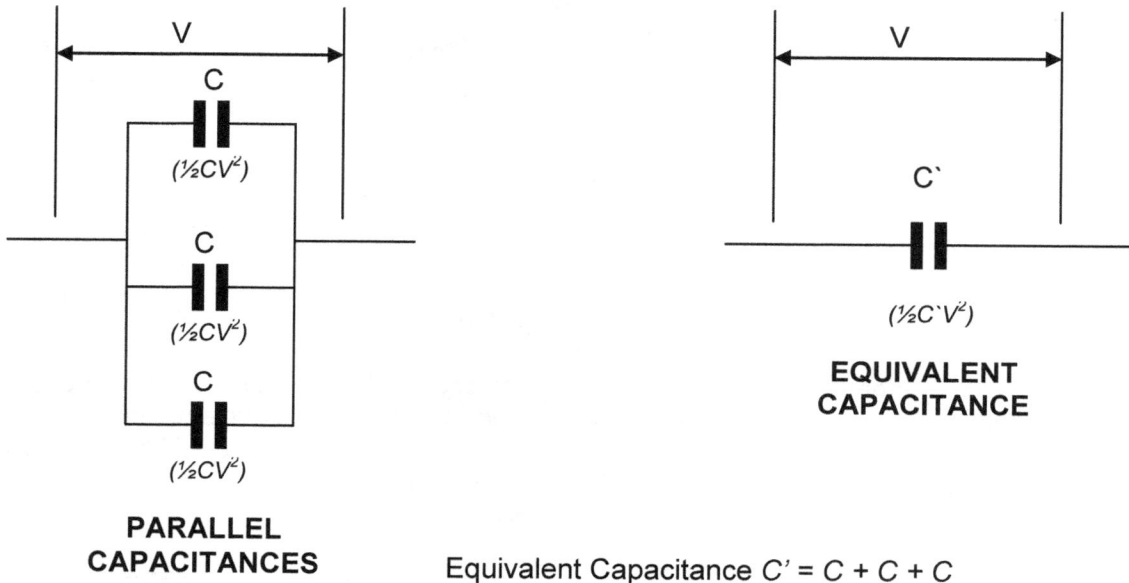

PARALLEL CAPACITANCES

Equivalent Capacitance $C' = C + C + C$

EQUIVALENT CAPACITANCE

FIGURE 9.9
CAPACITORS IN PARALLEL

If C' is the capacitance of the equivalent capacitor which stores the same total energy, then this energy will be $\frac{1}{2}C'V^2$

So
$$\frac{1}{2}C'V^2 = \frac{3}{2}CV^2$$

or
$$C' = 3C$$

To generalise, the capacitance of a single capacitor equivalent to n in parallel is n times that of each individual capacitor, assuming that they are of equal value.

Series

If a voltage V is applied to, say, three series capacitors each of capacitance C, then one-third of the applied voltage will appear across each (Figure 9.10). Therefore the energy stored by each is

$$\frac{1}{2}C\left(\frac{V}{3}\right)^2 = \frac{1}{18}CV^2$$

The **total** energy of the three is therefore three times, or $\frac{1}{6}CV^2$

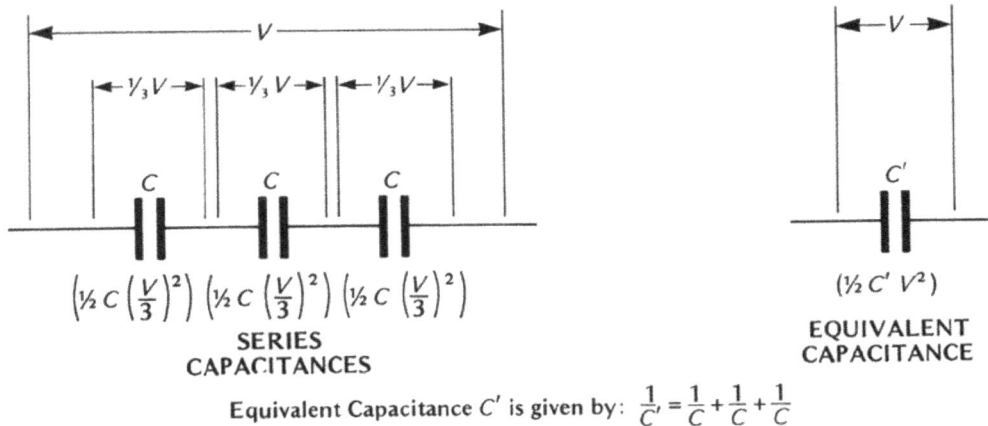

$$\left(\frac{1}{2}C\left(\frac{V}{3}\right)^2\right)\left(\frac{1}{2}C\left(\frac{V}{3}\right)^2\right)\left(\frac{1}{2}C\left(\frac{V}{3}\right)^2\right) \qquad (\frac{1}{2}C'\,V^2)$$

SERIES   EQUIVALENT
CAPACITANCES   CAPACITANCE

Equivalent Capacitance C' is given by: $\frac{1}{C'} = \frac{1}{C} + \frac{1}{C} + \frac{1}{C}$

FIGURE 9.10
CAPACITORS IN SERIES

If C' is the capacitance of the equivalent capacitor which stores the same total energy, then this energy will be $\frac{1}{2}C'V^2$

So
$$\frac{1}{2}C'V^2 = \frac{1}{6}CV^2$$

or
$$C' = \frac{1}{3}C$$

or
$$\frac{1}{C'} = 3\frac{1}{C}$$

To generalise, the reciprocal capacitance of a single capacitor equivalent to n in series is n times the reciprocal capacitance of each individual capacitor, assuming that they are of equal value. It should be noted that the equivalent capacitance of series capacitors is smaller than that of the individual elements.

Summary

The total capacitance of capacitors in parallel add directly:

e.g.
$$C' = C_1 + C_2 + C_3 + \cdots$$

but that of capacitors in series add inversely:

e.g.
$$\frac{1}{C'} = \frac{1}{C_1} + \frac{1}{C_2} + \frac{1}{C_3} + \cdots$$

(Compare the sum of series and parallel resistances in Chapter 5. The formulae for series and parallel are interchanged from those above.)

CHAPTER 10 D.C. POWER

10.1   POWER UNITS

Power is defined as the rate of using (or providing) energy and, in the electrical world, is measured in the unit 'watt' (W).

In the mechanical world of hydraulics, power is the product of pressure and volume flow. In modern SI units pressure is measured in newtons per square metre (N/m^2) and volume flow in cubic metres per second. The product of these two

$$\frac{N}{m^2} \times \frac{m^3}{s} = \frac{Nm}{s} = \frac{J}{s} = W$$

The electrical equivalent of pressure is the volt, and the electrical equivalent of hydraulic flow is the ampere, so that the power in watts is the product of voltage and current - that is

$$W = V \times I$$

where V is measured in volts and I in amperes; W is then in watts.

Applying Ohm's Law $V = IR$ or $I = V/R$ to this formula:

$$W = IR \times I = I^2 R$$

or

$$W = V \times V/R = V^2/R$$

These are two alternative forms for power when only I and R, or when only V and R, are known.

In earlier years the power output of electric motors was measured in 'horsepower' to align them with mechanical engine practice. Many motor nameplates are still marked in 'hp', but more and more are now being marked in kilowatts (kW).

The kilowatt used in this case is the equivalent of the mechanical power output of the motor. Horsepower and kilowatts are directly related:

1 hp = 0.746kW ($= \frac{3}{4}$ kW approximately)

The horsepower in this case is British horsepower. (Some Continental countries use a slightly different unit, known as the 'metric horsepower', which is equivalent to 0.735kW, a difference of about 1½%.)

Special care is needed when referring to the power of motors. Motors are rated by their mechanical output (hp or kW), but, because no motor is 100% efficient, its electrical power input, also measured in kW, is always greater than its mechanical power output. Because both may be measured in kilowatts, confusion can easily arise. When it is desired to distinguish between mechanical output and electrical input, suffixes 'm' and 'e' are often used: thus output is kW_m and input kW_e. Their ratio is the efficiency of the motor, thus:

$$\frac{kW_m}{kW_e} = \text{efficiency}$$

Power - that is, the rate of producing, absorbing or transmitting energy - is in the SI system always measured in watts or, more usually, kilowatts. It occurs in fields other than electricity and mechanics. For example energy can be produced thermally in boilers or reactors, or chemically in batteries or by burning fuel. The power being developed is still measured in kilowatts and would be distinguished by suffixes 'kW_{th}' or 'kW_{ch}'.

CHAPTER 11 PRINCIPLES OF D.C. MEASUREMENT

11.1 HOT-WIRE INSTRUMENTS

Electrical quantities such as pressure (voltage) and flow (current), being invisible to the eye, can only be measured indirectly by observing their effect on other things such as a mechanical indicating system. Even in the mechanical field steam or hydraulic pressure is indicated on a pressure gauge, and flow rate or flow volume on a flowmeter.

In the electrical world the first 'indirect' means of measurement used the heating effect of a current passing through a wire. In Chapter 4 it was shown that, if a current I amperes flows through a conductor having a resistance of R ohms, then heat is generated in that conductor at the rate of I^2R watts.

This heat in the wire raises its temperature until the rate of radiation away of the heat just balances the rate of heat generation. At that point the temperature will stabilise, and it will then be a measure of the current (strictly, the square of the current) which was causing the heating.

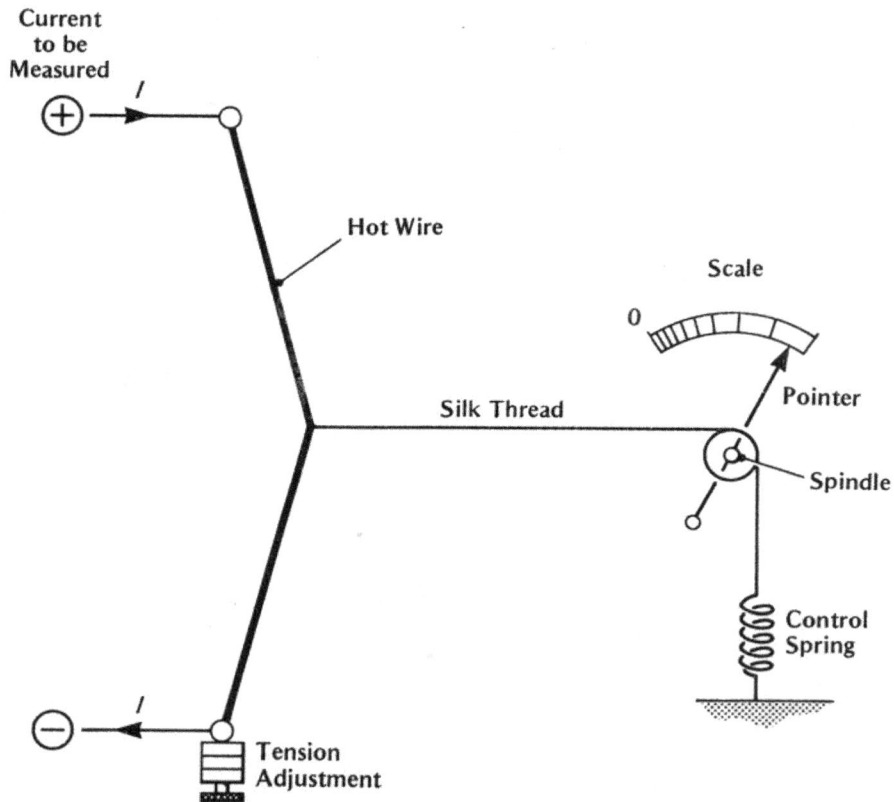

**FIGURE 11.1
HOT-WIRE AMMETER**

Use of this is made in the 'hot-wire ammeter' shown in Figure 11.1. A short length of platinum-iridium wire is connected electrically in series with the circuit whose current flow is to be measured. The wire is tensioned until it is fairly taut, and from its centre is taken a silk thread tensioned by a spring and whose movement actuates a spindle to which a pointer is fixed.

When no current flows through the main wire its tautness keeps it almost straight, and the centre thread is at its left-most position. When current flows the temperature of the wire rises, and the wire becomes longer and slacker. The slack is taken up by the spring pulling on the centre thread and moving it to the right. In doing so the spindle is rotated clockwise, and the pointer moves over a scale.

Since any given current causes a given I^2R heating rate and therefore a given elongation of the wire and a given deflection of the pointer, for each current there is a definite position of the pointer, which depends on the square of the current. The scale can be calibrated in current units (amperes), though it will not be an even scale, being crowded at the lower end.

The instrument described is the original 'hot-wire ampere-meter' (or ammeter) and was used generally until about the turn of the century. (It is of interest that it will also indicate alternating currents, as explained later.)

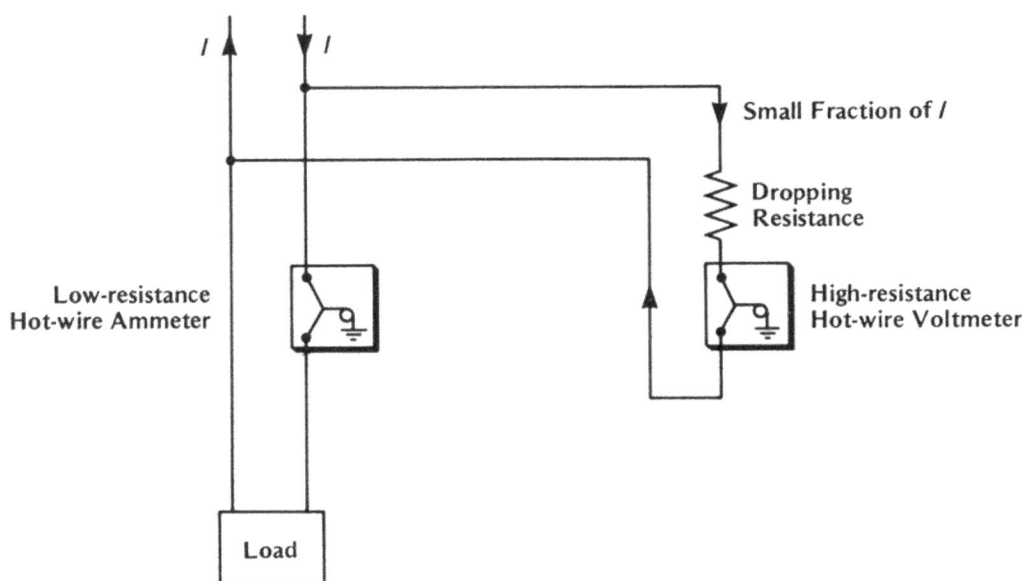

FIGURE 11.2
HOT-WIRE AMMETER AND VOLTMETER

A similar instrument is used for indicating voltage, since, by Ohm's Law $V = IR$ or $I = V/R$

If R is the resistance of the hot wire (considered fixed), then the current indicated by it will be proportional to V. In Figure 11.2 a hot-wire ammeter in series with the circuit to be measured is shown on the left, and on the right is a parallel circuit operating at the same voltage. A second hot-wire instrument is connected across this parallel circuit, but one with a high resistance in series with it. As explained above, the current, small because of the series resistance, drawn by this second instrument will still be proportional to V, and so the instrument acts as a voltmeter and can be scaled directly in volts. It too will have an uneven scale.

11.2   MOVING-IRON INSTRUMENTS

Whereas the hot-wire instruments described above depend on the heating effect of a current for its indirect method of indicating, another class of instruments makes use of the **magnetic** effect of the current discovered by Oersted. Its development was largely due to Lord Kelvin.

41

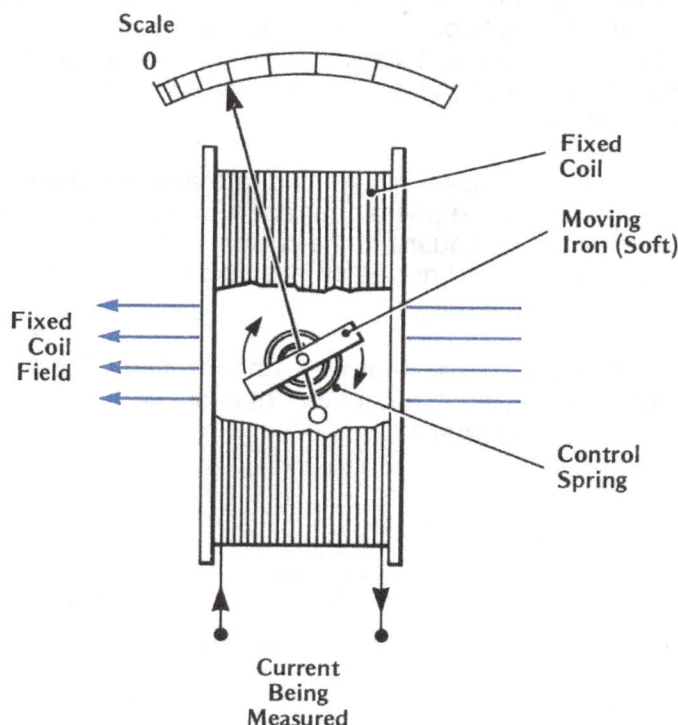

FIGURE 11.3
MOVING-IRON INSTRUMENT

Figure 11.3 shows a fixed coil through which the current to be measured passes. The coil is wound with wire of sufficient section to carry the current without overheating.

As current flows through the coil it gives rise to a magnetic field along its axis. In one form of the instrument a small piece of soft iron is pivoted so that it lies normally well across the coil's axis and is held there by a spiral control spring.

When current flows through the coil the soft iron is magnetised by induction and tries to align itself along the axis of the coil's magnetic field, but it is restrained from completely doing so by the back-pull of the spring. It takes up a balance position where the magnetic pull of the coil is just counterbalanced by the spring.

The moving-iron 'armature' is pivoted on a spindle carrying a pointer which moves over the instrument scale. If the instrument has a low-resistance coil and is used in series with the circuit whose current is to be measured, it acts as an ammeter. If however it has a high-resistance coil and is used in parallel with the circuit, it acts as a voltmeter. There is otherwise no basic difference between the movements.

Some such instruments are provided with two fixed coils in parallel planes and arranged to assist each other magnetically. The moving-iron armature is then placed symmetrically on their common axis. In other makes the iron may be pivoted differently, but it always moves towards the coils' magnetic axis against some form of restraint. Various types of damping are usually added.

Because the magnetic pull of a coil on a piece of iron magnetised by induction is proportional to the **square** of the current, the movement of a moving-iron instrument is proportional to I^2 just as was the case with the hot-wire instrument where the heating depended on I^2. The scale of a moving-iron instrument, calibrated in amperes or volts, is therefore uneven and is crowded towards the lower end. Indeed it is the crowding of the scale that makes it possible to tell a moving-iron from a moving-coil instrument at a glance.

11.3   MOVING-COIL INSTRUMENTS

The moving-coil type of instrument is magnetic in principle and depends on the interaction between a coil and a magnet. As shown in Figure 11.4, the magnet is fixed and takes the form of a permanent magnet; the 'armature' is a small pivoted coil to which connections are made by light flexibles. Normally the axis of the coil lies well across the axis of the permanent magnet field and is held there by a spiral control spring.

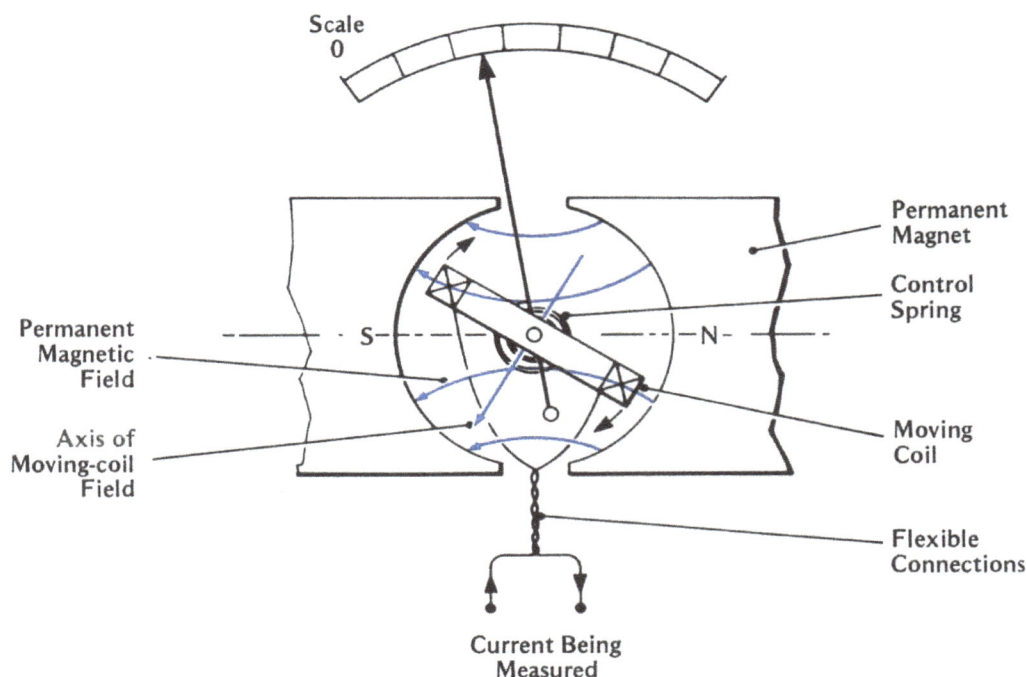

FIGURE 11.4
MOVING-COIL INSTRUMENT

The circuit current to be measured (or a known proportion of it) flows through the coil, causing it to produce an electromagnetic field along its axis. This field reacts with the field of the fixed permanent magnet and causes the coil to try to align its own axis with it, but it is restrained from completely doing so by the back-pull of the spring. It takes up a balance position where the magnetic pull on the moving coil is just counterbalanced by the spring.

The coil is pivoted on a spindle carrying a pointer which moves over the instrument scale. If the instrument has a coil of fairly low resistance and is used in series with the circuit whose current is to be measured, it acts as an ammeter. But here it is necessary to point out that the moving coil is necessarily of light construction, and, with its flexible connections, there is a limit to the size of wire that can be used. For heavier currents therefore such an ammeter would be used with a shunt (see para. 11.5). If the coil has a high resistance and is used in parallel with the circuit, it acts as a voltmeter. There is otherwise no basic difference between the movements.

In a moving coil operating in a permanent magnet field, the magnetic pull on the coil depends only on the current in that coil, not on the square of the current as in the moving-iron type where the magnetisation is induced. Therefore the movement of the coil is almost linear with the current, and the scale of a moving-coil instrument is fairly even over its whole range, not crowded at its lower end. It is also reversible - that is, negative currents cause negative movement - and, if a 'set-up zero' is used, they can be scaled to show the reverse currents; a centre-zero ammeter is an example.

In order that a moving-coil instrument shall retain its accuracy, it is important that the permanent magnet retain its original strength and not weaken with the passage of time or with vibration or shock. Therefore the magnets used in these instruments are made from specially chosen iron of high retentivity. They then undergo a thorough artificial 'ageing' process and are given rigorous tests to prove that they are truly magnetically stable. It should seldom be necessary to change a magnet.

Moving-coil instruments can only be used with d.c.; they will not function on a.c., as explained in Chapter 35.

11.4   DYNAMOMETER INSTRUMENTS

The dynamometer instrument is really a special case of the moving coil. There is a moving coil as already described, but the permanent magnet is replaced by a fixed coil (usually a pair, for reasons of symmetry, since they produce a nearly uniform field). (Figure 11.5.)

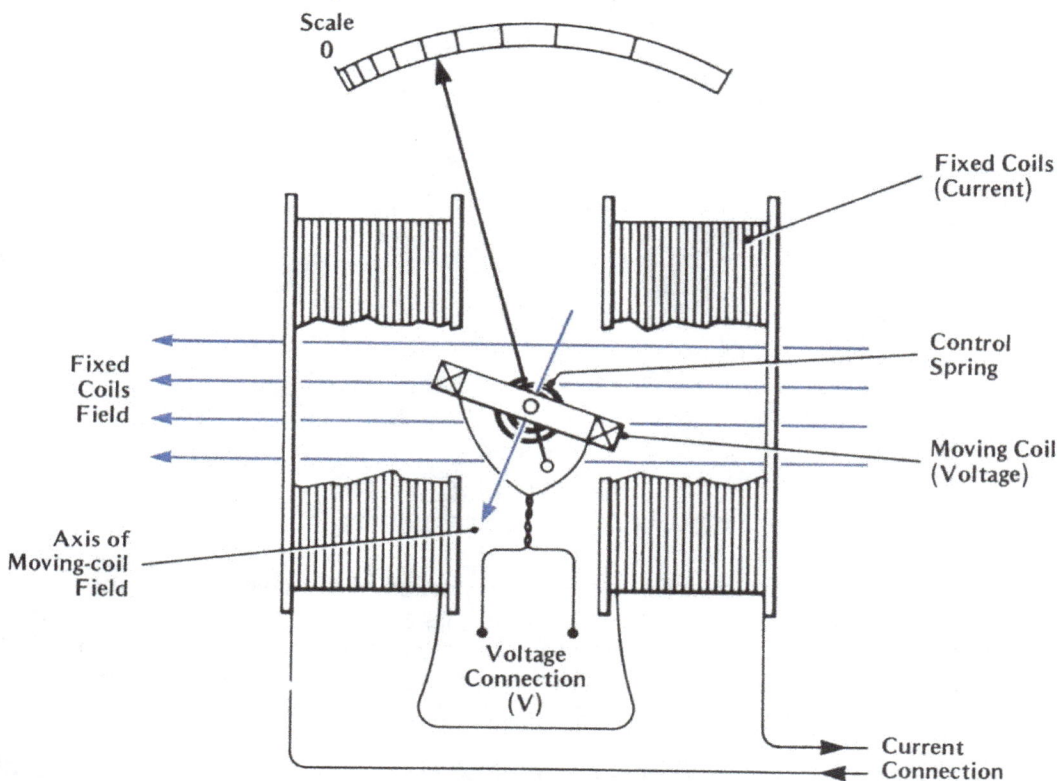

FIGURE 11.5
DYNAMOMETER INSTRUMENT CONNECTED AS WATTMETER

If such an instrument is used as an ammeter or voltmeter, both fixed and moving coils are in series, and the torque depends on the current in each. So the **total** torque depends once again on the square of the current, giving an uneven scale. Being more costly than moving-iron, they therefore have little advantage over the moving-iron type. They are also inferior to the simple moving-coil instruments and are not much used in this application.

However, one application is important. The actual torque on the moving coil depends, as stated, on both the current in the fixed coils and that in the moving coil; it is therefore proportional to their product. If the fixed coils are connected in series with the circuit (so carrying the current) and the moving coil is connected in parallel with the circuit (so carrying a current proportional to the voltage), the instrument will indicate $I \times V$, or d.c. **watts.** This dynamometer instrument, so connected, then becomes a **wattmeter.** (It is of interest that this instrument will also register a.c. watts, as explained in Chapter 35.)

Since the dynamometer instrument indicates the product of V and I, then if I reverses, the torque also reverses and the instrument indicates backwards. If its zero is 'set up', it can be scaled to indicate both normal (forward) and reverse power that is to say, it is **directional**. Whenever it is desired to take direction into account (such as for some relays in protective schemes), a dynamometer or 'wattmetric' type of movement is always used.

If the currents to be measured are too high for the coils of the instrument, the series (fixed coil) current can be taken through a shunt (see para. 11.5 below), thereby using only a known proportion of the circuit current. Similarly the voltage coil (moving coil) can be fed through a 'dropping' resistance of known value, so that only a known fraction of the circuit voltage is applied.

11.5   SHUNT-CONNECTED INSTRUMENTS

Most instruments receive an input from the circuit whose current is to be measured, but, due to their construction, they can in most cases accept only a small, but accurately known, proportion of that current. It is therefore necessary to divide the circuit current into two parts: the main part carrying the majority of the current, and the other part a known fraction of it - in many cases only an infinitesimal fraction, perhaps one-thousandth part.

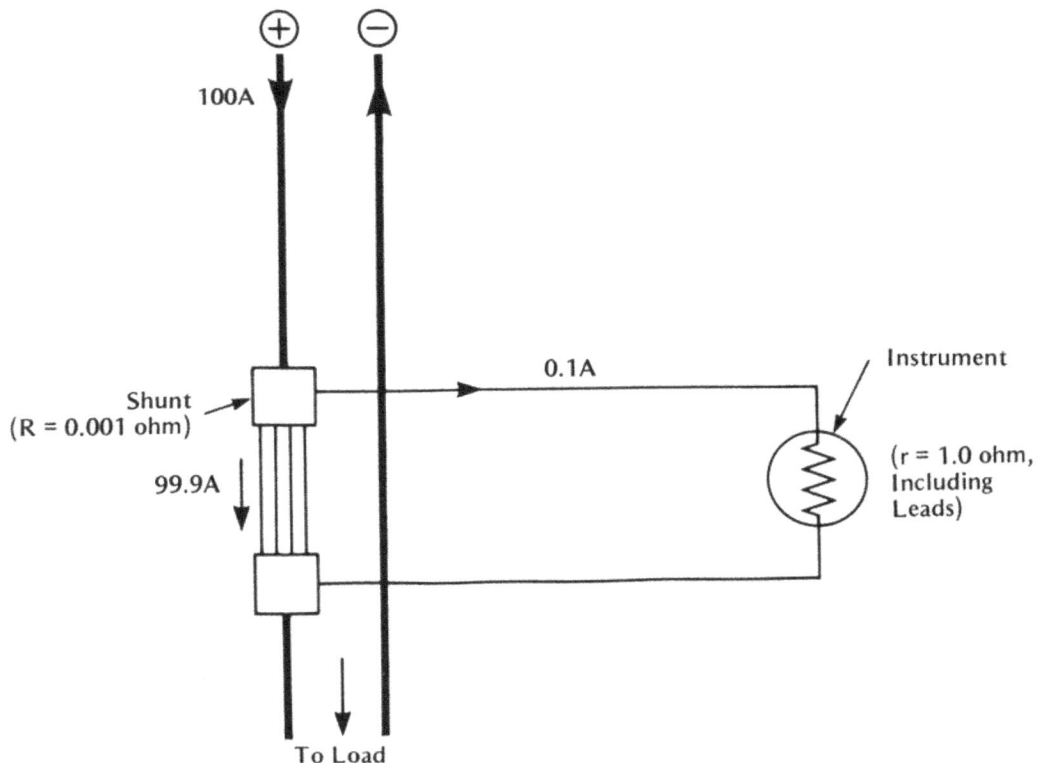

FIGURE 11.6
INSTRUMENT WITH SHUNT

In Figure 11.6 the line current is *I* and is shown in heavy line. It passes through a 'shunt' which usually for the heaviest currents takes the form of two brass blocks connected together by strips of conducting metal of very small, but not negligible, resistance. In parallel with this shunt is the instrument's current coil with a relatively high resistance, perhaps many thousand times that of the shunt. The line current will divide in inverse proportion to the resistances. For example, if the shunt resistance were 0.001 ohm and that of the instrument coil 1.0 ohm, a current of, say, 100A would divide approximately 99.9A through the shunt and 0.1A through the coil. The instrument would then actually measure only 0.1A, but it would be scaled one thousand times up to read '100A'.

When instruments receive their current input through a shunt, it is important that they be calibrated by the maker with their own shunt and with the actual connecting leads with which they will eventually be installed. Because of the great disparity between the resistance of the shunt itself and of the parallel instrument circuit, small errors in the latter could cause considerable errors in the instrument reading; hence the need to calibrate all together.

When an instrument is being installed with its shunt, the leads must never be altered or shortened; any extra length must be left in and flaked out. Shunt-operated instruments should all be marked with their associated shunt serial number and used with no other. For example, in the case quoted, one-tenth of an ohm cut out of the shunt leads would lead to a 10% error in the instrument's reading.

11.6   TERMINOLOGY

The words used with indicating instruments and similar devices are often misused, and below is a summary of the terms and their correct usage:

'Indicating Instrument' or simply **'Instrument'**. An indicating device to show the instantaneous value of the quantity being measured. Note that, although the terms 'voltmeter', 'ammeter', 'wattmeter' etc. are generally used, the generic term 'meters' should never be used for indicating instruments.

'Integrating Meter'. A device for integrating a quantity over a period of time. Usually associated with watts and vars, the total quantity shown by dial or digitally as watt-hours (Wh) or var-hours (varh). Such integrating devices are known collectively as 'meters' (as distinct from 'instruments' above).

'Recording Instrument' or **'Recorder'**. A device for continuously recording the instantaneous value of a quantity, the record being made by pen-on-paper or similar device, or digitally on paper or tape.

'Set-up Zero'. When an instrument is carrying no current, the position taken up by its pointer under the influence of the control spring is its 'zero', and it is usually at the extreme left-end of the scale. However, instruments can be arranged so that the zero is not at the extreme end but at some distance to the right; it is then said to have a 'set-up zero'. This allows the instrument to register, in part at least, in the negative or reverse direction. It can only be applied to moving-coil or dynamometer instruments. A centre-zero ammeter showing battery charge and discharge is an example of a set-up zero.

Each group of devices has a separate British Standards symbol.

A — Indicating Instrument

Wh — Integrating Meter

V — Recorder

The letter within shows the quantity being measured, for example A (amperes), Wh (watthours), V (volts), etc.

CHAPTER 12 USEFUL FORMULAE: CHAPTERS 1 - 11

D.C. CIRCUITS AND OHM'S LAW

Ohm's Law $\qquad V = IR$

or $\qquad I = V/R$

or $\qquad R = V/I$

Heating rate $\qquad = I^2R$ watts (W)

INDUCTANCE

Back emf $\qquad E = -L\frac{di}{dt}$

Stored energy $\qquad = \frac{1}{2}LI^2$ joules (J)

Time constant $\qquad = L/R$ seconds

CAPACITANCE

Stored energy $\qquad = \frac{1}{2}CV^2$ joules (J)

Time constant $\qquad = CR$ seconds

Capacitances in series $\qquad 1/C' = 1/C_1 + 1/C_2 + 1/C_3 + \cdots$

Capacitances in parallel $\qquad C' = C_1 + C_2 + C_3 + \cdots$

RESISTANCES IN SERIES AND PARALLEL

Resistances in series $\qquad R' = R_1 + R_2 + R_3 + \cdots$

Resistances in parallel $\qquad 1/R = 1/R_1 + 1/R_2 + 1/R_3 + \cdots$

D.C. POWER

Power $\qquad VI$ watts

or $\qquad I^2R$ watts

or $\qquad V^2/R$ watts

CHAPTER 13 QUESTIONS AND ANSWERS: CHAPTERS 1 – 11

13.1   QUESTIONS

1.   What is it in an atomic structure which differentiates between inert gases and some metals? Name some examples.

2.   What charge does an electron carry?

3.   How do you explain an electric current in electron terms? What is its direction?

4.   What is the difference between a primary and a secondary cell?

5.   Describe the elements of a 'dry' battery.

6.   What are the main differences between lead-acid and alkaline secondary cells? What are the advantages of alkaline cells?

7.   Why must a lead-acid cell never be left in a discharged state?

8.   Why must battery rooms be ventilated?

9.   About what specific gravity reading would you expect a fully charged lead-acid cell to show?

10.   Approximately how much of the charge energy of a battery would you expect to recover in use?

11.   A battery is rated 250Ah at the 5-hour rate. For how long could you draw 7A from it continuously? If you drew 70A continuously, what effect would you expect on the battery's performance?

12.   What effect would the discharging of a nickel-cadmium cell have on the electrolyte's specific gravity?

13.   What is the difference between natural magnetism and electromagnetism? Sketch an electromagnetic coil, illustrating the field direction.

14.   Name some uses for an electromagnet.

15.   What is the equivalent resistance of several resistors in series?

16.   What is the equivalent resistance of several resistors in parallel?

17.   Write down Ohm's Law in its three variations, naming the units used.

18.   What is the rate of heat production when current flows in a conductor of known resistance? How can the heat rate be reduced?

19.   What is understood by Faraday's 'Law of Electromagnetic Induction'? Illustrate your answer by a sketch.

20.   How do you use Fleming's Right-hand Rule for generated emf?

21.   What force acts on a conductor carrying current in a magnetic field? Illustrate your answer with a sketch.

22. Why does the current rise slowly when a voltage is applied to an inductive circuit? What do you understand by 'back-emf', and what is it proportional to?

23. For an inductor of inductance L henrys, write down the formula connecting L, the back-emf E and the rate of change of current.

24. In the inductor of Q.23, if the current flowing at any instant is I, what is the energy stored in the inductor? In what unit?

25. What do you understand by the 'time constant' of an inductive (R-L) circuit?

26. What happens if you suddenly switch off an inductive circuit? How can any ill-effects be mitigated?

27. In what unit is capacitance now measured? Is it a practical unit?

28. What do you understand by the 'dielectric' of a capacitor? How does it function?

29. A capacitor of capacitance C is charged to a voltage V. What is the energy stored in it? In what unit?

30. Describe an electrolytic capacitor. How does it compare with a normal type?

31. Why must an electrolytic capacitor never be reverse-charged? How is this to be avoided?

32. What is the equivalent capacitance of a number of capacitors in series?

33. What is the equivalent capacitance of a number of capacitors in parallel?

34. What is the time constant of a capacitive (R-C) circuit?

35. How is a capacitor/resistor combination used to provide a time delay to an operating circuit? Make a sketch.

36. What is the power transmitted in a d.c. circuit by a current of I amperes at a pressure of V volts? If the resistance of the circuit is R ohms, express this power in two alternative forms. What is the unit of the power?

37. What is the equivalent in kW_m of 20 horsepower?

38. What is the efficiency of a motor with full-load output of 50 hp and an electrical input of 45kW?

39. Name any three classes of d.c. measuring instrument, indicating their principal features and differences.

40. How can you tell a moving-iron from a moving-coil instrument simply by looking at it?

41. What difference is there between the movements of a moving-iron ammeter and a moving-iron voltmeter?

42. How does a dynamometer instrument operate as a wattmeter? Can it indicate reverse power flow?

43. What is a 'set-up zero' instrument, and when is this feature used?

13.2 ANSWERS
(Figures in brackets after each answer refer to the relevant paragraphs in the text).

1. Inert gases have a full outer ring of electrons - examples are helium, neon, argon. All atoms which have one electron only in their outer ring (except hydrogen) are metals and conduct electricity freely. Examples are lithium, sodium, potassium. (1.1)

2. It carries a single negative unit of charge. (1.1)

3. The passage of one electron from the outer ring of an atom to its neighbour, and the ensuing chain reaction, constitutes a continuous flow of electrons and so an 'electrical current'. The electrons, being negatively charged, migrate from the negative end to the positive and so move in a direction counter to the conventional flow of current from positive to negative. (1.1)

4. A primary cell relies on the electrolytic current which flows between two dissimilar metals in a conducting electrolyte when a return path is provided between the metals. The energy of the cell is derived from the energy released by the chemical reactions at the metal/electrolyte interfaces. (2.2)

5. A 'dry' battery is a primary cell which consists essentially of a carbon rod inside a zinc case, the intervening space being filled with stiff moist paste acting as an electrolyte. (2.2)

6. Lead-acid cells use lead plates which, when charged, change to spongy lead (negative) and lead peroxide (positive), and the sulphuric acid electrolyte becomes more dense. On discharge the process is reversed and the acid becomes weaker; this is used to indicate the state of charge. (2.3)

   An alkaline cell (nickel-iron or nickel-cadmium) has plates of nickel hydrate (positive) and iron (or cadmium) oxide (negative). The electrolyte is a solution of potassium hydroxide. On charging, the negative plates revert to pure metal and the positive plates become heavily oxidised, but the density of the electrolyte is not affected. (2.3)

   Advantages of the alkaline cell are: longer life; greater reliability; less maintenance, lighter weight, greater robustness against vibration and shock; good high-rate discharge performance; ability to accept high rates of charge; better charge retention; rapid voltage recovery after heavy discharge; and immunity from harm if over-discharged. (2.3)

7. On discharge a lead-acid cell forms lead sulphate on both plates. If left too long the sulphate hardens and the cell is said to be 'sulphated'; it will not readily recharge and becomes unfit for further use. (2.3)

8. On completion of a charge the current, having no more chemical changes to perform, starts to electrolyse the water in the electrolyte, giving off oxygen and hydrogen at the plates - the cell starts 'gassing'. This is a very explosive mixture and, if allowed to accumulate, presents a serious hazard. For this reason battery rooms must be continuously ventilated, especially when charging is in progress. (2.3)

9. Approximately 1.250 (varies between makes). (2.3)

10. About 70%. $\left(\text{Efficiency} = \dfrac{\text{output}}{\text{input}} = \dfrac{1}{1.4} \right)$. (2.3)

11. 35.7 hours. If discharge rate were 70A, time would theoretically be 3.6h, but, as 70A exceeds the 5-hour rate (50A), this time would not be achieved; it would be appreciably less. (2.3)

12. None. In an alkaline battery the electrolyte is a 'catalyst' and is not affected chemically. (2.3)

13. Magnetic ores are found in nature and their magnetism can be imparted to iron, which then becomes a natural, or permanent, magnet without any outside assistance. In an electromagnet, magnetism is induced artificially by an 'exciting coil' through which an electric current is passed. The electromagnetism can be controlled by controlling the electric current. (Sketch as Figure 3.2)

14. As a solenoid to actuate valves and other mechanical devices, and to operate switchgear. Used with a crane to pick up scrap-metal. On a small scale to actuate relays and control devices. Also to actuate the movements of instruments and measuring devices. (3.2)

15. Equivalent resistance (series) $\qquad R' = R_1 + R_2 + R_3 + \cdots$ (5.1)

16. Equivalent resistance (parallel) $\qquad {}^1\!/_R = {}^1\!/_{R_1} + {}^1\!/_{R_2} + {}^1\!/_{R_3} + \cdots$ (5.2)

17. Ohm's Law $\qquad V = IR$

    or $\qquad I = {}^V\!/_R$

    or $\qquad R = {}^V\!/_I$

    where V is in volts, I in amperes and R in ohms. (4.2)

18. Heat production rate is: I^2R.

    If the current cannot be altered, the heat rate can be reduced by using a conductor with a lower R - either by using one with a larger cross-section and therefore with a lower resistance, or by using two or more conductors in parallel, which has the same effect. (4.3)

19. If a conductor moves in a magnetic field so as to cut it, an electromotive force (emf) is induced along it which is proportional only to the strength of the field and to the speed of movement. (Sketch as Figure 6.1)

20. Forefinger in line with magnetic field (N to S);
    Thumb in direction of physical motion;
    Centre finger then points in direction of induced emf. (Figure 6.1)

21. If a conductor carries a current when in a magnetic field, it is subjected to a mechanical force mutually at right angles to the directions of the field and of the current. The magnitude of this force is proportional only to the strength of the field and the strength of the current. (7.1)

22. Changing a current in an inductance gives rise to a 'back-emf' of value $-L\frac{di}{dt}$; it opposes the applied voltage which is trying to change the current. L is the inductance in henrys, and $\frac{di}{dt}$ is the rate of change of current. Thus a given applied voltage can only change the current at a rate which will just balance that voltage. The current will therefore rise only slowly; faster with a higher applied voltage, but slower with a greater inductance in henrys. (8.2)

23. $E = -L\frac{di}{dt}$. (8.2)

24. Energy = $\frac{1}{2}LI^2$ joules. (8.2)

25. The time constant of an inductive circuit containing resistance is the time taken from its original state at switch-on to $\left(1 - \frac{1}{e}\right)$, or about 63%, of its final value. Its value is $\frac{L}{R}$, and it can be deduced graphically by drawing the tangent to the current curve at switch-on to cut the final steady-state line. (see Figure 8.4)

26. If an inductive circuit is suddenly switched off, the rate of change of current (downwards) is theoretically infinite so an infinitely large back-emf $-L\frac{di}{dt}$ would be induced in it and would appear across the switch contacts. Before that stage was reached the contact gap would break down and a severe arc would form, burning the contacts and perhaps endangering personnel both by fire and by very high voltage. The entire energy in the coil $(\frac{1}{2}LI^2)$ would have to be dissipated in the arc.

    The effect can be mitigated by shunting the coil with a permanent resistance. The voltage, and so the arcing at the switch contacts, can be much reduced, and the stored energy in the coil can dissipate itself harmlessly by causing the coil current to circulate through the resistance until it is damped out. (8.3)

27. Farad. This is a very large unit -too large to be practical - so a unit one-millionth of this is generally used: the microfarad (μF). (9.1)

28. The dielectric is the insulating material separating the two plates of a capacitor. It takes up the electric strain (like a spring) when the plates are charged and stores the energy, releasing it again when the plates are connected together and the capacitor is discharged. (9.1)

29. Energy = $\frac{1}{2}CV^2$ joules (9.2)

30. An electrolytic capacitor consists of a continuous spiral of aluminium foil with a very thin film of aluminium oxide which acts as a dielectric. The electrolyte itself (ammonium borate) acts as the second 'plate' and makes contact with the metal case. The very thin oxide film allows the 'plates' to be very close and so to increase the capacitance greatly as compared with that of a normal capacitor of the same size. (9.2)

31. The very thin oxide film on the aluminium foil of an electrolytic capacitor was originally formed chemically by an electrolytic process, and if the polarity were reversed, that process would be reversed and the oxide film destroyed.

    To avoid risk of reverse polarity connection the positive terminal of an electrolytic capacitor is always marked. (9.2)

32. Equivalent capacitance (series) $\frac{1}{C} = \frac{1}{C_1} + \frac{1}{C_2} + \frac{1}{C_3} + \cdots$ (9.4)

33. Equivalent capacitance (parallel) $C' = C_1 + C_2 + C_3 + \cdots$ (9.4)

34. Time constant $(\tau) = CR.$ (9.3)

35. When a voltage is applied to an R-C circuit the circuit voltage rises at a rate depending on the time constant. If a voltage-sensitive relay is included in the circuit the operation of the relay will be delayed until the circuit voltage reaches the relay operating voltage. The time delay can be made variable by using a variable resistor.
(Figure 9.8)

36. Power transmitted: VI watts (W)

    Alternatively I^2R watts

    or $V^2/_R$ watts (10.1)

37. 1 hp = 0. 746kW$_m$

    $\therefore$ 20 hp = 0.746 x 20

    = 14.92kW$_m$ (approx. 15, = $^3/_4$ x 20) (10.1)

38. Efficiency is $\dfrac{\text{output}}{\text{input}} = \dfrac{50 \text{ x } 0.746\text{kW}}{45\text{kW}} = 0.83$

    $\therefore$ Efficiency of motor is 83%. (10.1)

39. D.C. measuring instruments may be:

    Hot-wire
    Moving-iron
    Moving-coil
    Dynamometer

    The hot-wire instrument depends on the heating and elongation of a wire carrying the current. A moving-iron instrument has a fixed coil which carries the current and which magnetises a pivoted piece of soft-iron, drawing it into alignment against a spring. The moving-coil instrument has a fixed permanent magnet and a small pivoted coil carrying the current. The coil tries to align its axis with the axis of the magnet against the pull of a spring. A dynamometer instrument has both fixed and moving coils which may carry the same current in series or currents from different sources. (11.1 - 11.4)

40. A moving-iron instrument, operating on the square of the current, has a 'square-law' scale which is crowded at the lower end. A moving-coil instrument, operating directly on the current, has a near-linear scale, which displays nearly equal divisions over its whole length. The appearance of the scale therefore immediately shows whether the instrument is moving-iron or moving-coil. (11.2 - 11 .3)

41. No essential difference in the movements. The ammeter has a low-resistance fixed coil and the voltmeter a high-resistance one. (11.2)

42. In a dynamometer-type wattmeter the fixed coil is connected in the current line (or to a shunt in the current line), and the moving coil is connected across the voltage supply (with or without a dropping resistor). The movement of the pointer will indicate watts. (11.4)

43.   An instrument with a 'set-up' zero is one where its natural, or rest, position is not at the end of the scale but some distance along it. The part to the left of the zero is used to indicate when the measured quantity goes negative, such as reverse current or reverse power. A centre-zero ammeter is a special case.

Setting up of the zero can only be applied to moving-coil or dynamometer instruments; it cannot be applied to the moving-iron type.   (11.6)

CHAPTER 14 PRINCIPLES OF A.C. GENERATION

14.1   FARADAY'S LAW OF ELECTROMAGNETIC INDUCTION

In Chapter 6 it is explained how Faraday was led to propound his 'Law of Electromagnetic Induction'.

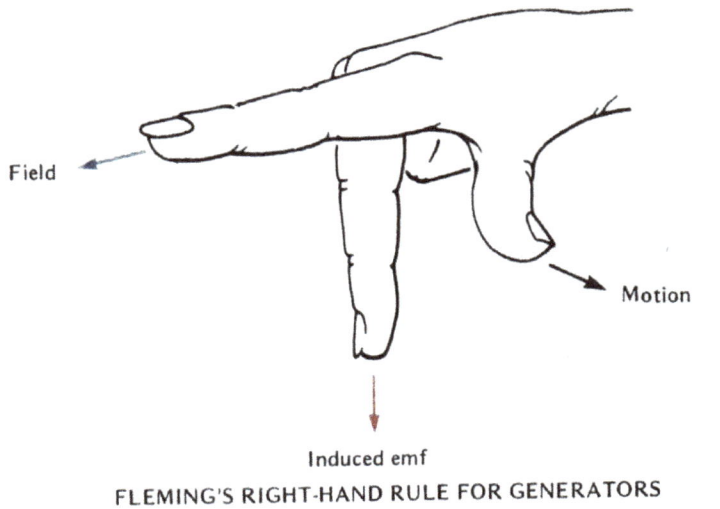

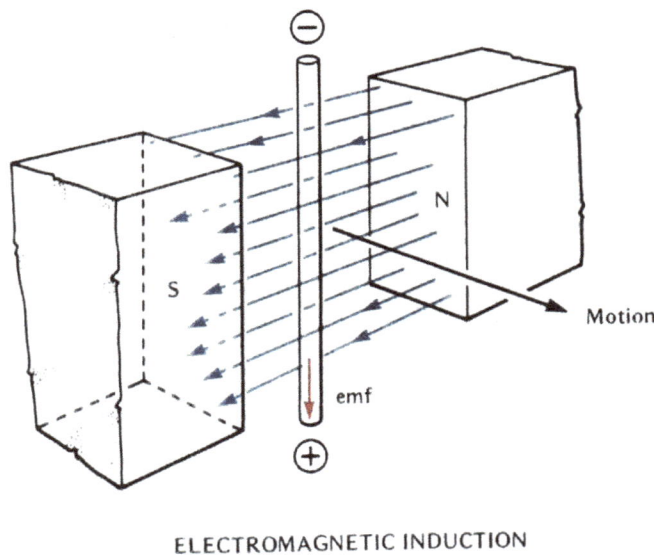

Induced emf

FLEMING'S RIGHT-HAND RULE FOR GENERATORS

ELECTROMAGNETIC INDUCTION

FIGURE 14.1
FARADAY'S LAW OF ELECTROMAGNETIC INDUCTION

This law states that, if a conductor is moved in a magnetic field, then an 'electromotive force' (emf) - or, simply, a voltage - is induced in that conductor, as shown in Figure 14.1. It follows that, if the ends of the conductor are connected to an external load, then an electric current, driven by that voltage, will flow from the conductor, through the load and back again. The set of conductors in which the voltage is induced is called the 'armature'.

Whereas Oersted showed that an electric current in a wire gives rise to an artificial magnetic field, Faraday showed the opposite - that if a wire moves in a magnetic field an artificial charge, or voltage, will be created in that wire. Faraday also showed that the magnitude of the voltage induced in the moving conductor depends on the strength of the magnetic field and the speed of movement, and on nothing else.

These two laws form the whole basis of electrical power generation, both a.c. and d.c. Starting with a magnetic field, either a natural magnet or an artificial electromagnet of Oersted's type, a conductor or a number of conductors are caused to move past it, from which the current is extracted as they are moving. First however look at one other rule which determines how this is achieved and the directions of movement and induced voltage.

Figure 14.1 also shows 'Fleming's Right-hand Rule for Generators'. If the right hand is held with the thumb, forefinger and centre finger extended mutually at right angles, then, with the magnetic field in the direction (North to South) pointed by the forefinger and the motion of the conductor in the direction indicated by the thumb, the centre finger will point in the direction in which the emf (i.e. voltage) is induced in that conductor (and in which current will flow when connected to a load).

The following paragraphs show how this can be put into practice.

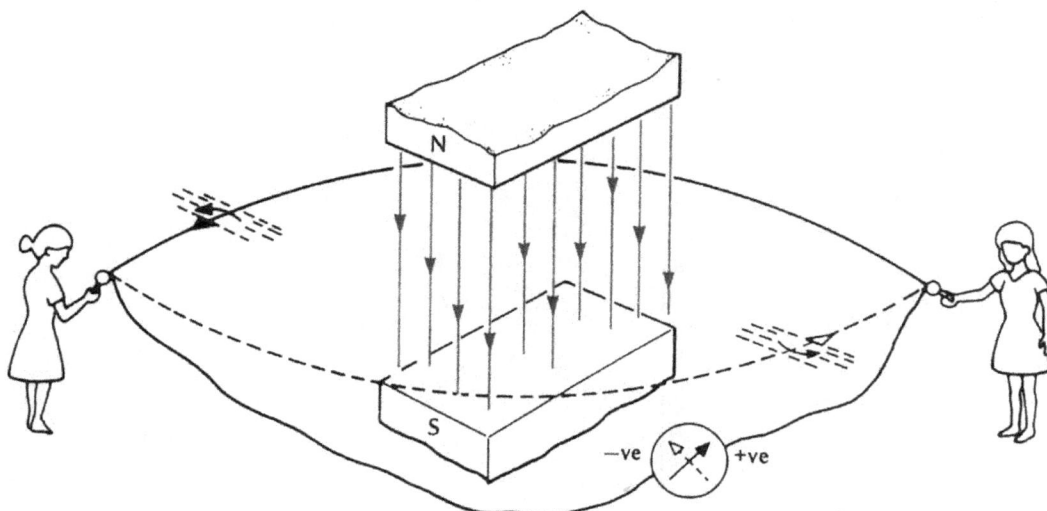

(a) ELECTROMAGNETIC INDUCTION

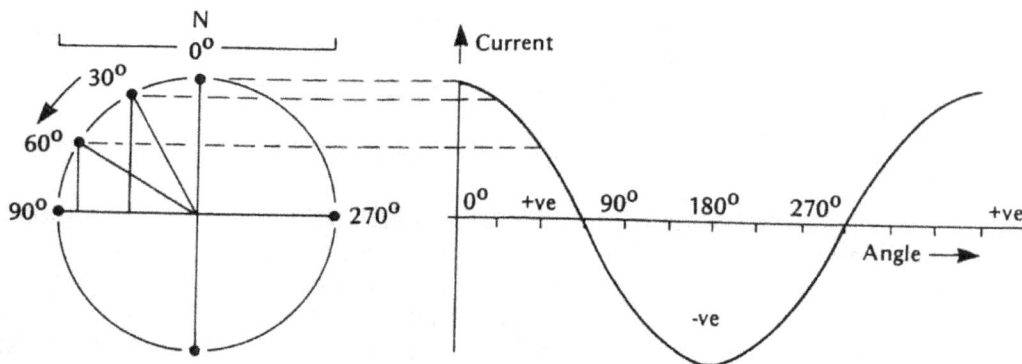

(b) ALTERNATING CURRENT

FIGURE 14.2
APPLICATION OF FARADAY PRINCIPLE

14.2   A.C. GENERATORS

Consider the scene (Figure 14.2(a)) where two girls are swinging a skipping rope. Suppose the 'rope' is a copper wire with its ends connected to a voltmeter, and suppose the rope swings between the poles of a large magnet - north pole overhead and south pole in the ground. There is then a downward magnetic field all over the rope.

If the rope is swinging anti-clockwise as seen from the left, as it passes the 12 o'clock position it is moving at its fastest past the N magnetic pole, and, by Fleming's Right-hand Rule, a voltage will be induced from right to left. The voltmeter will swing one way - say to the right.

As the rope moves on it moves less quickly across the magnetic field until, by 9 o'clock, it is not crossing it at all. No voltage will be induced, and the voltmeter indication falls to zero.

After 9 o'clock as the rope continues its swing it begins to move through the field in the other direction. The right-hand rule says that the voltage is induced the other way (left to right), and the voltmeter needle swings to the other side. At 6 o'clock the rope is moving at its fastest past the S-pole, and the voltmeter reaches its maximum left swing.

So on, past 3 o'clock, where the induced voltage is again zero, back to 12 o'clock where the rope is once more moving at its fastest past the N-pole, and the voltmeter needle swings back to its maximum reading to the right.

If the voltage indicated by the voltmeter is plotted against the rope's position (considered as 360° for one revolution) it takes a waveform (Figure 14.2(b)) - maximum positive at 12 o'clock (0° and 360°), maximum negative at 6 o'clock (180°) and zero at 9 o'clock and 3 o'clock (90° and 270°). The shape of the curve is that of a pure sine-wave (or more strictly in this case a pure cosine-wave).

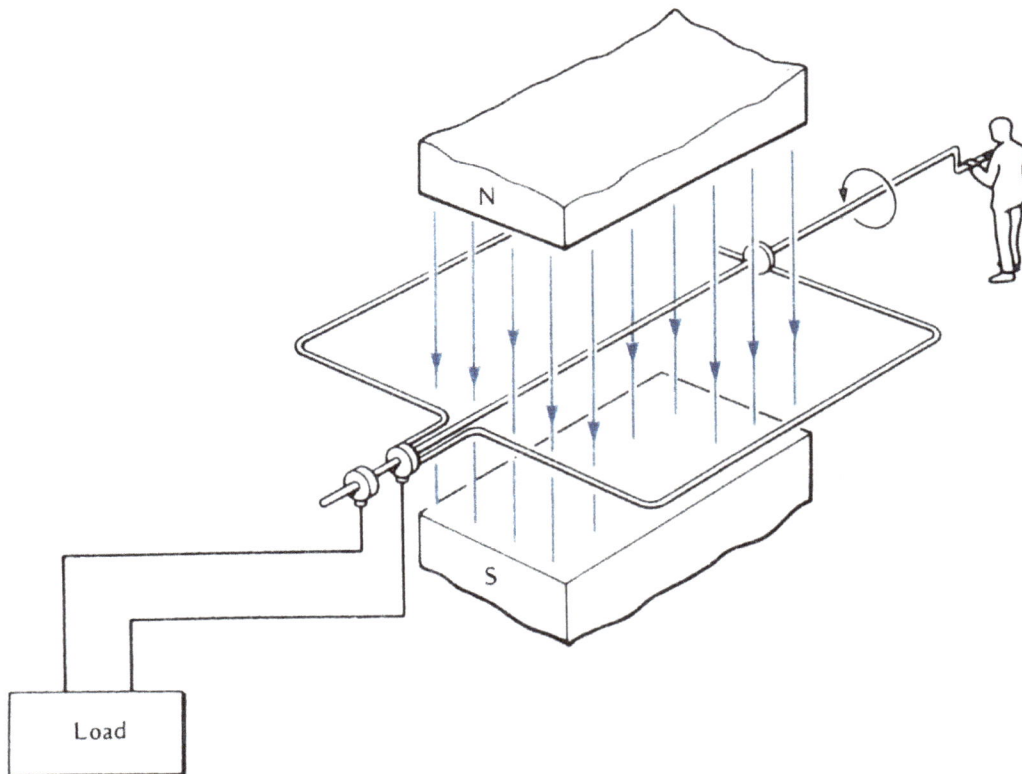

FIGURE 14.3
A.C. GENERATION - FIXED FIELD

Suppose, instead of the skipping rope, there were a loop of stiff wire on a shaft which can be turned, as shown in Figure 14.3. Suppose each end of the wire is connected to a slipring, insulated from the shaft, upon which brushes bear which are connected to a load or voltmeter as before.

As the shaft is turned, one bar passes the N-pole as the other passes the S-pole. Voltage is induced one way in one of them and the opposite way in the other. But as they are in series the two voltages add up and appear as a double voltage at the sliprings, and so at the voltmeter.

Faraday's theory required only that the conductor should be moving through a magnetic field - that is, that there should be relative motion between conductor and field. It would work just as well if the magnetic field moved past the conductor.

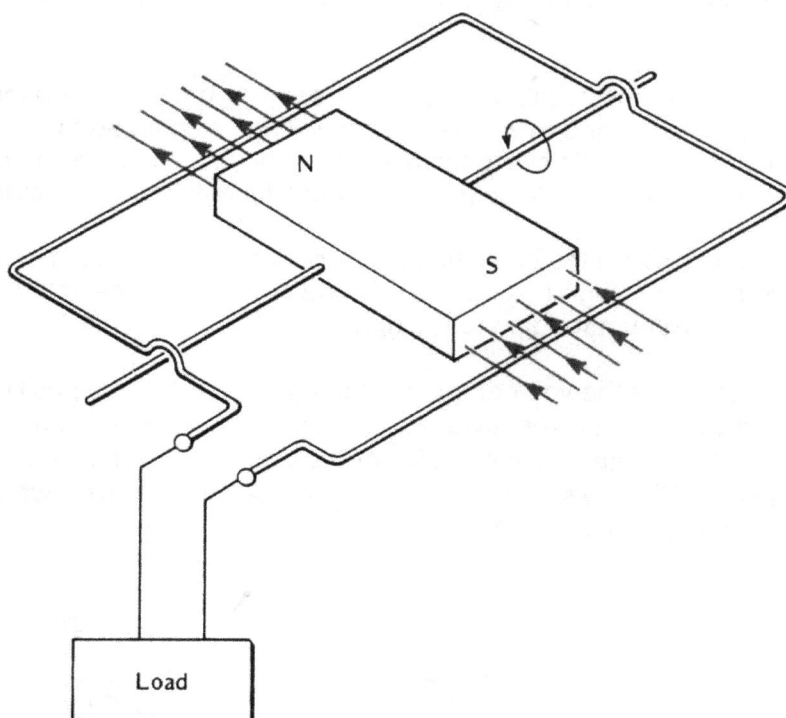

FIGURE 14.4
A.C. GENERATION - ROTATING FIELD (PERMANENT MAGNET)

In the arrangement shown in Figure 14.4 this is just what is happening. The stiff wire loop is fixed, and the permanent magnet is rotated past it and inside it. As a pole passes a fixed conductor a maximum voltage is induced in it, opposite voltages on opposite sides, and they add up to give a double voltage at the terminals or at the voltmeter. Only in this case no sliprings or brushes are needed - a great advantage for many reasons, not least that it eases maintenance.

So far we have only considered a permanent magnet as producing the magnetic field. But far better results can be achieved by using an electromagnet, as in Figure 14.5, which can produce much stronger fields and therefore much higher induced voltages. In that case however d.c. power must be provided to the coil which magnetises it. This can come from a battery or other d.c. source, but a pair of sliprings must be reintroduced to bring the battery current to the moving magnetising coil - called the 'field coil'. This coil is said to 'excite' the field, and the whole process is called 'excitation'.

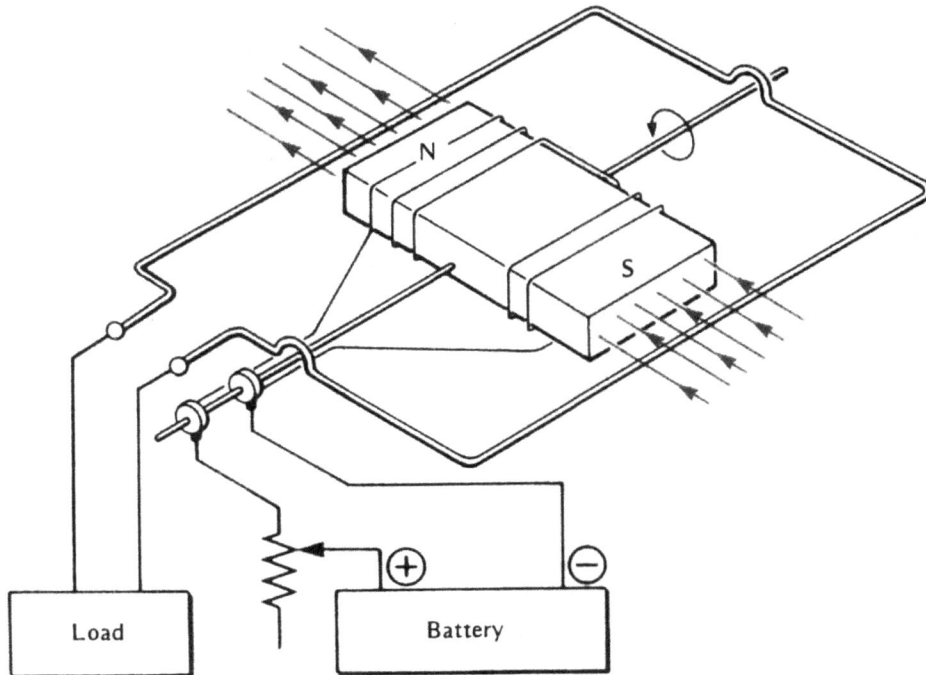

FIGURE 14.5
A.C. GENERATION - ROTATING FIELD (ELECTROMAGNET)

Because the field magnet is not permanent but is an electromagnet, it is possible to vary the coil current by a resistance and so to vary the strength of the magnetic field itself. This in turn will vary the amount of the induced voltage. Here then is a simple method of controlling the main machine's voltage by varying the excitation.

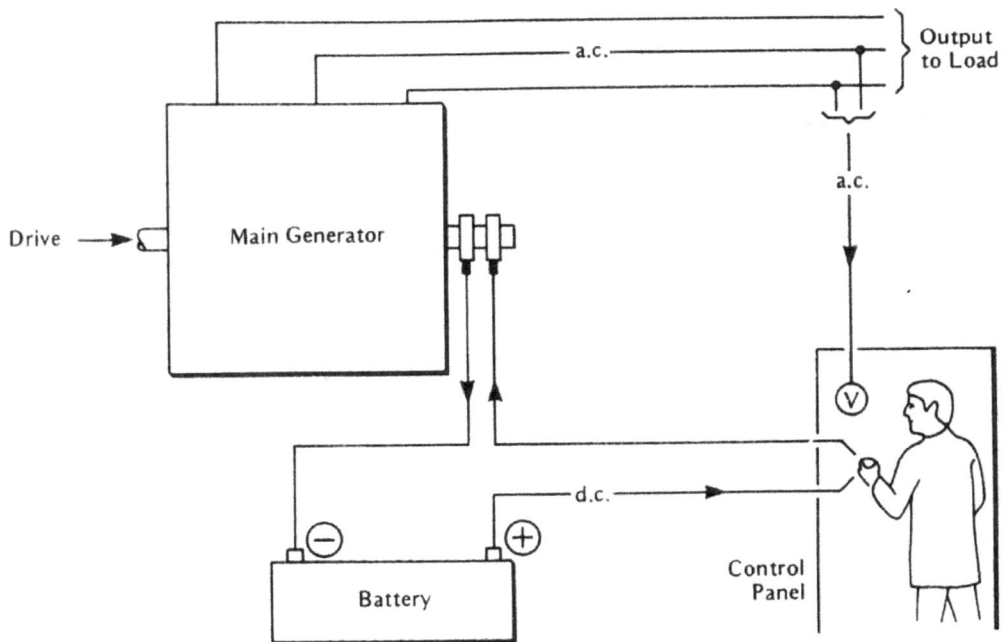

FIGURE 14.6
A.C. GENERATION - VOLTAGE CONTROL

Figure 14.6 shows in principle how this is done. Top left is the generator, and on its right are the sliprings carrying the exciting current to the field coils. Connected to the generator's output lines is a voltmeter, which can be seen by the operator. He knows what voltage he wants to see. If it is not right, he adjusts the resistance controlling the level of field current from the battery. As he adjusts this current up or down, so the voltmeter will indicate a rise or fall of output voltage. He continues until the voltmeter reads the voltage desired.

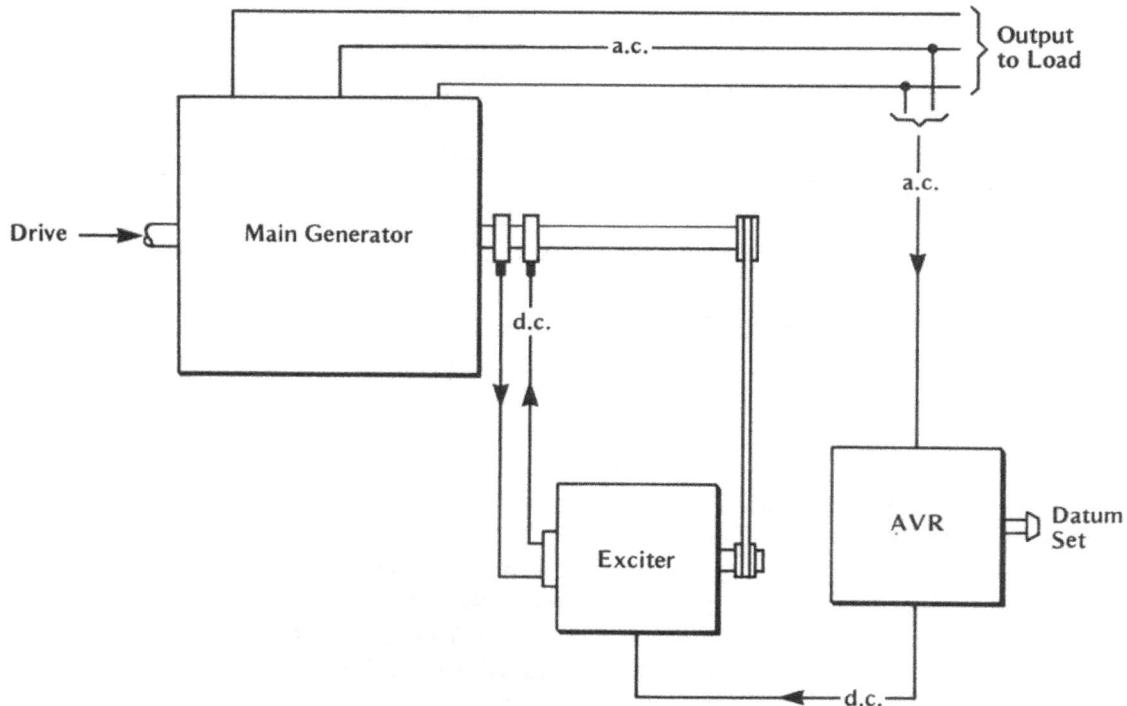

FIGURE 14.7
A.C. GENERATION - AUTOMATIC VOLTAGE CONTROL

Figure 14.6 showed an operator intervening between the voltage output and the adjustment needed to correct it. The next stage is to make it automatic. An electronic device called an 'Automatic Voltage Regulator' (AVR for short) senses the output voltage and compares it with a datum which has previously been set on by hand. It decides whether the output voltage is correct, too high or too low. This is shown in Figure 14.7.

In this case the battery which has hitherto been providing the d.c. power for the field coils is replaced by a d.c. generator, called an 'Exciter', driven mechanically by the main generator shaft. If the AVR has decided that the generator's output voltage is too high or too low, it signals the exciter to decrease or increase its current to the main generator's field coils, so bringing the main generator's output voltage down, or up, until it is correct. All platform and onshore generators work on this principle.

The AVR acts on the machine's voltage much as a governor acts on its speed. Both can have their datum set on by hand, after which the voltage (or speed) should be held automatically between close limits, no matter how the loading varies. If either device fails, it can be put into manual control.

The final stage in the development of an a.c. generator is shown in Figure 14.8. The exciter, instead of being belt- or gear-driven by the main shaft and requiring sliprings to bring the field coil current into the generator, is now moved up to the main machine itself and shares

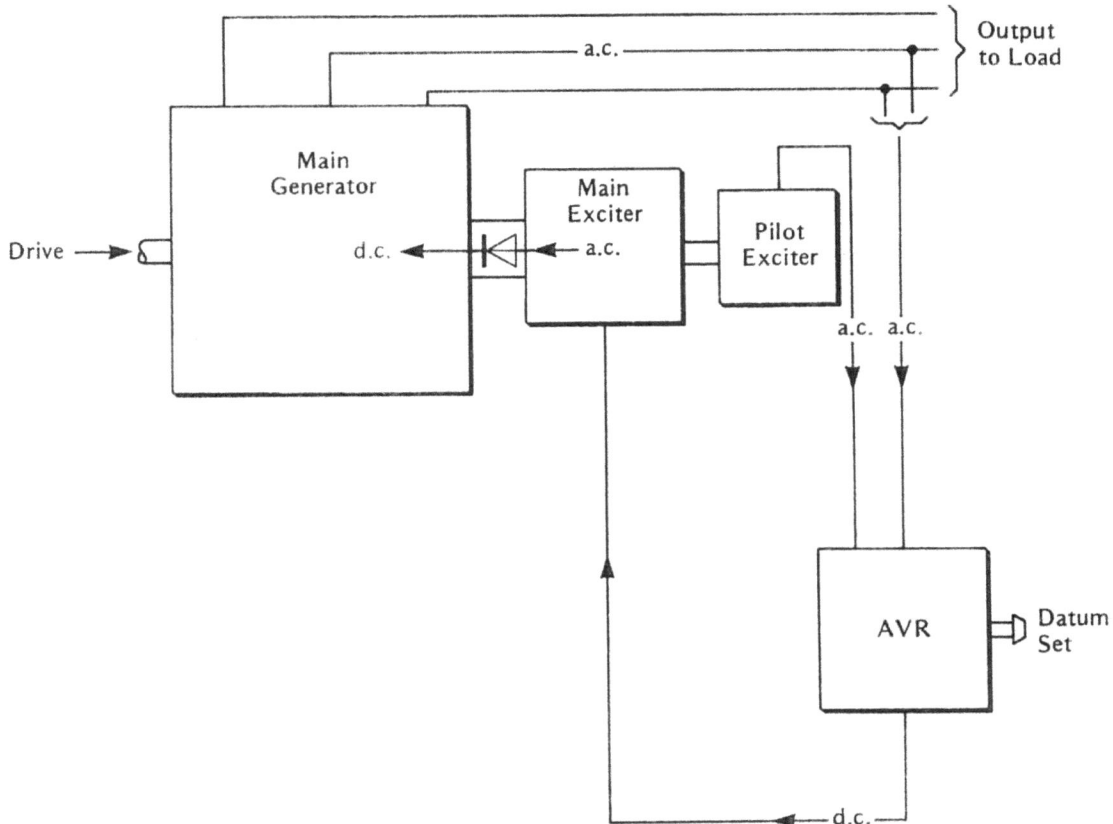

FIGURE 14.8
A.C. GENERATION - BRUSH LESS EXCITATION

a common shaft. A small second, or 'pilot', exciter is sometimes added to excite the main exciter. The AVR, as before, signals the main exciter whether to raise or lower its output to the field coils, to which it is connected by cables through the hollow shaft.

For mechanical reasons it is not possible with this arrangement to use a d.c. generator as an exciter, as the brushgear would have to rotate. Instead the exciter is another, but smaller, a.c, generator working on the principle of Figure 14.3 - that is, with stationary field and rotating armature. The a.c. output from this armature is taken through static 'rectifiers' rotating with the shaft; they convert the a.c. into d.c. and pass it on to the rotating field winding of the main generator.

This arrangement gives better and tighter control over the voltage, but its prime advantage is that it totally dispenses with the sliprings and brushes. It is known as 'Brushless Excitation' and is now used almost exclusively on all platforms and many shore-side generators.

The actual construction of an a.c. generator and of its driving engine is described in Part 2 Electrical Power Generation.

14.3 3-PHASE A.C. GENERATORS

Figure 14.4 showed how the generator developed into a magnet or 'field' rotating inside a fixed loop. As the poles passed each side of the loop, an alternating voltage was induced in it which was conveyed to the terminals, and thence to any connected load.

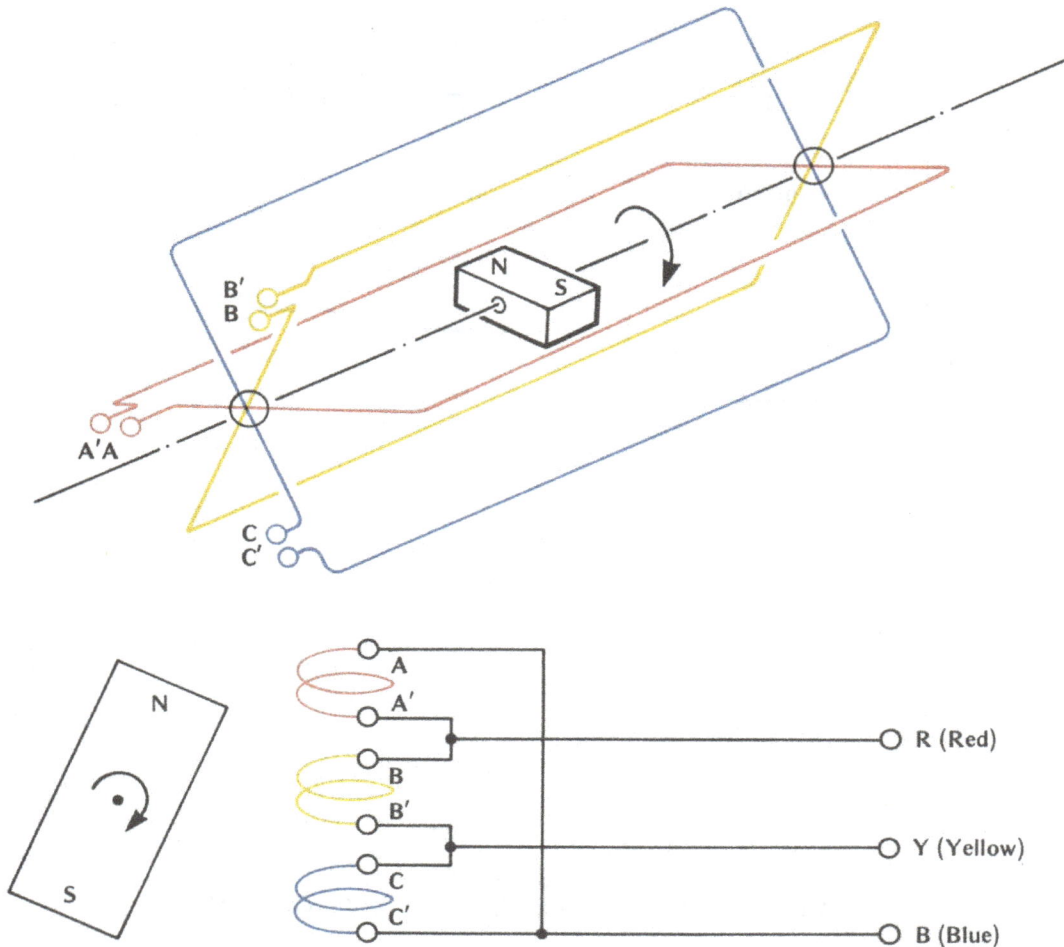

FIGURE 14.9
3-PHASE GENERATION - WINDINGS

In practice such a system, though used in small installations, is not very economical, and with most large installations three separate loops are provided, equally spaced at 120° around the shaft as shown at the top of Figure 14.9 where they are distinguished by different colours. The same field rotates inside all three loops, so the same voltage is generated in each, and at the same frequency, but, because of the 120° spacing, the voltage in each loop rises to its peak one-third of a cycle, or 120°, later than the one before it. There are in fact three generators in a single machine. The voltage from each of the three loops is brought out to separate pairs of terminals A, A'; B, B'; and C,C'.

At first sight it would appear that six wires would be needed to remove the current from the three loops, but actually each loop shares a wire with its neighbour, as shown diagrammatically at the bottom of the figure. Thus the red and yellow loops share wire R; yellow and blue loops share wire Y, and blue and red loops (if examined closely) share wire B. So only three wires are actually needed to convey the power away from the generator. They are variously identified as A, B, C or L1, L2, L3 or U, V, W (in Continental machines) or are simply coloured red, yellow and blue.

Such a machine is called a '3-phase' generator, and if you look at the overhead pylons all over the country you will see that each carries one, or sometimes two, sets of three wires.

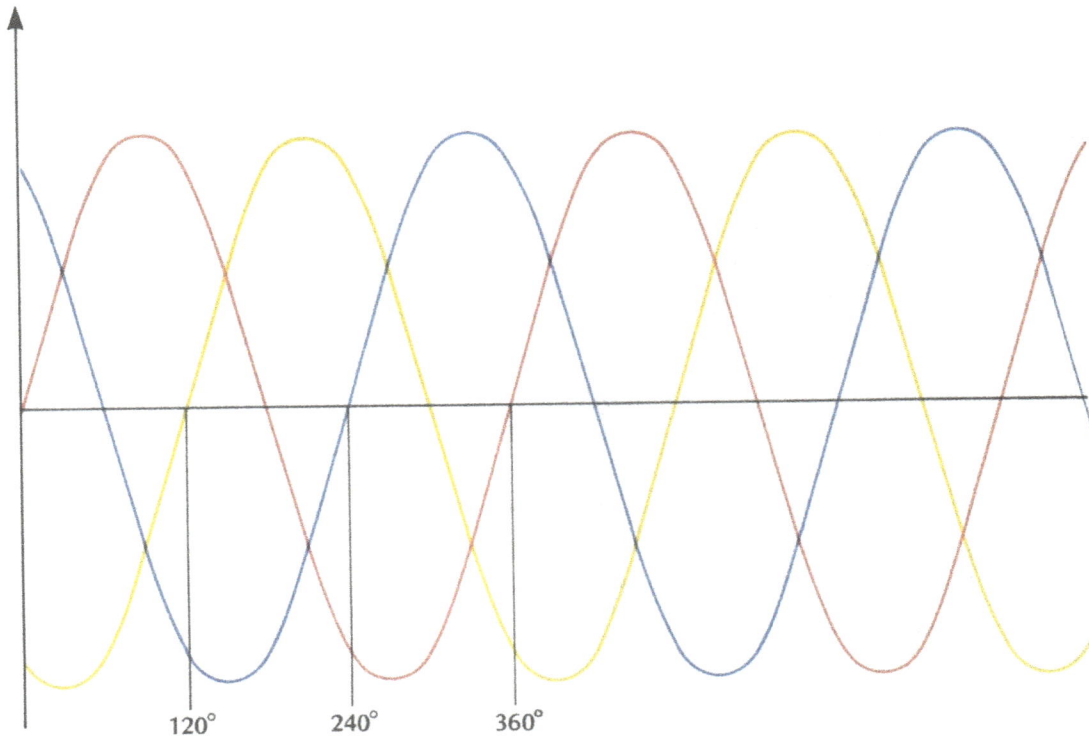

FIGURE 14.10
3-PHASE GENERATION - WAVEFORMS

The voltage waveform of a 3-phase system is simply three sine-waves similar to those of Figure 14.2, but each displaced one-third of a cycle, or 120° behind the other. This is shown in Figure 14.10.

The advantages of using a 3-phase alternating current may be compared with those of using an engine with many cylinders. With a single cylinder power comes from the engine in spurts, but with the addition of more cylinders the spurts of power overlap, producing a steadier level of power. In the same way the overlapping phases of a 3-phase circuit maintain a smoother and higher level of power. Power in large installations - and certainly on all platforms and shore networks - is always produced and transmitted in 3-phase systems because the generators and power lines are less expensive than those for a single-phase system carrying the same power, and also because they are best adapted to motors (as is shown in Part 5).

14.4   D.C. GENERATORS

Reverting to Figure 14.3, a stiff wire loop is rotating between the poles of a permanent magnet, and with its two ends connected each to a separate slipring on which brushes bear. As the loop rotates, each side has induced in it a voltage which changes direction at every half rotation - that is, as that side passes a N- or a S-pole. Each slipring, and so each brush, receives alternately a forward and a reversed voltage, so that the output from the brushes to the load is 'alternating'.

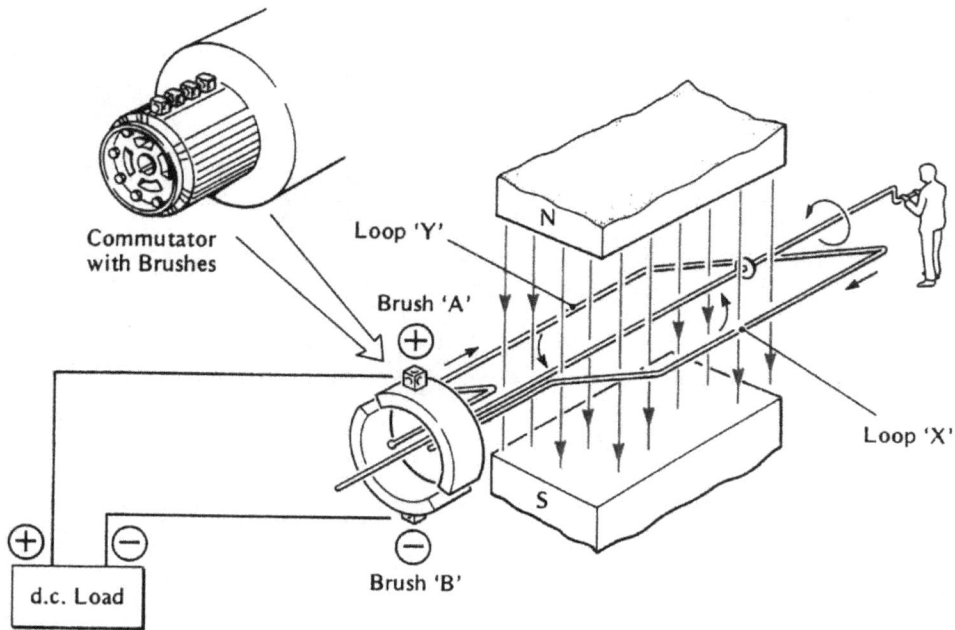

FIGURE 14.11
DIRECT CURRENT GENERATOR

Suppose now, instead of the two sliprings, there had been a single slipring split into two and with a conductor-end connected to each half, but still with two brushes, which will be called 'A' and 'B'. This is shown in Figure 14.11.

As the side 'X' of the loop is passing the N-pole, by Fleming's Rule the voltage (V) induced in that loop is towards the brushes, and in loop 'Y', which is passing the S-pole, it is away from the brushes. As they are in series a voltage (2V) appears across the two halves of the slipring, that half connected to loop 'X' being positive and that to loop 'Y' negative, and so brush 'A' also is positive and brush 'B' negative.

Half a revolution later the voltages in loops 'X' and 'Y' are reversed, and the half slipring connected to loop 'X' is negative and to loop 'Y' positive.

But the split slipring itself has now turned half a revolution, so loop 'Y' (positive) is now with brush 'A', and loop 'X' (negative) with brush 'B'. So brush 'A' is still positive and brush 'B' is still negative, that is to say there has been no change from half a revolution before. Brush 'A' remains positive and brush 'B' negative throughout. The split slipring acts as a reversing switch, and the voltage does not alternate but remains unidirectional at all times.

Such a machine is a 'Direct Current Generator' (in the old days called a 'Dynamo') and is identified by having a split slipring, the two parts being insulated from each other but the brushes passing over both.

In practice of course such a generator would have not one loop but many, so the slipring would be split not into two parts but into several. It would appear something like that shown on the top-left of Figure 14.11 and is called a 'Commutator'. It is the unique indication of all d.c. machines.

D.C. generators are not much used today as power sources except for special applications, principally because the commutator limits the voltage which can be generated, and it also poses maintenance problems.

CHAPTER 15 STAR AND DELTA CONNECTIONS

15.1   3-PHASE CONNECTIONS

In Chapter 14 was described how an a.c. generator is usually equipped with three separate windings, disposed at equal 120° space intervals around the stator. These windings produce identical alternating voltages, but each is timed to peak one-third of a cycle (120° time degrees) after its predecessor. The three windings are distinguished by calling them 'red', 'yellow' and 'blue', or sometimes A, B and C or U, V and W. The red-yellow-blue notation will be used in this book.

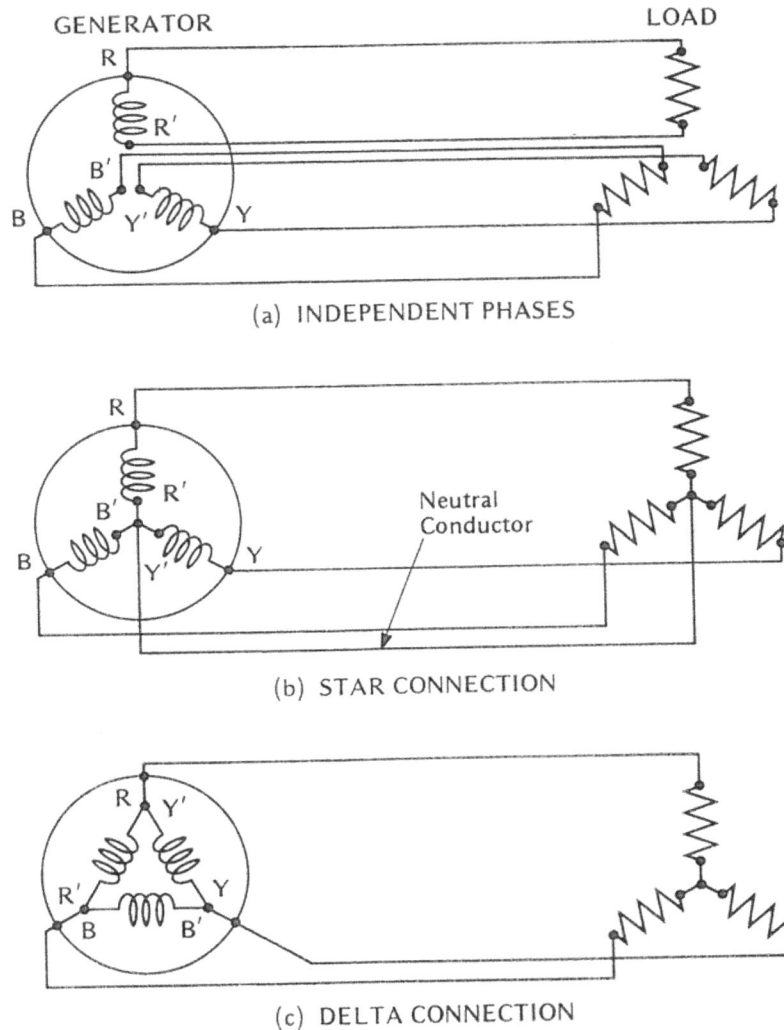

(a) INDEPENDENT PHASES

(b) STAR CONNECTION

(c) DELTA CONNECTION

FIGURE 15.1
STAR AND DELTA CONNECTIONS

Each winding has two ends, called R & R', Y & Y' and B & B'. Normally each winding supplies one element of a 3-phase load. If the three loads are identical - that is, if they each draw the same current - the load is said to be 'balanced'. Unless otherwise stated in the description which follows, loads will always be assumed to be balanced.

The simplest way of connecting the generator windings to the loads is to arrange that each is connected to its own load independently of the others, as shown in Figure 15.1(a). Since each winding has two terminals, this will require six wires to convey the power from all three phases.

As will be shown later, such a form of connection, though perfectly correct, is wasteful of copper and cable material and is seldom, if ever, used.

15.2 STAR CONNECTION

What is done is to make each phase share one conductor with the others. One way of doing this is to connect the inner ends of all three phase windings together - that is to say, terminals R', Y' and B' are commoned - and a fourth wire is taken from this common point to a similar common point of the three loads; this is shown in Figure 15.1(b). The wires connected to the three outer terminals R, Y and B each carry the outward phase currents into their respective loads, but all use the common return path to R', Y' and B'. Such an arrangement is called a 'star connection' (American 'wye'). The common point is called the 'star point', and the fourth or common wire is called the 'neutral' conductor.

Star connection of generators and transformer secondaries is widely used both ashore and on platforms.

15.3 NEUTRAL CURRENT

Look now more closely at the current which flows in the neutral wire; it is the sum of the red, yellow and blue return currents.

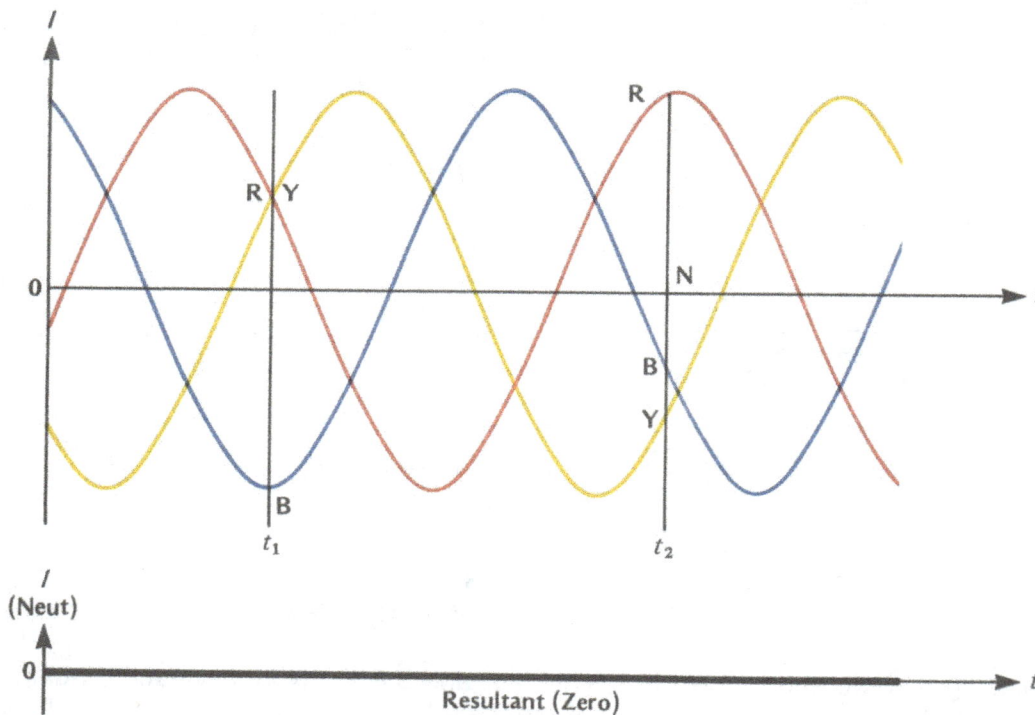

FIGURE 15.2
NEUTRAL CURRENT

In a balanced system the currents in the three phases are equal in magnitude but are displaced one-third of a cycle, or 120° time degrees, from each other. Three such balanced phase currents are shown in the upper part of Figure 15.2, and it will be easily seen that yellow current lags 120° on red, and blue current 120° on yellow. It follows that, on completion of one full cycle, that red once more lags 120° on blue.

Since the neutral return current is the sum of all three phase currents at all instants, the neutral current wave can be obtained by adding the three values of the phase currents at every instant, taking account of their signs. Suppose we take the instant t_1 where blue is at a negative peak (B). At that moment red and yellow are both positive at half peak value, yellow rising (Y) and red falling (R). So the sum of the three is -1 +½ +½, which is zero. If another arbitrary point is taken, say at t_2, and the three values of current at that instant are measured taking account of their signs (NR + NY + NB), it will be found that, once again, they add up to zero.

The surprising conclusion is that, although the neutral wire is carrying the sum of the three phase currents, it is actually carrying nothing at all. (Note that it was assumed that the loads were balanced; if they had not been, this conclusion would be no longer valid.)

15.4   3- AND 4-WIRE SYSTEMS

If the neutral conductor carries no current, why have it at all, or waste money on expensive cable?

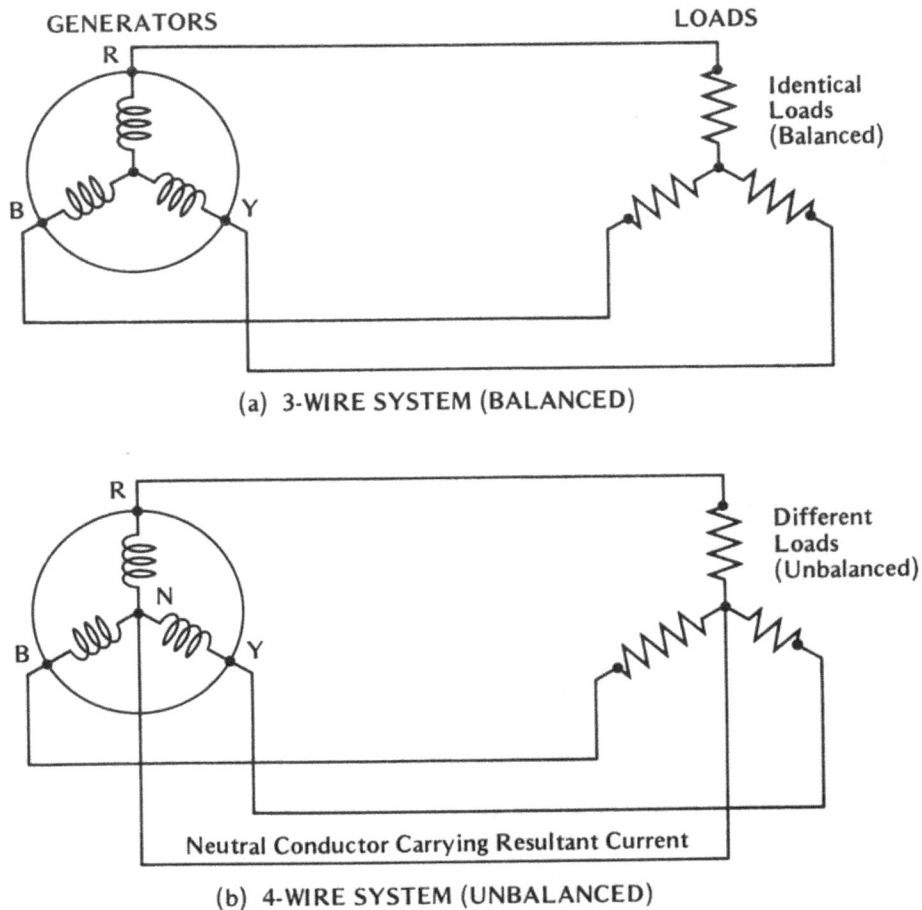

(a) 3-WIRE SYSTEM (BALANCED)

(b) 4-WIRE SYSTEM (UNBALANCED)

FIGURE 15.3
STAR CONNECTION 3- AND 4-WIRE SYSTEMS

It is in fact usually left out, at least in high-voltage systems where the loads are always regarded as nearly balanced and also in some low-voltage systems where balance may be assumed (for example motors). The neutral conductor is entirely dispensed with, and all the power is transmitted by the three phase conductors only. Such a system is known as '3-wire' distribution and is shown in Figure 15.3(a).

Where balance cannot be assumed, particularly in low-voltage systems where there may be many single-phase loads, the neutral current is not zero, and the neutral wire must be retained. This is shown in Figure 15.3(b) and is known as '4-wire' distribution. It is generally used on platform and shore-side low-voltage distribution systems.

Where there are many single-phase loads which cause unbalance (for example lighting circuits), they are connected between one phase and the neutral which is available on a 4-wire system. Every effort is made at the design stage to distribute the single-phase loads as evenly as possible between each of the phases and neutral so as to reduce to a minimum any unbalance caused. As a result, although the neutral does not carry zero current and therefore cannot be dispensed with, the current which it does carry is relatively small compared with the phase currents. If the cables from a transformer feeding a low-voltage system are examined, it will usually be found that, although each phase may require perhaps four cables in parallel for each phase to carry the large phase currents, there may be only two, or even one, neutral cable.

15.5   DELTA CONNECTION

Referring to Figure 15.1(c) it will be seen that there is another way by which the generator windings can share conductors. In this case, instead of sharing a common return conductor as with star connection, each winding shares a conductor with its neighbour at both ends. That is to say, R is commoned with Y', Y with B' and B with R'. There are only three conductors leaving the generator and carrying power to the loads. Because, for convenience of drawing, such an arrangement is usually shown in a triangular form, it is called a 'delta connection' (American 'mesh'). There is no star-point in this case and therefore no possibility of any neutral connection. Distribution from a delta-connected source or to a delta-connected load must therefore always be 3-wire.

Figure 15.1 shows the **loads** all star-connected, even when supplied from a delta-connected generator. The loads themselves however may equally well have been delta-connected, although this is unusual. In that case there would be no neutral conductor.

From Figure 15.1(b) it will be seen that, with star connection, the line current from generator to load must be the same as the generator winding's phase current, since they are in series. With a delta connection however (Figure 15.1(c)) the line current is divided between two phases. This gives a slight cost advantage where heavy currents are involved and the copper section is large.

From Figure 15.1(c) it will be seen that, in a delta-connected circuit, the voltage between lines is the same as the voltage across one winding of the generator, since they are in parallel.

For various reasons, chiefly because of the availability of the neutral for earthing purposes (see Part 4 Electrical Systems), star connection is almost always used with generators and with the secondary sides of distribution transformers (the primaries however are usually delta-connected).

One notable exception is the generator system of many platform drilling installations where they are separate from the platform system. They are seldom earthed, and drilling practice customarily uses delta-connected generators and transformer secondaries.

The relationships between the phase and line voltages, and between the phase and line currents, in star- and delta-connected systems are explained in Chapter 20.

15.6   DRAWING OF STAR- AND DELTA-CONNECTED APPARATUS

In Figures 15.1 and 15.3, star connection was shown by a 'star' or 'Y' arrangement of windings or loads, and delta connection by a triangular arrangement. This is an advantage for instructional purposes, but for distribution drawings another way is used for showing these connections (Figure 15.4).

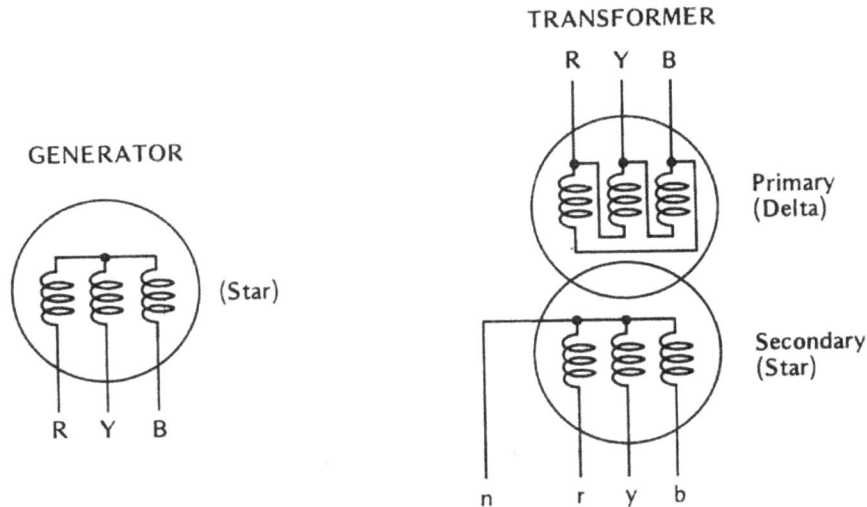

FIGURE 15.4
PRESENTATION OF STAR AND DELTA CONNECTIONS
IN DISTRIBUTION DIAGRAMS

With this presentation no attempt is made to 'angle' the windings, but they are drawn parallel, in red-yellow-blue order. On the left is a star-connected generator, and on the right a delta-primary/star-secondary transformer is shown in which the neutral has been brought out as a fourth wire. It can be seen how this arrangement simplifies the drawing of distribution circuits. It is mentioned here because the reader is bound to come across this method on system drawings and should recognise the star and delta connections when seen.

CHAPTER 16 PRINCIPLES OF ALTERNATING AND ROTATING MAGNETIC FIELDS

16.1   ALTERNATING MAGNETIC FIELDS

It is shown earlier how Oersted's discovery that a magnetic field surrounds every current-carrying conductor led to the construction of coils, with or without an iron core, which concentrated the magnetic field surrounding each single piece of wire in each turn of the coil to give a strong magnetic field mainly along the axis of the coil.

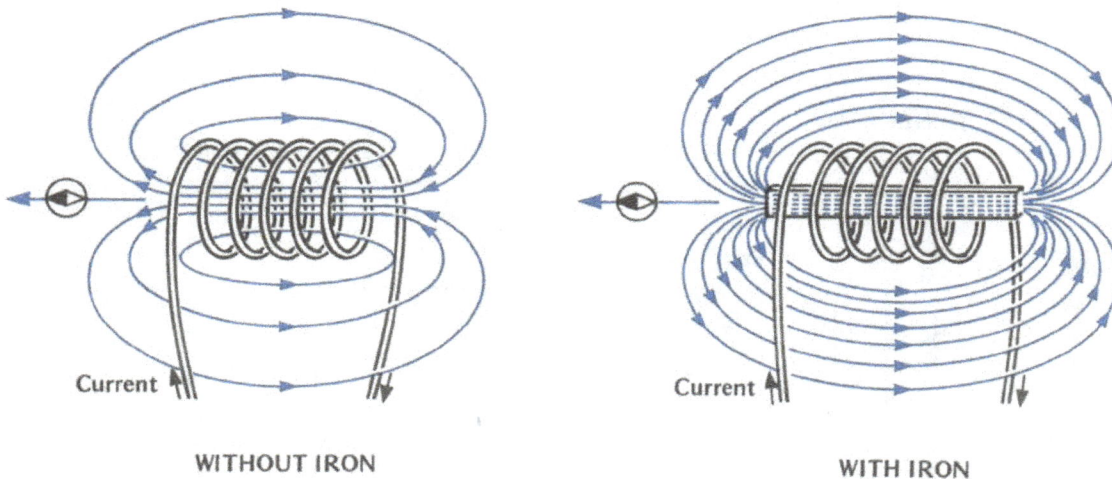

WITHOUT IRON   WITH IRON

FIGURE 16.1
MAGNETIC FIELD AROUND A COILED CONDUCTOR

The word 'mainly' is used advisedly, because there is also some return field **outside** the coil, and not all the inside field is wholly concentrated in the centre. When iron is present the concentration is greater, but some of the field 'gets away', even with iron. This is known as 'leakage field', as distinct from the useful axial field which it is the coil's purpose to produce.

Taking Oersted's Law literally, any current in a wire gives rise to a field around it, and the strength of that field is proportional to the strength of the current producing it. This applies to the **direction** of the current also. A current flowing in one direction will produce a certain field; the same magnitude of current flowing in the opposite direction will produce a field of the same strength but in the opposite direction.

Thus in Figure 16.1, with an alternating current flowing, if the current direction in one half-cycle is as shown, the field direction at the same time is also as shown. Half a cycle later the current will be flowing in the opposite direction; so too the field will be in the opposite direction.

Therefore an alternating current will produce a corresponding alternating field, whose strength at any instant is proportional to the magnitude of the current at that instant and which varies 'in phase' with the current - that is, it rises, falls and reverses in exact synchronism with it.

16.2   ROTATING MAGNETIC FIELD

A rotating magnetic field is an essential part of all electrical machines. In an a.c. generator it is provided by the rotor carrying d.c.-excited magnetic poles, whose effect is just as if a large permanent magnet were rotating inside the machine.

A rotating field is also necessary for the operation of an induction motor (see Part 5 Electric Motors), where it must be produced by the stator which, of course, is not moving. The problem then is: how to produce a moving, rotating magnetic field from stationary windings.

Fortunately this is possible if the supply is 3-phase. Three separate windings are provided in the stator slots, and their axes are arranged to be displaced in space by 120° as explained in Chapter 14. Also the three alternating line voltages of a 3-phase system are displaced 120° in time. It is this combination which makes possible a rotating magnetic field. This can be proved mathematically, but the following description explains it using first principles.

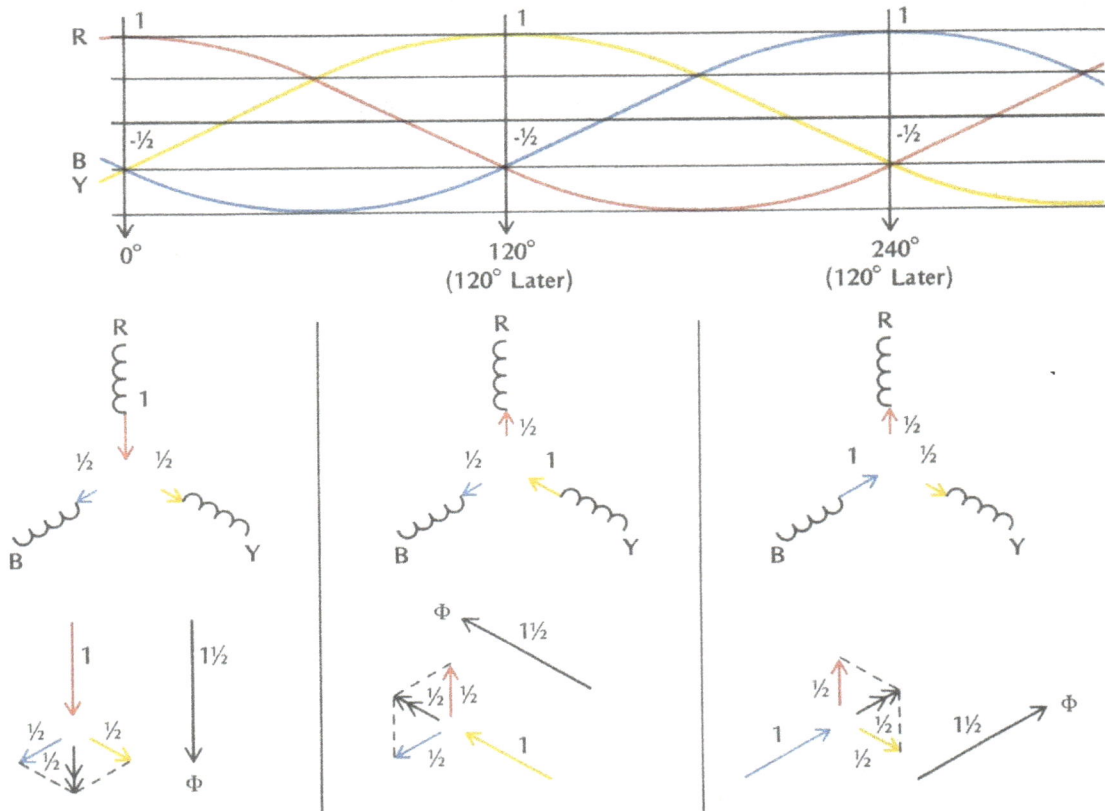

FIGURE 16.2
ROTATING MAGNETIC FIELD

In the centre of Figure 16.2 are shown the three stator windings of an induction motor, each drawn as a simple coil. The winding which is connected to each phase is distinguished by its colour - red, yellow or blue. The three windings are distributed 120° apart in space around the stator.

At the top is shown one cycle of currents of the 3-phase supply, each current wave being coloured. The 120° displacement in time is clearly seen.

The current in each of the three windings gives rise to a magnetic field along its axis; these are shown by coloured arrows for each field near the centre of the figure. As the current alternates, so also does the direction of the field along its axis. The convention has been adopted that, when the current is positive (above the line in the upper part of the figure), the arrow points **towards** the centre, and when it is negative, it points away from the centre.

Looking at the left-hand centre figure, taken at the instant where $t = 0$ (left-hand edge of the current waves), R phase current is at its maximum positive, and Y and B phases are negative and both equal in magnitude to half the maximum. Therefore the R phase flux is shown as '1' with its arrow pointing towards the centre, and Y and B phase fluxes are both shown as '½' with their arrows pointing away from the centre.

These arrows can be regarded as flux vectors, and in the bottom figure they are resolved. The R phase flux is at 12 o'clock and pointing downwards, and the Y and B phase fluxes are respectively at 4 and 8 o'clock, half the length of R and pointing outwards in those directions. The latter two, by normal vector addition, combine to give a single flux of length ½ at 6 o'clock in a downward direction.

The R phase (length 1) and the combined Y and B phases (length ½) are both in the same direction and add together to give a single resultant flux of length 1½ in a 6 o'clock direction.

If exactly the same procedure is adopted for a time one-third of cycle (120°) later, Y phase is now at a maximum positive and B and R phases at half that value negative. B and R now combine to add to Y, giving a resultant flux of magnitude 1½ in the 10 o'clock direction.

Similarly at a time one-third of a cycle still later (240°), B phase is at a maximum positive and R and Y phases at half that value negative. R and Y now combine to add to B, giving a resultant flux of magnitude 1½ in the 2 o'clock direction.

One-third of a cycle still later (360°) the conditions are the same as at 0° ($t = 0$), and the process repeats with each cycle.

As time advances and the current cycle changes from 0°, through 120° and 240°, back to 0° and then repeats, the resultant flux constantly changes direction (but not magnitude, which remains constant at 1½) and swings from 6 o'clock, through 10 and 2 o'clock, back to 6 o'clock and then starts again. Although only three different points in the cycle have been described, the same treatment at intermediate points would show the resultant flux in intermediate positions. For example, if the currents were taken at 60° and the fluxes combined as before, the resultant flux would come out at 8 o'clock.

This is therefore a continuous process, and the resultant flux direction moves continuously round the stator, remaining constant in magnitude and completing one revolution per cycle of supply. A rotating magnetic field has been produced from a set of stationary stator windings.

Note: the statement that the flux completes one revolution per cycle of supply is only true when there is one winding per phase as shown in Figure 16.2; this is a so-called '2-pole' machine (by analogy with d.c.), and with a 60Hz supply the field rotates at 60 rev/s or 3 600 rev/mm (with 50Hz at 50 rev/s or 3 000 rev/mm). If there were **two** windings per phase ('4-pole' machine), a similar treatment would show that the field rotates only half a revolution per cycle, so that it makes one revolution per two cycles; at 60Hz this is 30 rev/s or 1 800 rev/mm (at 50Hz, 25 rev/s or 1 500 rev/mm). Similarly a '6-pole' machine (three windings per phase) would at 60Hz have 1 200 rev/mm (50Hz, 1 000 rev/mm), and so on.

16.3   ROTATING FIELD FROM SINGLE PHASE

What has been described (para. 16.2) is the production of a rotating field from a 3-phase supply. It can however be done from a single-phase supply, although it is seldom found except in the smallest motors which have only a single-phase supply to run on.

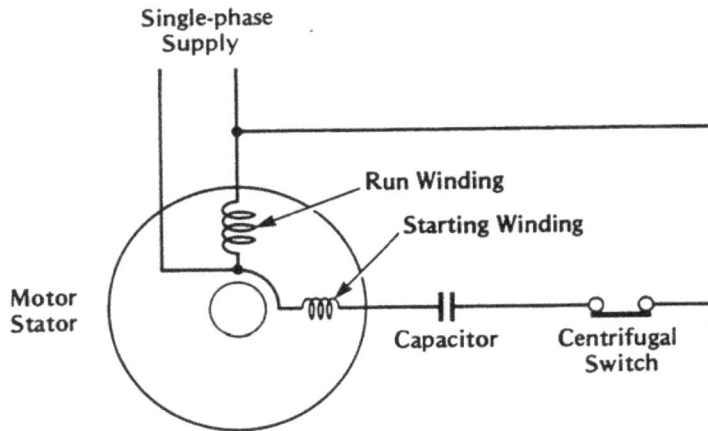

FIGURE 16.3
ROTATING FIELD FROM SINGLE-PHASE SUPPLY
(APPLIED TO A CAPACITOR MOTOR)

It is achieved by giving the motor two separate windings displaced in space by 90° One, the 'running' winding, is fed direct from the single-phase supply, and the other, the 'starting winding', is fed from the same supply but through a series capacitor. This causes the current in that winding to **lead** nearly 90° on the current in the other, so there is the situation where the two windings are displaced 90° in space and are fed from supplies 90° displaced in time. By an exactly similar treatment to that used for the 3-phase case, it will be found that this too will produce a rotating field, enabling the motor to start and run. Such a motor is referred to as a 'Capacitor Motor' or 'Split Field Motor'. It is further described in Part 5 Electric Motors.

On most such motors a centrifugal switch cuts out the capacitor-fed starting winding when the motor is up to speed; it will then continue to run with only its single-phase 'running' winding connected. Very small motors often dispense with the centrifugal switch and leave the capacitor in circuit all the time.

CHAPTER 17 SIMPLE HARMONIC MOTION

17.1   ALTERNATING QUANTITIES

An alternating current or voltage is 'periodic' - that is to say, it repeats itself exactly after each period or 'cycle'. Ideally an alternating quantity can be represented by a pure sine-wave (see Figure 17.1(a)). In practice such waves are seldom quite pure and suffer some distortion. With rectifier equipments, which are dealt with in Chapter 33, the distortion can be considerable (see Figure 17.1(b)). But with this exception most alternating currents and voltages are reasonably pure, and for the purposes of this and the following paragraphs they will be assumed to have a pure sine-wave form.

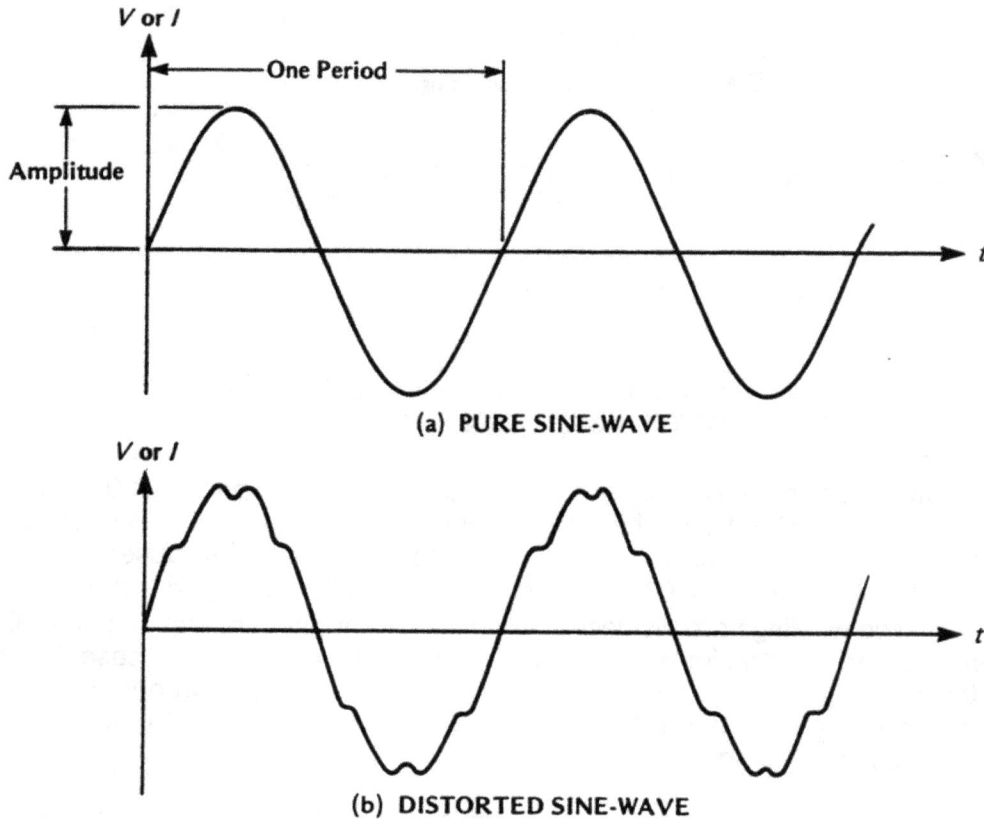

(a) PURE SINE-WAVE

(b) DISTORTED SINE-WAVE

FIGURE 17.1
ALTERNATING QUANTITIES

The vertical distance from the centreline to the peaks on either side is called the 'amplitude'. Sometimes the expression 'peak-to-peak value' may be seen; this is double the amplitude and is a term which should be avoided, as it could lead to confusion.

If an alternating quantity repeats itself f times per second, f is called the 'frequency' of the quantity and is measured in 'hertz' (Hz) (formerly in 'cycles per second' or c/s). The time of one period or 'cycle' is then $1/f$ seconds. For example, if the frequency of an electrical system is 50Hz, then the time for one cycle is one-fiftieth of a second.

17.2 SIMPLE HARMONIC MOTION

An alternating quantity can be developed as shown in Figure 17.2. Here there is a bar OP of length 'A' rotating with a constant angular velocity about O.

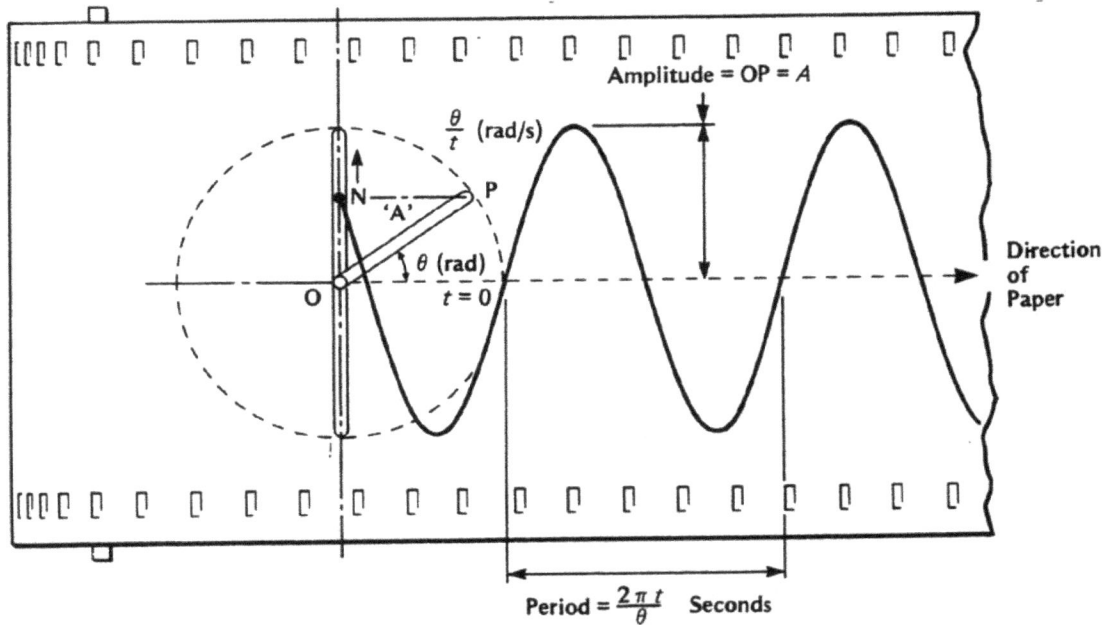

FIGURE 17.2
SIMPLE HARMONIC MOTION

If the angular displacement θ (radians) of the bar OP from its starting horizontal position at $t = 0$ occupies a time t (seconds), then the angular velocity of OP is:

$$\frac{\theta}{t} \text{ radians per second (or 'rad/s')}$$

If the point N on the vertical line through O is the projection (or 'shadow') of P, then N moves vertically up and down about O in what is called 'simple harmonic motion', and the length ON is in fact a pure alternating quantity. Its maximum length is equal to OP either side of O; this is its amplitude A. The value of ON at any instant is OP $\sin \theta$,

$$\text{or} \qquad ON = A \sin \theta \qquad\qquad\qquad \dots (i)$$

The point N will complete one full cycle in the time it takes for P to complete one full circle: that is, in the time that the bar OP takes to rotate through 2π radians (360°) at its constant angular velocity of $\frac{\theta}{t}$ radians per second.

The time for one complete cycle is therefore $\frac{2\pi}{\theta/t}$ (or $\frac{2\pi t}{\theta}$) seconds.

Put another way, the number of complete cycles in one second is $\frac{\theta}{2\pi t}$. But the number of cycles per second has already been defined as the 'frequency' f, hence

$$f = \frac{\theta}{2\pi t}$$

or $$\theta = 2\pi t$$

The general expression for simple harmonic motion can now be written from expression (i):

$$ON = A \sin 2\pi f t \qquad\qquad .... \text{(ii)}$$

Suppose N were the pen of a chart recorder, and the paper were moving from left to right at a steady speed. Then N would leave a trace-as shown in Figure 17.2 which is a pure sine-wave. It could represent any alternating quantity of amplitude A and frequency f. The value of that quantity at any instant of time is given by $A \sin 2\pi f t$.

Any alternating quantity can thus be completely specified by its amplitude and its frequency, as in the expression (ii) above. If the amplitude is used in this manner, it represents the 'peak' value of the quantity.

The expression (ii) above now contains variable quantities which can be measured by practical instruments:

t the time in seconds
f the frequency in hertz
A the amplitude of the wave (volts, amperes, etc.)

In electrical engineering practice the actual amplitude A - that is, the peak value - is not what is usually measured. It is however a function of a particular value of the wave which **can** be measured, the so-called 'root mean square' or 'rms' value. This concept is discussed in Chapter 18.

It should be noted that the reason for the term 'sine-wave' being given to the waveform generated in this unique manner is simply that its value at any instant is a function of the displacement angle θ and is, in fact, $A \sin \theta$ from equation (i).

CHAPTER 18 PEAK, RMS AND MEAN VALUES

Figure 18.1(a) shows, at the top, a current/time curve for a pure alternating quantity. Figure 18.1(b) is the corresponding (current)^2 /time curve for the same quantity, and, shown dotted through it, is the average value of the (current)^2 - that is, where areas above and below the dotted line are equal. Projected back onto the current/time curve of Figure 18.1(a), and shown chain-dotted, is the square root of this average (current)^2. Since the height of the peaks of the curve of Figure 18.1(b) is A^2, its mean value must be ½A^2. The square root of this average quantity is $\sqrt{\frac{1}{2}A^2}$ or $\frac{1}{\sqrt{2}}A$ (= 0.707A). This then is the 'root mean square ('rms') of the quantity whose peak value is A.

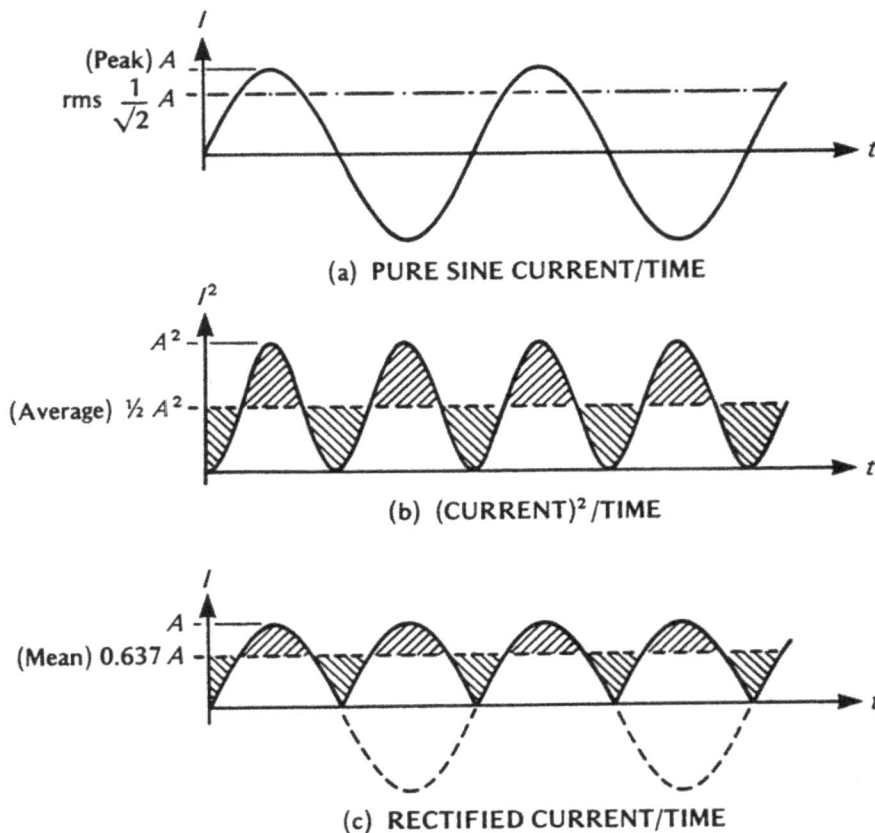

(a) PURE SINE CURRENT/TIME

(b) (CURRENT)^2/TIME

(c) RECTIFIED CURRENT/TIME

FIGURE 18.1
PEAK, RMS AND MEAN QUANTITIES

Since the heating of a conductor due to its resistance is at any instant I^2R, it is proportional to the square of the current, and the total heat developed during a cycle is proportional to the average of the square of the current over that cycle - that is, to the 'mean square'. So, as an a.c. ammeter reads the root mean square current, that reading, squared and multiplied by the resistance R, gives the net heating rate in an a.c. circuit. So the rms current in a circuit causes the same heating rate as would a d.c. current of the same value. In other words, when speaking of 'I^2R' losses in a circuit, it applies equally to d.c. as to a.c. so long as we are speaking (as is usually the case) of rms currents.

All a.c. ammeters and voltmeters are calibrated to indicate rms values, and current- and voltage-operated relays are set to operate at certain predetermined rms values.

For certain applications, notably in rectifier work (see Chapter 33), **mean** (or average) values of the rectified a.c. over the period are required rather than rms values. Figure 18.1(c) shows the average value of the 'rectified' current curve of Figure 18.1(a), where areas above and below the average line are equal. For a pure sine-wave its value is 0.637 of the amplitude (i.e. less than the rms).

Care must be taken not to confuse peak, rms and mean values of an alternating quantity. Since almost all instruments indicate, and relays operate on, rms quantities, these are the ones that are generally used and are intended to be understood when voltages and currents are discussed. However, it must be remembered that in the mathematical expression for an alternating quantity, $A \sin 2\pi f t$, 'A' is here the amplitude or peak value.

If it is necessary to distinguish between peak, rms and mean values of an alternating quantity 'A', they can be written respectively $\hat{A}$, A and A_m.

Whereas rms current values are important when considering the thermal current-carrying capacity of cables and items of plant, the peak values of current become important when considering the mechanical strength of such items under conditions of short-circuit when currents may be very high. The peak values of voltage are especially significant when considering questions of insulation.

Examples

If a voltmeter reads 33 000V, this is an rms value, and the peak value is √2 times this, namely about 47 000V. This is the voltage against which the system must be insulated, since it occurs twice every cycle.

If an ammeter reads 2 000A, this is an rms value, and the peak current is √2 times this, namely 2 800A. The peak value, which occurs twice every cycle, does not affect the heating, which is based on the rms value, but it does affect electromagnetic forces in the neighbourhood of the conductor, and therefore the mechanical bracing of the current-carrying conductors. This is described further in Part 10 Electrical Protection.

If an ammeter reads 40A (rms) in a circuit of resistance 5 ohms, then the heating rate, or losses, would be 40^2 x 5 watts, or 8 000W, or 8kW. This would be equally true if it were a d.c. circuit and the d.c. ammeter read 40A.

CHAPTER 19 PYTHAGORAS AND TRIGONOMETRICAL FUNCTIONS

19.1   GENERAL

It is necessary here, while dealing with fundamentals, to look again at two important mathematical concepts which many will remember from their schooldays. Both are widely used in calculations associated with electrical engineering, especially when dealing with alternating currents. The two concepts are Pythagoras' Theorem and, associated with it, the basic trigonometrical functions. At the end of this chapter are some examples showing how these functions are used.

19.2   PYTHAGORAS' THEOREM

If ABC is a right-angled triangle with the right angle at C, the side AB opposite the right angle is called the 'hypotenuse' (Figure 19.1).

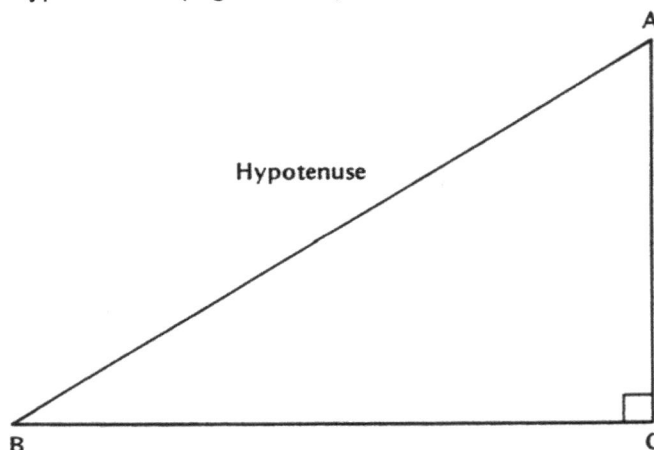

FIGURE 19.1
PYTHAGORAS' THEOREM

Pythagoras' Theorem states that 'the square on the hypotenuse is equal to the sum of the squares on the other two sides'. Put into mathematical form:

$$AB^2 = AC^2 + BC^2 \qquad ....(i)$$

For example, if AC = 4 cm and BC = 3 cm, then:

$$
\begin{aligned}
AB^2 &= 4^2 + 3^2 \\
&= 16 + 9 \\
&= 25 \\
\\
\therefore \quad AB &= \sqrt{25} \\
&= 5 \text{ cm}
\end{aligned}
$$

Thus, if the 'other two' sides are both known, the length of the hypotenuse can be calculated.

Expression (i) can also be written:

$$AC^2 = AB^2 - BC^2 \qquad \dots\text{(ii)}$$
$$\text{or} \quad BC^2 = AB^2 - AC^2 \qquad \dots\text{(iii)}$$

For example, if AB = 13 cm and BC = 5 cm:

$$
\begin{aligned}
AC^2 &= 13^2 - 5^2 \text{ from equation (ii)} \\
&= 169 - 25 \\
&= 144
\end{aligned}
$$

$$
\begin{aligned}
\therefore \quad AC &= \sqrt{144} \\
&= 12 \text{ cm}
\end{aligned}
$$

Similarly, if AB and AC are known, BC can be calculated from equation (iii).

To sum up: in a right-angled triangle, if the lengths of any two of the sides are known, that of the third side can be calculated by using one of the three forms (equations (i), (ii) or (iii)) of Pythagoras' Theorem:

$$AB^2 = AC^2 + BC^2$$

19.3 TRIGONOMETRICAL FUNCTIONS

The word 'trigonometry' is derived from the Greek, meaning 'three-cornered', and, as may be expected, it deals with triangles and the relationships between their sides and angles. For the sake of this discussion it will be limited to right-angled triangles and so stays within the realm of Pythagoras.

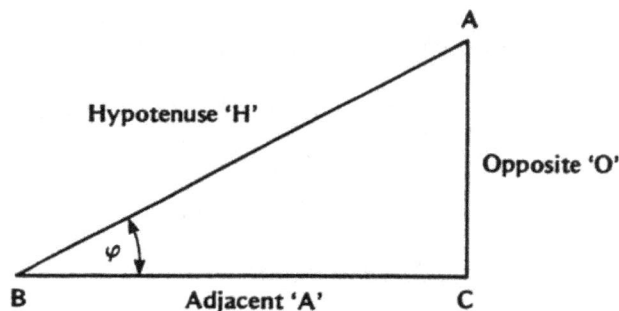

FIGURE 19.2
TRIGONOMETRICAL FUNCTIONS

Figure 19.2 shows a right-angled triangle with the right angle at C and with AB as its hypotenuse. The angle ABC will be called 'φ'; the side BC adjoining φ is called the 'adjacent side' (symbol 'A'), and the side AC opposite φ is called the 'opposite side' (symbol 'O'). Keeping to the symbol notation, the hypotenuse AB will be called 'H'. Then, by Pythagoras:

$$AB^2 = AC^2 + BC^2$$

or, using the symbol notation,

$$H^2 = A^2 + O^2$$

This is the Pythagoras relationship between the hypotenuse and the other two sides.

So far only the relationship between the triangle's three sides has been discussed. Look now at the angle φ and consider three more definitions:

(a) Ratio $\dfrac{O}{H}$ is called the 'sine' of the angle φ (symbol 'sin φ')

(b) Ratio $\dfrac{A}{H}$ is called the 'cosine' of the angle φ (symbol 'cos φ')

(c) Ratio $\dfrac{O}{A}$ is called the 'tangent' of the angle φ (symbol 'tan φ')

Each of these three is the ratio of a pair of sides, the ratio of different pairs having different names, but all being functions of the angle φ. Tables are available of sines, cosines and tangents from which, if the ratio is known, the angle φ can be read off directly. So, if any two sides of a right-angled triangle are known, not only can the third side be calculated by Pythagoras, but either of the two acute angles can be determined by use of the appropriate table. (Nowadays 'scientific' pocket calculators are used instead of tables.)

Relations (a), (b) and (c) above can also be written:

$$O = H \sin \varphi \qquad \qquad \text{....(iv)}$$
$$A = H \cos \varphi \qquad \qquad \text{....(v)}$$
$$O = A \tan \varphi \qquad \qquad \text{....(vi)}$$

It is in these forms that the relationships are most useful.

Example

In case of the triangle of Figure 19.2, suppose the opposite side 'O' had length 10cm and the adjacent side 'A' had length 15cm. Then by Pythagoras:

$$
\begin{aligned}
H^2 \;&=\; 10^2 + 15^2 \\
&=\; 100 - 225 \\
&=\; 325
\end{aligned}
$$

$$
\begin{aligned}
\therefore \quad H \;&=\; \sqrt{325} \\
&=\; 18.03 \text{ cm.}
\end{aligned}
$$

Also:

$$
\begin{aligned}
\tan \varphi \;&=\; \frac{O}{A} \\
&=\; \frac{10}{15} \\
&=\; 0.667
\end{aligned}
$$

From the tangent tables or pocket calculator:
$$\varphi \;=\; 33°42'$$

(Note that this can be calculated without working out H first.)

But if H is known:

$$\sin \varphi = \frac{O}{H} \qquad \text{or} \qquad \cos \varphi = \frac{A}{H}$$

$$= \frac{10}{18.03} \qquad\qquad\qquad = \frac{15}{18.03}$$

$$= 0.555 \qquad\qquad\qquad\qquad = 0.832$$

From the **sine** tables or pocket calculator:

$$\varphi = 33°42'$$

From the **cosine** tables or pocket calculator:

$$\varphi = 33°42'$$

It will be seen that, whichever of the three functions is used, the calculation leads to the same answer for $\varphi = 33°42'$.

When it is stated that $\sin \varphi = \dfrac{O}{H}$, it could equally well be said that φ is the angle whose sine is $\dfrac{O}{H}$. This may be written mathematically as:

$$\varphi = \sin^{-1}\frac{O}{H} \qquad \text{or} \qquad \arcsin\frac{O}{H}$$

The expression 'sin^{-1}' or 'arcsin' means simply 'the angle whose sine is ...', and one or other of these expressions will be found on most scientific pocket calculators. It is important to distinguish between 'sin' (the sine of the keyed-in angle) and 'sin^{-1}' (the angle whose sine is the keyed-in figure).

Thus: $\qquad\qquad\qquad \sin 30° = 0.5$, but $\sin^{-1} 0.5 = 30°$

19.3.1 Use of Trigonometric Functions
In many a.c. electrical power problems much use is made of Pythagoras and of the trigonometrical functions sine, cosine and tangent. Chapters 21 to 23 show that currents in purely resistive circuits are in phase with the applied voltage, but currents in purely reactive circuits lag or lead 90° on the applied voltage. We therefore have all the elements of a right-angled triangle to which Pythagoras and the simple trigonometrical functions can be applied.

The principal uses of these methods in electrical circuits are for COMBINING and for RESOLVING. These two terms are explained below.

19.3.2 Combining
Anticipating for a moment the use of vector notation described more fully in Chapter 20, Figure 19.3 shows an applied voltage V supplying a purely resistive circuit R in parallel with a purely reactive (inductive) circuit X. The resistive circuit draws a current I_R which is in phase with the voltage, whereas the reactive circuit draws a current I_x which lags 90° on the voltage. These two currents are so shown in the figure on the right.

The diagram shows that two separate currents are flowing in the two limbs, although of course only one current can come out of the voltage source, and that must clearly be the combination of the two separate currents. It is therefore required to **combine** the two currents I_R and I_x to give a single equivalent or 'resultant' current I.

FIGURE 19.3
COMBINED CURRENTS

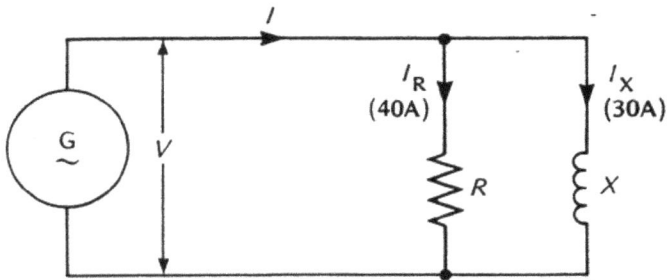

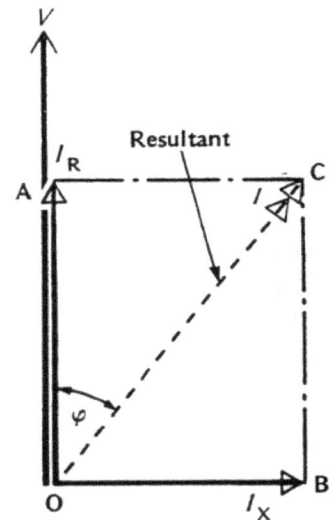

FIGURE 6.3

In the diagram on the right of Figure 19.3, OA represents the current I_R in phase with the voltage V, and OB the current I_x lagging 90° on V. Complete the rectangle OACB and draw the diagonal OC. Then OC represents the resultant or equivalent current I to the same scale as I_R and I_x

By Pythagoras:

$$
\begin{aligned}
OC^2 &= BC^2 + OB^2 \\
&= OA^2 + OB^2 \\
&= I_R^2 + I_x^2 \\
\therefore \quad OC &= \sqrt{I_R^2 + I_x^2}
\end{aligned}
$$

That is to say, the resultant current is the square root of the sum of the squares of the individual resistive and reactive currents. Putting numbers to these, if $I_R = 40A$ and $I_x = 30A$, then:

$$
\begin{aligned}
I^2 &= 40^2 + 30^2 \\
&= 1\,600 + 900 \\
&= 2\,500
\end{aligned}
$$

$$\therefore \quad I = 50A$$

Thus the resultant current equivalent to the 40A and 30A with this arrangement is not 70A but 50A. This is an example of **combining**. (Compare Chapter 20, Figure 20.5.)

This combining of currents can also be used to determine the phase angle (φ) of the resultant current with respect to the applied voltage.

For, as already defined,

$$
\begin{aligned}
\tan \varphi &= \frac{AC}{OA} \\
&= \frac{OB}{OA} \\
&= \frac{I_x}{I_R}
\end{aligned}
$$

Expressed mathematically:

$$\varphi = \tan^{-1} \frac{I_x}{I_R}$$

That is to say, the phase angle φ is the angle whose tangent is the ratio of the reactive current (I_x) to the resistive current (I_R). It will be shown later that this is the same as the ratio of the actual reactance (X) to the resistance (R); these two terms are explained in Chapters 21 to 23.

19.3.3 Resolving

When the actual current is known and also its angle of lag (or lead) φ, as shown in Figure 19.4, it is often required to break down, or 'resolve', that current into components which are in phase with the applied voltage and at right angles to that voltage - that is the reverse process to 'combining' which has just been described.

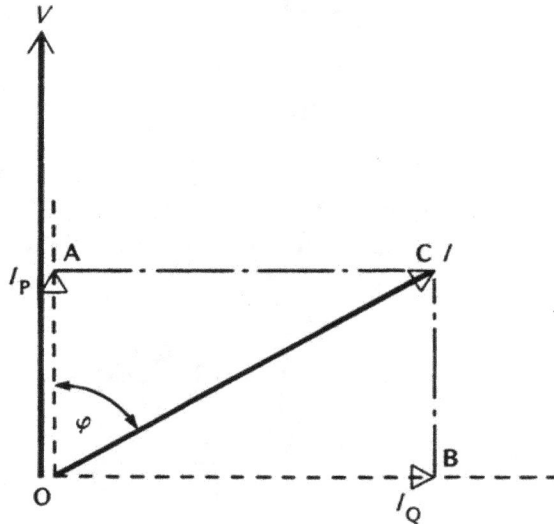

FIGURE 19.4
RESOLVED CURRENTS

All that it is necessary to do is to apply the trigonometrical functions cos φ and sin φ to the current. In the figure, V is the applied voltage and I the total current which flows. The angle between I and V is φ. If perpendiculars are dropped from C to the horizontal (CB) and to the vertical (CA), then OA is the in-phase component I_P, of the current I (that is, in the same phase as the voltage), and OB is the right-angled (or 'quadrature') component I_Q of the current I (that is, at right angles to the voltage).

Remembering equations (v) and (iv),
$$BC = OA = OC \cos \varphi$$
$$OB = AC = OC \sin \varphi$$

But OC is the total current I, and OA and OB are the in-phase and quadrature components I_P and I_Q respectively.

$$\therefore \quad I_P \quad = \quad I \cos \varphi$$

$$I_Q \quad = \quad I \sin \varphi$$

Thus, if the total current and phase angle are known, the in-phase and quadrature components can be immediately calculated. (Usually the actual angle φ is not given, but the power factor, cos φ, is. I_P then makes direct use of it, but to find sin φ it is necessary to use the cosine tables to find φ, then the sine tables to find sin φ to evaluate I_Q. A good scientific calculator can do this transition in one operation.) In finding I_P, and I_Q from I in this manner, the current I is said to be **resolved** into in-phase and quadrature components.

Example

A circuit has a current of 100A at a lagging power factor of 0.8 on the applied voltage. Find the in-phase and quadrature components of this current.

Power factor is 0.8	=	cos φ
From the cosine tables (or calculator) φ	=	36°52'
From the sine tables (or calculator) sin φ	=	0.60

∴
In-phase current	=	100 cos φ = 80A
Quadrature current	=	100 sin φ = 60A

(Note that, if these were recombined by Pythagoras,

(Total current)^2	=	$80^2 + 60^2$
	=	6 400 + 3 600
	=	10 000

∴
Total current	=	$\sqrt{10\,000}$
	=	100A

which is the total current that we started with. This shows that combining and resolving are merely reverse processes.)

CHAPTER 20 VECTORS (OR PHASORS)

20.1 VECTOR DIAGRAMS

In alternating current practice the quantities such as voltages, currents, impedances, fluxes and so on have not only numerical values but also a time or phase relationship with each other. Although their numerical values may follow a pure sine-wave form, the instants they reach their peaks may be different, one following (or 'lagging') on the other. This time difference is expressed not in seconds but as a fraction of the time for one period, or 'cycle', which is taken to be 360°. Thus a quantity whose peak value follows one-quarter of a cycle behind another's is said to lag 90° on it. (If there is likely to be confusion between any other degrees - say the rotation of a rotor - the lag may be expressed in 'electrical degrees'.) A lag of half a cycle is 180°, and of one-tenth of a cycle 36°. If the peak value of a quantity occurs **before** that of the reference quantity, it is said to 'lead' on it, the amount of lead being similarly expressed in (electrical) degrees. The difference angle, usually called 'φ', is the 'phase relationship' between the quantities.

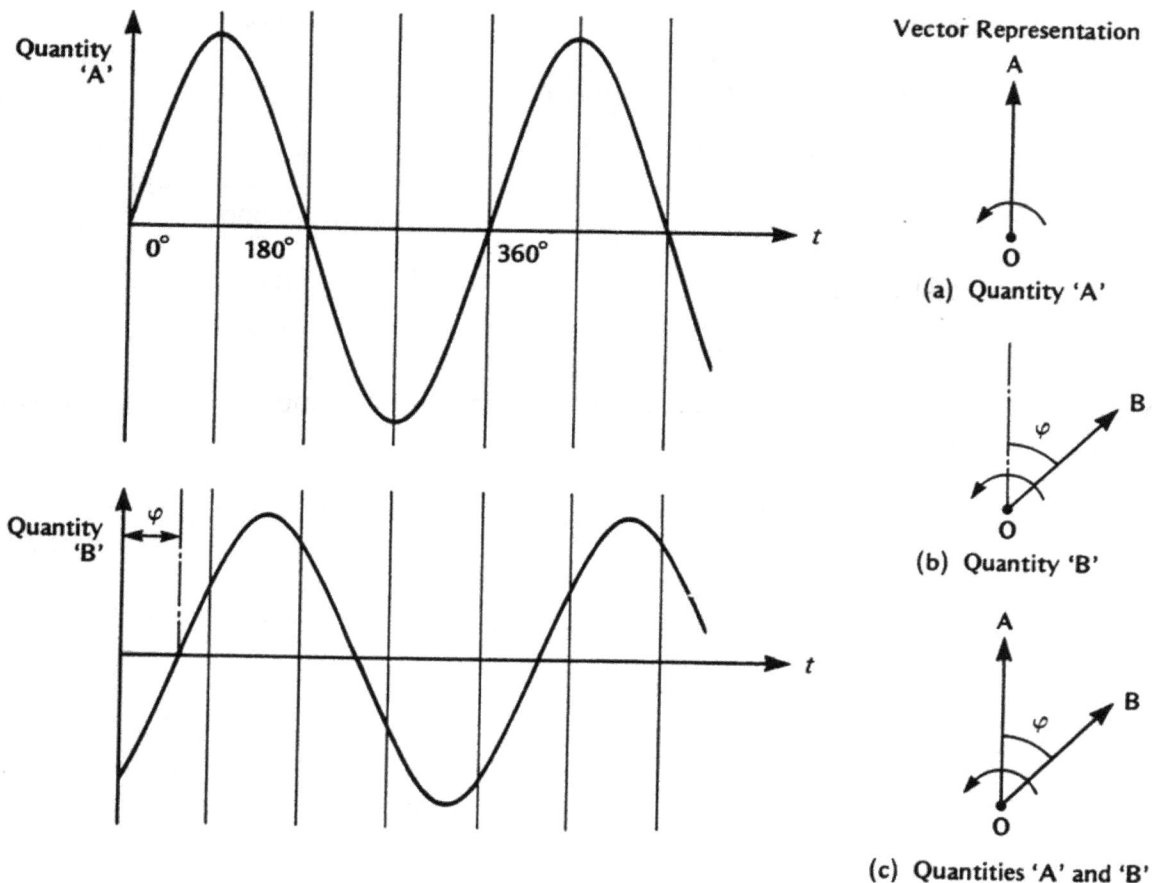

FIGURE 20.1
VECTOR DIAGRAM

This relationship is easily expressed in diagram form. In any relationship one of the partners is considered to be the 'reference' or datum against which the other is measured. In Figure 20.1(a) suppose the line OA represents the reference quantity, which may be an alternating voltage at a certain point or any other quantity. Its length represents the numerical peak value of the reference voltage, and an arrow is added to show the direction which is taken as the reference datum. OA is said to be a 'vector', having both length and direction. (Note: the term 'phasor' is now often used in such cases instead of 'vector'. However in this book the word 'vector' will be used exclusively.)

The line OA is by convention considered to be-rotating anti-clockwise at a speed which will cause it to complete one full revolution in the time of one period or cycle; thus its speed is determined solely by the system frequency. Figure 20.1(a) is simply a flash 'photograph' of the line as it passes through the datum direction, here taken to be 12 o'clock (though some prefer to use 3 o'clock as datum; it makes no difference which is used).

If now another alternating quantity of a different peak value and lagging in phase on OA exists (it might be another voltage at some other point, or a current, or a flux), it could be fully represented by another vector line OB, where the length OB represents the numerical peak value of the second quantity, and the angle φ between OB and OA is the angle in electrical degrees by which OB lags on OA, as shown in Figure 20.1(b). Since both are considered to be rotating anti-clockwise, OB is **following** OA and therefore truly lagging on it, and Figure 20.1(c) is the flash photograph of both vectors at the same instant when OA is passing through the 12 o'clock datum direction. If OB were leading on OA, it would be on the left of OA and so rotating ahead of it

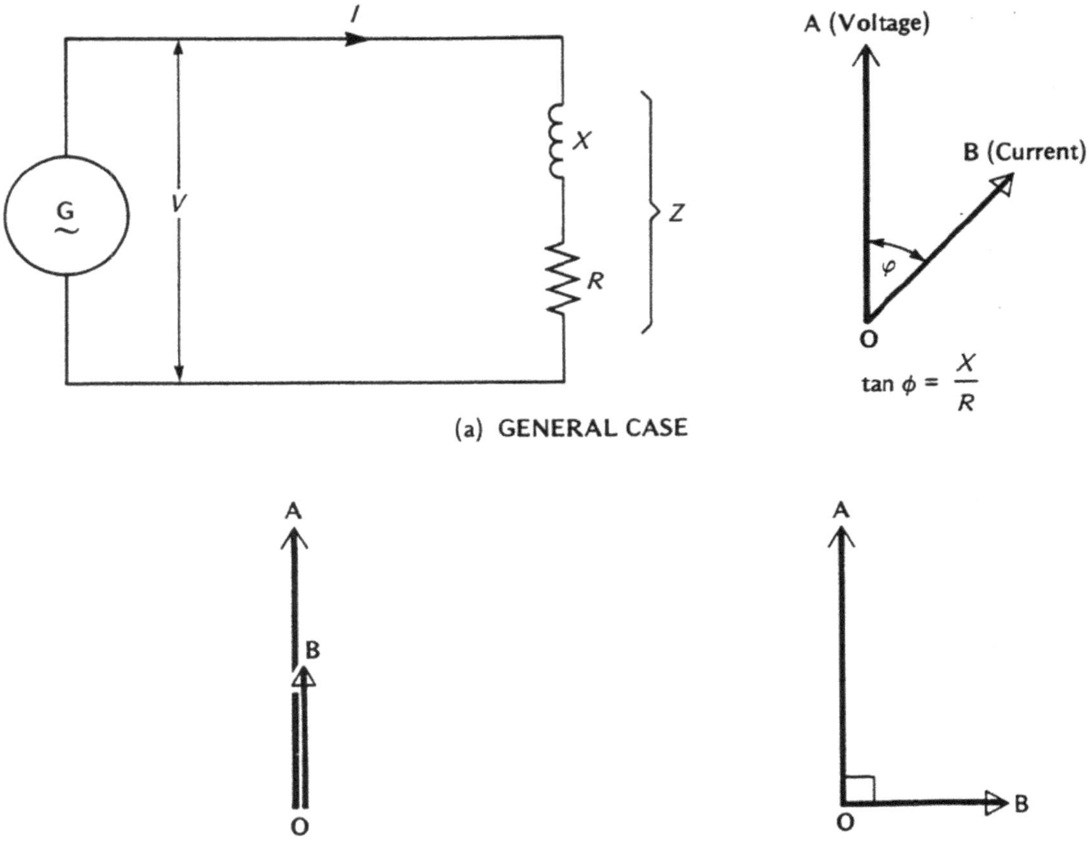

(a) GENERAL CASE

$$\tan \phi = \frac{X}{R}$$

(b) RESISTANCE ONLY
$(X = 0, \tan \phi = 0, \phi = 0)$

(c) REACTANCE ONLY
$(R = 0, \tan \phi = \infty, \phi = 90^\circ)$

FIGURE 20.2
CURRENT/VOLTAGE RELATION

A particular and important case is when OA represents the voltage applied to a circuit with an impedance having both resistance and reactance (see Chapters 21, 22 and 23), and OB represents the current that consequently flows in it. In general the current will lag on the voltage by an angle which depends on the relative amounts of resistance and reactance (see Figure 20.2(a)). As explained in Chapter 22, the angle φ of lag is given by

Part 1 Electrical Theory

the expression

$$\tan \varphi = \frac{X}{R}$$

where X and R are respectively the reactance ($= 2\pi f L$) and the resistance in ohms. It can be seen at once that if there is only resistance and no reactance, $X = 0$ and so $\tan \varphi = 0$, which makes $\varphi = 0$. This means that there is no angle between current and voltage, and the current is said to be 'in phase' with the voltage (see Figure 20.2(b)).

If on the other hand there were only reactance and no resistance, $R = 0$ and $\tan \varphi$ becomes infinite, which makes $\varphi = 90°$. In that case the current lags 90° on the voltage and is said to be 'in quadrature' with the voltage (see Figure 20.2(c)). This is a very common situation when dealing with generators and transformers under fault conditions, as discussed in Part 10 Electrical Protection.

Both the above, $R = 0$ or $X = 0$, are extreme cases. The general case is that shown in Figure 20.2(a), where the phase angle φ is somewhere between 0° and 90°.

20.2   ADDING AND SUBTRACTING VECTORS

Vectors represent quantities, such as voltages and currents, and they can be added and subtracted just like ordinary numbers.

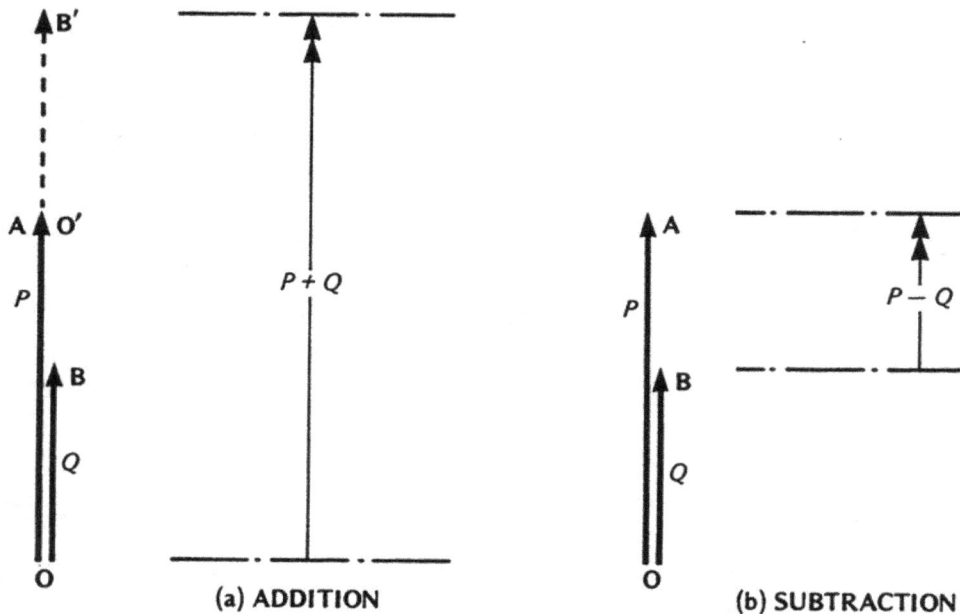

(a) ADDITION

(b) SUBTRACTION

FIGURE 20.3
VECTORS IN PHASE

OA or OB are two vector quantities P and Q in phase with each other. In Figure 20.3(a) it is required to **add** them together. It is clear that, if OB were picked up and put down over OA, the base (O) of OB being placed on the arrow-tip of OA to the position O'B', then the combined length OB' represents the sum of OA and OB and so is the quantity $P + Q$. If OA were 4 cm and OB 3 cm, $P + Q$ would be 7 cm.

In Figure 20.3(b) it is required to **subtract** OB from OA. If a line were drawn joining the two arrow-tips, its length would be BA; it represents the difference between OA and OB and so is the quantity P - Q. The direction of this difference line is **from** the arrow-tip of the subtracted quantity **to** that of the quantity it was taken from - that is from B to A.

Where the quantities OA and OB were in phase, as above, the process was simple, as shown in Figure 20.3.

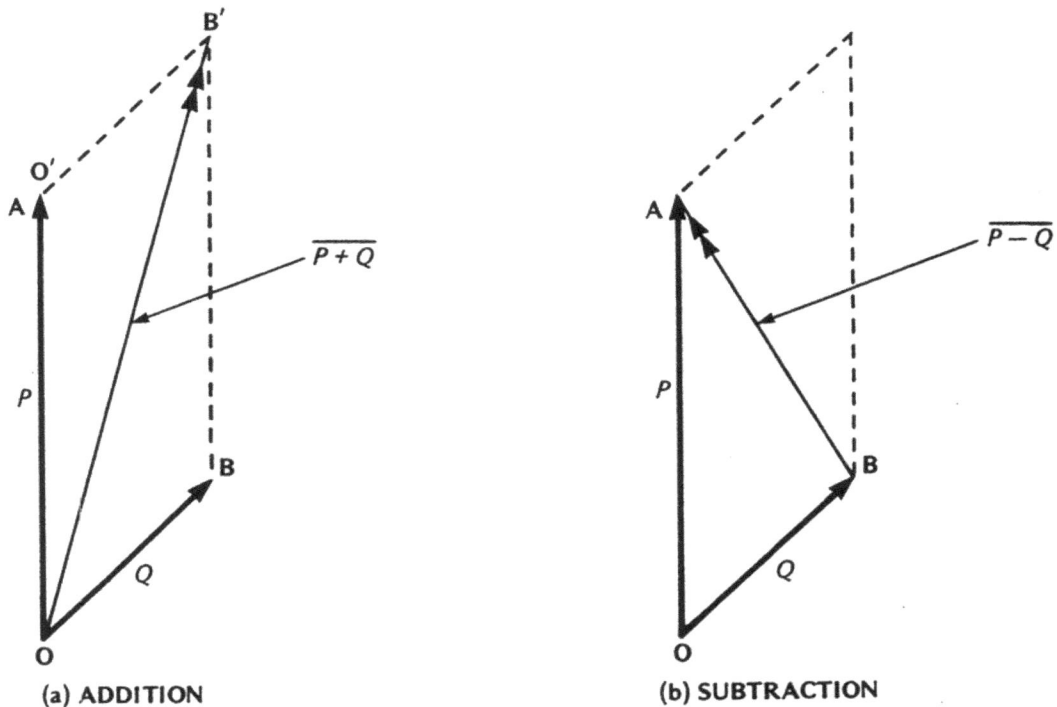

(a) ADDITION

(b) SUBTRACTION

FIGURE 20.4
VECTORS NOT IN PHASE

The process is exactly the same however where the quantities are not in phase. To add the two vectors OA and OB in Figure 20.4(a), simply pick up vector OB and place it, without altering its direction, on top of OA in position O'B'. OA and O'B' are added together as in Figure 20.4(a), but the route now goes round a corner. The line OB' joining O to the new point B' is then the **vector** sum of OA and OB; it is written $\overline{P + Q}$ to distinguish it from the numerical sum $P + Q$. It is clear that the vector sum $\overline{P + Q}$ is in general less than the numerical sum $P + Q$.

Some prefer to use the construction where the parallelogram on OA or OB is completed. The diagonal through O is then the addition vector representing the sum $\overline{P + Q}$.

To subtract two vector quantities which are not in phase, the process is exactly similar to that shown in Figure 20.3(b). Simply join the tips of the two vector arrows, as shown in Figure 20.4(b). The vector BA is then the difference between the vectors OA and OB and is written $\overline{P - Q}$.

Since OB is being subtracted from OA, the direction of the difference vector is **from** B **to** A, as shown in Figure 20.4(b). If OA had been subtracted from OB, the direction would have been from A to B.

If the parallelogram on OA and OB is drawn as before, the difference vector is then the **other** diagonal - that is the diagonal **across** O.

20.3   EXAMPLE: ADDITION OF ACTIVE AND REACTIVE ELEMENTS

Chapters 21 to 23 show that a current in a purely resistive circuit is in phase with the applied voltage, whereas that in a purely inductive circuit lags 90° on the voltage. Vector methods can be used to handle such currents - for example to add them.

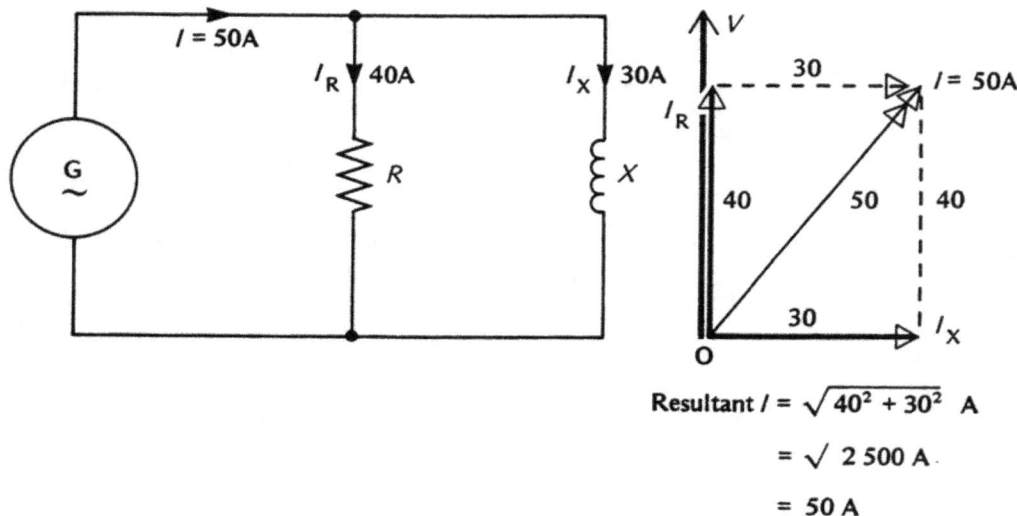

$$\text{Resultant } I = \sqrt{40^2 + 30^2} \text{ A}$$

$$= \sqrt{2\,500} \text{ A}$$

$$= 50 \text{ A}$$

FIGURE 20.5
EXAMPLE OF THE VECTOR ADDITION OF CURRENTS

In Figure 20.5 an a.c. generator feeds a resistance and a reactance in parallel which draw 40A and 30A respectively. The resistance current I_R is in phase with the applied voltage, but the reactance current I_x lags 90° on the voltage (see Chapter 22). The combined current is now not 70A but is 50A $\left(= \sqrt{40^2 + 30^2}\right)$ because the currents add **vectorially**.

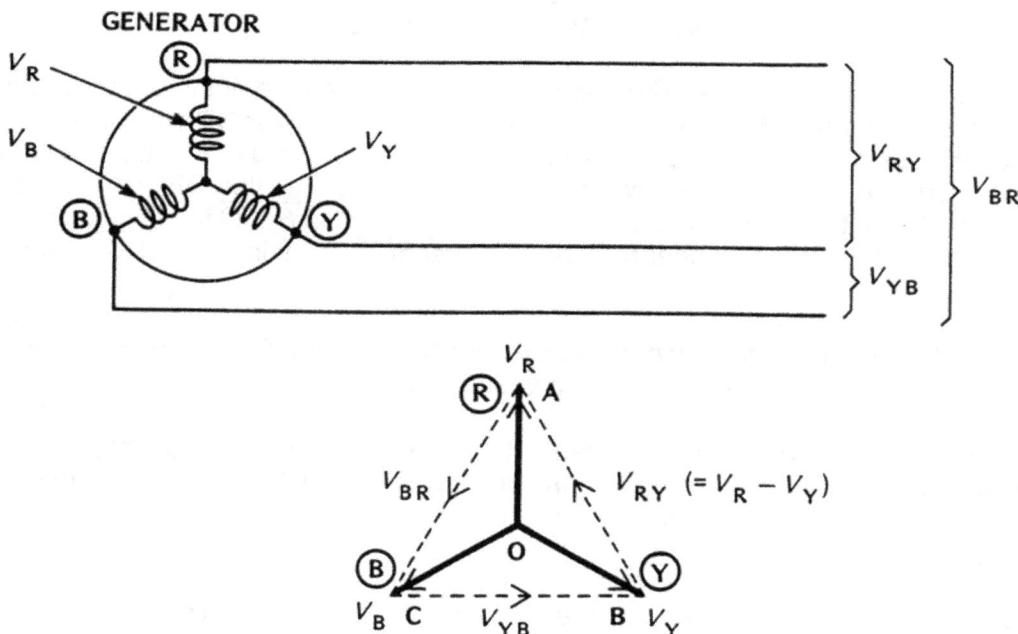

FIGURE 20.6
STAR CONNECTION VOLTAGES

20.4   PHASE AND LINE VOLTAGES AND CURRENTS

An important application of vectors is for determining the relationship between the phase and the line-to-line voltages and currents in a 3-phase system.

Consider a 3-phase system supplied from a generator as shown at the top of Figure 20.6. Suppose that the generator is star-connected. Then, since the three voltages are equal in magnitude and are spread 120° apart in time, the three vectors V_R, V_Y and V_B in the lower part of the figure represent the three **phase** voltages as developed in the three windings of the generator. They are the voltages between each of the three terminals and the star (or neutral) point.

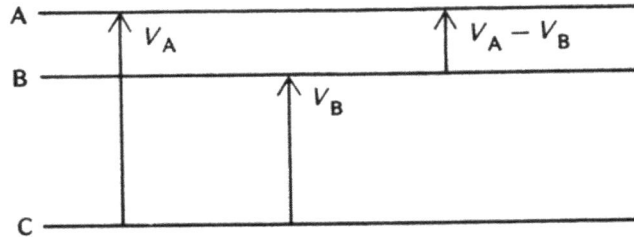

FIGURE 20.7
VOLTAGE BETWEEN LINES

Consider the simple system of Figure 20.7. Two lines A and B carry voltages V_A and V_B measured with respect to a common line (or earth) C. Clearly the voltage between lines A and B is the difference between the individual line voltages V_A and V_B.

Reverting to Figure 20.6, by the same token the voltage between any two lines, say R and Y, is the difference between their individual voltages V_R and V_Y measured with respect to their common neutral point.

. V_R and V_Y, are vectors, and it has already been shown that the difference between two vector quantities is obtained by joining their arrow-tips. So the voltage between lines R and Y, which we will call ' V_{RY}', is the difference between vectors V_R and V_Y, and is therefore represented by the line AB. Similarly line BC represents the voltage between lines Y and B, which is V_{YB}, and line CA represents the voltage between lines B and R, which is V_{BR}.

Thus in Figure 20.6 the sides of the triangle represent the line-to-line voltages, and the 'spokes' represent the three phase (or line-to-neutral) voltages.

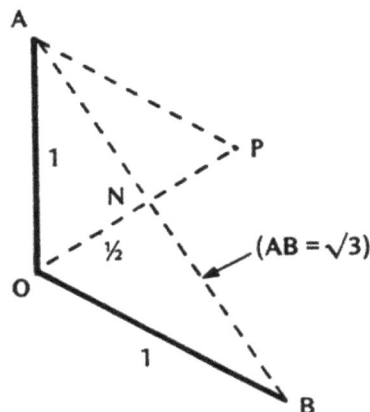

FIGURE 20.8
PHASE AND LINE VOLTAGES

Consider more closely the geometry of the triangle OAB. This is reproduced in Figure 20.8, and an equilateral triangle OAP is constructed on line OA. Let OP cut AB at point N.

If OA (= OB) is assumed to be one unit of length, then ON (which is half OP) will be one-half unit of length. By Pythagoras:

$$AN^2 = OA^2 - ON^2$$
$$= 1^2 - \left(\frac{1}{2}\right)^2$$
$$= \frac{3}{4}$$

so that

$$AN = \sqrt{\frac{3}{4}}$$

$$= \frac{\sqrt{3}}{2} \text{units}$$

Now

$$AB = 2AN$$
$$= 2 \times \frac{\sqrt{3}}{2}$$
$$= \sqrt{3} \text{units}$$

This means that the triangle's sides are √3 times as long as the spokes. Hence, with a star connection:

line voltage = √3 (phase voltage)

This is a very important result which comes into numerous system calculations, especially the calculation of 3-phase power which is explained in Chapter 31.

It should also be noticed from Figure 20.6 that the current in each **line** is the same as the current in the corresponding generator phase winding, since both are in series. So, with a star connection:

line current = phase current

A similar method is used to determine the individual phase currents (that is, the actual machine currents) in a delta-connected machine or transformer when only the line currents are known.

Figure 20.9 (top) shows a delta-connected generator. Suppose it is feeding a **balanced** load - that is where each of the three load elements has the same impedance and the same power factor. This load itself may be star- or delta-connected, as indicated. If I_{RY}, I_{YB} and I_{BR} are the currents in the three phase windings of the machine, and if I_R, I_Y and I_B are the three **line** currents, then applying Kerchoff's Law to the machine terminal R (which says that the total current entering any point is balanced by the current leaving it):

$$I_{BR} \text{ (in)} = I_{RY} + I_R \text{ (out)}$$
$$\text{or} \quad I_R = I_{BR} - I_{RY} \qquad \qquad \text{....(i)}$$

where all these are vector quantities.

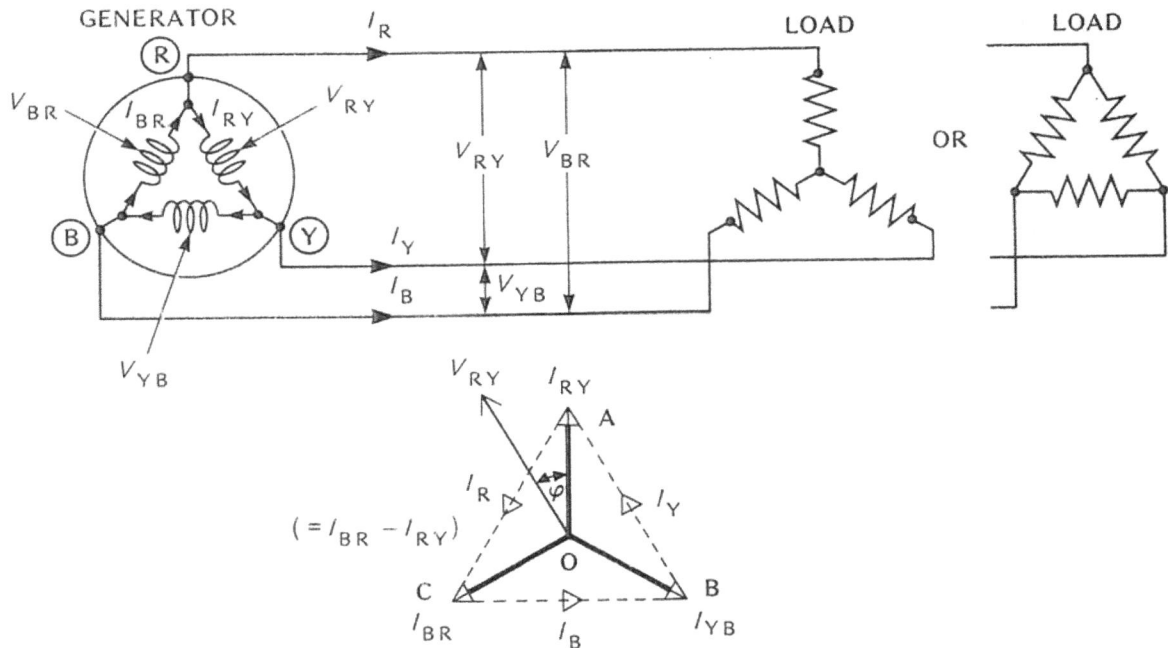

FIGURE 20.9
DELTA CONNECTION CURRENTS

The three balanced currents are shown at the bottom of Figure 20.9. The vectors I_{BR} and I_{RY} are two of them, and their difference (see equation (i)) is obtained, as usual, by joining the arrow-tips, A and C. The line AC then represents the line-current vector I_R from equation (i).

Using the same geometry as was used in Figure 20.8 for the voltages, it is clear that $I_R = \sqrt{3}$ times I_{RY}. Similarly for I_Y and I_B

Thus, with a delta connection:

line current $= \sqrt{3}$ (phase current)

It should also be noticed from Figure 20.9 that the voltage across each pair of lines is the same as that of the corresponding generator phase winding, since both are in parallel. So with a delta connection:

line voltage $=$ phase voltage

20.5   SUMMARY

A.   **Star Connection**

line voltage $= \sqrt{3}$ (phase voltage)
line current $=$ phase current

B.   **Delta Connection**

line current $= \sqrt{3}$ (phase current)
line voltage $=$ phase voltage

CHAPTER 21 RESISTANCE

21.1   CURRENT AND VOLTAGE

Earlier in the book resistance in a circuit was likened to friction in a mechanical system. It opposed any force attempting to cause motion, and it absorbed energy which showed in the form of heat.

Unlike inductance which caused a slow build-up of current, or unlike capacitance which caused a slow build-up of charge, resistance has an instantaneous effect on the current in a circuit. In a d.c. circuit Ohm's Law states that:

$$\frac{V}{I} = R$$

where V is the d.c. voltage applied and I is the current (in amperes) caused by that voltage. R is then defined as the 'resistance' of the circuit and is measured in the unit 'ohm'.

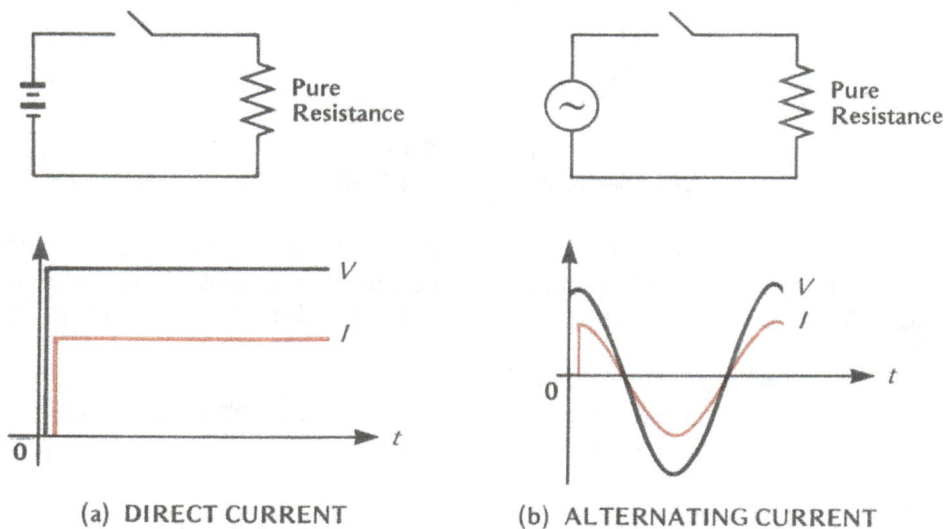

(a) DIRECT CURRENT   (b) ALTERNATING CURRENT

FIGURE 21.1
RESISTIVE CURRENTS, D.C. AND A.C.

If Ohm's Law is written in the form $\frac{V}{R} = I$, the current (in amperes) in a d.c. circuit is equal to the voltage divided by the resistance (in ohms) and starts to flow virtually instantaneously the moment that the voltage is applied (see Figure 21.1(a)). For any given circuit or sample the resistance is fixed (though it differs between samples), so that the current too is constant and proportional to the voltage.

The same argument applies to a.c. when it flows through a resistance. Since $I = \frac{V}{R}$ at all times, and R is fixed, the current at any instant is directly proportional to the voltage **at that instant**. As in an a.c. system the voltage is changing periodically, so also will the current change periodically and will bear a fixed ratio to the voltage at all times, as shown in Figure 21.1(b).

Because of this fixed ratio the current will reach its peaks at the same instants as the voltage, and it will also pass through zero at the same instants as the voltage. This is shown clearly in Figure 21.1. The current is then said to be 'in phase' with the voltage.

It follows that Ohm's Law applies not only to d.c. but also to a.c. so long as the a.c. circuit consists only of resistance, and in the a.c. case the resulting current is in phase with the voltage.

The effect of inductance and capacitance on the current in an a.c. circuit, and the consequent modification of Ohm's Law, is dealt with in Chapters 22 to 24.

21.2   HEATING

It was stated earlier that, whenever a d.c. current is forced by pressure of voltage to flow through a conductor which has resistance (and all conductors do, even metals), heat is generated in that conductor. The rate of heat generation is proportional to the resistance and to the **square** of the current (in amperes squared). That is to say, the heat generated is I^2R, and, since it represents a continuing loss of energy, it is expressed in the energy-rate unit 'watts' (W).

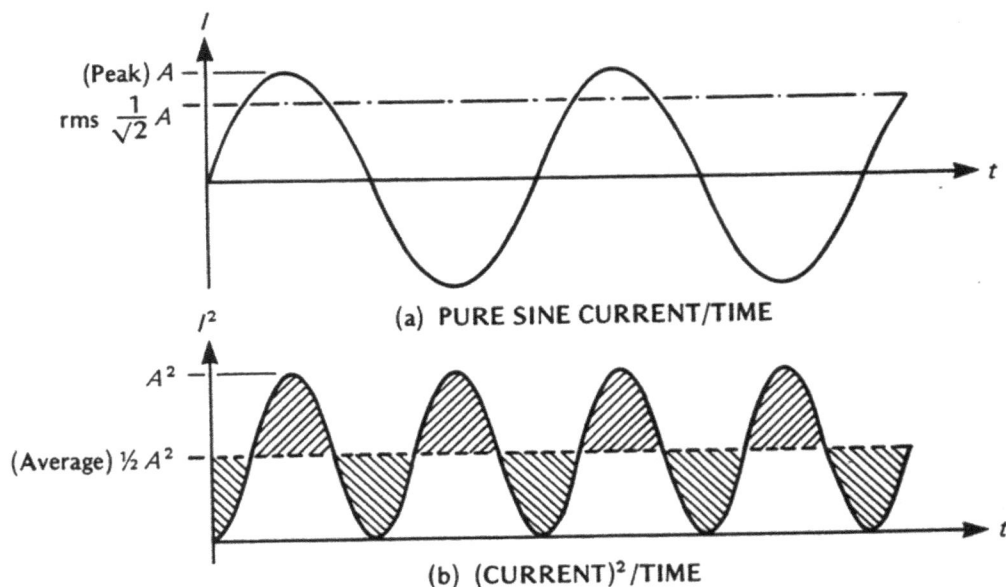

FIGURE 21.2
CURRENT HEATING EFFECT

Consider now the heating effect of an alternating current when flowing through a resistance R. In Figure 21.2(a) is a pure sine-wave current trace with amplitude (or peak value) 'A'. In Figure 21.2(b) is the corresponding 'current squared' wave, whose amplitude must be A^2. Since the square of a quantity, whether positive or negative, is always positive, the current-squared wave is wholly above the line.

The rate of heat generation depends on the resistance and the square of the current, so that the height of this curve at any instant indicates the heating rate I^2R at that instant, and the area below the curve is the total heat generated over a given period. The middle line (shown dotted) is then the **average** rate of heat generation. It therefore represents the 'mean square' current, and its height is $\frac{1}{2}A^2$.

It is shown in Chapter 18 that currents in a.c. systems are measured not by their amplitudes or peak values but by their 'root mean square' values, which were shown to be the square root of the mean square current, or 'root mean square' (or 'rms' current for short). This has the value $\sqrt{\frac{1}{2}A^2}$ or $\frac{1}{\sqrt{2}}A$

So long as the current measured (I) js the rms current (which it normally is), then its square is the 'mean square' current which, when multiplied by R, determines the average heating rate. Consequently with a.c. the average heating rate is given by the expression I^2R (where I is the rms current), which is the same expression as used for d.c.

CHAPTER 22 INDUCTIVE REACTANCE

22.1   CURRENT IN A PURELY INDUCTIVE CIRCUIT

The effect of inductance on the behaviour of the d.c. 'magnetising current', especially when combined with resistance, was examined earlier. As a d.c. voltage was applied and current started to build up, a 'back-emf' was induced in the inductor coil, and it opposed the applied voltage in a decaying manner until the current settled down to a steady value given by Ohm's Law.

In this chapter we shall examine the behaviour of an inductive circuit (initially a 'purely inductive' one without resistance) when an a.c. voltage is applied to it. It will be seen to be apparently very different from its d.c. behaviour.

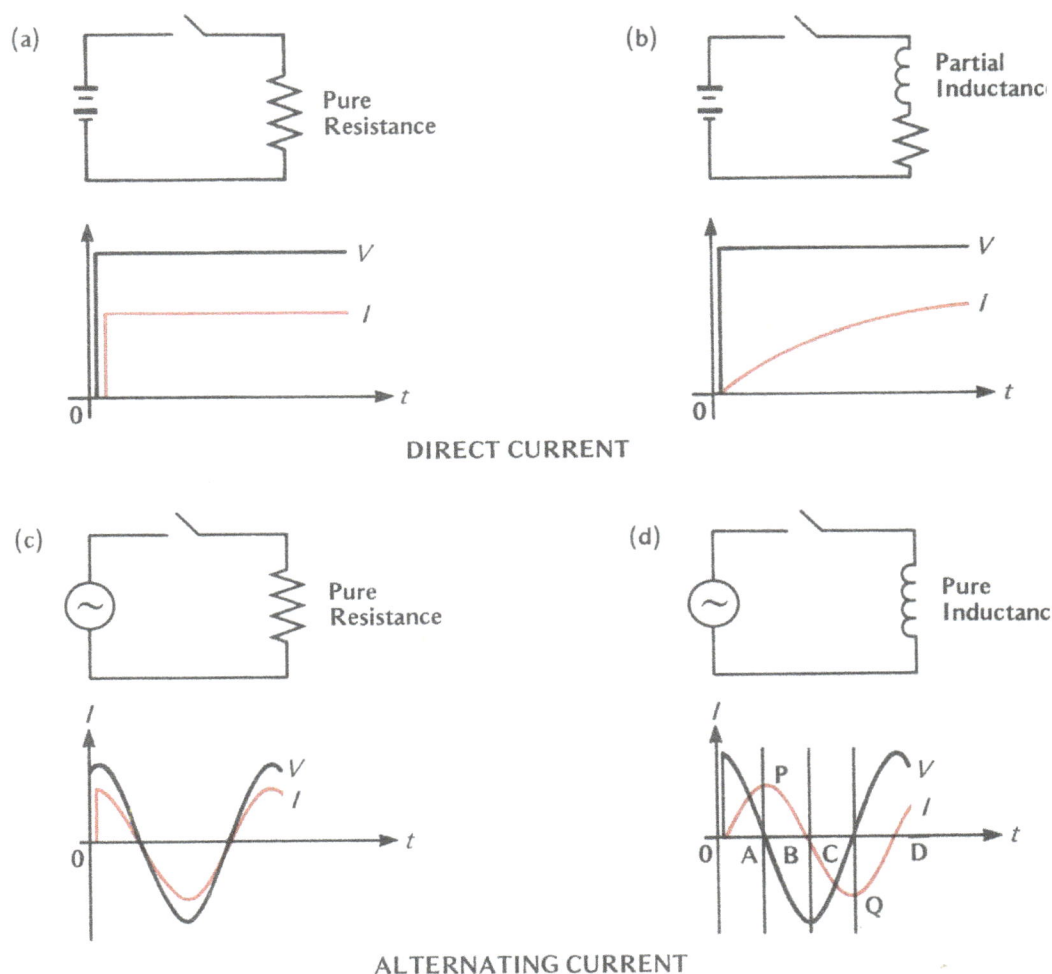

FIGURE 22.1
CURRENT IN D.C. AND A.C. CIRCUITS

In the upper part of Figure 22.1 a **d.c.** voltage is suddenly applied by a switch (a) to a resistive circuit, and (b) to an inductive circuit with both resistance and inductance.

In the case of the resistive circuit (a) the current rises immediately to the value determined by Ohm's Law, and it stays at that value. In the case of the inductive circuit (b) the current rises fairly slowly, since the applied voltage is overcoming the 'back-emf', or inertia, of the system to make the current grow, as explained earlier. Eventually the current will settle down at the steady d.c. level determined by Ohm's Law $I = \frac{V}{R}$. At this point the current ceases to grow and there is no more inertia to overcome.

Consider now the bottom two figures, where an **a.c.** voltage is suddenly applied by a switch (c) to a purely resistive circuit and (d) to a purely inductive circuit.

In the case of the resistive circuit (c) the voltage has only resistance to overcome, and the current wave follows the voltage wave exactly. As explained in Chapter 21, the amount at any instant is then determined by Ohm's Law. Since the current peaks and valleys coincide with the peaks and valleys of the voltage wave, the current is then 'in phase' with the voltage.

In the case of the inductive circuit (d) the process needs to be followed rather more carefully. Suppose the switch is closed at the instant when the voltage wave is at a positive peak. Because the load is inductive, the first application of voltage will cause the current to rise sluggishly, as it did in case (b). It will continue to rise in this manner under the positive drive of the voltage wave up to time 'A', by which time the voltage wave has fallen to zero. There is then no more drive, and the current ceases to rise, as shown at point 'P'.

After this the voltage becomes increasingly negative, opposing the current flow and causing it to reduce. At time 'B' the voltage is at its negative maximum and the current is reducing at its fastest rate, passing through zero and becoming negative. Between times 'B' and 'C' the voltage is still negative, so the current continues to become increasingly negative. At time 'C' the voltage wave has returned to zero and the negative drive has gone, so the current wave levels out at a negative peak 'Q'.

After time 'C' the voltage becomes positive again, now opposing the current's negative flow. The current therefore becomes less negative until, at time 'D', it has returned to zero. The voltage however continues positive, so the current continues its rise to become positive again. The conditions at time 'D' are the same as they were at the start time (time 'O'), and the whole cycle begins again.

It can be seen from (d) that the current wave is 'late' compared with the voltage wave by one-quarter of a cycle. It is said to 'lag', and, if one cycle is considered as 360°, it lags by 90°.

In the case shown in (d) the circuit was considered purely inductive (i.e. no resistance). In practice there will always be some resistance, however small.

To sum up: in purely resistive circuits the current will follow the voltage exactly, whether d.c. or a.c. In the a.c. case moreover the current will be in phase with the voltage. In inductive circuits however the current will follow the voltage sluggishly. In the d.c. case it will rise slowly to its final settled value, whereas in the a.c. case the alternating current wave will not rise slowly but will lag by up to 90° on the voltage wave.

Figure 22.2 shows, for comparison, the currents (dotted curves) which would be caused by applying an alternating voltage *(V)* as in curve (a)

> (b)   to a purely resistive circuit
>
> (c)   to a purely inductive circuit.

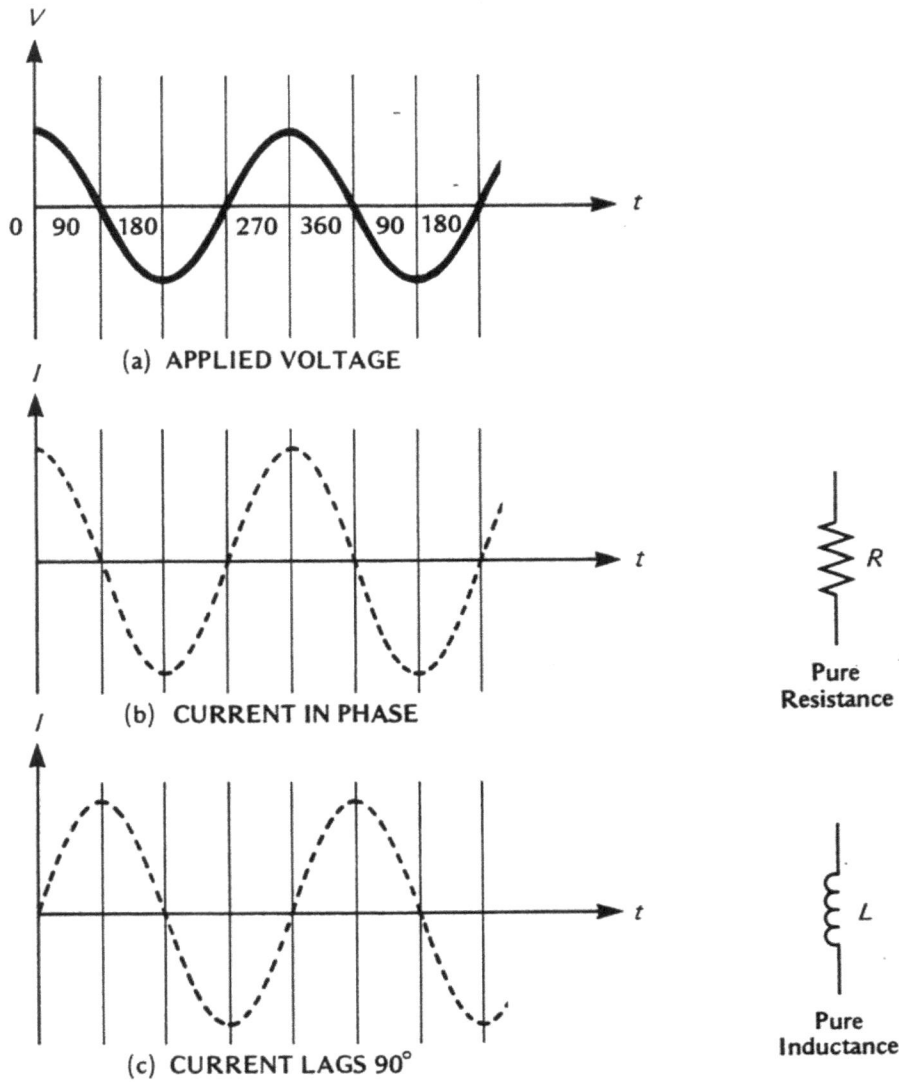

FIGURE 22.2
PURELY RESISTIVE AND PURELY INDUCTIVE CURRENTS

It is shown in Chapter 21 that the current in the purely resistive circuit follows the applied voltage exactly - its peaks, valleys and zeros occur always at the same instants. The current is then 'in phase' with the voltage.

With a purely inductive circuit however the current wave **lags** on the voltage wave in time by one-quarter of a cycle, or 90°. Each current peak occurs 90° **after** the previous voltage peak.

For the situation where the circuit is not **purely** inductive but contains also some resistance (the general case) see Chapter 24.

22.2   INDUCTIVE REACTANCE

In the absence of any resistance in the circuit, the applied voltage (V) is at any instant opposed only by the back-emf (E) which is being induced in the inductor by the changing current, so that at all times:

$$V = -E$$

The back-emf in an inductor depends on its inductance L (in henrys) and on the **rate of change** of the current through it, namely:

$$E = -L\frac{di}{dt}$$

The inductance L is a property of the inductor itself, but the rate of change of current depends directly on both the amplitude of the current itself (I) and on the speed of change. In an a.c. system, that means on the frequency (f), or number of changes per second. That is to say:

$$V(= -E) \propto f, L \text{ and } I$$

In fact the actual relationship is: $\qquad\qquad V = 2\pi f L I$

Rewriting this: $\qquad\qquad\qquad\qquad \dfrac{V}{I} = 2\pi f L$

by comparison with Ohm's Law for resistance

$$\frac{V}{I} = R$$

the expression $2\pi f L$ is the equivalent in an inductive circuit to resistance and, like resistance, is also measured in ohms. It is called the 'inductive reactance' of the circuit (or sometimes just 'reactance') and has the symbol $'X'$. If necessary to distinguish it from the capacitive reactance (see Chapter 23), the symbol $'X_L'$ may be used.

$$\therefore \text{ Inductive reactance } X_L = 2\pi f L$$

where X_L is in ohms, f in hertz and L in henrys.

See also the end of Chapter 23 for certain notes on the relationship between inductive and capacitive reactance.

CHAPTER 23 CAPACITIVE REACTANCE

23.1 CURRENT IN A PURELY CAPACITIVE CIRCUIT

The effect of capacitance on the behaviour of the d.c. 'charging current', especially when combined with resistance, was examined earlier. As a d.c. voltage was applied and a charging current set up, the growing charge on the capacitor increasingly opposed the applied voltage until the charging current decayed and ceased.

This chapter examines the behaviour of a capacitive circuit (initially a 'purely capacitive' one without resistance) when an **a.c.** voltage is applied to it. It will be seen to be apparently very different from its d.c. behaviour.

This can best be explained, without resorting to mathematics, in the following way.

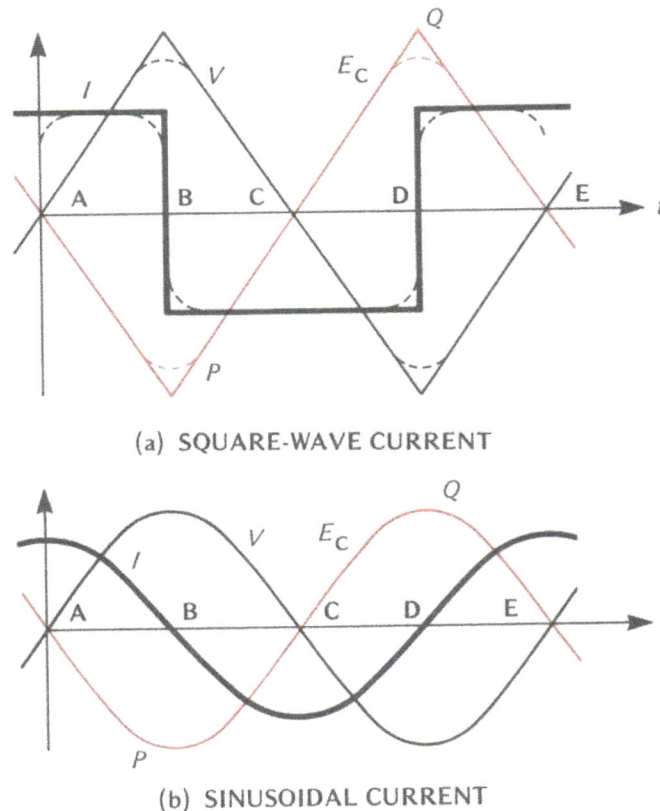

(a) SQUARE-WAVE CURRENT

(b) SINUSOIDAL CURRENT

FIGURE 23.1
PURELY CAPACITIVE CIRCUIT

Suppose an a.c. charging current, instead of being the classical 'sine-wave' shape, is square shaped - that is, as the heavy lines in Figure 23.1(a). It is constant and positive during the first quarter cycle (from A to B); constant and negative during the second and third quarter cycle (B to C and C to D); and constant and positive again during the fourth quarter cycle (from D to E).

Then during the period A to B the capacitor is charging at a constant rate, its charge voltage growing steadily and opposing the applied voltage (V). Its charge voltage (E_C) is therefore represented by the red line, increasing uniformly and negatively between A and B to point P.

During the next quarter cycle (B to C) the charging current has reversed and is constant; it will steadily discharge the capacitor until at instant C it is again without charge. The red

101

line rises uniformly from point P to zero.

During the next quarter cycle (C to D) the charging current is still uniform and negative, so the capacitor will charge up negatively. This **assists** the applied voltage, and its charge voltage is therefore represented by the red line above the axis, increasing uniformly and positively between C and D to point Q.

Finally, during the fourth quarter cycle (D to E) the charging current becomes positive again and, being uniform, discharges the negatively charged capacitor steadily. The capacitor voltage E_C (red line) therefore falls uniformly from Q to zero during this quarter cycle.

The applied voltage is opposed only by the capacitor's charge voltage and is therefore equal and opposite to the red E_C curve - it is the full-line curve V. (Compare the purely inductive case (Chapter 22) where it is only the 'back-emf' E of the inductor which opposes the applied voltage V.)

For this explanation the charging current wave was assumed to be square-topped. This is of course not the typical case; normally the current wave would be a sine curve. If the square-topped and angled lines of Figure 23.1(a) are 'rounded off', implying a gradual rather than a sudden change, then we approach the sine-wave curves of Figure 23.1(b). The charging current curve I is in heavy line, and the capacitor voltage curve E_C is as shown in red.

The applied voltage is opposed only by the capacitor's charge voltage and is therefore equal and opposite to the red E_C curve - it is the light full-line curve V as in Figure 23.1(a).

If Figure 23.1(b) is examined closely, it will be seen that the heavy current curve I is **ahead** in time of the applied voltage curve V, the current is said to **lead** the applied voltage (as distinct from the inductive case where it lags). In the case described it leads by one-quarter of a cycle, or 90°.

To sum up: if an alternating voltage is applied to a purely capacitive circuit, the alternating current which will flow will lead the applied voltage by 90°

Figure 23.2 shows, for comparison, the currents (dotted curves) which would be caused by applying an alternating voltage *(V)* as in curve (a)

      (b)   to a purely resistive circuit

      (c)   to a purely capacitive circuit.

It is shown in Chapter 21 that the current in the purely resistive circuit follows the applied voltage exactly - its peaks, valleys and zeros occur always at the same instants. The current is then 'in phase' with the voltage.

With a purely capacitive circuit however the current wave **leads** the voltage wave in time by one-quarter of a cycle, or 90°. Each current peak occurs 90° **before** the next voltage peak. (Compare the purely inductive case of Chapter 22 where the current wave **lags** by 90°. Particularly compare Figure 23.2 with Figure 22.2.)

For the situation where the circuit is not **purely** capacitive but contains also some resistance, see Chapter 24.

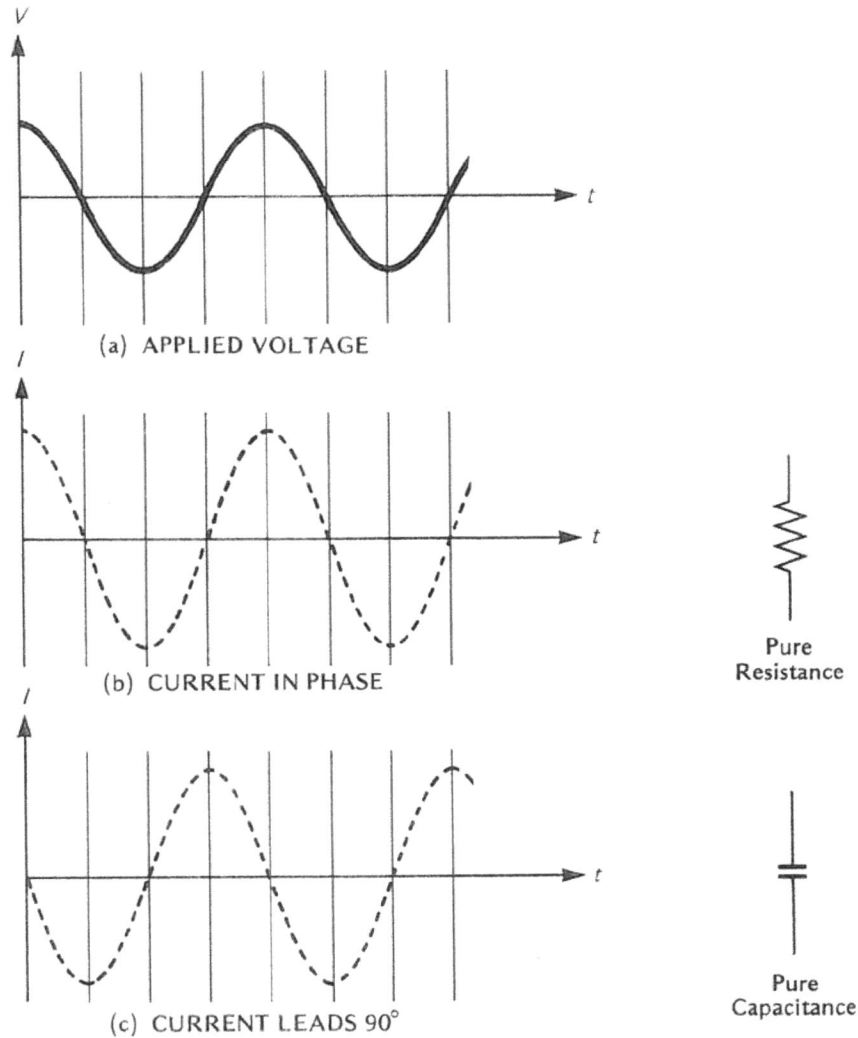

FIGURE 23.2
PURE RESISTIVE AND PURE CAPACITIVE CURRENTS

23.2 CAPACITIVE REACTANCE

From Figure 23.1(b) it is seen that the charge E_C on the capacitor is at all times equal and opposite to the value of the applied voltage V at any given instant. But the charging current I of a capacitor depends on its capacitance C and on the applied voltage V. (Think of a tank filling with water under pressure. The rate of flow into the tank will depend on the depth of water already there (its contents at the moment), and the pressure of the water entering.)

But the value of current depends also on the frequency with which the alterations take place; for obviously if the frequency is doubled, the same charge or discharge must take place in **half** the time, so the value of current throughout the cycle is doubled. Therefore I is proportional to the capacitance (C), to the applied voltage (V) and to the frequency (f) - that is $I \propto f \times C \times V$. In fact the actual relationship is:

$$I = 2\pi f C V$$

If it is turned and written:

$$\frac{V}{I} = \frac{1}{2\pi fC}$$

by comparison with Ohm's Law for resistance

$$\frac{V}{I} = R$$

the expression $\frac{1}{2\pi fC}$ is the equivalent in a capacitive circuit to resistance and, like resistance, is also measured in ohms. It is called the 'capacitive reactance' of the circuit and has the symbol 'X'. If necessary to distinguish it from the inductive reactance (see Chapter 22), the symbol 'X_C' may be used.

$$\therefore \text{ Capacitive reactance} \qquad X_C = \frac{1}{2\pi fC}$$

where X_C is in ohms, f in hertz and C in farads.

23.3 RELATIONSHIP BETWEEN CAPACITIVE AND INDUCTIVE REACTANCE

Chapters 22 and 23 developed the concept of two kinds of reactance - inductive (X_L) and capacitive (X_C). Both are measured in ohms and both act to limit the current, when an a.c. voltage is applied, as if they were resistances behaving as the resistance in Ohm's Law.

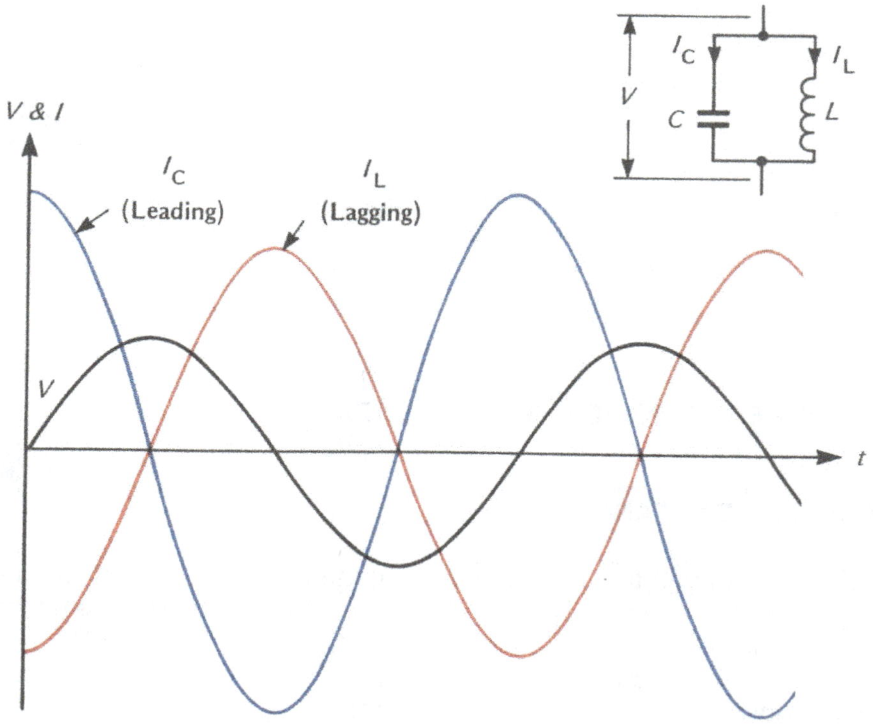

FIGURE 23.3
CAPACITIVE AND INDUCTIVE CURRENTS

Their essential difference is that an inductive reactance draws a current which lags 90° on the applied a.c. voltage, whereas a capacitive reactance draws a current which leads 90° on the voltage. The two, I_L (red) and I_C (blue) respectively, of different magnitudes, are shown in Figure 23.3 related to the applied voltage V.

It will be seen that, since each current wave is displaced 90° either side of the voltage wave, there is 180° between them - that is, the inductive and capacitive current waves are exactly opposite to each other in phase - they are said to be 'anti-phase'. One can be regarded merely as the negative of the other, or, put slightly differently, X_L and X_C can be regarded as having opposite signs.

Convention regards inductive reactance (being far the more common in power engineering) as positive, which makes capacitive reactance negative. If it were stated that two circuits had '20 ohms' and '-15 ohms' reactance, it would be inferred that the former was inductive and the latter capacitive.

Taking this concept a little further, since the inductive or capacitive loads here considered are regarded as 'drawing' currents - that is, taking currents in - a capacitive load (which draws a leading current) could be equally regarded as supplying - that is, giving **out** - a lagging current. This idea will not be considered further here, but it is used in the application of power-factor correction, which is dealt with in Part 5 Electric Motors.

CHAPTER 24 IMPEDANCE; OHM'S LAW FOR A.C.

24.1 PURE RESISTANCE, INDUCTANCE OR CAPACITANCE

It is shown in Chapters 21, 22 and 23 that opposition to the flow of current when an alternating voltage is applied to a circuit can be caused by any one or more of the following:

- resistance (R)
- inductive reactance (X_L)
- capacitive reactance (X_C)

All are measured in ohms, and, provided that only one of the three is present at any one time - that is, that the resistance or reactance is 'pure' - the current which flows is governed by Ohm's Law, namely:

$$I = \frac{V}{R} \text{ or } \frac{V}{X_L} \text{ or } \frac{V}{X_C}$$

where V and I are rms values.

In this chapter will be considered how the current is affected if more than one of these factors are present.

24.2 RESISTANCE PLUS INDUCTIVE REACTANCE

Suppose that there is a series circuit containing a resistance R and an inductance L. If the frequency f is known, the inductive reactance $X_L = 2\pi f L$, so X_L also is known.

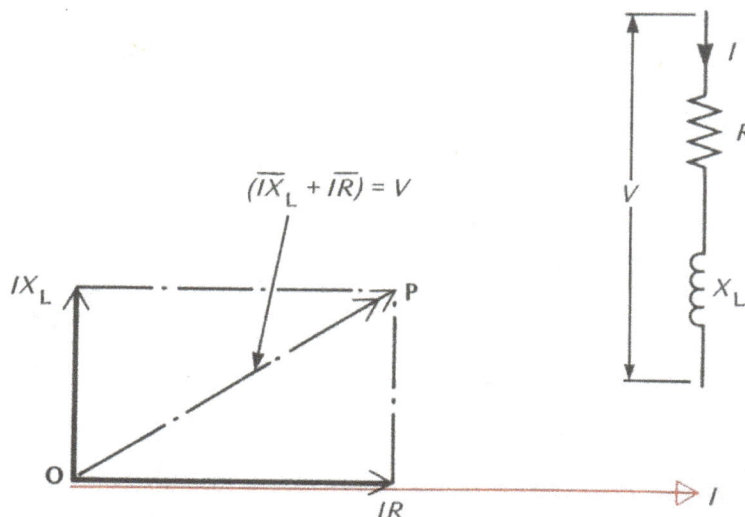

FIGURE 24.1
RESISTIVE/INDUCTIVE CIRCUIT

If the pair of elements R and X_L are fed in series from an a.c. supply voltage V, they will have a common current I. This is shown as the red vector I in Figure 24. 1.

Since current I passes through the pure resistance R, it will be in phase with the voltage IR developed across it, the magnitude of this voltage being determined by Ohm's Law $V = IR$. This is shown as voltage vector IR (full line) in Figure 24.1, where it is in phase with the current I.

The same current I also passes through the pure inductive reactance X_L. The current I through X_L will lag 90° on the voltage IX_L across it, the magnitude of this voltage being determined by Ohm's Law $V = IX_L$ for a pure inductance. As the current I lags 90° on voltage IX_L, voltage IX_L **leads** on the current I and is shown as voltage vector IX_L (full line) in Figure 24.1.

Thus the total voltage across R and X_L is the **vector** sum of IR and IX_L, which has been shown in Chapter 20 as being the diagonal of the rectangle formed by IR and IX_L - that is, the line OP. It can be written:

$$\overline{OP} = \overline{IR} + \overline{IX_L}$$

$$= I(\overline{R} + \overline{X_L})$$

This combined voltage is of course the same as the applied voltage V, so that:

$$V = I(\overline{R} + \overline{X_L})$$

or
$$\frac{V}{I} = \overline{R} + \overline{X_L} \quad \text{(where } V \text{ and } I \text{ are rms values)}$$

Compare this with Ohm's Law for a simple resistance, which is:

$$\frac{V}{I} = R$$

This shows that, for a combined resistive/inductive circuit, Ohm's Law applies if the **vector** sum of R and X_L is substituted for R. This vector sum is called the 'impedance' of the circuit; it has the symbol 'Z' and is measured in ohms.

Numerically) from the right-angled triangle formed by the diagonal OP and the vectors IR and IX_L,

$$Z^2 = R^2 + X_L{}^2$$

or
$$Z = \sqrt{R^2 + X_L{}^2}$$

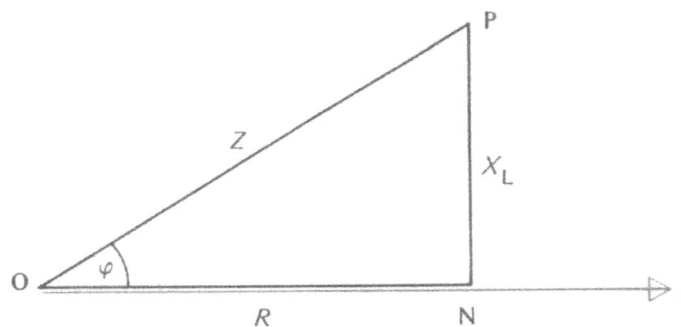

FIGURE 24.2
IMPEDANCE TRIANGLE

This triangle OPN is called the 'impedance triangle' (Figure 24.2) of the circuit and enables the impedance to be calculated, or measured directly, if R and X_L are both known.

24.3 RESISTANCE PLUS CAPACITIVE REACTANCE

Suppose that there is a series circuit containing a resistance R and a capacitance C. If the frequency f is known, the capacitive reactance $X_C = \frac{1}{2\pi f C}$, so X_C also is known.

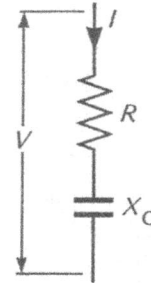

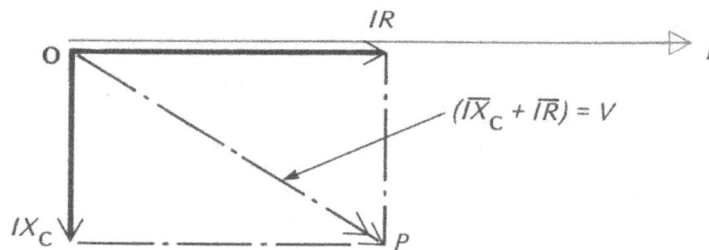

FIGURE 24.3
RESISTIVE/CAPACITIVE CIRCUIT

If the pair of elements R and X_C are fed in series from an a.c. supply voltage V, they will have a common current I. This is shown as the red vector I in Figure 24.3.

The argument from here on is exactly as for the inductive case above, except that the capacitive reactance X_C, being regarded as negative, is drawn downwards. This has no effect on the magnitudes of the various quantities, and the circuit's impedance Z is still the vector sum of $\overline{R}$ and $\overline{X_C}$ and has the magnitude given by $Z = \sqrt{R^2 + X_C^2}$. The impedance triangle OPN is similar but inverted.

Ohm's Law for a resistive/capacitive circuit is, as for the resistive/inductive case,

$\frac{V}{I} = Z$ ohms, where $Z = \sqrt{R^2 + X_C^2}$.

24.4 GENERAL CASE - OHM'S LAW FOR A.C.

For a circuit containing resistance and reactance, whether inductive or capacitive reactance, or both, Ohm's Law applies in the form:

$$\frac{V}{I} = Z \text{ ohms} \qquad \qquad ....(i)$$

where V and I are rms values and

$$Z = \sqrt{R^2 + X^2} \qquad \qquad ....(ii)$$

The reactance X is, in the general case, the sum of all reactances in the circuit, whether inductive or capacitive, remembering that capacitive reactances are regarded as negative.

In the special case where the inductive and capacitive reactances are numerically equal (but opposite), $X_L = -X_C$, or $X_L + X_C = 0$. In that case the expression (ii) for Z reduces to:

$$Z = \sqrt{R^2 + 0} = R$$

Z has then its minimum value and behaves as a simple resistance. With Z at a minimum, equation (i) shows that I is at its maximum - we have a condition known as 'resonance'.

24.5 PHASE ANGLE

In Figure 24.2 the angle between OP and ON (the Z and R vectors) is called the 'phase angle' and is given the Greek symbol 'φ' (phi, for 'phase'). From the trigonometry of the impedance triangle:

$$\tan \varphi = \frac{X}{R}; \quad \cos \varphi = \frac{R}{Z}; \sin \varphi = \frac{X}{Z}$$

so that, if any two of R, X and Z are known, the phase angle φ can be determined. Also, if X is capacitive ($= X_C$), it is by convention negative, so that φ too is negative and below the line. If φ above the line represents a lagging phase angle, below the line it represents a leading phase angle.

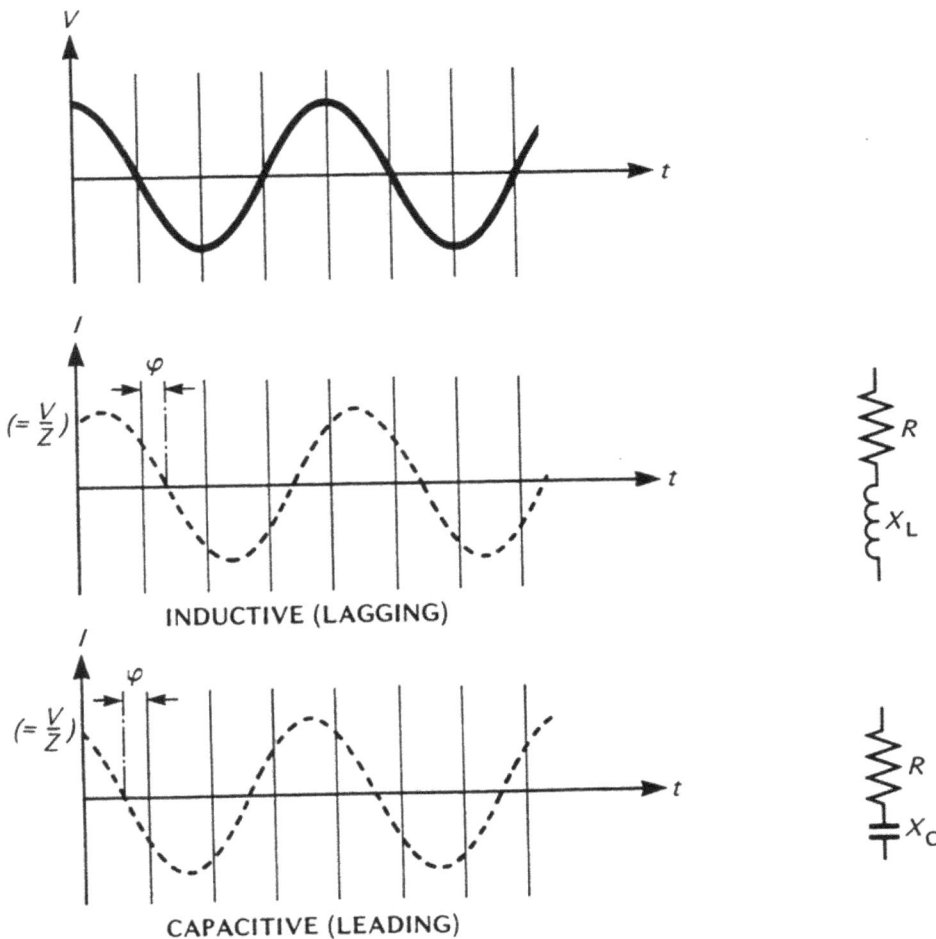

FIGURE 24.4
IMPEDANCE - GENERAL CASE

In the special cases (a) where there is only resistance and no reactance, $X = 0$ and therefore $\varphi = 0$; (b) where there is only reactance and no resistance, $R = 0$ and therefore $\varphi = 90°$. That is to say, in these two special cases of pure resistance and pure reactance the phase angles are 0° and 90° (lagging or leading) respectively, which are precisely the situations shown in Figures 21.1, 22.2 and 23.2 respectively, where the current is in phase or 90° lagging, or 90° leading, on the voltage.

In the general case where there is both resistance and reactance, the phase angle lies somewhere between 0° and +90° for inductive circuits, and between 0° and -90° for capacitive circuits.

Figure 24.4 shows the two general cases of a partly inductive and a partly capacitive circuit, where it will be seen that the current lags, or leads, on the voltage by an angle which is less than 90° and which is, in fact, the phase angle φ.

CHAPTER 25 IMPEDANCES IN SERIES AND PARALLEL

In Chapter 5, the behaviour of resistances in series and in parallel was considered when they carried direct current.

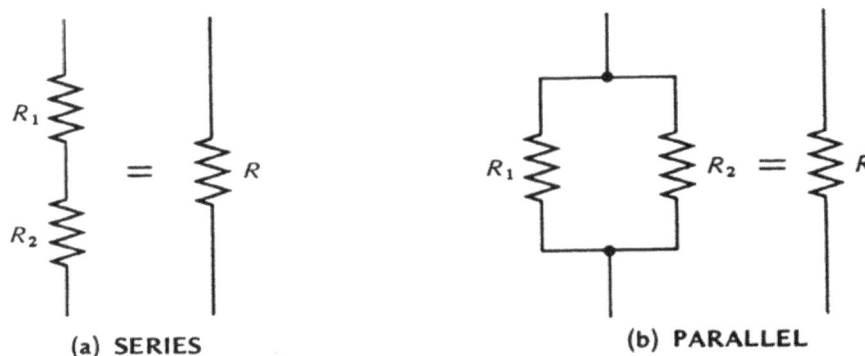

(a) SERIES   **(b) PARALLEL**

FIGURE 25.1
SERIES AND PARALLEL RESISTANCES

The conclusion was reached that, when a number of resistances R_1, R_2, R_3 ......etc were placed in series, they behaved as a single equivalent resistance R, where:

$$R = R_1 + R_2 + R_3 + \cdots \qquad \qquad \text{.... (i)}$$

When however a number of resistances R_1, R_2, R_3 .... etc. were placed in parallel they behaved as a single equivalent resistance R, where R is given by:

$$^1/_R = {}^1/_{R_1} + {}^1/_{R_2} + {}^1/_{R_3} + \cdots \qquad \qquad \text{.... (ii)}$$

That is to say, each resistance element must first be inverted, the inverses added together, and the sum so obtained inverted again to obtain the equivalent resistance R.

Exactly the same rules apply when the resistances carry alternating current. When a current flows in a pure resistance it was shown in Chapter 21 that it is in phase with the applied voltage. In a network containing nothing but resistances, all currents are in phase with all voltages and therefore with each other. It was shown in Chapter 20, Figure 20.3, that quantities which are in phase simply add or subtract numerically and so behave just as in the d.c. case. Therefore Figure 25.1 and equations (i) and (ii) apply equally to the a.c. case.

If the network consisted only of pure **reactances**, then all currents would lag (or lead) 90° on the applied voltages and so would be in phase **with each other**. They would therefore add and subtract numerically just as if they were resistances, and the series and parallel counterparts of equations (i) and (ii) for the equivalent reactance in an all-reactance network are:

Series: $X = X_1 + X_2 + X_3 + \cdots$ $\qquad$ .... (iii)

Parallel: $^1/_X = {}^1/_{X_1} + {}^1/_{X_2} + {}^1/_{X_3} + \cdots$ $\qquad$ .... (iv)

(Note that, when a reactance is capacitive, it is regarded as negative. Provided that care is taken with the sign of X, the above expressions apply equally to inductive or capacitive reactances, or to a mixture of both.)

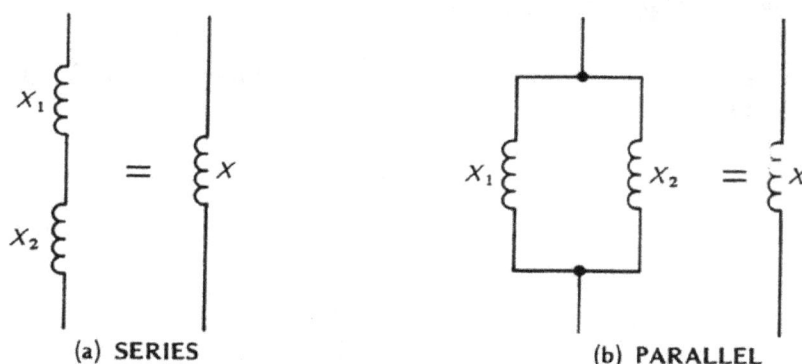

(a) SERIES   (b) PARALLEL

FIGURE 25.2
PARALLEL REACTANCES

When both resistance and reactance are present together, the problem is less simple. It was shown in Chapter 20, Figure 20.4, that quantities which are not in phase can only be added or subtracted vectorially. When two parallel limbs of a circuit are pure resistance and pure reactance (Figure 25.2(a)), the current in the resistive limb is in phase with the common applied voltage, whereas that in the reactive limb lags 90° on that voltage; the currents are therefore 90° out of phase with each other.

Similarly, when a resistance is in series with a reactance (Figure 25.2(b), the common current is in phase with the voltage across the resistive element but lags 90° on the voltage across the reactive one. The former therefore has a voltage in phase with the common current, and the latter has a voltage which leads 90° on that current. The two voltages, which are thus 90° out of phase with each other, together add up to the total applied voltage V. They can therefore only be added vectorially.

Figure 25.3(a) (bottom) is the 'impedance triangle' for the series case; I is the common current, R is in phase with it and X is at 90°. Z is the total impedance given by the vector addition of R and X - that is, $\overline{Z} = \overline{R} + \overline{X}$, or numerically:

$$Z = \sqrt{R^2 + X^2} \qquad\qquad .... \text{(iv)}$$

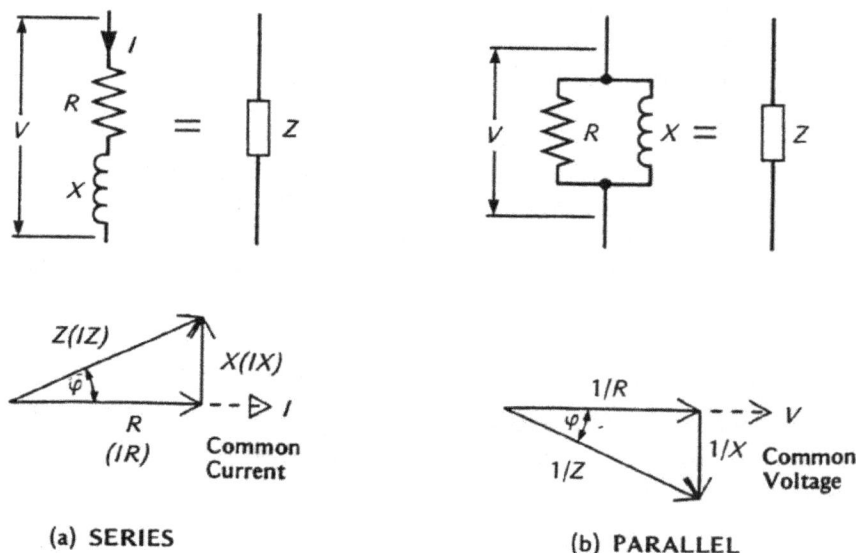

(a) SERIES   (b) PARALLEL

FIGURE 25.3
SERIES AND PARALLEL IMPEDANCES

If all three sides of the triangle are multiplied by I, it becomes a voltage triangle, IR being the voltage developed across R by the common current (in phase with I), and IX being the voltage developed across X by that current and leading 90° on I, and IZ being the applied voltage across the whole impedance.

If there are several resistances and several reactances, R and X in expression (v) are respectively the sums of all the resistances and all the reactances as given by the series expressions (i) and (iii).

The phase angle so of the equivalent impedance Z is given by:

$$\tan \varphi = \frac{X}{R}$$

and the power factor (cos φ) is $\frac{R}{Z}$.

Figure 25.3(b) (bottom) is the 'impedance triangle' for the parallel case. We have already seen from expressions (ii) and (iv) that with parallel circuits it is the **inverses** of resistances or reactances which must be added together. This is also the case when they are mixed, except that the addition must be vectorial. (Note that in this case the impedance triangle is drawn **downwards**, since the current in the reactance limb must **lag** 90° on the common voltage.)

The impedance triangle is therefore drawn not for the actual resistance and reactance but for their inverses $1/R$ and $1/X$. They can then be added vectorially to give the inverse of the equivalent impedance Z. Thus:

$$\overline{1/Z} = \overline{1/R} + \overline{1/X}$$

or numerically
$$\frac{1}{Z} = \sqrt{\left(\frac{1}{R}\right)^2 + \left(\frac{1}{X}\right)^2} \qquad \dots\text{(vi)}$$

$$= \sqrt{\frac{X^2 + R^2}{R^2 \times X^2}}$$

whence
$$Z = \frac{RX}{\sqrt{R^2 + X^2}} \qquad \dots\text{(viii)}$$

As before, if there are several resistance and several reactance limbs, $\frac{1}{R}$ and $\frac{1}{X}$ in expression (vi) are the sums of the inverses of all the resistances and all the reactances as given by the parallel expressions (ii) and (iv).

The phase angle φ of the equivalent impedance Z is given by:

$$\tan \varphi = \frac{1/X}{1/R} = R/X \text{ (the opposite of the series case)}$$

and the power factor (cos φ) is $\frac{1/R}{1/Z} = Z/R$ (again the opposite of the series case).

CHAPTER 26 USEFUL FORMULAE: CHAPTERS 14 - 25

A.C. WAVEFORM

If peak amplitude is $\hat{A}$ and rms value is A, then:

$$\hat{A} = \sqrt{2}A$$

$$A = \frac{1}{\sqrt{2}}\hat{A}$$

STAR AND DELTA CONNECTIONS

If V_P is phase voltage and V_L line voltage, and if I_P, is phase current and I_L line current, then in a 3-phase system with

Star connection:
$$V_L = \sqrt{3}V_p \quad \text{or} \quad V_p = \frac{1}{\sqrt{3}}V_L$$

$$I_L = I_P$$

Delta connection:
$$I_L = \sqrt{3}I_p \quad \text{or} \quad I_p = \frac{1}{\sqrt{3}}I_L$$

$$V_L = V_P$$

PYTHAGORAS AND TRIGONOMETRICAL FUNCTIONS

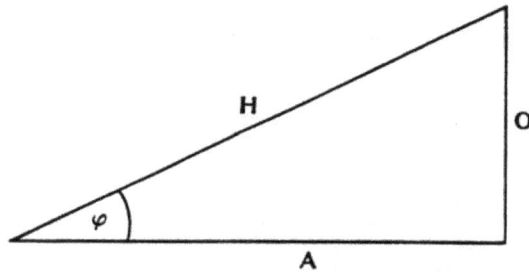

$$H^2 = O^2 + A^2 \quad \text{or} \quad H = \sqrt{O^2 + A^2}$$

$$O^2 = H^2 - A^2 \quad \text{or} \quad O = \sqrt{H^2 - A^2}$$

$$A^2 = H^2 - O^2 \quad \text{or} \quad A = \sqrt{H^2 - O^2}$$

$$\sin\varphi = \frac{O}{H} \quad \text{or} \quad O = H\sin\varphi$$

$$\cos\varphi = \frac{A}{H} \quad \text{or} \quad A = H\cos\varphi$$

$$\tan\varphi = \frac{O}{A} \quad \text{or} \quad O = A\tan\varphi$$

VECTORS

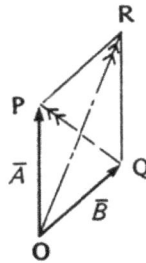

To add vectors $\overline{A}$ and $\overline{B}$:
Complete parallelogram OPRQ. The sum $\overline{A+B}$ is given by the diagonal OR through O.

To subtract vector $\overline{B}$ from $\overline{A}$:
Complete parallelogram OPRQ. The difference $\overline{A-B}$ is given by the **other** diagonal across O (or by joining the arrow-tips). The direction is from $\overline{B}$ to $\overline{A}$.

RESISTANCE, REACTANCE, IMPEDANCE AND OHM'S LAW FOR A.C.

Heating due to current I (rms) through resistance $R = I^2 R$

Inductive reactance X_L in a circuit of inductance L henrys at a frequency $f = 2\pi f L$ ohms

Capacitive reactance X_C in a circuit of capacitance C farads at a frequency $f = \frac{1}{2\pi f C}$.

In a circuit containing resistance R and reactance X, the total impedance (Z) is given by:
$$Z^2 = R^2 + X^2 \quad \text{or} \quad Z = \sqrt{R^2 + X^2}$$

Ohm's Law for a.c.: $\quad V = IZ \quad \text{or} \quad I = \frac{V}{Z} \quad \text{or} \quad Z = \frac{V}{I}$

IMPEDANCES IN SERIES AND PARALLEL

Reactances in series: $\quad X = X_1 + X_2 + X_3 + \ldots$

Reactances in parallel: $\quad \frac{1}{X} = \frac{1}{X_1} + \frac{1}{X_2} + \frac{1}{X_3} + \ldots$

CHAPTER 27 QUESTIONS AND ANSWERS: CHAPTERS 14 - 25

27.1 QUESTIONS

1.   Show by a sketch the principle of Faraday's Law of Electromagnetic Induction and explain how it is used to achieve electrical generation.

2.   Explain the difference between a permanent magnet and an electromagnetic field system.

3.   What is the name given to that part of a generator in which voltage is induced?

4.   What is the essential difference between a fixed field and a rotating field generator? What are the advantages of the latter?

5.   In a fixed field generator, how is the current taken from the rotating conductors?

6.   How is a rotating field generator excited?

7.   What do you understand by 'brushless excitation'? What are its advantages over other forms? Make a sketch diagram of an a.c. generator with brushless excitation.

8.   How is the output voltage of a generator controlled?

9.   What distinguishes a d.c. generator from an a.c.?

10.   Describe briefly the construction of a 3-phase generator.

11.   What are the two main methods of interconnecting the three 3-phase machine windings?

12.   In a star-connected generator feeding a balanced load what current flows in the neutral (star-point) connection?

13.   Where would you expect to find a 3-wire, and where a 4-wire, distribution on a 3-phase system?

14.   What is the effect of including iron inside a coil of wire?

15.   If a single-phase a.c. supply is connected to such a coil, what effect would it have on the magnetic field (a) inside, and (b) outside the coil?

16.   How is a rotating magnetic field produced statically from a 3-phase system?

17.   A 50Hz, 3-phase supply is applied to a '2-pole' a.c. motor. What is the speed of rotation of the field?

18.   If the motor in Q.17 were '4-pole' and the frequency of the supply were 60Hz, what would then be the speed of rotation of the field?

19.   Can a rotating field be produced from a single-phase supply? If so, describe one method.

20.	What do you understand by 'simple harmonic motion'? Sketch a voltage sine-wave and identify its parts by symbol and name.

21.	What is the 'amplitude' of an alternating quantity? How do you express it in the mathematical form for such a quantity?

22.	In practice are electrical alternating quantities always of pure sine-wave form? From what sources would you expect to find distortion?

23.	What is the relationship between an rms quantity and its peak value? Define what is meant by an rms current.

24.	Write down the relationship between the sides of a right-angled triangle. If the angle at one corner is φ, the hypotenuse is 'H' and the sides adjacent to and opposite that corner are 'A' and 'O' respectively, write down the values of cos φ, sin φ and tan φ.

25.	In a right-angled triangle one angle is 40° and the hypotenuse is 15cm. What are the lengths of (a) the opposite and (b) the adjacent side?

26.	What is a 'vector', and how does it differ from a simple measurement? Sketch a vector system where the voltage is 100V and the current 80A lagging 30° on the voltage.

27.	What is the vector sum $\overline{A} + \overline{B}$?

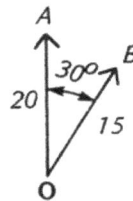

28.	In the figure of Q.27, what is the vector difference $\overline{A} - \overline{B}$?

29.	What is the line voltage of a star-connected 3-phase system whose individual phase voltages are 3.81kV? What would the line voltage be if these same windings were delta-connected?

30.	If a 3-phase, delta-connected generator supplied a load with balanced line current of 200A in each line, what current flows in each phase-winding of the generator?

31.	What is the greatest voltage experienced under normal conditions between the line conductors of a cable (and therefore the greatest strain on the insulation between them) if line-voltage voltmeter reads 132kV?

32.	What is the phase relationship between current and voltage in an a.c. circuit containing only pure resistance? If a current of 10A (rms) flows in an a.c. circuit of resistance 4 ohms, at what rate is heat produced, and what unit is used to express it?

33.	What is the effect of inductive reactance on the current in an a.c. circuit? If that circuit had an inductance of 0.1 henrys and the system frequency were 60Hz, what would be the reactance? What unit?

34. What is the effect of capacitive reactance on the current in an a.c. circuit? If that circuit had a capacitance of 100μF (100 microfarads) and the system frequency were 50Hz, what would be the reactance? And how would it differ in principle from the inductive reactance of Q.33?

35. If a single-phase circuit has an inductive reactance of 30 ohms and a resistance of 40 ohms, what is the total impedance? (State the unit.) If 250V (rms) a.c. is applied to this circuit, what current would flow?

36. Three inductive reactances, 12 ohms, 20 ohms and 30 ohms are placed in parallel. What is the equivalent single reactance?

37. An inductive reactance of 50 ohms and a capacitive reactance of 25 ohms are placed in parallel. What is the equivalent single reactance?

38. A generator of 250V, 60Hz, single-phase feeds two inductive loads of 0.1H and 0.4H and a capacitor of 10μF, all in parallel. What are the three individual currents, and what is the total current supplied by the generator?

39. An inductor of 1.013H is placed in series with a capacitor of 10μF and a resistor of 5 ohms. A voltage is applied overall at 50Hz. What is the equivalent impedance? What is noteworthy about it?

40. Write down three alternative versions of Ohm's Law for a.c.

41. What is the impedance of a 50Hz series circuit consisting of a resistance of 40 ohms and an inductance of 0.2H?

42. What current does a single-phase, 220V, 60Hz generator supply to a **parallel** circuit consisting of a resistance of 15 ohms and a capacitance of 100μF?

43. What do you understand by an 'impedance triangle' of a circuit? How do you use it?

44. A single-phase circuit has a resistance of 20 ohms and a series capacitive reactance of -15 ohms. A voltage of 250V is applied. What **total** current flows, and by what angle does its vector differ from the voltage vector? Does it lag or lead?

45. A single-phase circuit has a resistance of 20 ohms and a **series** inductive circuit of +15 ohms. A voltage of 250V is applied. What current flows and by what angle does its vector differ from the voltage vector? Does it lag or lead? What are the individual voltages across the resistive and the inductive elements? Why do they not add up to 250V?

27.2 ANSWERS

(Figures in brackets after each answer refer to the relevant paragraphs in the text.)

1. Sketch as Figure 5.1 in Chapter 5. By arranging a machine so that conductors move successively past a fixed magnetic field, or that a magnetic field moves past all conductors, and by interconnecting the conductors; the electromotive forces (emf) in each are added together to produce a useful voltage at the terminals of the generator. (14.1)

2. In a permanent magnet generator the field is produced by natural magnets or by magnets artificially made. Once made, their magnetisation is fixed and cannot be altered. In an electromagnetic generator the field is produced by an iron core 'excited' by a d.c.-fed winding; the strength of magnetisation can be controlled by varying the exciting field current. (14.2)

3. 'Armature'. (14.1)

4. In a fixed field system the magnets, whether permanent or electromagnetic, are stationary and the armature moves past them. In a rotating field system the armature is stationary and the magnets move past its conductors. (14.2)

5. By sliprings on the armature shaft. (14.2)

6. By an external d.c. supply through sliprings, taken from a battery, a d.c. generator or an a.c./d.c. rectifier. (In a 'brushless' system the sliprings are dispensed with.) (14.2)

7. A form of a.c. generator excitation where an a.c. exciter is mounted on the main rotor shaft and its a.c. output passed through rotating diode rectifiers to the main field. Voltage control is on the exciter's stationary field. Reduces maintenance by omitting sliprings and brushgear as compared with other types. Sketch as in Chapter 14, Figure 14.8. (14.2)

8. By controlling the exciting field current - either manually or automatically by an automatic voltage regulator ('AVR'). (14.2)

9. A d.c. generator has a commutator, whereas an a.c. generator has sliprings or, in the case of the brushless type, no connections to the rotating part at all. (14.4)

10. A 3-phase generator has three separate armature windings and a common field system, the axes of the three windings being spaced at 120° intervals around the machine. Each winding may be brought out to two separate terminals (six in all), or they may be star- or delta-connected inside the machine, resulting in four or three terminals. (14.3)

11. Star or delta. (15.2, 15.5)

12. Nil. The three phase return currents, when combined vectorially, add up to zero.(15.3)

13. High.voltage systems are generally star-connected 3-wire because the loads are all 3-phase and effectively balanced. Low-voltage systems are usually star-connected 4-wire because the many single-phase loads produce appreciable unbalance and thereby give rise to resultant current in the neutral conductor. (15.4)

14. The presence of iron intensifies any magnetic field because of its 'permeability'. (Figure 16.1)

15. (a) The axial field inside the coil would be intensified. It is at all times in phase with the coil current. (16.1)

    (b) The field outside the coil is in the reverse direction to that inside and is not so concentrated. It too is at all times in phase with the coil current but opposite in direction. (16.1)

16. By arranging three field windings at equal 120° intervals around the core of a stator and feeding each with one phase of the 3-phase supply, so there is both 120° space displacement and also 120° time displacement between each. (16.2)

17. 3 000 rev/mm. (16.2)

18. 1 800 rev/min. (16.2)

19. Yes, by splitting the field current into two parts mutually at right angles to one another. One part is fed direct, the other through a capacitor large enough to cause the current to lead on the other part by almost 90°. A motor fed this way is termed a 'capacitor' or 'split field' motor. (Figure 16.3)

20. If a point rotates, say at the end of an arm, at constant angular velocity about a centre, then the 'shadow' (that is, the projection) of that point on a plane surface parallel to the axis of rotation describes a to-and-fro motion known as 'simple harmonic motion'. (17.2)

21. Amplitude is the peak value (referred to the central axis) of a sinusoidal quantity. It is the term 'A' in the expression $A \sin 2\pi ft$. (17.2)

22. No, but in most cases it is assumed. Distortion can arise through magnetic saturation (for example of transformers) and more particularly through rectifying equipment. (17.1)

23. rms $= \dfrac{1}{\sqrt{2}} \hat{A}$ where $\hat{A}$ is the peak value (or amplitude) of the quantity.

    An rms current is the square root of the average (or mean) value of the (current)^2 wave. The rms value of an a.c. current causes the same rate of heating in a given resistance as would a d.c. current of the same numerical value. (Chapter 18)

24. $H^2 = A^2 + O^2$

    $\sin \varphi = \dfrac{O}{H}$

    $\cos \varphi = \dfrac{A}{H}$

    $\tan \varphi = \dfrac{O}{A}$ (19.3)

25. (a) Opposite side is 15 sin 40° = 9.64cm.
    (b) Adjacent side is 15 cos 40° = 11.49cm. (19.3)

26. A vector represents not only a numerical value (its length) but also a **direction**.

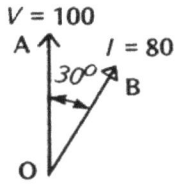

(20.1)

27.

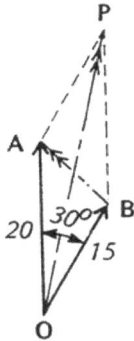

Complete the parallelogram $OAPB$

Diagonal $\overline{OP}$ represents the vector sum of $\overline{A}$ and $\overline{B}$.

By measurement of $OP, \overline{A} + \overline{B}$ = 3.38 units

(20.2)

28. Complete the parallelogram OAPB as in Q.27.
Joining the arrow-tips of $\overline{A}$ and $\overline{B}$ (or taking the other diagonal BA) represents the difference $\overline{A} - \overline{B}$

By measurement $\overline{A} - \overline{B}$ = 10.3 units (20.2)

29. $\sqrt{3} \times 3.81 = 6.6kV.$
If delta-connected, line voltage would be the same as the phase voltage, namely 3.81kV. (20.4)

30. $200 \div \sqrt{3} = 115.4A.$ (20.5)

31. $132 \times \sqrt{2} = 187kV.$ (20.5)

32. In a purely resistive a.c. circuit the current is always in phase with the voltage. Heating rate is I^2R watts, where I is the rms current. If I = 10A and R = 4 ohms

$$I^2R = 10^2 \times 4 = 400 \text{ watts.}$$ (21.1)

33. Inductive reactance causes the current to lag on the voltage. If no resistance is present, it lags 90°

Reactance $X_L = 2\pi fL$

As f = 60Hz and L = 0.1H
$$X_L = 2\pi \times 60 \times 0.1 = 37.7 \text{ ohms}$$ (22.1)

34. Capacitive reactance causes the current to lead on the voltage. If no resistance is present, it leads 90°

    Reactance $X_C = \frac{1}{2\pi f C}$

    As $f = 50\text{Hz}$ and $C = 100\mu F = 100 \times 10^{-6}\,F = 0.0001\,F$

    $$X_C = \frac{1}{2\pi \times 50 \times 0.0001} = 31.8 \text{ ohms}$$

    As this is a capacitive reactance it would differ from the inductive reactance of Q.33 by being negative and could be written '-31.8 ohms'. (23.1)

35. $Z = \sqrt{R^2 + X^2} = \sqrt{40^2 + 30^2} = \sqrt{1600 + 900} = \sqrt{2500} = 50$ ohms

    $V = IZ$ or $I = {V}/{Z} = {250}/{50} = 5A$ (24.2)

36. $\frac{1}{X} = \frac{1}{12} + \frac{1}{20} + \frac{1}{30} = 0.083 + 0.05 + 0.033 = 0.166$

    $\therefore \quad X = 6.02$ ohms (Chapter 25)

37. $\frac{1}{X} = \frac{1}{50} - \frac{1}{25} = 0.02 - 0.04 = -0.02$

    $\therefore \quad X = -50$ ohms

    (Note: the answer is negative, so that the equivalent single reactance of 50 ohms is capacitive.) (Chapter 25)

38. At 60Hz, 0.1H has reactance $2\pi f L = 2\pi \times 60 \times 0.1 = 37.7$ ohms
    At 60Hz, 0.4H has reactance $2\pi f L = 2\pi \times 60 \times 0.4 = 150.8$ ohms

    At 60Hz, 10μF has reactance $\frac{1}{2\pi f C} = \frac{1}{2\pi \times 60 \times 10^{-5}} = -265.2$ ohms

    Equivalent reactance X is given by $\frac{1}{X} = \frac{1}{37.1} + \frac{1}{150.8} - \frac{1}{265.2} = 0.0294$

    whence $\quad X = 34.0$ ohms ($= Z$, since no resistance)

    If 250V applied, current $= \frac{V}{Z} = \frac{250}{340} = 7.35A$ (Chapter 25)

39.

At 50Hz, $X_L = 2\pi \times 50 \times 1.013 = +318.3$ ohms

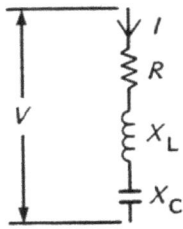

At 50Hz, $X_C = \dfrac{1}{2\pi \times 50 \times 10^{-5}} = -318.3$ ohms

For the two reactances $X = X_L + X_C = 318.3 - 318.3 = 0$ ohms.

They act as a short-circuit, leaving only R in circuit:

i.e. $Z = \sqrt{R^2 + X^2} = \sqrt{5^2 + 0} = 5$ ohms

This is the condition of 'resonance', where Z is at its minimum and therefore I at its maximum. It behaves as a purely resistive circuit, with I in phase with V. (Chapter 25)

40. $V = IZ$ or $I = \dfrac{V}{Z}$ or $Z = \dfrac{V}{I}$ \hfill (24.4)

41. At 50Hz 0.2H has reactance $2\pi f L = 2\pi \times 50 \times 0.2 = 62.8$ ohms $(=X)$

$$Z = \sqrt{R^2 + X^2} = \sqrt{40^2 + 62.8^2} = 74.4 \text{ ohms} \qquad (24.2)$$

42. At 60Hz 100μF has reactance $\dfrac{1}{2\pi f C} = \dfrac{1}{2\pi \times 60 \times 10^{-4}} = 26.5$ ohms $(=X)$

As circuit is parallel, $\dfrac{1}{Z} = \sqrt{\dfrac{1}{R^2} + \dfrac{1}{X^2}} = \sqrt{\dfrac{1}{15^2} + \dfrac{1}{26.5^2}} = 0.00587$

$\therefore \qquad Z = 170$ ohms

Current $= \dfrac{V}{Z}$ $\therefore I = \dfrac{220}{170} = 1.29A$ \hfill (24.2)

43. If any two of the quantities resistance (R), reactance (X) and impedance (Z) of a series circuit are known, the third can be deduced by Pythagoras from the right-angled triangle formed by R, X and Z, called the 'impedance triangle'. This also enables angles, and so phase relationships, to be measured or calculated. If each side of the triangle is multiplied by the common current I (so not altering the shape of the triangle), the three sides then give the voltages across each element and overall.

If the circuit is a parallel one the impedance triangle is formed by the inverse quantities $\dfrac{1}{R}$, $\dfrac{1}{X}$ and $\dfrac{1}{Z}$. If each side is multiplied by the common voltage V (so not altering the shape of the triangle), the three sides then give the currents in each element and overall. \hfill (24.2)

44. Impedance Z \hfill $= \sqrt{R^2 + X^2} = \sqrt{20^2 + (-15)^2} = 25$ ohms

Current $= \dfrac{V}{Z}$ \hfill $\therefore I = \dfrac{250}{25} = 10A$

Phase angle \hfill $= \tan^{-1}\dfrac{X}{R} = \tan^{-1}\dfrac{-15}{25} = \tan^{-1} -0.6$

\hfill $= -31.0°$ leading (negative) \hfill (24.2)

45. Impedance Z

$$= \sqrt{R^2 + X^2} = \sqrt{20^2 + 15^2} = 25 \text{ ohms}$$

Current $= \dfrac{V}{Z}$

$$\therefore I = \frac{250}{25} = 10A$$

Phase angle

$$= \tan^{-1}\frac{X}{R} = \tan^{-1}\frac{+15}{25} = \tan^{-1} +0.6$$

$$= +31.0° \text{ lagging (positive).}$$

Individual voltage across R is $IR = 10 \times 20 = 200V$
Individual voltage across X is $IX = 10 \times 15 = 150V$

Numerical total is 350V, but this differs from the actual applied voltage (250V) because the two individual voltages must be added **vectorially**: thus

$$\overline{IR} + \overline{IX} = \sqrt{200^2 + 150^2} = 250V \text{ (the actual applied voltage)} \qquad (24.2)$$

CHAPTER 28 ACTIVE POWER

28.1   POWER

The purpose of most electrical systems is to generate real electrical power and to convey it to those consumer installations which will use it.

True 'power' may be used to provide a mechanical drive, or to provide heating and lighting, or to energise control and communication systems such as instrumentation or radio and telephone installations. All of these things consume energy, and that energy is absorbed at a stated rate: the **rate** of consuming energy is defined as the true power consumption of that system. In electrical installations it is measured in the unit 'watt' (W). For practical power purposes this unit is generally found to be too small, and more often used is the kilowatt (kW), one thousand watts, or even the megawatt (MW), one million watts.

Electric power is most usually obtained from an electric generator, but this in turn receives its power from the engine which drives it (termed the 'prime mover'). It may be of many types: a steam or gas engine, a diesel, a steam or gas-turbine, a water-wheel or even a windmill, though in some installations only gas-turbines or diesel engines are used. With the exception of water- and wind-driven sets, the energy delivered by these engines to the generator is derived from the fuel which they burn - that is to say, the energy source is ultimately a chemical one.

What has been said so far refers to **energy** and to the rate at which it is delivered (the power). When an electric generator is delivering this energy it is at the same time usually delivering also another type of 'false energy' which too is required by certain consumer equipment. To distinguish between them the true power, which represents real energy, is called 'active power' (sometimes also called 'wattful power'). The other kind, which is the rate of delivering 'false energy', is termed 'reactive power' (sometimes also called 'wattless power 'or 'blind power').

Reactive power is dealt with separately in Chapter 29.

28.2   ELECTRICAL POWER

Voltage is a pressure, and current is a flow. In mechanical engineering, power - the rate of doing work - is the product of pressure and volume flow; in electrical circuits, power is the product of voltage and current - that is, power $= V \times I$. If V is measured in volts and I in amperes, their product is the power in watts (W).

In d.c. this presents no problem. Both V and I are steady quantities and their product is a direct measure of the power in watts. Indicating instruments - wattmeters - are made which do this multiplication internally.

With a.c. rather more care must be taken. The same rule applies: namely that the watts **at any instant** are the product of the volts and the amperes **at that instant**, but these quantities are constantly changing as the voltage and current alternate. It is therefore necessary to look at this product instant by instant to see whether it has any average value.

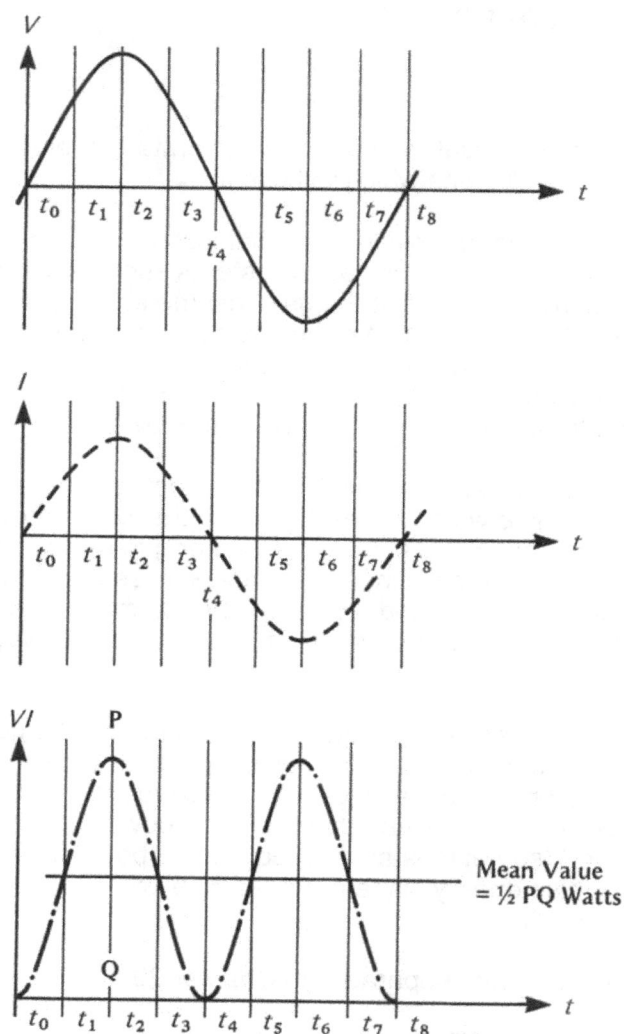

FIGURE 28.1
A.C. POWER - PURE RESISTIVE LOAD

Consider an a.c. voltage feeding a purely resistive load. If the top wave of Figure 28.1 represents the alternating voltage, the second wave represents the current, which, as has been shown for a resistive circuit, is in phase with the voltage. Thus both voltage and current have their positive parts together, and also their negative parts together.

The power at any instant is the product of the voltage and current at that instant. Clearly at times t_0, t_4 and t_8 both waves are at zero, so their product is also zero. At any time, say t_1, in the first half-cycle voltage and current are both positive, so their product is also positive and is greatest at time t_2, where both are at their maximum.

At any time, say t_5, in the second half-cycle voltage and current are both **negative**, so their product is again positive and is greatest at time t_6, where both are at their negative peaks.

The power wave is therefore the third in Figure 28.1. It is of double frequency (i.e., two peaks for every one voltage peak) and is wholly above the line (positive). It represents pulses of power, always positive, and the average value of that power is midway between the power peaks and valleys.

Numerically, the mean power level is half the peak-peak value (PQ) of the power curve, and this itself is the product of the peak voltage and the peak current. If $\hat{v}$ is the peak voltage and $\hat{\imath}$ the peak current, then mean power is:

$$\frac{1}{2}\hat{v}\hat{\imath} \qquad\qquad \text{...(i)}$$

But it was shown earlier in Chapter 18 that peaks of voltage or current are $\sqrt{2}$ times their rms values: i.e. $\hat{v} = \sqrt{2}V$ and $\hat{\imath} = \sqrt{2}I$ where V and I are rms values.

So from (i) above the mean power (symbol $'P'$) is given by:

$$P = \frac{1}{2} \times \sqrt{2}V \times \sqrt{2}I = \frac{1}{2} \times 2VI = VI \ \text{(watts)}$$

That is to say, in the purely resistive case the mean power is the true or active power *(P)* and is measured in watts. It is the product of the **rms voltage** and **rms current** (in amperes), exactly similar to the d.c. case.

The dynamometer wattmeter, described earlier and also in Chapter 36, does this multiplication automatically for a.c. as well as d.c., and it will indicate the average watts being transmitted.

CHAPTER 29 REACTIVE POWER

29.1 INDUCTIVE CASE

In Chapter 28 it was shown that the true power transmitted where the load is purely resistive is given by:

$$P = VI \text{ watts}$$

where V and I are the rms values of voltage and current.

Consider now the purely inductive case, where there is no resistance, as represented by Figure 29.1.

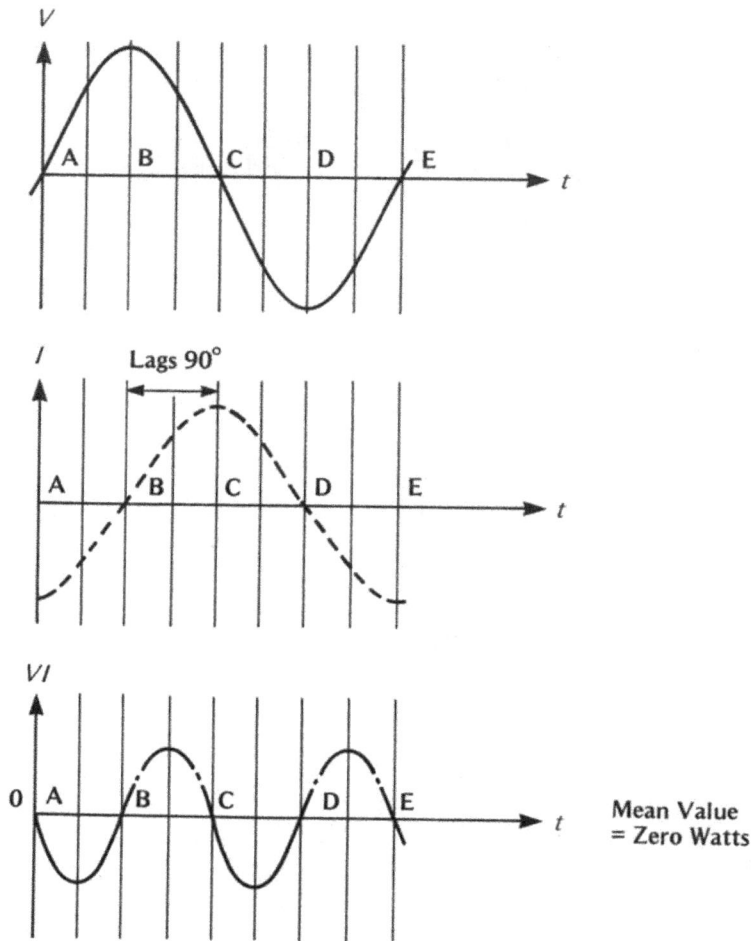

FIGURE 29.1
A.C. POWER - PURE INDUCTIVE LOAD

It has already been shown in Chapter 22, that, if the top wave represents the alternating voltage, the second wave represents the current, which now lags one-quarter of a cycle, or 90°, behind the voltage.

Using the same method as in Chapter 28, multiply the voltage and current at each instant. We now have:

In first quarter cycle (A - B) voltage is positive current is negative } product negative

In second quarter cycle (B - C) voltage is positive current is positive } product positive

In third quarter cycle (C - D) voltage is negative current is positive } product negative

In fourth quarter cycle (D - E) voltage is negative current is negative } product positive

after which the cycle repeats.

At points A, B, C, D and E either the voltage or the current is zero, so that their product at all these points is zero.

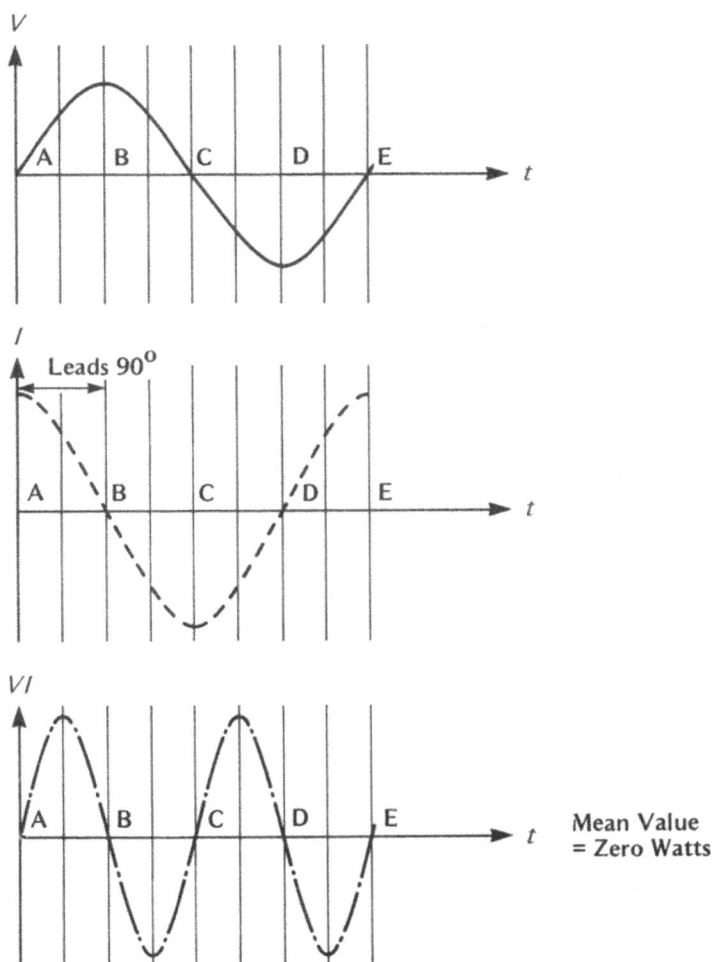

FIGURE 29.2
A.C. POWER - PURE CAPACITIVE LOAD

If the product (power) curve is now drawn, it will be as the third wave of Figure 29.1. It will be, as before, of double frequency but now it is symmetrical about the zero line, and therefore the average power will be zero. Power is put in at each positive part and taken out again at each negative part, giving a net power transmission of NIL.

29.2   CAPACITIVE CASE

It has been shown above that a current lagging 90° on the voltage, in the purely inductive case, gives rise to a double-frequency power wave which is symmetrical about the axis and therefore has no net or average power.

The purely capacitive case is quite similar except that the current wave leads the voltage by 90°, as shown in Figure 29.2.

Exactly the same treatment as that given in para. 29.1 will produce the power wave at the bottom of Figure 29.2. Compared with that of Figure 29.1 it will be reversed in sign, but it will still be symmetrical about the axis and therefore will have no net or average power.

29.3   THE 'VAR' UNIT

The conclusions of both paras. 29.1 and 29.2 - namely that no net power is passed in either a purely inductive or purely capacitive circuit - lead to a certain re-thinking of the power rules if we are used only to d.c. If the d.c. voltmeter reading is multiplied by the d.c. ammeter reading, the result is the d.c. power in watts. In the a.c. case however this is only true if the load is purely resistive (as in Chapter 28), which it seldom is.

In the reactive cases (inductive or capacitive) the voltmeter and ammeter readings can still be multiplied together, but they do not now represent true power in watts. Yet the product is a perfectly good figure. What then is it? It represents a 'false power' (the Germans call it 'blind power') and it is measured in a unit called the 'var' (short for volt-ampere-reactive). It is called 'reactive power', or sometimes 'wattless power' with symbol 'Q', and is a measure of the energy stored (but not consumed) in a magnetised system. Since a platform or shore installation consists of a vast number of transformers, motors, etc. which all need to be magnetised, the demand for vars is considerable, as will be shown by the varmeter (also a dynamometer instrument) now installed on most switchboards.

It is convenient for system operators to consider the true, or useful, or 'active' power (P) in watts quite separately from the 'reactive' or wattless power (Q) in vars, though in practice both are present together, travel down the same cables and are produced by the same generator. Separate wattmeters and varmeters are usually installed at switchboards.

CHAPTER 30 POWER FACTOR

30.1 PURELY RESISTIVE OR REACTIVE CIRCUITS

In Chapter 28 it was shown that in a purely resistive circuit, where the load current is completely in phase with the applied voltage, the power output takes the form of a double-frequency wave which is wholly on the positive side of the zero axis. The wave has an average or net value equal to half its peak-to-peak height; this average value represents the net power transmitted and is equal to $V \times I$, where V is the rms value of the applied voltage and I the rms current in amperes.

In Chapter 29 it was shown that in a purely reactive circuit, whether inductive or capacitive and where the current lags or leads 90° on the applied voltage, the power output takes the form of a double-frequency wave which is wholly symmetrical about the zero axis and therefore has a mean or net value of zero. That is to say, in purely reactive circuits no net power is transmitted; the power going in during one half-cycle is returned during the next.

It was further shown that, although no **active** power (watts) was transmitted, the product of rms volts and rms amperes is still a perfectly good figure which represents the magnetic energy stored, but not consumed, in the system. This product of volts and amperes does not represent true power (watts) but is given the name 'vars'. It is termed 'reactive power' as distinct from the true or active power measured in watts. Active and reactive power can be separately indicated on switchboard wattmeter and varmeter instruments.

30.2 GENERAL (INDUCTIVE) CASE

We have considered until now only power in purely resistive and purely reactive (inductive and capacitive) circuits. The general case occurs when a circuit is partly resistive and partly reactive, which is much more common.

Figure 30.1 shows the general case of a resistive/inductive circuit. The resistive part of the load draws in-phase current, and the reactive part a current lagging 90°. Between them they draw a single current somewhere between in-phase (0° lag) and 90° lag, as shown on the second curve of the figure. The actual phase angle between current and voltage is usually written 'φ' (Greek 'phi' for 'phase'); it is considered positive when current is lagging and negative when leading.

If the same process is used, as before, of multiplying the voltage by the current at each instant of time, the power wave so produced (bottom of the figure) will again be double-frequency but will now be neither wholly symmetrical about the zero axis nor wholly asymmetrical above it. It will be **partly** asymmetrical, and its average value (half-way between its upper and lower peaks) will be positive and will lie somewhere between zero and the half-way value shown in Figure 28.1. This means that, in the general case, the average active power (watts) will always be less than the maximum value which occurs in the purely resistive case, where the net power was shown to be VI watts (V and I being rms values).

Because in the d.c. days power was always the simple product of V and I, with the advent of a.c. people continued with this outlook and preferred still to think of power as the product of V and I (rms values) but to insert a 'correcting factor' to make it apply to a.c. This correcting factor was given the name 'power factor' ('pf').

For the general case therefore:

$$P = V \times I \times \text{(power factor) (watts)} \qquad \ldots\text{(i)}$$

the power factor being in general less than 1. In the special case where the circuit is resistive only, the power factor equals 1, and where the circuit is reactive only it equals zero.

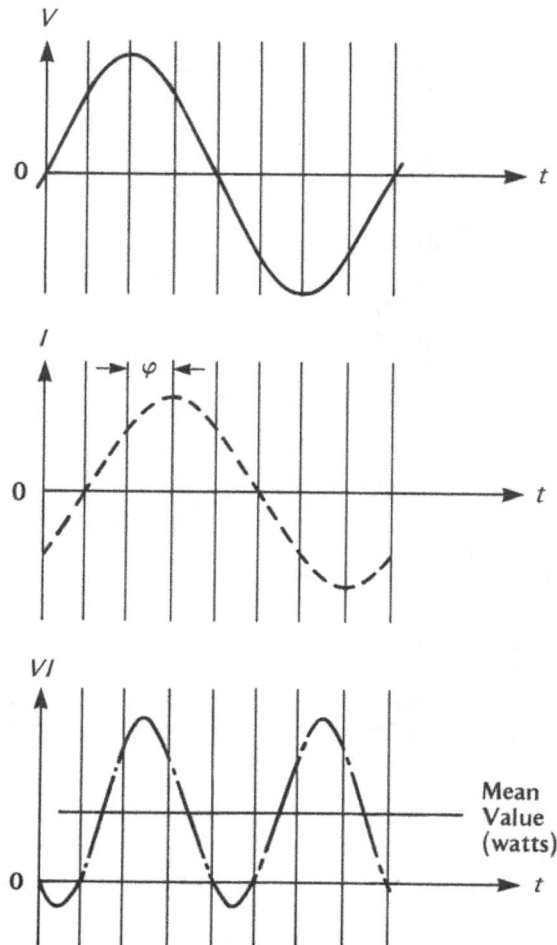

FIGURE 30.1
A.C. POWER - GENERAL CASE

Reverting to the 'impedance triangle' shown in Figure 30.2 the angle between the impedance vector Z and the current vector I is 'φ'. Now Z is the overall impedance across which the voltage is applied, so the voltage vector V lies along Z, just as the current vector I lies along R, and φ is then also the angle between voltage and current - that is, the 'phase angle'.

By Ohm's Law for a.c.: $\frac{V}{I} = Z$ or $V = IZ$

and the uncorrected power (volts x amperes) is $V \times I$; or, substituting for V, it is $IZ \times I$ or I^2Z

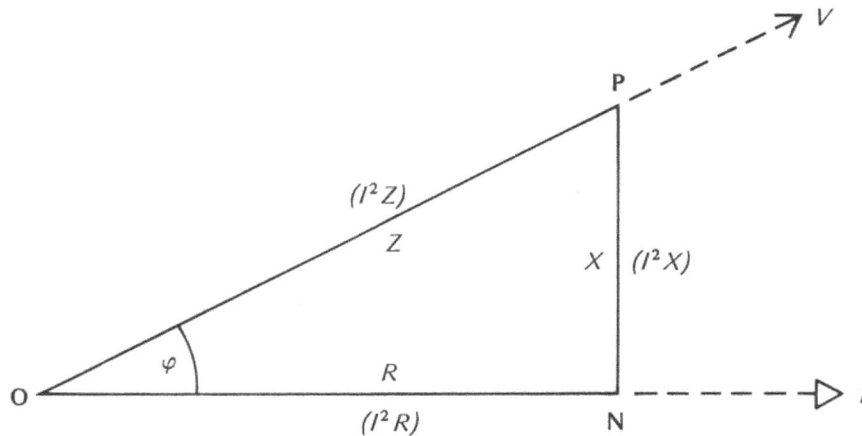

FIGURE 30.2
IMPEDANCE AND POWER TRIANGLE

In the impedance triangle of Figure 30.2 multiply all three sides by the same quantity I^2 (which will not alter its shape). The hypotenuse is now I^2Z and the horizontal side I^2R. I^2Z has just been shown to be the uncorrected power ($V \times I$); I^2R is the active power absorbed by the resistance; and I^2X is the reactive power in the inductance. The impedance triangle has now become a power triangle.

But by ordinary trigonometry ON = OP cos φ. Therefore

active power (watts) = uncorrected power (VI) × $\cos \varphi$

This shows that cos φ is in fact the 'power factor' of equation (i).

Therefore in an a.c. system, where the phase angle between applied voltage and load current is φ, the active power is obtained by the formula:

$$P = VI \cos \varphi \text{ (watts)}$$

where V and I are rms values, and cos φ is the power factor.

Switchboard instruments are provided which show the power factor direct. They used to be marked 'POWER FACTOR' or simply 'PF', but modern instruments are now generally marked 'COS φ'. The direct uncorrected product 'VI', referred to above as 'uncorrected power', is more properly called 'apparent power'. It is given the symbol 'S' and is measured in volt-amperes (VA).

It should be noted in the impedance triangle that

$$\begin{aligned} S^2 &= P^2 + Q^2 \\ &= V^2I^2\cos^2\varphi + V^2I^2\sin^2\varphi \\ &= V^2I^2(\cos^2\varphi + \sin^2\varphi) \\ &= V^2I^2 \end{aligned}$$

and therefore $\qquad S\text{(volt-amps)} = \sqrt{watts^2 + var^2}$

30.3 GENERAL (CAPACITIVE) CASE

It has already been shown that the phase angle φ between load current and voltage is considered to be positive when the current is lagging, and negative when it is leading.

Since cos (- φ) is the same as cos (+ φ), the power factor does not change sign when φ goes negative, and cos φ is still the power factor even of leading loads.

Therefore the statement above for inductive loads applies equally to capacitive loads and is indeed general for all a.c. loads, namely that, where the phase angle between applied voltage and load current is φ, whether lagging or leading, the active power is obtained from the formula:

$$P = VI \cos \varphi \text{ (watts)}$$

where V and I are rms values and cos φ is the power factor $\left(= \frac{R}{Z} \right)$

30.4 POWER-FACTOR METERS

Power-factor meters, which are basically only φ-indicators calibrated to read cos φ, always show the power factor as a positive number whether the current is lagging or leading, but they are arranged to indicate lagging and leading power factors in opposite directions (φ positive and negative respectively). This can be seen on both examples of Figure 30.3, where lagging pf's are to the left and leading to the right.

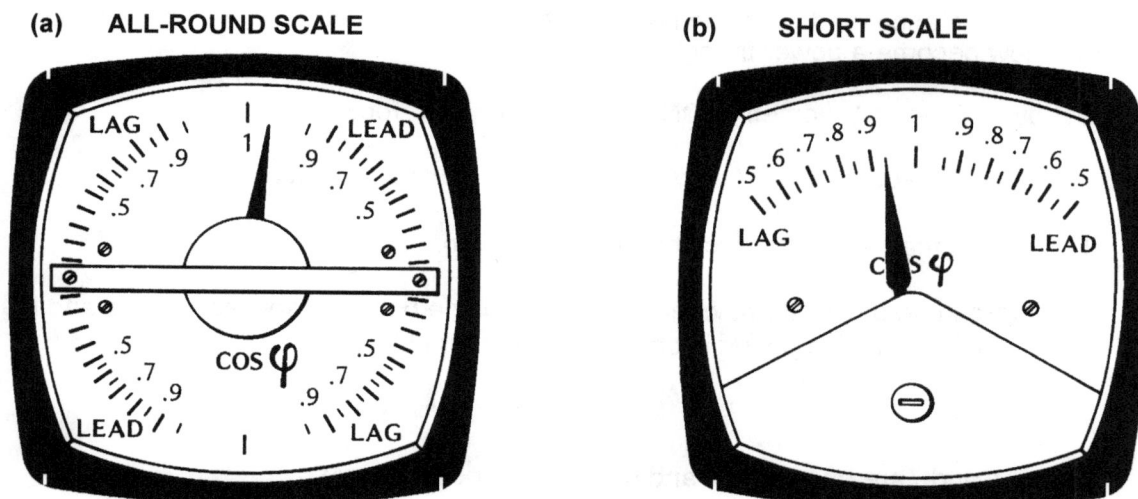

(a)   ALL-ROUND SCALE

(b)   SHORT SCALE

**FIGURE 30.3
POWER-FACTOR METERS**

The older type of power-factor meter, shown in Figure 30.3(a), has an all-round (360°) scale. The upper two quadrants are the ones normally used, but if power can flow in either direction (for example in a ring main or interconnector) the upper two quadrants are used for the forward direction and the lower two for the reverse direction of flow. These directions are sometimes marked 'Export' and 'Import'.

Where only one direction of power flow is involved, a 'short-scale' instrument is nowadays more generally used, as shown in Figure 30.3(b). Lagging pf is to the left, as before, and leading to the right, but the scales are limited from 1 down to about 0.5 in either direction. Sometimes the '1' point, instead of being in the centre, is biased one way to give a longer lagging scale and a shorter leading scale. This type of power-factor meter, which is transducer-operated (see Chapter 36), is almost universally used on platform switchboards.

30.5   REACTIVE POWER FACTOR

It was shown above that in the impedance triangle of Figure 30.2, if all three sides are multiplied by I^2, the hypotenuse OP represents the uncorrected power ($V \times I$) and that ON is the active power (I^2R) in watts,

$$\text{Where ON} = \text{OP cos } \varphi$$

$$\text{or active power } (P) = VI \text{ cos } \varphi \text{ watts.}$$

A similar argument applies to the reactive power, represented by NP (I^2X). This is measured in 'vars'. But by trigonometry:

$$\text{NP} = \text{OP sin } \varphi$$

$$\text{or reactive power } (Q) = VI \text{ sin } \varphi \text{ vars.}$$

The expression 'sin φ' is often called the 'reactive power factor' (as distinct from cos φ, the active or ordinary 'power factor') and is equal to $\frac{X}{Z}$.

30.6   SUMMARY

To sum up: in a circuit with applied rms voltage V and rms load current I, whose resistance is R, reactance X and impedance Z ohms, the expression:

$V \times I$	is the 'apparent power' (in volt-amps)
$VI \cos \varphi$	is the active power (in watts)
$VI \sin \varphi$	is the reactive power (in vars)

Cos φ is the power factor and sin φ the reactive power factor.

CHAPTER 31 THREE-PHASE POWER

31.1   GENERAL

Consider a 3-phase system supplied from a generator as shown at the top of Figure 31.1. Suppose the generator is star-connected and that each terminal, R, Y and B, is connected to the corresponding terminal of a 3-phase, star-connected load.

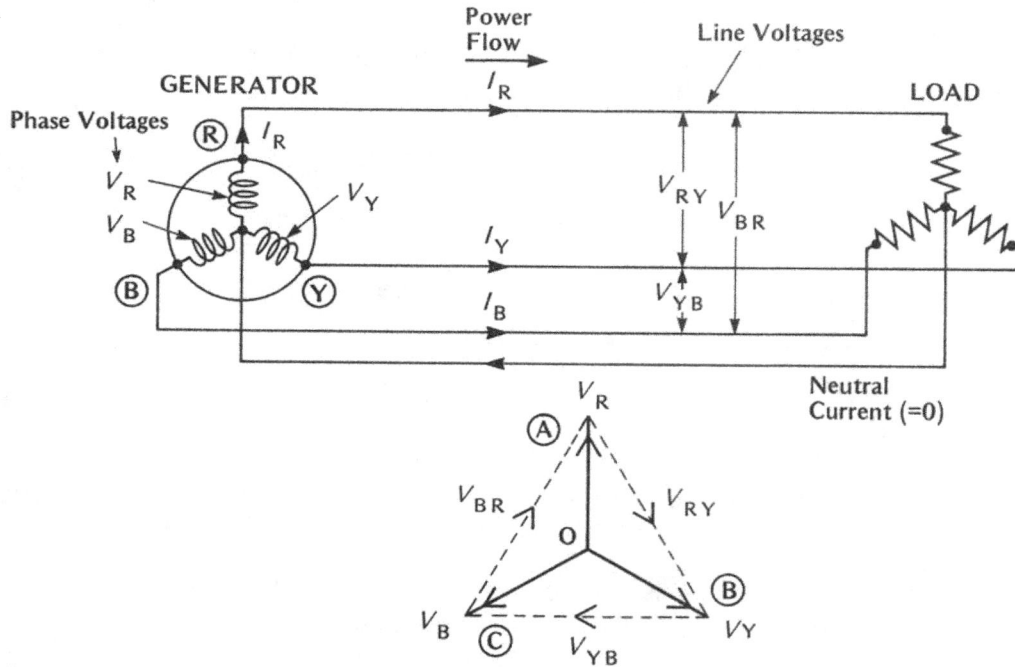

FIGURE 31.1
3-PHASE SYSTEM -STAR

Although it has been shown that three wires are sufficient in this instance, nevertheless for the sake of this description the neutral point of the generator is shown connected to the neutral point of the load; it is known that, under these balanced conditions, no current will actually flow in this fourth wire (see Chapter 15).

Then, since the three phase voltages are equal in magnitude and are spread 120° apart in time, the three vectors V_R, V_Y and V_B in the lower part of the figure represent the three phase voltages of the system - that is, the voltages generated between the neutral point and each of the three terminals R, Y and B.

31.2   3-PHASE POWER

If in Figure 31.1 the three phases are considered quite separately, each generator winding has phase voltage (e.g. V_R) developed in it. If the **line** current is I_L, then, if the loading on that phase is purely resistive, the power transmitted by that phase alone is:

$$V_R \times I_L$$

Similarly the power transmitted by the two other phases alone is:

$$V_Y \times I_L \quad \text{and} \quad V_B \times I_L$$

Since $V_R = V_Y = V_B$ = phase voltage, the total power in all three phases is:

$$3\,V_R\,I_L$$

If the load were not purely resistive but had a power factor cos φ, then the active power transmitted by each phase would be:

$$V_R\,I_L\,\cos φ$$

and the total power for all three phases would then be:

$$3V_R\,I_L\,\cos φ \qquad\qquad ....(i)$$

If V_L is the line-to-line voltage (always indicated by switchboard voltmeters), it has already been shown that $V_L = \sqrt{3}V_R$, or $V_R = \frac{1}{\sqrt{3}}V_L$

Therefore from equation (i) the total 3-phase active power *(P)* is:

$$P = 3 \times \frac{V_L}{\sqrt{3}}I_L\cos\varphi$$

$$= \sqrt{3}V_L I_L\cos\varphi \qquad\qquad ....(ii)$$

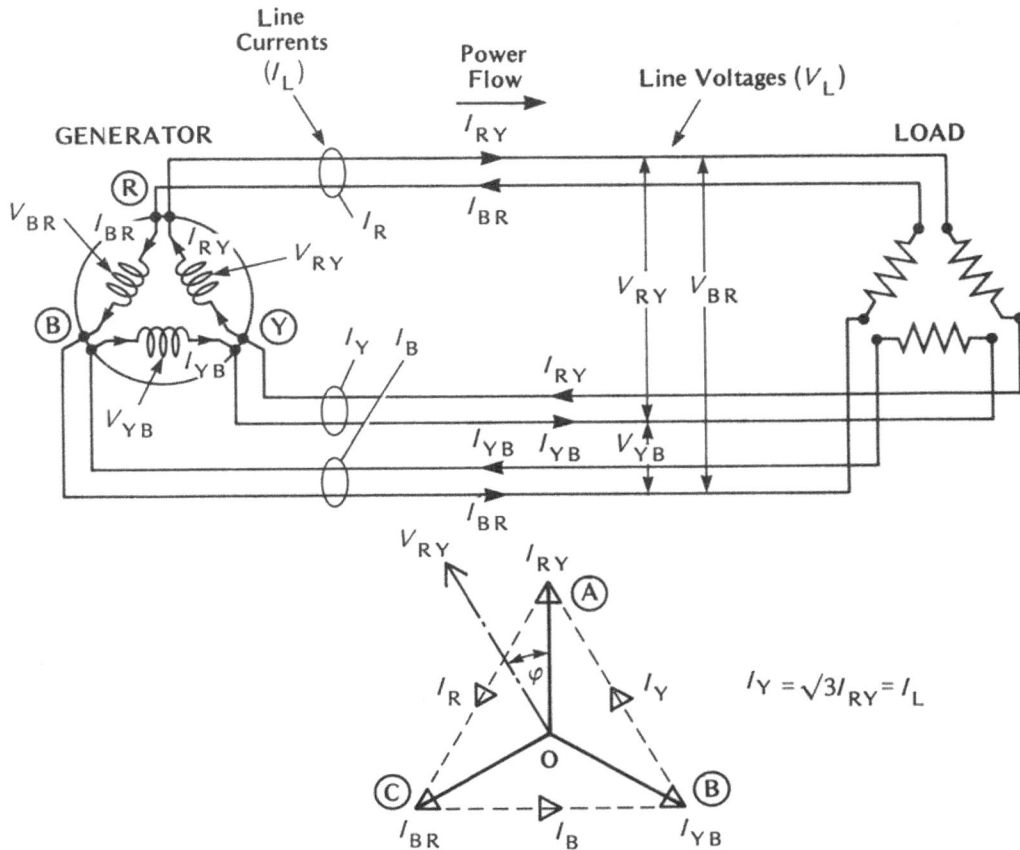

FIGURE 31.2
3-PHASE SYSTEM – DELTA

Figure 31.2 shows a 3-phase system feeding a balanced load, but this time the generator is delta-connected. As before, the generator terminals, R, Y and B, are connected to the corresponding terminals of a balanced 3-phase, delta-connected load, but for the sake of this description suppose that each generator winding is independently connected to its corresponding load element, so that six wires are used, as shown in the diagram.

If the currents in the generator windings (i.e. the phase currents) are I_{RY}, I_{YB} and I_{BR}, and the voltages developed in each generator phase are V_{RY}, V_{YB} and V_{BR}, then, considering the winding RY alone, there is a voltage V_{RY} causing a current I_{RY} to flow from terminal R, through one element of the load, returning to terminal Y and through the generator phase winding YR back to terminal R. Similarly for the other two generator windings.

Thus the two wires leaving terminal R carry respectively an outward current I_{RY} and a return current I_{BR} from a different load element.

In practice, of course, there are not two wires from each terminal, but each pair is commoned into a single wire at each terminal R, Y and B. The single wire from terminal R thus carries the difference current $I_{RY} - I_{BR}$, which we will call 'I_R'. Similarly the single wire from terminal Y carries the difference current $I_{YB} - I_{RY}$, which will be 'I_Y' and that from terminal B carries $I_{BR} - I_{YB}$, which will be 'I_B'.

The vector diagram at the bottom of Figure 31.2 shows the generator's three balanced phase currents I_{RY}, I_{YB} and I_{BR}. The differences are obtained, as explained earlier', by joining the arrow-tips, so that line AC equals $I_{RY} - I_{BR}$, which, as explained above, is the net **line** current I_R leaving terminal R. Similarly CB represents the net line current I_Y, leaving terminal Y, and BA the net line current I_B leaving terminal B. That is to say, the 'spokes' of the diagram represent the three phase currents in the generator, and the three sides of the triangle represent the three line currents I_R, I_Y and I_B. Since these are all numerically equal, they will be called I_L (for line current).

It has been shown, from the geometry of Figure 20.6 in Chapter 20, that the length of the triangle's sides is √3 times the length of its spokes, so that:

$$I_L(= I_R = I_Y = I_B) = \sqrt{3} \times I_{RY} \text{ (or } I_{YB} \text{ or } I_{BR})$$

that is, line currents $= \sqrt{3} \times$ any phase current

The active power delivered by any one phase (say phase RY) alone is:

$$V_{RY} I_{RY} \cos \varphi \text{ watts} \qquad \qquad \text{....(iii)}$$

where cos φ is the power factor of the load (indicated on the vector diagram by I_{RY} lagging by an angle φ on the corresponding voltage V_{RY}). But it has just been shown that
$$I_L = \sqrt{3} \times I_{RY} \text{ or } I_{RY} = \frac{1}{\sqrt{3}}I_L$$

Therefore, from equation (iii), the power delivered by phase RY alone is:

$$V_{RY} \frac{1}{\sqrt{3}} I_L \cos \varphi$$

and the power delivered by all three phases (P) is three times this, namely:

$$3V_{RY} \frac{1}{\sqrt{3}} I_L \cos \varphi$$

$$\therefore \quad P = \sqrt{3}V_{RL}I_L \cos \varphi \qquad \qquad \text{...(iv)}$$

With a delta connection any line-to-line voltage (V_L) is the same as the corresponding generator phase voltage, since both are in parallel, so that $V_L = V_{RY}$. Therefore, from equation (iv), the total active power transmitted is:

$$P = \sqrt{3}V_L I_L \cos \varphi \text{ watts}$$

If this is compared with equation (ii) for a star-connected generator, it will be found to be exactly the same.

Therefore, if in a balanced 3-phase system the line voltage is V_L (volts) and the line current I_L (amperes), and if the load power factor is cos φ, then the active power (P) delivered is:

$$P = \sqrt{3}V_L I_L \cos \varphi \text{ watts}$$

whether the power source is star or delta connected.

Example

In a 6.6kV system a 3-phase load draws 200A line current at a power factor (cos φ) 0.8.

$$\therefore \quad \text{Total power} = \sqrt{3} \times 6\,600 \times 200 \times 0.8\text{W}$$
$$= 285\,780\text{W}$$
$$= 286\text{kW approximately.}$$

CHAPTER 32 SWITCHING ON - ASYMMETRY

32.1   SWITCHING ON - SINGLE-PHASE

When a current is switched on in a d.c. inductive circuit, it rises from zero and gradually approaches its steady value, the rate depending on the inductance and resistance of the circuit, as explained earlier. The behaviour when an a.c. circuit is switched on is however quite different, and there is no time delay in the build-up of current.

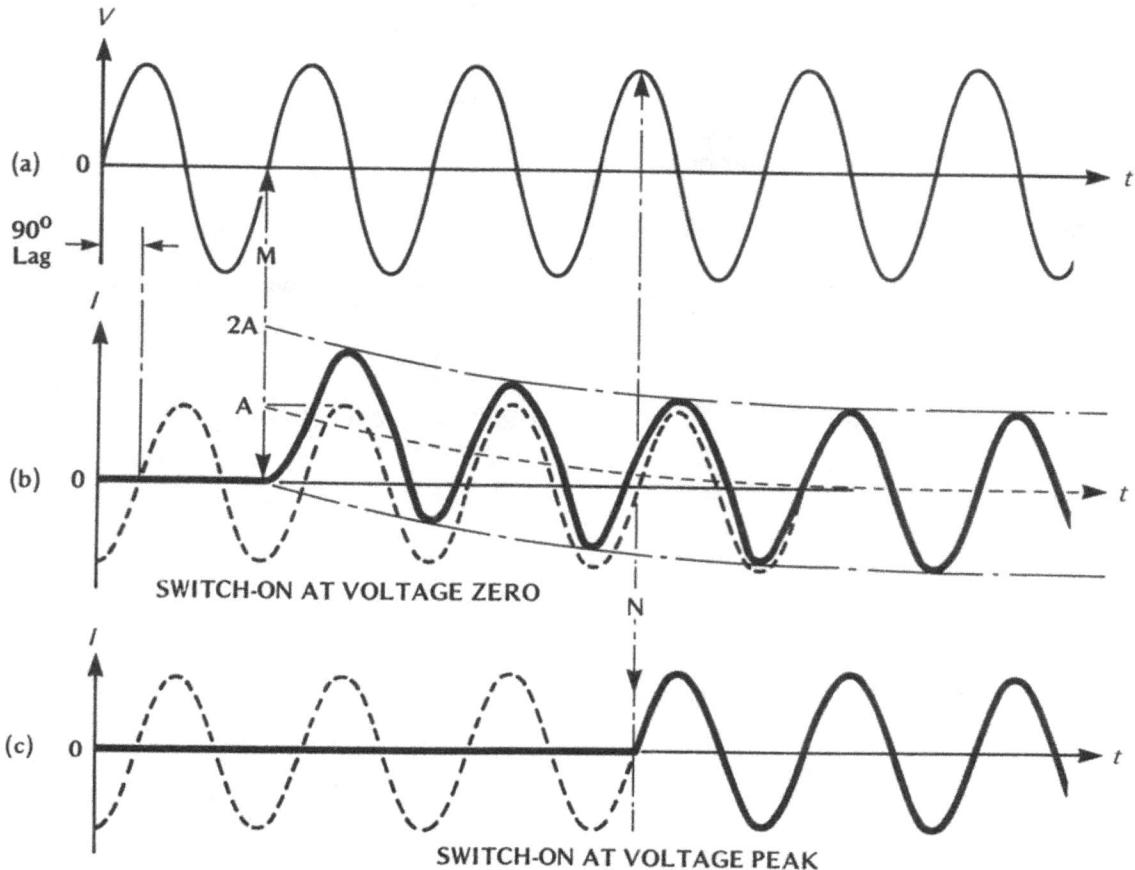

FIGURE 32.1
SWITCHING ON - 90° LAG

Figure 32.1(a) shows the voltage in an a.c. circuit having inductance only and little resistance. This is the condition for a 90° lag, and the current peaks occur at the same instants as the voltage zeros. Figure 32.1(b) shows the corresponding current wave both before and after the current is switched on, which is assumed to occur at the point M at one of the voltage zeros. The dotted wave in Figure 32.1(b) is what the current **should be** doing after switching on - namely lagging 90° behind the voltage.

But this would entail the current jumping suddenly from zero at point M just before the switch-on to its maximum an instant after - which it clearly cannot do. What happens is that the current grows so that the whole current waveform moves bodily upward (in this case) by an amount equal to its amplitude, as shown in full line in Figure 32.1(b). If this displaced wave is examined closely, it will be seen that there is no jump at the moment of switch-on (M) - it is zero immediately before and immediately after - and it also lags 90° on the voltage wave, its peaks coming opposite the voltage zeros.

It is completely asymmetrical at the instant of switching on, but it gradually regains symmetry a few cycles later; the more resistance that is present, the quicker symmetry is restored. Where, as here, the displacement is complete and equal to the a.c. amplitude, the wave is said to be '100% asymmetrical'.

If, instead, the current had been switched on at the point N at one of the voltage **peaks**, as in Figure 32.1(c), the effect would be different. At a voltage peak the current with a 90° lag would in any case be zero so there would be no need for any sudden change at the moment of switch-on - it would be zero both immediately before and immediately after the switching. There would therefore be no asymmetry to compensate for a jump, and the current would start and remain symmetrical throughout; it would be '0% asymmetrical'.

With 100% asymmetry the current peak is about double the symmetrical peak, and this itself is √2 times the symmetrical rms value. So the asymmetrical peak is 2√2 or 2.83 times the rms. In practice, because the current has already started to regain symmetry by the time the first peak appears, the 'doubling factor' is usually taken to be 2.55.

Figures 32.1(b) and 32.1(c) are the two extreme cases with a 90° lag; switching at the instant of voltage zero and at the instant of a voltage peak. The general case would be somewhere in between, where there would be partial asymmetry, something between 0% and 100%.

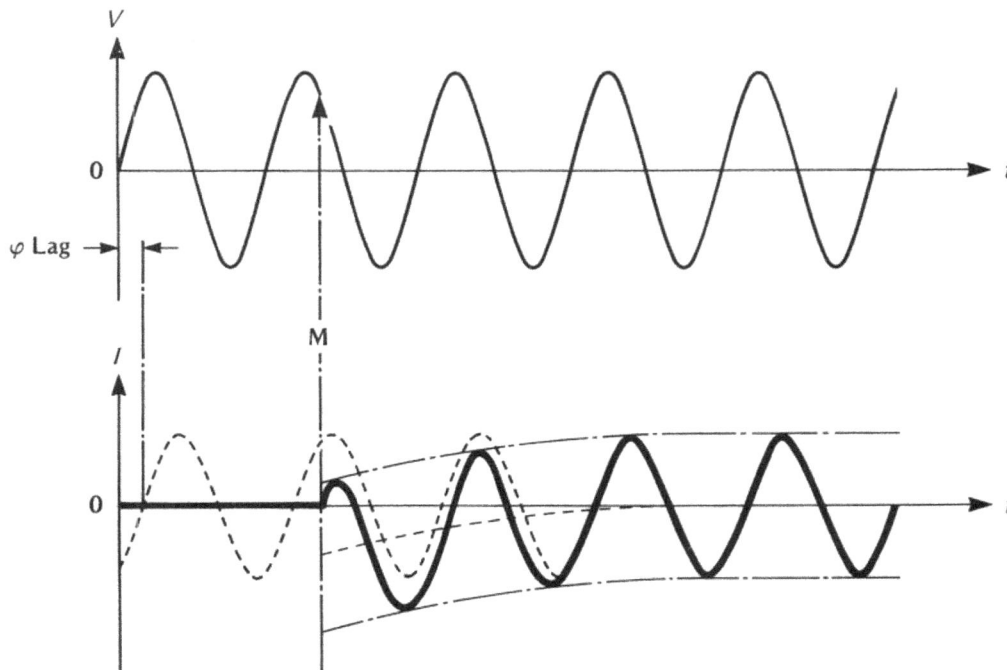

FIGURE 32.2
SWITCHING ON – GENERAL CASE (PARTIAL ASYMMETRY)

With the general case of a circuit having both inductance and resistance the power factor would be higher than zero and the current lag would be less than 90° (see Figure 32.2). In this case the current has to jump from zero before the switch-on to a point something less than peak value immediately after. Therefore the asymmetry to compensate for this jump is less than a full amplitude (100%), as shown in Figure 32.2. There is then 'partial asymmetry', between 0% and 100%.

32.2   SWITCHING ON – 3 PHASE

All the above discussion has been about a single-phase voltage being switched onto a circuit. However, most switching on platform and shore equipment is 3-phase. In a 3-phase circuit the voltage phases are 120° apart so that, even if at the instant of switching one of them occurs at a voltage zero or voltage peak, the other two will not be so, and they will be at voltage points somewhere between zero and peak. Therefore, even if one phase of current is wholly asymmetrical or wholly symmetrical, the other two will be partially asymmetrical. This is shown in Figure 32.3 where red-phase current is 100% asymmetrical; the other two in that case will be 50% asymmetrical in the opposite direction.

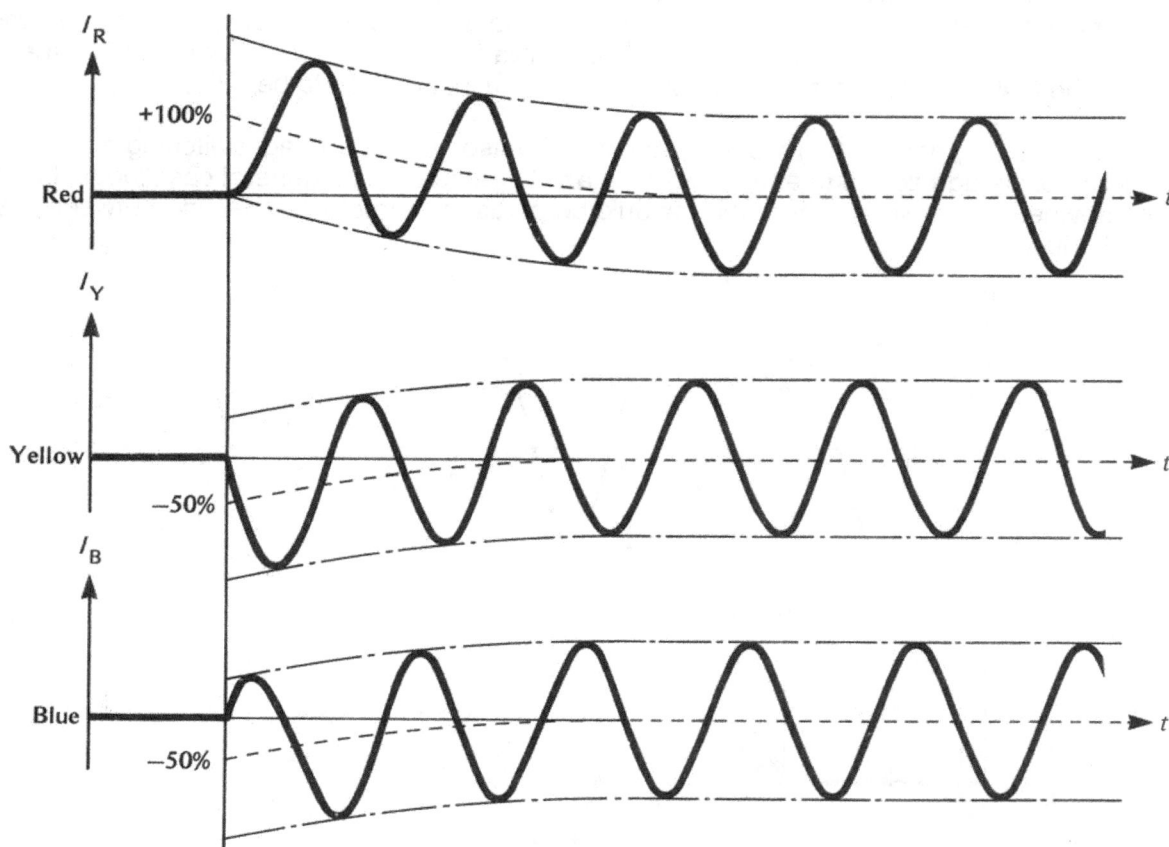

Switch-on (at Red Voltage Zero, Rising)

FIGURE 32.3
SWITCHING ON - 3-PHASE (PF ZERO)

The problem of asymmetry on switching on is an important one in platform and shore networks, especially under conditions of sudden short-circuit. Whatever the calculated short-circuit current is, the peak current which flows in the first cycle of short-circuit, where the power factor may be typically about 0.15, will be about 2.55 times the calculated rms current, and this could put damaging strain upon busbars and other distribution equipment unless it is allowed for in design. Switchgear which operates very quickly - that is, within the first few cycles of a short-circuit - will have to handle and clear this excessive asymmetrical current in at least one pole, and as the instant of short-circuit would be entirely random, it might occur in any pole. For this reason all main switchgear is given two breaking current ratings: an rms symmetrical breaking capacity and a peak asymmetrical breaking capacity, which is about 2.55 times the symmetrical; this is not usually given on the nameplate.

CHAPTER 33 PRINCIPLES OF RECTIFICATION

33.1 DIODES

An element which is constantly used in control circuits, especially d.c. circuits, is the diode. It is a device which allows current to flow freely in one direction but presents a high resistance to it if it flows in the other. It does not necessarily completely stop the reverse flow, but in many applications it is regarded as blocking it completely. It acts in much the same way as a mechanical non-return valve such as is fitted to a motor-car tyre; air may pass freely in, but it cannot get out.

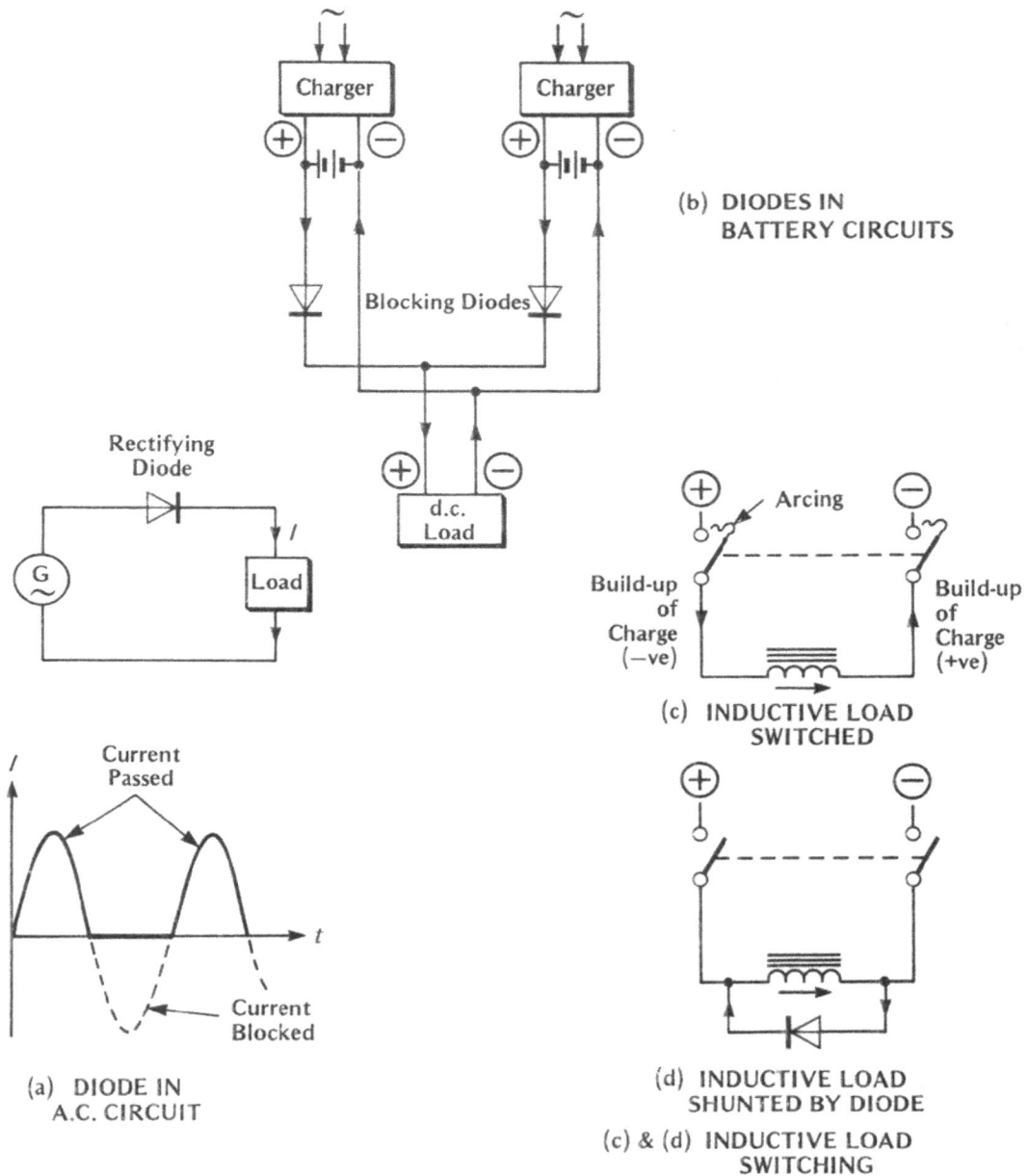

FIGURE 33.1
SOME USES OF DIODES

The first diode was the original Fleming thermionic tube, the forerunner of the electronic valve. Here the electrons passed easily from the heated filament or cathode to the positive anode, but they could not return to the cathode even if the anode were made negative. It was this one-way action and its likeness to the non-return valve which gave its name to the Fleming tube: the 'valve'.

Nowadays the same action can be obtained from solid-state material, and in a much simpler, cheaper and more compact manner. Solid-state diodes are widely used in electronic and control circuits, chiefly for their one-way blocking facility. The action of a diode on an a.c. circuit and in some d.c. circuits is shown in Figure 33.1.

In Figure 33.1(a) an a.c. generator is feeding a load through such a diode. In the positive halves of the cycle the current passes freely, giving a half-sine waveform. In the negative halves, where the current would normally flow back, it cannot do so because it is blocked by the diode. So the current waveform is a series of positive half-sine waves, with zero value in the gaps in between, and the current is unidirectional, though by no means constant. It might be called 'direct current', but that would be misleading. If the load were a lamp, there would be bad flicker, but if it were a battery to be charged, the pulsing nature would not matter and the battery would receive charging current in a series of pulses, and no discharge would take place in the negative parts of the cycle. It would therefore receive a net charge.

In Figure 33.1(b) two batteries are shown feeding a common d.c. load in parallel. Each has a charger, and a diode is placed in each battery output. So long as both batteries have approximately the same voltage, each will be contributing to the load, though perhaps not quite equally. But if one battery were discharged its voltage might be such that not only would it contribute nothing to the load but the healthy battery might try to feed current into it. This would also occur if one charger failed. The diodes prevent this by blocking any reverse current into either battery.

In Figure 33.1(c) is another classic example of the use of a diode. A highly inductive load - for example a solenoid - is being fed from a d.c. source with a main switch or contactor. When the switch is opened the current cannot immediately stop in the inductor, and it carries on to 'pile up' on the switch contacts. This causes a very high voltage to appear there and usually severe arcing.

Suppose now a diode is placed in reverse to shunt the inductor as shown in Figure 33.1(d). It will not pass any current while the switch is closed because it is placed so as to block it. If now the switch is opened, the inductive current in the coil, instead of 'piling up' to put a charge on the switch contacts, will have an easy path through the diode and back into the coil. It will circulate in this manner until eventually it is damped out by the resistance of the coil/diode circuit. But it will have prevented high voltage and arcing at the switch contacts. Here the diode is used as an arc-suppressor.

33.2   RECTIFICATION - SINGLE-PHASE

It has already been shown in Figure 33.1(a) how a diode can change an alternating into a unidirectional current - this action is called 'rectifying' - and it was mentioned that such an arrangement could be used, for example, to charge a battery.

Figure 33.2(a) repeats Figure 33.1(a). This arrangement is wasteful of time, as useful current flows for only half the available time. It is called 'half-wave' rectification. If the unidirectional current pulses are 'smoothed' to give a mean direct current, the d.c. level will be the line (shown dotted) where the areas above and below it (shaded) are equal. It is in fact 0.318 times the amplitude, or 0.45 (= $\sqrt{2}$ x 0.318) times the rms value of the current.

This can be improved by the arrangement of Figure 33.2(b), where four diodes are connected in the form of a bridge. It turns the negative half-wave into a positive instead of blocking it, so that each half-cycle has its quota of unidirectional current. This arrangement is called 'full-wave' rectification. It is more efficient and gives less flicker if used for lighting. The 'smoothed' mean d.c. level is higher than in the half-wave case, as shown by the dotted line in Figure 33.2(b). It is in fact 0.635 times the amplitude, or 0.90 (= $\sqrt{2}$ x 0.635) times the rms value of the current. Thus an a.c. current of rms value 10A (peak 14.1A) will be converted to 9A d.c. (apart from losses).

A full-wave bridge is sometimes drawn in the alternative manner shown in the centre of Figure 33.2(b).

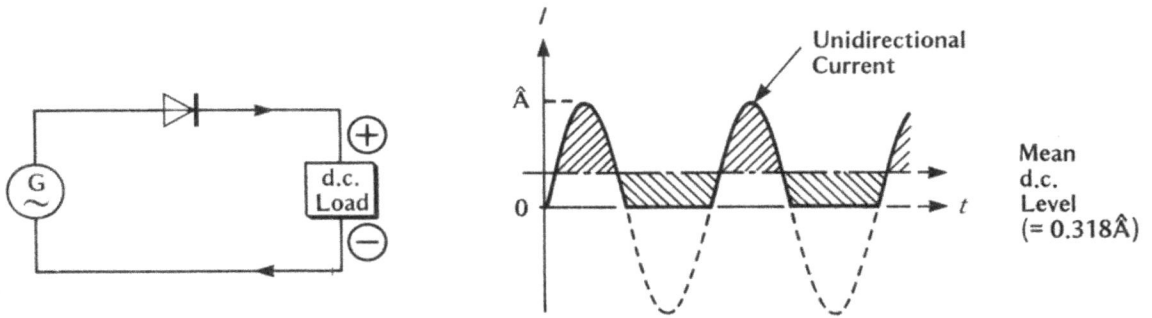

(a) HALF-WAVE RECTIFIER

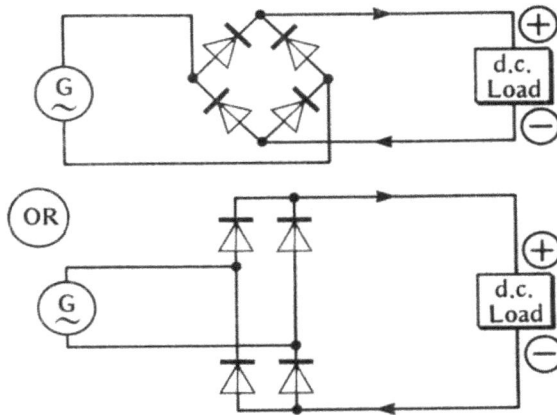

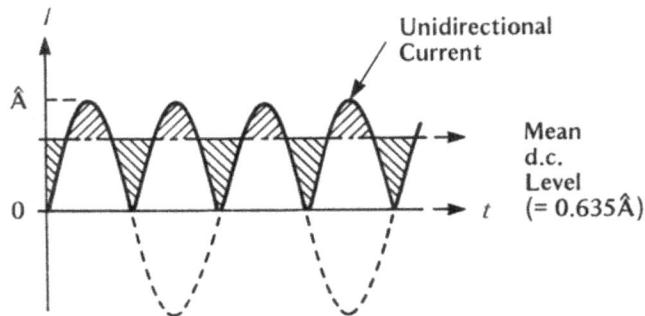

(b) FULL-WAVE RECTIFIER

FIGURE 33.2
DIODES USED AS RECTIFIERS

33.3   RECTIFICATION – 3-PHASE

The idea can be extended to 3-phase, as shown in Figure 33.3.

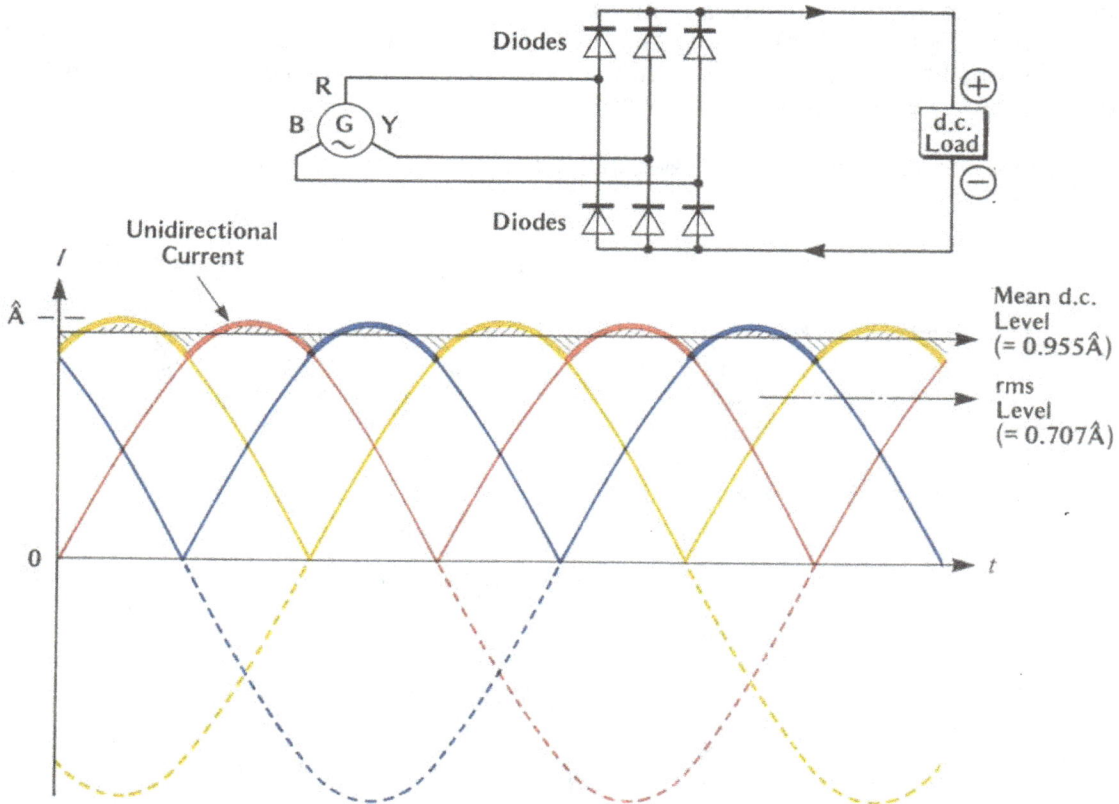

FIGURE 33.3
3-PHASE FULL-WAVE RECTIFIER

Here a six-diode bridge is connected to receive a 3-phase supply and to produce a unidirectional output. The arrangement shown is full-wave, and it reverses the three negative halves each cycle to produce a unidirectional current with six peaks each cycle. This is much less 'peaky' than the single-phase case, and it is more readily smoothed to produce a good, low-ripple direct current.

As before, the smoothed mean d.c. level is the line, shown dotted, where the shaded areas above and below it are equal. The level is much higher than even the full-wave single-phase case, being equal to 0.955 times the amplitude, or **1.35** ($\sqrt{2}$ x 0.955) times the rms value. Thus an a.c. current of rms value 10A (peak 14.1A) will be converted to 13.5A d.c. (apart from losses). It should be particularly noted that with 3-phase full-wave rectification the d.c. level is **higher** than the rms a.c. value.

33.4   CONTROLLED RECTIFICATION

It has been shown that the d.c. output from a 3-phase full-wave rectifier with six diodes is fixed at approximately 1.35 times the rms a.c. input. The d.c. voltage output can be controlled by substituting three **thyristors** for three of the diodes - see Figure 33.4. A thyristor is a solid-state device like a diode but with a third electrode which prevents the device passing even forward current until the third electrode is 'triggered'.

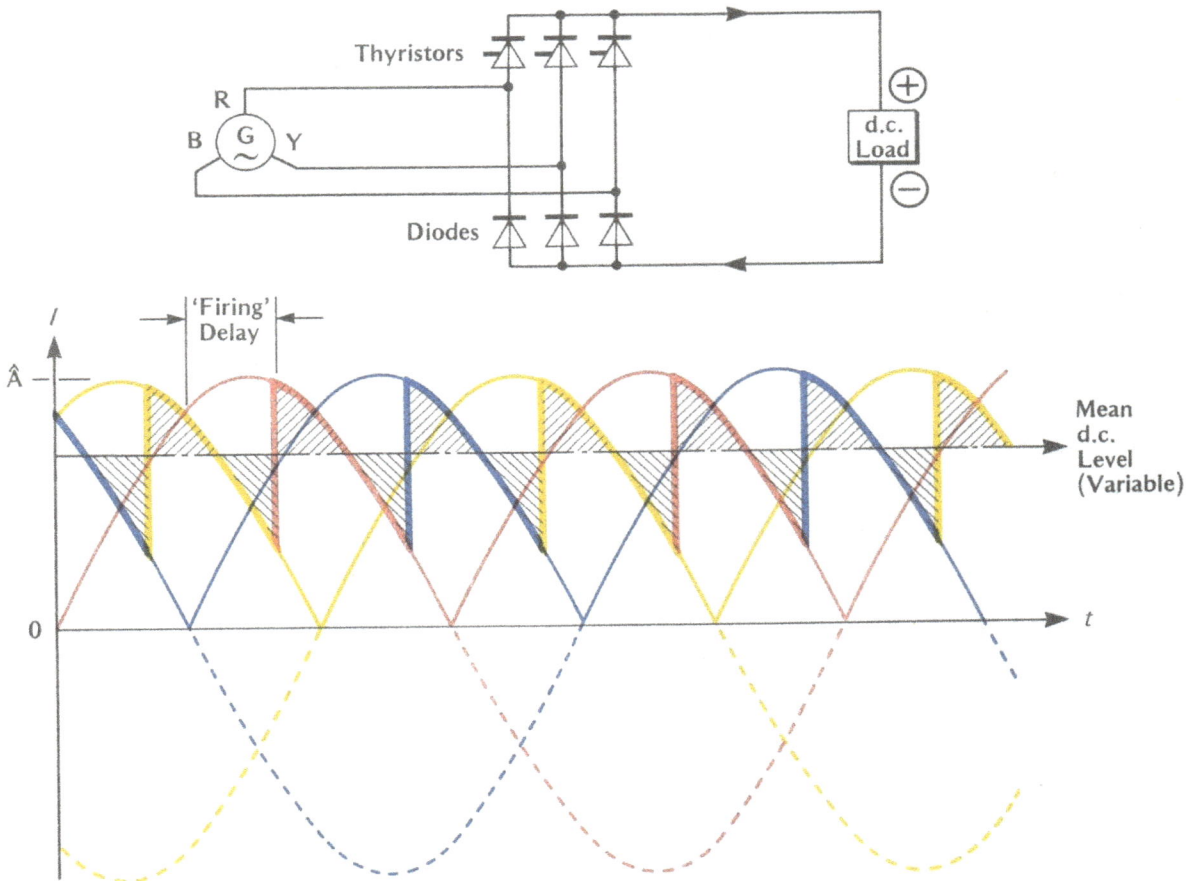

FIGURE 33.4
CONTROLLED RECTIFIER (3-PHASE FULL-WAVE)

Since the thyristor will not conduct until signalled to do so on the third electrode, the 'firing' can be deliberately delayed. An electronic circuit provides a firing pulse with a variable delay, so that the waveform appears as in Figure 33.4. The mean d.c. level - that is, the line where areas above and below it are equal - will be different with differing delay times, so that the bridge can be used to give different d.c. output levels simply by controlling the electronic delay circuit.

33.5   SUMMARY

Rectifiers, which in modern practice are normally solid-state diodes or thyristors, are used in many applications of power work. They all have a 'one-way' or blocking function, like a non-return valve, which can be used to convert a.c. into d.c. as well as preventing reverse flow of current, providing a means of spark suppression and similar applications. They are described more fully in Part 7 Control Devices.

When diodes are used for a.c.-to-d.c. conversion, the d.c. level is fixed and depends entirely on the level of the a.c. applied, but if thyristors are used, the d.c. level can be controlled within broad limits. In this form they are widely used for battery charging and, in the drilling world, for applying a variable d.c. voltage to the drilling motors in order to regulate their speed; they are there referred to by their earlier name 'silicon controlled rectifiers' (SCR).

The d.c. levels when rectified from a.c. are as follows:

Type of Rectification	D.C. Level	
	Fraction of a.c. peak	Fraction of a.c. rms
Single-phase, half-wave	0.318	0.45
Single-phase, full-wave	0.635	0.90
Three-phase, full-wave	0.955	1 .35

Thus a single-phase a.c. rms supply of 250V will give:

112V d.c. with half-wave rectification
225V d.c. with full-wave rectification

and a 3-phase a.c. rms supply of 440V will give:

594V d.c. with full-wave rectification

CHAPTER 34 TRANSMISSION THEORY

34.1 TRANSMISSION AND DISTRIBUTION

Both ashore and on platforms electricity is generated in bulk in a relatively few generator sets of high capacity. It is transmitted, still in bulk, to a number of distribution centres, called 'substations', where it is further divided into smaller parcels and sent on either to the individual consumers or, in large shore networks, to a number of smaller substations local to consumer centres for further subdivision.

In shore networks the transporting of bulk power from the generating station to the principal substations is referred to as 'transmission', whereas the further spreading of power from those points is referred to as 'distribution'. As will be explained below, transmission is usually at a much higher voltage than distribution. In the shore networks generation and transmission are usually the responsibility of a Central Electricity Generating Board (CEGB), but distribution will be in the hands of the various Area Boards who pass the power on to the consumers.

On platforms the distance between generators and substations is so short that the passing of bulk power between the two, even though it is at high voltage, is not regarded separately as 'transmission', and the whole network is considered as distribution.

It will be noticed that on most platforms the power is generated at high voltage, 6 600V or 4 160V. In shore installations such as refineries and fractionating plants power is also brought in in bulk from the Area Board, at high voltage, typically at 11 000V.

Why, it may be asked, is it done this way when the bulk of power-consuming equipments require only 415V or 440V?

The answer lies in one word - 'current'.

34.2 WHY USE HIGH VOLTAGE?

This question is best answered by considering a d.c. situation, although the argument applies equally to a.c.

It is one of the basic facts of electrical life that power (measured in watts) is the product of voltage and current (measured in amperes). This is the exact equivalent of what happens in the mechanical world: e.g., hydraulic power transmitted is the product of pressure and oil volume flow.

Putting figures to this (and assuming d.c. for the moment), a 200 hp motor is equivalent to 150kW or 150 000 watts. If it is supplied at 440V, then the current that it draws is

$$\frac{150\ 000}{440}$$

$$= 340A$$

But suppose the motor were rated **2 000 hp**, equivalent to 1 500kW or 1 500 000 watts, and that it is still supplied at 440V, the current would then be ten times as great, namely

$$\frac{1\ 500\ 000}{440}$$

$$= 3\ 400A$$

The copper windings of a machine, and the cores of the connecting cables, require a cross-section of something more than 6.5 cm^2 for every 1 000 amperes that they carry. So in this case the windings would need to be more than 22.6 cm^2 in section, as would the cores of the connecting cables. Such a size is simply not practical. The machines would be enormous and heavy, and the cables would be as stiff as pipes, to say nothing of cost. Further, the very heavy starting currents of these motors, up to five times normal (see Part 5 Electric Motors), would present an additional problem.

If however the voltage were increased, say, ten times to 4 400V, for the same power the current needed would be

$$\frac{1\ 500\ 000}{4\ 400}$$

$$= 340A$$

This is no more than the current taken by a 200 hp motor fed at 440V. It is quite practical, requiring as it does only a 2 cm^2 section of winding or cable core. True, the higher voltage would call for thicker insulation, but this adds little to the weight and size of the machine.

So the principle is that, where powers are such that the currents become unmanageable, the operating voltage is raised so as to reduce the currents again to manageable levels. This applies particularly to transmission lines, where the power is handled in bulk and the currents are consequently heavy.

The current limit of a 440V motor is reached at about 400 hp (300kW), above which a higher voltage must be used. For a 440V generator the practical limit is about 2 000kW.

In theory the higher voltage used need only be enough to reduce the current to a manageable level, but in practice the voltage steps are fewer and coarser. This is because British and International Standards have laid down certain standard operating voltages, and manufacturers design their equipment to these standards. At the lower end of the scale the standard voltages of a.c. systems, and the US equivalents, are:

UK/European	US
380/415/440V	440V
3 300V	4 160V
6 600V	
11 000V	13 800V

On onshore installations in the EU 11kV, 6.6kV and 3.3kV may all be found, with distribution at 415V. On most platforms generation is normally at 6.6kV and distribution at 440V, though some of them generate at the US standard of 416kV. Motors rated above about 400 hp operate directly at high voltage. Exceptionally drilling equipment operates at 600V.

With the coming of SI units the horsepower (equal to 0.746 or approximately ¾kW) has been superseded by the kilowatt not only in electrical plant but increasingly in mechanical plant also. Nevertheless many motor rating plates will still be found stamped in hp.

Examples

A motor is rated at 50 hp. What is its output in kilowatts?

$$50hp = 50 \times ¾$$
$$= 37.5kW$$

A diesel engine has an output of 1 200 hp. What is the equivalent in kilowatts?

$$1\ 200\ hp = 1\ 200 \times ¾$$
$$= 900kW.$$

CHAPTER 35 PRINCIPLES OF A.C. MEASUREMENT

35.1   USE OF D.C. INSTRUMENTS FOR A.C.

In Chapter 11, are described a number of different types of instrument used for d.c. measurements. They are the hot-wire, moving-iron, moving-coil and dynamometer types and are there described and illustrated in detail.

Some of them may also be used for a.c. measurements, and each is considered below in that application.

35.2   HOT-WIRE INSTRUMENTS

This instrument depends for its action on the heating and stretching of a wire due to the passing of current through it. The heating with a d.c. current is at the rate I^2R, where R is the resistance of the internal hot wire. But with a.c. the heating is also at the rate I^2R provided that I is the rms current -indeed, it has been shown that rms current is defined as that a.c. current which produces the same heating as a d.c. current of the same numerical value.

Consequently a hot-wire instrument to which a.c. is applied will correctly indicate the rms value of the applied quantity.

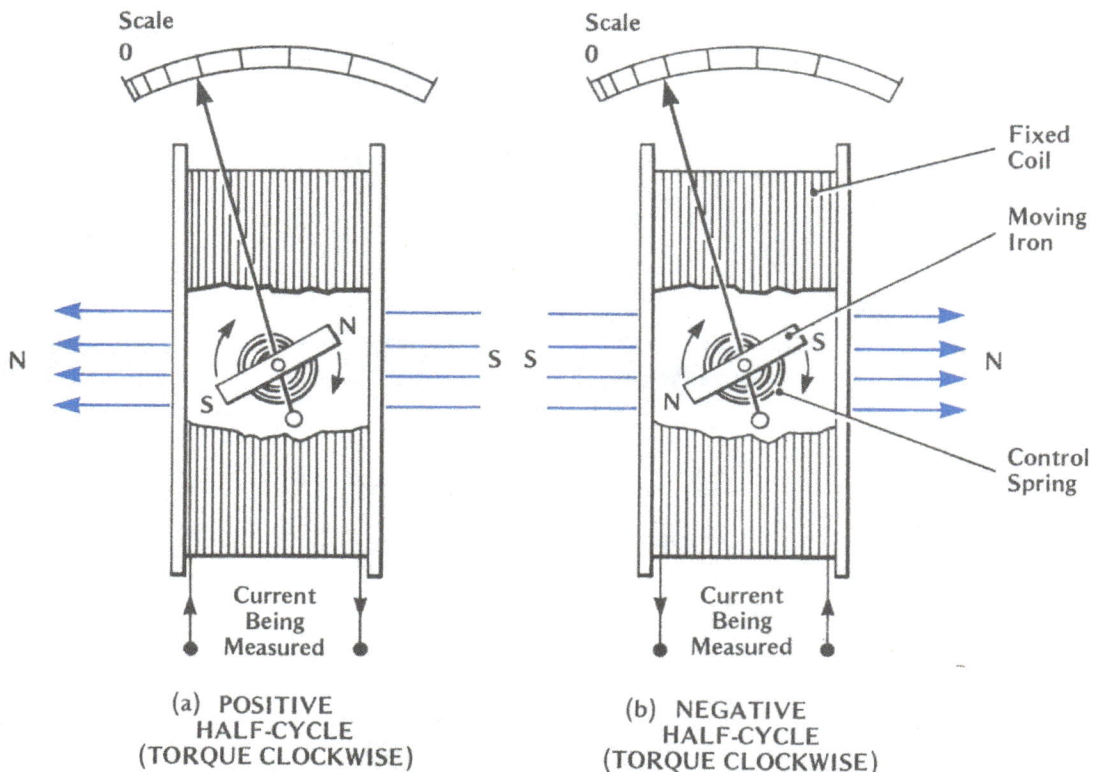

(a) POSITIVE
HALF-CYCLE
(TORQUE CLOCKWISE)

(b) NEGATIVE
HALF-CYCLE
(TORQUE CLOCKWISE)

FIGURE 35.1
MOVING-IRON INSTRUMENT MEASURING A.C.

35.3   MOVING-IRON INSTRUMENTS

A moving-iron instrument will also correctly indicate the rms value of an applied a.c. quantity.

Figure 35.1 shows a basic moving-iron instrument. As explained earlier the fixed coil induces in the moving-iron poles of opposite polarity to the flux which induces them. Thus, in Figure 35.1(a), which is assumed to be the state during a positive half-cycle of current in the coil, the coil's field is from right to left. A N-pole will be induced on the right tip of the moving iron, and a S-pole on its left. Each pole will be attracted towards the axis of the coil, so giving a clockwise torque.

On the next half-cycle the coil field will be reversed, as shown in Figure 35.1(b). The field will now be from left to right, and a N-pole will be induced on the left tip of the moving iron and a S-pole on the right. Each will be attracted to the axis of the coil, and the torque is again clockwise. There is thus no reversal of torque as between positive and negative half-cycles, and the pull will be always in the same direction.

The magnitude of the magnetic pull between a coil and its induced magnetic pole is proportional to the product of the coil's flux (and so the current in it) and of the strength of the induced pole. But that strength is itself proportional to the flux causing it, so the total pull - and hence the torque on the moving iron - is proportional to the **square** of the coil current.

Therefore, like the hot-wire type, the moving-iron instrument responds not simply to the current but to the square of the current. Its scale will be uneven and crowded towards the lower end, and it will indicate the rms value of the current being measured. Moving-iron instruments are relatively cheap and are widely used ashore and in platforms on a.c. switchboards. They can be instantly recognised by their scales.

35.4   MOVING-COIL INSTRUMENTS

Moving-coil instruments cannot be used with a.c. for the following reason.

Figure 35.2 is a reproduction of the corresponding moving-coil figures in Chapter 11. As explained there, a permanent magnet provides a constant field in which a moving coil rotates.

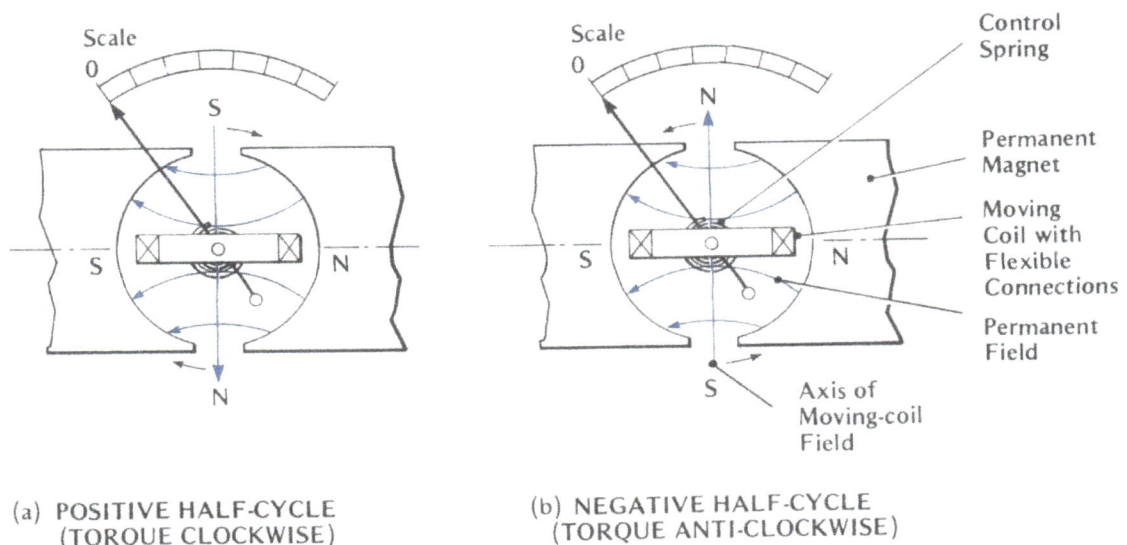

(a) POSITIVE HALF-CYCLE
(TORQUE CLOCKWISE)

(b) NEGATIVE HALF-CYCLE
(TORQUE ANTI-CLOCKWISE)

FIGURE 35.2
MOVING-COIL INSTRUMENT ATTEMPTING TO MEASURE A.C.

In the figure the N-pole is assumed to be on the right, and the field in the gap is therefore from right to left.

The current to be measured flows through the moving coil and gives rise to its own flux. This flux reacts with the permanent field and causes a torque on the moving coil. The coil turns against a control spring so as to try to align its own axis with that of the permanent magnet. This is shown in Figure 35.2(a), which is assumed to be the state during a positive half-cycle of current in the moving coil. The coil's 'S' side is attracted towards the magnet's N-pole, and its 'N' side to the magnet's S-pole, so producing a clockwise torque.

Half a cycle later the coil current is reversed, as shown in Figure 35.2(b), but the direction of the permanent magnet field is unchanged. The coil's 'N' side is now on top and is repelled by the magnet's N-pole, just as its 'S' side is repelled by the magnet's S-pole. The two combine to produce an anti-clockwise torque.

The direction of torque thus reverses with every half-cycle, and, because of the inertia of the movement, no motion whatever takes place (though there might be a buzz). For this reason moving-coil instruments cannot be used directly on a.c., although they can be used with transducers (see para. 35.6).

35.5   DYNAMOMETER INSTRUMENTS

A dynamometer instrument consists of fixed coils and a moving coil.

When used as an ammeter or voltmeter (rarely done), the fixed and moving coils are in series or parallel respectively, and therefore both fixed and moving fluxes reverse together with each half-cycle. Therefore there is no change in the direction of torque, and the pull is always one way, as shown in Figure 35.3.

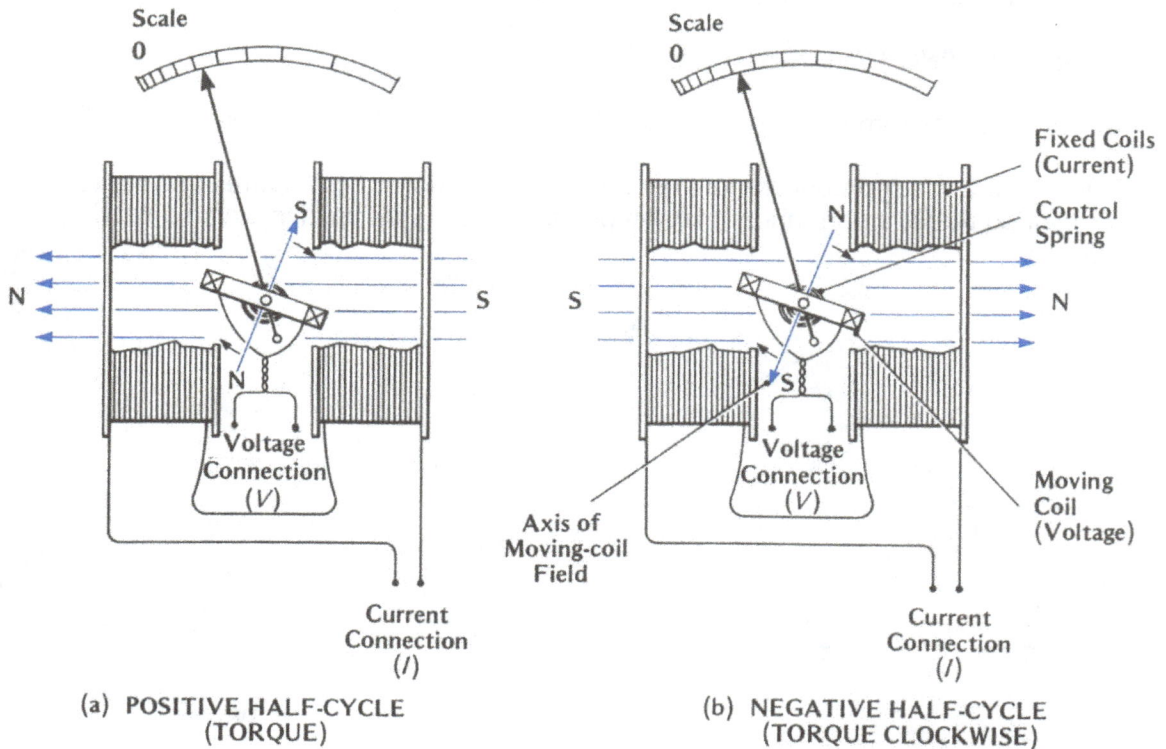

(a) POSITIVE HALF-CYCLE (TORQUE)

(b) NEGATIVE HALF-CYCLE (TORQUE CLOCKWISE)

FIGURE 35.3
DYNAMOMETER INSTRUMENT USED AS AN A.C. WATTMETER

Moreover, since the torque depends on the product of the currents in both the fixed and moving coils, and since these are either equal (in an ammeter, where they are in series) or proportional (in a voltmeter, where they are in parallel), the torque is proportional to the square of the coil current. The instrument will therefore have an uneven scale and will indicate rms current or voltage.

Dynamometer instruments can consequently be used as ammeters or voltmeters on an a.c. system, where they will indicate rms values, although, as already said, this is not often done because of cost.

The dynamometer instrument is however principally used as a wattmeter, where the fixed coils carry the line current and the moving coil the line voltage. Here again when going from a positive to negative half-cycle, both change sign together, so that there is no reversal of torque.

The magnitude of the torque depends on the product of the voltage (moving) field strength, of the current (fixed) field strength and of the cosine of the phase angle between them. The torque therefore is proportional to $VI\cos\varphi$. But $\cos\varphi$ is the power factor, so the torque is proportional to the active power in watts. The dynamometer instrument can consequently be used as an a.c. wattmeter.

This instrument can also be adapted for 3-phase working, where, by suitable connections between the phases, it can be made to indicate the **total** watts in a balanced or an unbalanced 3-phase system. By suitable reconnections between the phases the same instrument can be made to indicate $VI\sin\varphi$ - that is, reactive power, and so is used as a varmeter.

35.6   TRANSDUCER-OPERATED INSTRUMENTS

There are in big installations many disadvantages in carrying the signals (especially currents) from the sensing devices over long distances to a control centre. Also wattmeters and similar instruments of the dynamometer type are expensive. Much cost can be saved and the instrumentation simplified if all a.c. measurements could be converted to simple equivalent d.c. voltages proportional to the quantities sensed, and the d.c. signals so derived distributed to all control points in parallel and displayed on simple d.c. moving-coil voltmeters. The scale would of course not be in d.c. volts, but it would be calibrated in whatever unit the original sensing device was measuring. This may be volts, amperes, watts, hertz (frequency), power factor or any other quantity.

The sensing device, most usually a current transformer or voltage transformer (see para. 35.8), feeds its signal into a solid-state electronic circuit where the signals are processed, converted to d.c. exactly proportional to the quantity sensed and passed out as a variable-voltage d.c. signal to be displayed on voltmeters wherever desired. Such an electronic device is called a 'transducer'; it is relatively cheap and may be incorporated inside the instrument, or it may be in a separate box near to the sensing point when remote instruments need to be connected to the d.c. side.

35.7   INDUCTION (EDDY-CURRENT) INSTRUMENTS

There is another class of instrument which works on a totally different principle and which operates only with alternating current. They are 'induction' instruments, also called 'eddy-current' type, as shown in Figure 35.4.

The movement consists of a thin copper or aluminium disc which is caused to rotate between the poles of a special electromagnet. As it rotates it winds up a spiral control spring which opposes the rotation increasingly as the disc moves.

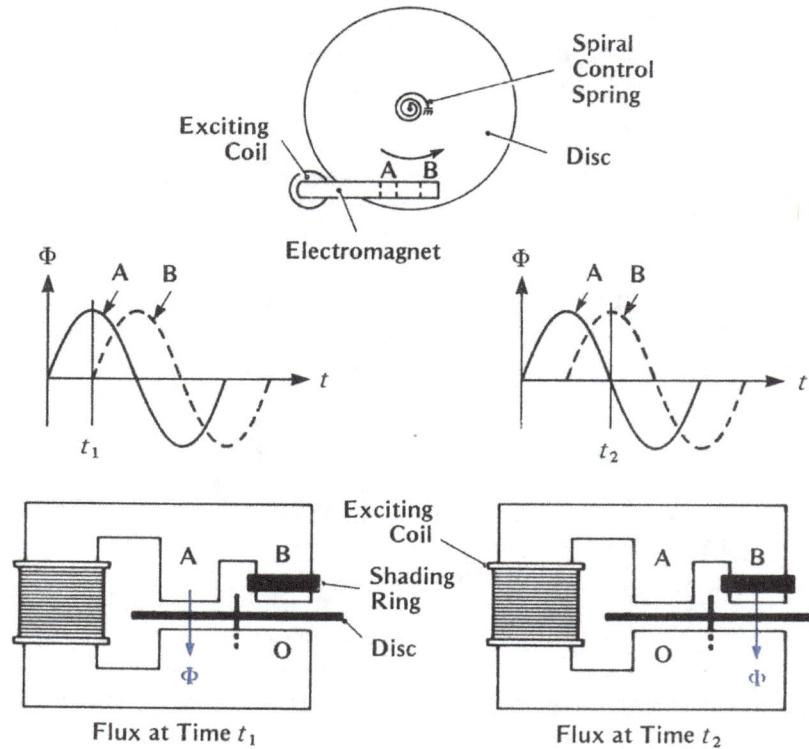

FIGURE 35.4
INDUCTION MOVEMENT

The electromagnet is of a special shape, as shown in Figure 35.4. It is wound with an exciting coil: the coil is connected in series with the line current to be measured if the instrument is to be an ammeter, or in parallel with the line voltage if it is to be a voltmeter.

The pole on one side of the disc is split into two parts, 'A' and 'B', and one of the parts is surrounded by a bare copper ring, called a 'shading ring'. The alternating flux due to the exciting coil passes freely through the disc from pole 'A', but the parallel flux in pole 'B' sets up an alternating emf, and so a current, in the closed shading ring which produces its own flux. The combined effect of these two fluxes in pole 'B' is to produce a net flux which lags nearly 90° on that in pole 'A'.

There is thus a situation where the fluxes in poles 'A' and 'B' are separated in both space and time - the classic requirement for a travelling magnetic field. On the left of Figure 35.4 is shown the situation at the instant when the current in the coil is at a positive peak (time t_1). The flux in pole 'A' is maximum (downwards, say), and the flux in pole 'B' is zero.

One-quarter of a cycle later (time t_2, right-hand side of Figure 35.4) the flux in pole 'A' is zero and the flux in pole 'B', which lags 90° on that in pole 'A', is now maximum positive. So the peak flux has moved from 'A' to 'B' in one-quarter of a cycle - it has, in fact, **travelled** from 'A' to 'B'.

As a flux wave travels, it induces in the metal disc a mass of so-called 'eddy currents' - local whirls of current within the disc which react with the travelling field and, as explained earlier for the interaction of currents in a magnetic field, produce a mechanical force on the outside of the disc, causing it to try to follow the travelling field. The disc therefore undergoes a torque which is proportional to the travelling flux and so to the square of the current in the coil.

Under this torque the disc starts to rotate, and as it does so it begins to wind up the spiral spring. This continues until the torque exerted by the spring exactly balances the driving torque due to the travelling field of the magnet; the disc then comes to rest.

The rest position is thus an indication of the current in the coil, and the disc actuates a pointer which moves over a scale, which is calibrated in amperes or volts. Since the torque is proportional to the square of the current, this type of instrument, like the moving-iron type, has a non-linear scale and is calibrated to read rms values.

One distinguishing feature of the induction-type instrument is that the disc has a far greater range of movement than is possible in the moving-iron type, and the scale is therefore very long, or open, covering almost the whole of the face of the instrument.

Induction instruments are now not much used as ammeters or voltmeters, as the moving-iron type is much cheaper, but the method is used with integrating meters such as the kWh meter found in domestic and other installations. In that application there are two coils, one voltage and one current, on either side of the disc. There is no spiral spring but instead a brake magnet which controls the disc speed. The torque is then proportional to the product of the voltage flux, of the current flux and of the cosine of the phase angle between them, namely:

$$\text{torque} \propto VI \cos \varphi$$

Since cos φ is also the power factor

$$\text{torque} \propto P, \text{ the active power.}$$

This instrument's disc therefore moves, controlled by the brake magnet, at a speed proportional to the active power, watts. Having no control spring it does not stop but continues to rotate, operating a counter as it does so; this indicates the watt-hours (or kWh) which have been consumed by the circuit over any period of time. It is thus an **integrating** meter, in that it continuously adds up the energy consumed.

The induction, or eddy-current, movement is also widely used in overcurrent relays (see Part 7 Control Devices), where the method is the same as for the watt-hour meter as described, but the disc, after rotating through a preset angle, strikes an adjustable contact. The time which elapses before this happens depends on the speed of the disc, and so on the line current causing it. A variable time delay can therefore be set on this relay, depending on the current being measured and by varying the distance to be travelled by the disc before striking the contact.

35.8   INSTRUMENT TRANSFORMERS

The current, or voltage, in the operating coil of any type of a.c. instrument could be the actual line current or voltage of the system. However, in most a.c. systems the operating voltages are very high and the currents are large. Severe practical difficulties would arise if such voltages were applied to these small instruments or switchboards, or if they had to be designed to carry such heavy currents.

It is therefore universal practice to apply the operating voltage through a step-down 'voltage transformer' or to apply the operating current through a step-down 'current transformer'. In either case a much lower voltage or current, which is an exact proportion of the line voltage or current, is applied to the instrument. The scale however is calibrated to read the actual line values, not the ones actually applied.

These 'instrument transformers' are described more fully in Part 3 Electrical Power Distribution.

CHAPTER 36 USEFUL FORMULAE: CHAPTERS 28 - 35

ACTIVE, REACTIVE AND APPARENT POWER

Single-phase: If V is the rms voltage and I the rms current, and φ the phase angle between current and voltage, then:

$$\text{active power } P = VI \cos \varphi \text{ watts}$$

$$\text{reactive power } Q = VI \sin \varphi \text{ vars}$$

$$\text{apparent power } S = VI \text{ volt-amperes}$$

Three-phase: if V_L is the rms line voltage and I_L the rms line current and φ the phase angle between current and voltage for a balanced load, then:

$$\text{active power } P = \sqrt{3} \, V_L I_L \cos \varphi \text{ watts}$$

$$\text{reactive power } Q = \sqrt{3} \, V_L I_L \sin \varphi \text{ vars}$$

$$\text{apparent power } S = \sqrt{3} \, V_L I_L \text{ volt-amperes}$$

POWER FACTOR

If φ is the phase angle between a current and the voltage causing it, then $\cos \varphi$ is the power factor. This, when multiplied by the product of the rms volts and amperes, gives the active power *(P)* in watts:

thus: $\qquad P = VI \cos \varphi$ watts for single-phase

or $\qquad P = \sqrt{3} \, V_L I_L \cos \varphi$ watts for 3-phase

If $\sin \varphi$ is substituted for $\cos \varphi$, this is sometimes termed the 'reactive power factor' and, when used instead of $\cos \varphi$ as above, gives the reactive power (Q) in vars.

If R, X and Z are respectively the series resistance, reactance and impedance of a circuit,

$$\text{power factor } (\cos \varphi) = \frac{R}{Z} \text{ or } = \frac{\text{watts}}{\text{volt - amps}}$$

$$\text{reactive power factor } (\sin \varphi) = \frac{X}{Z} \text{ or } = \frac{\text{vars}}{\text{volt - amps}}$$

SWITCHING ON - ASYMMETRY

If an a.c. circuit is switched on at the worst possible instant so that the current is initially 100% asymmetrical, the first current peak may reach approximately 2.55 times the symmetrical rms value of the current to which it eventually settles.

RECTIFICATION

Half-wave,
single-phase,
amplitude $\hat{A}$

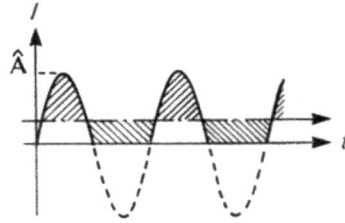

D.C. level $= 0.318\hat{A}$
or $= 0.45 \times$ rms

Full-wave
single-phase,
amplitude $\hat{A}$

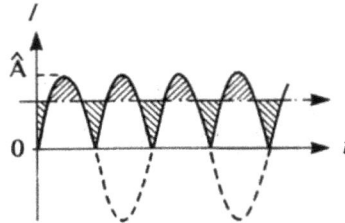

D.C. level $= 0.635\hat{A}$
or $= 0.90 \times$ rms

Full-wave,
3-phase,
amplitude $\hat{A}$

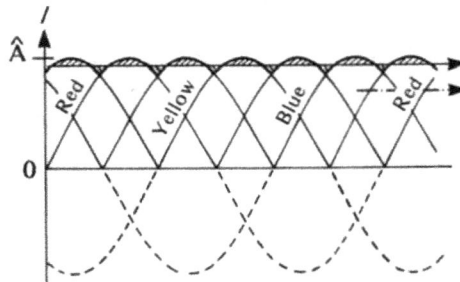

D.C. level $= 0.955\hat{A}$
or $= 1.35 \times$ rms

CHAPTER 37 QUESTIONS AND ANSWERS: CHAPTERS 28 - 35

37.1   QUESTIONS

1.   What do you understand by 'active power'? What type of circuit absorbs only active power? What is the relationship between current and voltage in a circuit passing only active power?

2.   What do you understand by 'reactive power'? What is the relationship between current and voltage in a circuit passing only reactive power? In what unit is reactive power measured?

3.   What is 'apparent power'? What is its relation to active and reactive power?

4.   In a 3-phase system with line-to-line voltage V_L and balanced line current I_L at a phase angle φ, write down expressions for (a) active power, (b) reactive power and (c) apparent power, giving the correct units.

5.   Describe the meaning of 'power factor'. A single-phase circuit, supplied at 250V has a resistance of 40 ohms and an inductive reactance of 30 ohms in series. What is (a) the power factor and (b) what active power flows?

6.   In a 250V single-phase system carrying a current of 50A at a power factor of 0.75 what is (a) the active power and (b) the reactive power? Give the units in both cases.

7.   In a 3-phase system the line voltage is 11kV and the line current 200A at a power factor of 0.8. What is the total active power transmitted?

8.   What is the power factor of a 50Hz single-phase circuit consisting of a 50 ohm resistor and a 0.1H inductor in series?

9.   In the circuit of Q.8 what is (a) the active power P and (b) the reactive power Q transmitted when a current of 25A is flowing?

10.

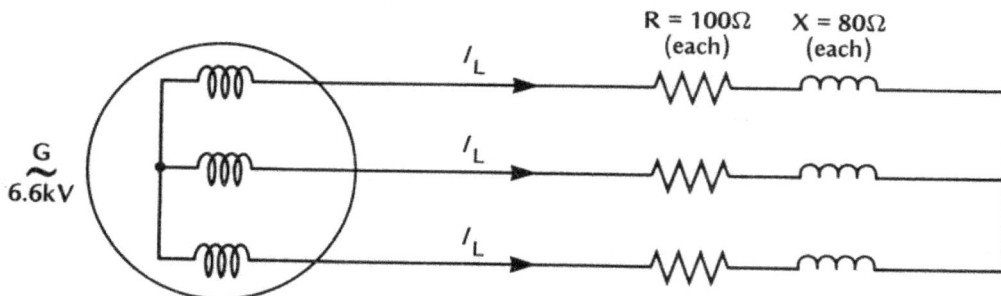

In the balanced 3-phase system shown in the figure determine:

(a)   the line currents I_L
(b)   the overall power factor
(c)   the total active power P
(d)   the total reactive power Q

11. What difference would it make if the generator in Q.10 were delta-connected?

12. A purely reactive (90° lagging) a.c. circuit is switched on at an instant when the voltage is passing through zero. Sketch the ensuing current wave.

13. When a balanced 3-phase system is switched on, why do the initial currents in the three phases show different degrees of asymmetry?

14. To what uses can a diode be put in simple power supply or control circuits?

15. How is an a.c. current converted into d.c.? What element is used to do this? If a single-phase a.c. current of 10A (rms) undergoes half-wave rectification, what d.c. current level is achieved?

16. With a 3-phase, full-wave rectifier what a.c. voltage (rms) must be applied in order to achieve 24V d.c. output?

17. How can a rectifier bridge be adapted to give a variable d.c. output?

18. Why is high voltage used to generate large bulk quantities of electricity?

19. A platform has a 6.6kV generation and a 440V distribution system. An oil transfer pump is driven by a motor rated at 1 200kW. At what voltage would you expect it to operate? Why?

20. A system has six motors with nameplate ratings 200 hp, 150 hp, 2 x 100 hp, 80 hp and 50 hp. What is the total full-load output in kW_m ?

21. What different types of instrument can be used to measure alternating voltages and currents? What voltage or current will they indicate? What do you expect to notice about their scales?

22. Why cannot a moving-coil instrument be used to indicate a.c. directly? How can it be made to do so?

23. What type of instrument would you use to indicate watts on a 3-phase a.c. system?

24. What type of instrument would you use to indicate vars on a 3-phase a.c. system?

25. Why are transducers often used in modern instrumentation?

37.2 ANSWERS

(Figures in brackets after each answer refer to the relevant paragraphs in the text.)

1. Active power represents true energy passing through a circuit or consumed by a load. It is measured in watts. Only those circuits with pure resistance (e.g. lighting and heating) absorb **only** active power. Other circuits, which include motors, absorb both active and reactive power. Current in a circuit passing only active power is in phase with the voltage. (28.1)

2. Reactive power represents 'false power' and in a network is a measure of the magnetisation of all inductive equipment. The current in a circuit passing only reactive power lags 90° on the voltage if the reactance is inductive, or leads 90° on the voltage if the reactance is capacitive. Reactive power is measured in 'vars'. (28.1, 29.3)

3. Apparent power is the product of voltage and current, disregarding any phase angle between them. Its value is $V \times I$ and is measured in 'volt-amperes' (VA), where V and I are rms quantities. Its symbol is S, and $S = \sqrt{watts^2 + vars^2}$ (30.2)

4. (a) active power $= \sqrt{3}\, V_L I_L \cos \varphi$ watts

    (b) reactive power $= \sqrt{3}\, V_L I_L \sin \varphi$ vars

    (c) apparent power $= \sqrt{3}\, V_L I_L$ volt-amperes. (31.2)

5. If in a circuit the current lags (or leads) by an angle φ, the expression 'cos φ' is the power factor. It is the factor which, when multiplied by the product of rms volts and amperes (the apparent power), gives the true active power being passed:
    $$P = VI\cos \varphi.$$

    (a) $\tan \varphi = \dfrac{X}{R} = \dfrac{30}{40} = 0.75, \quad \varphi = 36.7°$ and power factor $= \cos \varphi = 0.8$

    (b) active power $\quad z = \sqrt{R^2 + X^2} = \sqrt{40^2 + 30^2} = 50$ ohms
    $$I = \frac{V}{z} = \frac{250}{50} = 5A$$

    active power $\quad P = I^2 R = 5^2 \times 40 = 1\,000W$ (30.2)

6. (a) active power $\quad = VI \cos \varphi = 250 \times 50 \times 0.75$
    $P = 9\,375$ watts

    (b) reactive power $\quad = VI \sin \varphi$ (NB: $\cos \varphi = 0.75, \therefore \sin \varphi = 0.66$)
    $= 250 \times 50 \times 0.66$
    $\therefore \quad Q = 8\,250$ vars (30.2)

7. $$P = \sqrt{3}VI \cos \varphi = \sqrt{3} \times 11\,000 \times 200 \times 0.8$$
    $$= 3\,048\,409 \text{ watts}$$
    $$= 3.05 \text{ MW}$$
    (31.2)

8. 0.1 H at 50Hz has a reactance $X = 2\pi f L = 2\pi \times 50 \times 0.1 = 3.14$ ohms

$$\tan \varphi = \frac{X}{R} = \frac{31.4}{50} = 0.628$$

$$\therefore \quad \varphi = 32.1°$$

power factor $= \cos \varphi = 0.85$ (30.2)

9. Active power $P = I^2 R = 25^2 \times 50 = 31\,250$ watts (31.3 kW).

Reactive power $Q = I^2 X = 25^2 \times 31.4 = 19\,625$ vars (19.6 kvar).   (30.2)

10. Impedance **Z per phase** $= \sqrt{100^2 + 80^2} = 128$ ohms

(a) $\therefore$ current $= \dfrac{phase\ voltage}{Z}$

$$= \frac{6600/\sqrt{3}}{128} = 29.8A \text{ (in all lines)}$$

(b)   power factor $= \dfrac{R}{Z}$

$$= \frac{100}{128} = 0.78$$

(c)   total active power $P = \sqrt{3}V_L I_L \cos \varphi$ (Note: cos φ is power factor)

$$= \sqrt{3} \times 6600 \times 29.8 \times 0.78 = 265\,715 \text{ watts}$$
$$\text{(say 266kW)}$$

(d)   total reactive power $Q = \sqrt{3}V_L I_L \sin \varphi$ (Note, if cos φ = 0.78, sin φ = 0.63)

$$= \sqrt{3} \times 6600 \times 29.8 \times 0.63 = 214\,616 \text{ vars}$$
$$\text{(say 215 kvar)}$$

(Note: (c) and (d) could also have been evaluated: $P = 3I^2 R$ and $Q = 3I^2 X$ (31.2)

11. None. The same formula $P = \sqrt{3}V_L I_L \cos \varphi$ or $Q = \sqrt{3}V_L V_L \sin \varphi$ applies whatever the generator's phase connections.   (31.2)

12. See Chapter 32, Figure 32.1(b).

13. Because in a balanced 3-phase system the currents are separated 120° in time, at any given instant they will in general all be at different points on the wave, giving different degrees of partial asymmetry at the instant of switching. (See Chapter 32, Figure 32.3.)

---

14. A diode can be used:

(a) to rectify a.c. into d.c.

(b) to block the passage of current in an unwanted direction

(c) to provide a discharge path for an inductive element when its d.c. supply is switched off. (33.1)

15. By passing it through any device which allows current to travel only one way (a 'rectifier'). Nowadays most usually a solid-state diode element.

10A rms has a peak value $\sqrt{2} \times 10 = 14.1A$. The rectifying factor for single-phase half-wave is 0.318. Therefore d.c. level is $14.1 \times 0.318 = 4.5A$ d.c. (33.2)

16. With a 3-phase, full-wave rectifier

$$V_{dc} = 0.955\hat{V} \ \text{or} \ = 1.35V_{rms}$$

If $V_{dc} = 24V$, then $V_{rms} = \frac{24}{1.35} = 17.8V$ (rms). (33.5)

17. By replacing one of the diodes in each phase by a triode (thyristor) and arranging to delay (or advance) the firing of the thyristor as required. Delayed firing will cause a reduction in the mean d.c. output level. (33.4)

18. Large bulk generation involves considerable power, and power is proportional to the product of voltage and current. For a given power therefore at low voltage the current is very high - too high normally to be handled by machines, cables, etc. Raising the voltage lowers the current in inverse proportion for a given power. Therefore by raising the voltage level the currents can be kept down to manageable levels. (34.1)

19. On the 6.6kV system a 1 200kW motor is far outside the practical range of 440V motors because of the high current (about 2 000A) which it would draw. (34.2)

20. Total motor power is 680 hp. The equivalent mechanical output in kilowatt units is

$$680 \times 0.746 = 507kW_m$$ (34.2)

21. A.C. instruments may be hot-wire, moving-iron, dynamometer or induction (eddy-current) type, but not moving-coil except where transducers are used. All voltmeters and ammeters will indicate rms values, and their scales will normally tend to be crowded at the lower end. (35.3 to 35.7)

22. Because the moving coil would carry an alternating current which would react with the fixed magnetic field, causing it to suffer alternating torques with each cycle of current. There would therefore be no net movement of the coil. (35.4)

If it is desired to take advantage of the more open scale of a moving-coil instrument, it can be used on a.c. systems by using a transducer to convert the measured a.c. signal to d.c. and then applying that d.c. to a moving-coil voltmeter-type instrument, suitably scaled. (35.6)

23. Normally a dynamometer-type instrument would be used, with voltage connected to the moving coil and current to the fixed coils. (35.5)

24. An instrument constructed identically with the wattmeter, but with its external connections rearranged so that it indicates $VI \sin \varphi$.

25. Transducers reduce all a.c. signals, whether of voltage, current, power, frequency or power factor, to a proportional d.c. signal which can be made to indicate on a suitably scaled d.c. moving-coil voltmeter-type instrument. Many such instruments can be paralleled onto a single loop for remote indication. Transducers also avoid having long current-carrying connections (especially current transformer connections) between the sensor and the remote indicator. (35.6)

PART 2 ELECTRICAL POWER GENERATION

CHAPTER 1 INTRODUCTION TO GAS TURBINES

1.1   GENERAL

All generators send out energy, in the form of electrical power, and they have to be given the equivalent mechanical energy. This means that they have to be driven by an engine of some sort which derives its energy from fuel or some other natural source such as wind or water. The engine which drives an electric generator is called a 'prime mover and may take many forms.

In the early days steam-, gas- or oil-driven reciprocating engines were used. Later, steam turbines became more general, especially in large power stations. More recently gas turbines have come into use, especially on oil platforms where gas is produced as part of the production process usually in sufficient quantities to provide a source of fuel.
The modern form of oil engine, the diesel, is also much used, principally for standby or emergency plant when gas supplies to the gas turbines fail or are shut down. On a platform, of course, it is necessary to bunker diesel fuel for these engines. (See Chapter 6.)
As the various forms of gas turbine may not be familiar to some, a brief description of this type of engine and how it evolved is given overleaf. Firstly, however, it may be advantageous to recap the principles of operation of its predecessor, the steam turbine.

1.2   THE STEAM TURBINE

Fuel was burned under a boiler, whose water (shown blue) was turned into steam. This steam, at high pressure and speed, hit the inclined blades of a turbine wheel (yellow) and drove it round. In doing so it lost some of its pressure, but enough was left to drive a second wheel - and a third or fourth - on the same shaft. Finally the steam was exhausted into a condenser, turned back into water and returned to the boiler to be reheated and used again. This was the whole steam 'cycle' which is shown in Figure 1.1.

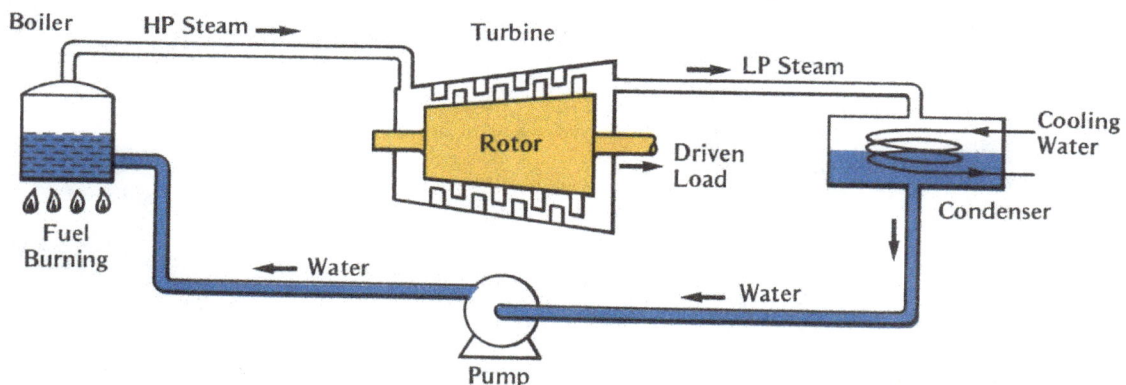

FIGURE 1.1
STEAM TURBINE SYSTEM

1.3   THE GAS TURBINE

Gas-turbine generators are found generally in offshore installations, where natural gas is available. Onshore installations normally have only small standby or emergency generators, and these are usually diesel driven. This chapter, therefore, applies to offshore installations only.

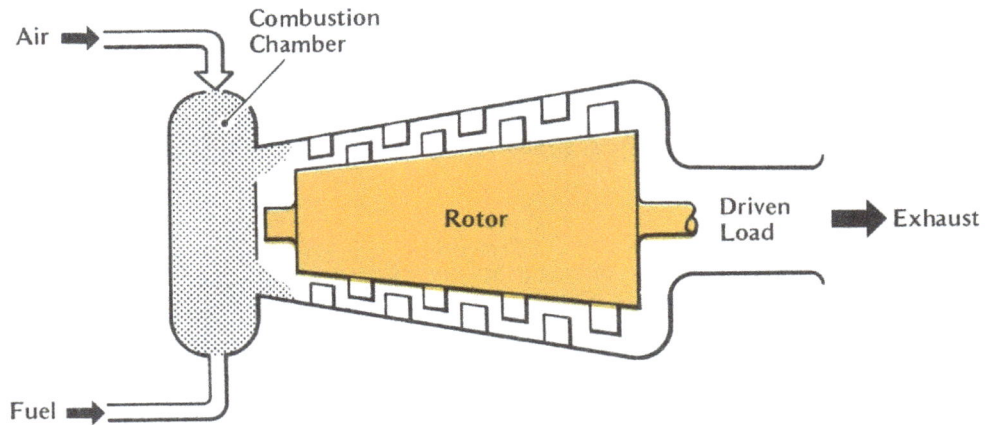

FIGURE 1.2
GAS TURBINE

The gas turbine works on a similar principle to that of the steam turbine except that there is no boiler or water: instead the fuel is burned in a combustion chamber at one end of the turbine where it produces a hot, high-pressure gas. This gas, in trying to expand, causes a reaction on each row of blades on a rotor (shown yellow), expanding and cooling as it does so and driving the blades round to produce mechanical power. By expanding in the confined volume of the turbine, the gas has to keep up and even increase its speed in order to pass through. The principle of the gas turbine is shown in Figure 1.2.
It should be noted that the gas is hottest at the combustion chamber or inlet end. As it expands in the turbine, it cools, and it should leave the exhaust end at a lower temperature, of the order of 650°C. Many turbines have instruments to measure exhaust temperature. If it is too high, it indicates some fault in the combustion, and the set is usually shut down to save the blades from damage.

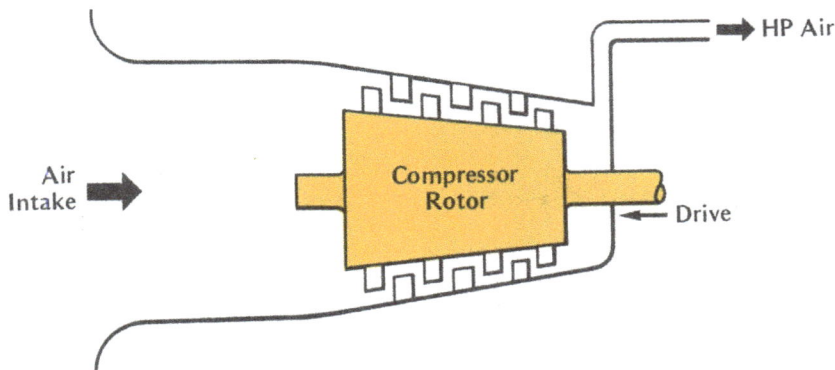

FIGURE 1.3
COMPRESSOR

In order for the fuel to burn, oxygen, in the form of air, is needed, and it must be at high pressure in order to enter the combustion chamber; therefore an air compressor is fitted integrally with the turbine. It is just like a turbine in reverse. Air is drawn in at the larger diameter end by the inclined blades acting as a suction fan. Once in, it is compressed by the blades of the compressor rotor (shown yellow) into a smaller volume, to be sucked in again by the next row of blades and compressed still further. Each stage of compression causes the air to become hotter. Eventually it emerges at the small diameter end as hot compressed air. The principle of the gas-turbine compressor is shown in Figure 1.3.

To provide the power to compress the air, the compressor must be driven mechanically. The turbine itself drives it. The gas-turbine shaft is coupled to the compressor shaft and constitutes the complete gas-turbine assembly. This looks like perpetual motion - which it is, provided that the fuel continues to be supplied. The combined turbine and compressor unit is shown in Figure 1.4.

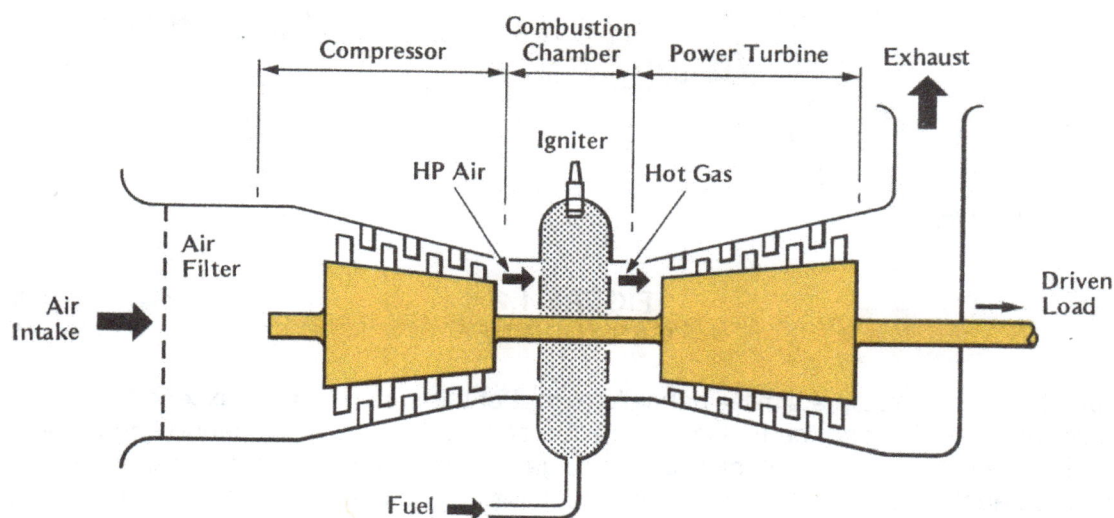

FIGURE 1.4
SINGLE SHAFT GAS TURBINE SET

In practice something like 80% of the power developed in the turbine from the combustion is needed to drive the compressor, leaving only about 20% 'payload' to drive the load. It can be seen that, if there is a drop of only about 5% in the combustion efficiency, so needing 85% of the output to drive the compressor, the effect is to reduce the payload from 20% to 15% - an effective reduction of load drive of 25%. Gas turbines are therefore very sensitive to combustion control.

In the gas turbine on an oil platform the power developed by the gas turbine, less that part of it needed to drive the compressor, is used to drive the load, which may be a machine such as a generator, air compressor or pump.

One of the great advantages of the gas turbine over other forms of prime mover is its high power/weight ratio.

An important point to note is that, unlike other types of engine, the gas turbine needs to take in up to 70 times the amount of air actually needed for combustion. The excess is for cooling. This means that gas turbines have very large intake ducting, usually provided with screens and filters to prevent the entry of sea birds and other sizeable particles. In freezing weather the screens can become iced up and restrict the flow of air. Therefore, anti-icing equipment and blow-in doors are often provided (see para. 1.13).

1.4   SINGLE SHAFT AND TWO SHAFT TURBINES

The type of gas turbine shown in Figure 1.4 is known as a 'single-shaft' type - that is, the power turbine, compressor and driven load are on a single, common shaft. The power delivered by the power turbine is divided between the compressor (about 70% to 80%) and the driven load (about 20% to 30%).

In some larger gas turbines the arrangement is different. A standard aircraft-type jet engine may be used, as shown in Figure 1.5, where the compressor turbine is only large enough to drive the compressor itself, with no driven load. But the exhaust gas which, in an aircraft, would go straight to jet is ducted to the input of a further power turbine on a separate shaft; its rotor is shown blue. This drives the load, and usually at a speed different from that of the compressor turbine.

This is known as a 'two-shaft' gas turbine and has the advantage that it can be used with an existing proved aero gas-turbine design with only minor modifications to the jet end. The power turbine is a completely separate design which need not even be in line with the compressor (though it usually is). The complete aero compressor/compressor-turbine unit is known as the 'gas generator', and the separate load-drive unit as the 'power turbine'.

In a single-shaft gas turbine the power turbine is usually coupled to the driven load (a generator or compressor) through a gearbox. The compressor and power turbine therefore run at the same fixed speed, which is the generator speed multiplied by the gear ratio. In a two-shaft turbine the compressor and the power turbine can, and do, run at different speeds. The power turbine is coupled to the generator and runs at governed speed, but the compressor speed varies with the loading. At light load it will be idling, but as loading increases it increases its own speed up to full load, when it will generally be running much faster than the power turbine.

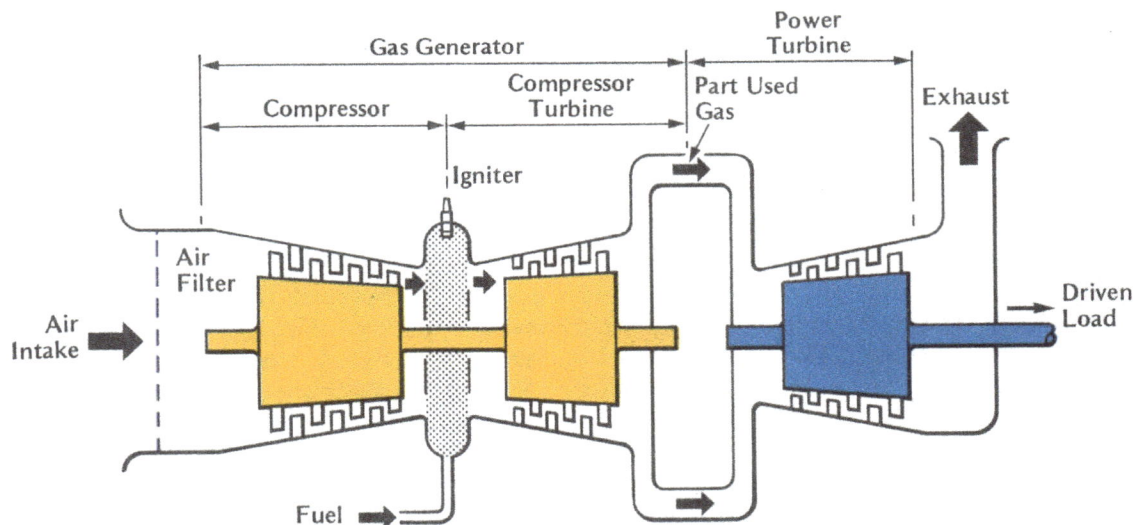

FIGURE 1.5
TWO SHAFT GAS TURBINE SET

1.5   FUEL

Like any other internal combustion engine, a gas turbine can burn gas or liquid fuel (usually diesel oil). Some turbines are designed for single-fuel burning - that is, for gas only or liquid only - whereas others may have been adapted for 'dual fuel'; they may be set to run on either fuel or, in some gas turbines, on a mixture of both.

On oil platforms where gas is available the turbines will be for gas only or dual fuel. If dual, they will normally run on gas, with liquid fuel as a fall-back if gas pressure should fail. If this happens, the changeover from gas to liquid is automatic; the turbine does not stop, but an alarm is given. When gas pressure is restored the change back must sometimes be done by hand in slow time, but on some sets the change back is automatic provided that 'Gas Fuel' had been selected originally and the fuel selector switch had not been moved.

There are exceptions to the arrangements described above. In some installations there is no automatic changeover, even from gas to liquid fuel.

Gas turbines can operate on a variety of fuels which range from crude oil (with some derating of the turbine) through to fuel gas.

1.6   SPEED CONTROL

The turbine speed is always controlled by a governor. In single-shaft sets the governor controls the shaft speed, but in two-shaft sets it controls only the power turbine speed - and so the speed of the driven load. The gas-generator shaft is free to take up its own speed, depending on the load, as explained below (see also Figures 1.6 and 1.7).

In single-shaft turbines the governor controls speed by regulating the gas control valve or the liquid fuel valve. In two-shaft sets the governor itself is driven from the power turbine shaft but regulates the fuel input to the gas generator. This runs at such a speed as to provide just enough gas to the power turbine to keep it at its correct speed. Thus, as load increases, the **gas generator** speeds up, but the power turbine stays at constant speed. The skilled operator can detect load changes by the note of a two-shaft machine, but not with a single-shaft.

Speed control is discussed in detail in Chapter 5.

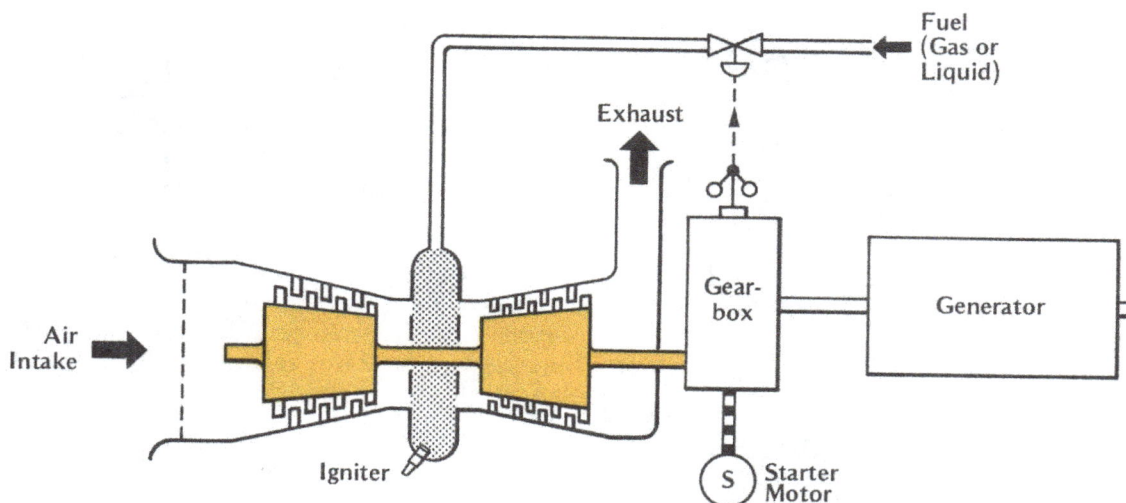

FIGURE 1.6
SINGLE SHAFT GAS TURBINE CONTROL

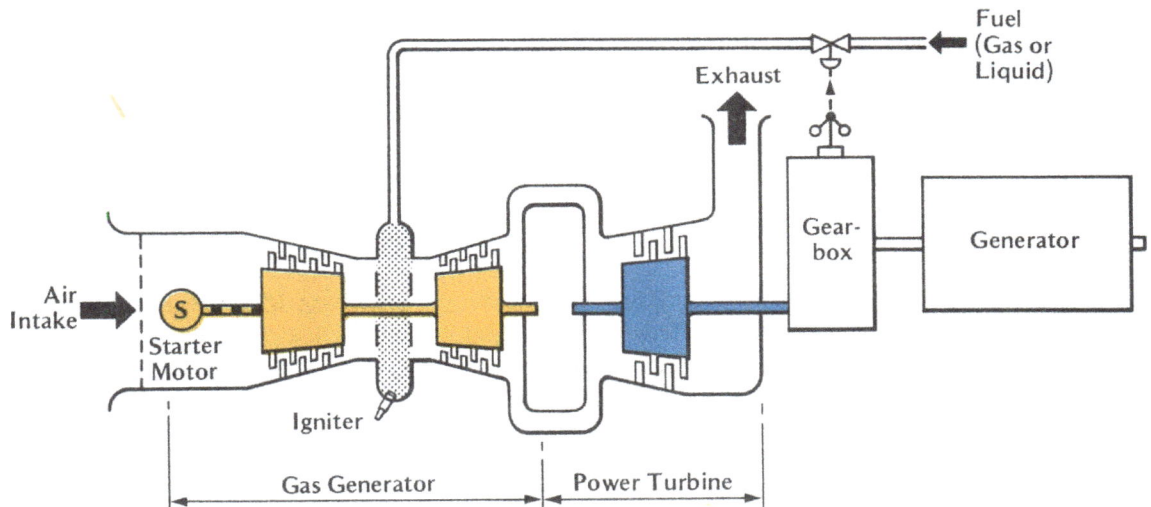

FIGURE 1.7
TWO SHAFT GAS TURBINE CONTROL

1.7   STARTING

Gas turbines must be started by an external starter. This may be electric, air-motor or even a diesel engine. For some sets a separate small gas turbine is used, which itself is electrically started. Electric starting requires a separate battery and charger.

Because in single-shaft machines not only the compressor and turbine but also the driven load (usually a generator) and gearbox must all be started together, the starting unit must be relatively heavy, and this precludes electric start, with battery, on any but the smallest machines. On the other hand in a two-shaft set the starter has only to spin up the compressor/turbine unit of the gas generator, so that the starter need only be quite small and is suited to an electric motor.

On some of the largest sets the d.c. power for the starting motor is taken not from a battery but is rectified from the set's general a.c. supplies. This requires that auxiliary a.c. supplies for the turbine set shall be available before the set can be started.

When the start button is pressed, but before the gas-turbine shaft actually begins to move, automatic circuits put into action a sequence programme which normally includes starting a lubrication pump to pre-lubricate the turbine, gearbox and generator bearings. In some machines the turbine hood is also purged with air to remove any gas present. When the lubricating oil pressure has reached a certain level, the start motor is actuated and begins to rotate and accelerate the turbine shaft (in the case of a two-shaft turbine, the gas-generator shaft only). When it reaches a certain speed, usually about 20%, fuel is admitted to the combustion chamber which is by now receiving some air from the compressor. Automatic ignition by spark-plug and torch follows, and the hot burning gas passes to the power turbine in the case of a single-shaft set, or to the compressor turbine of a two-shaft set.

The turbine gradually takes over the job of driving the compressor, and the starting motor steadily becomes off-loaded. When its load falls to a predetermined level, the electric start motor is switched off, or the mechanical start motor unclutched and stopped. The ignition is also switched off. As the set runs up, a mechanically driven lubricating oil pump begins to deliver lubrication to the bearings. When its pressure reaches a certain level, the

173

electrically driven pump stops automatically.

The turbine is now self-sustaining, and the speed continues to build up until it comes under governor control and settles at its correct level. In a two-shaft set the gas-generator speed continues to build up, and the hot burning gas from it passes on to the main power turbine, which then starts to move by itself without any mechanical starting. Its speed too builds up until it comes under governor control, when the fuel to the gas generator is cut back and it settles down to its no-load or 'idling' speed. The power turbine, however, is now running at its controlled speed.

1.8   STOPPING

To stop the set, fuel is simply cut off, and the set runs down steadily. Part of the stopping sequence includes the starting of the electrically driven lubricating oil pump so that bearing lubrication continues as the set runs down and the mechanical lubricating oil pump becomes less effective. The electric pump continues to run for some time after the set has actually stopped.

Some larger sets have a hydraulic ratchet arrangement which slowly turns the turbine rotor, after stopping, for up to 24 hours to prevent 'bowing' of the rotor due to uneven cooling.

All sets are arranged so that, when the 'Stop' button is pressed or when a shutdown signal is given for any other reason, the associated generator supply breaker is tripped, if not already open, to off-load the turbine. Sometimes automatic provision is made to off-load the set gradually (except with an emergency stop) before the supply breaker actually trips.

1.9   PROTECTION

The turbine (as distinct from the electric generator) has a number of protective devices to guard against malfunction both during starting and while running. During starting each stage of the sequence is monitored to ensure that it is completed within a certain time; if it is not so completed, the start is 'aborted', and the set, if already moving, is stopped and the appropriate alarm given.

During running other protection operates, including such obvious things as overspeed, excessive vibration, loss of lubricating oil pressure, high lubricating oil temperature and excess exhaust temperature (which indicates a combustion fault). All these and other malfunctions shut the set down (having also tripped the generator breaker) and give the appropriate alarm.

Alarms are visual and audible. The visual alarms are grouped into 'annunciator' lamp boxes on the control board, each lamp window being annotated with the fault it announces. At the onset of a fault the lamp flashes and a buzzer sounds. When the 'Accept' button is pressed the buzzer stops and the lamp burns steadily. It does not go out however until the fault has been cleared; even then a 'Reset' button must be pressed before the set can be started again. If a turbine stops during starting or shuts down during running, it should be possible for the operator to diagnose from the lamp indications the cause of the trouble.

On some sets the starting sequence lamps, which merely monitor the starting stages but do not indicate a malfunction, are segregated from the fault lamps. On most platforms the turbine malfunction alarms are repeated in the Electrical Control Room, though they are usually grouped to reduce the number of lamps there.

One of the most damaging things that can happen to a turbine is failure of lubrication. This can cause bearings to fail at high speed with probable catastrophic damage and danger to personnel. To guard against this all sets have not only the mechanically driven oil pumps and the a.c. electrically driven pumps for starting pre-lubrication, but also a d.c.-driven emergency pump fed from a battery, which cuts in automatically on failure of lubricating oil pressure. This is a vital piece of equipment, and it must be regularly tested to prove its proper functioning.

1.10   CONTROL

With every gas turbine there is a Local Control Panel, usually adjoining the control panel for the generator. The turbine control panel has instruments to indicate speed, temperatures and pressures at various points and fuel pressure, as well as controls for starting and stopping, for fuel selection and for setting the speed. There are a number of lamps in an annunciator panel to indicate malfunction, and others to indicate successful completion of each starting sequence step. On some larger sets the governor control occupies a complete panel on its own.

1.11   SPEED CONTROL

Control of speed by automatic governor is dealt with in Chapter 5.

1.12   WASHING

Air pollution, especially salt, can cause encrustation of the compressor blades, distorting their aerodynamic form and reducing the efficiency of compression. This shows up as higher exhaust temperature.

When this situation occurs (and high exhaust temperature indication is a pointer) the compressor must be 'washed'. Turbine manufacturers make arrangements for this to be done at reduced turbine speed using water, with or without detergent, or solid abrasives such as ground walnut husks or bran. The two methods are sometimes known as 'Crank Soak' (liquids) and 'Abrasive Cleaning' (solids).

Details of the methods are given in the manufacturers' Operations or Maintenance Manuals.

1.13   ANTI-ICING

In freezing weather the air intake filter screens can become iced up; in this state they can severely restrict the intake of air and cause serious combustion problems.

One way of dealing with this is to duct warm air from the engine to the area of the screens. On some makes of turbine this is taken from the exhaust ducting; on others it is bled off the later stages of the turbine compressor.

Blocked screens, whether due to icing or other causes, can bring about problems, and immediate steps should be taken to clear them. One feature which assists the air flow in the short term is the 'blow-in' door. This is a door, usually on the side of the intake and downstream of the screens, which is loosely hinged and is just kept closed by gravity. If the screens become blocked, the differential pressure across them increases, and the lower pressure inside the ducting sucks the door open, so allowing air to bypass the screens until action to clear them can be taken. The opening of a blow-in door gives an alarm to the operator. The door usually has a de-icing heater to prevent its becoming iced up.

CHAPTER 2 A.C. GENERATORS

2.1 GENERAL

The principle of a.c. generation is fully covered in Part 1, where it is developed from Faraday's Law of Electromagnetic Induction to the idea of a modern generator with a rotating field and a stationary armature. This chapter assumes familiarity with that concept and deals with the actual hardware. Chapter 3 discusses the various methods of excitation.

Figure 2.1 shows, in cutaway form, a typical a.c. generator in the 15-megawatt (20 000 hp) size range. The generator proper is enclosed in a box or 'hood'; this is both to exclude noise and to contain the closed ventilation system. It also assists purging before starting if gas has been present. The rotating parts are coloured yellow and the stator blue.

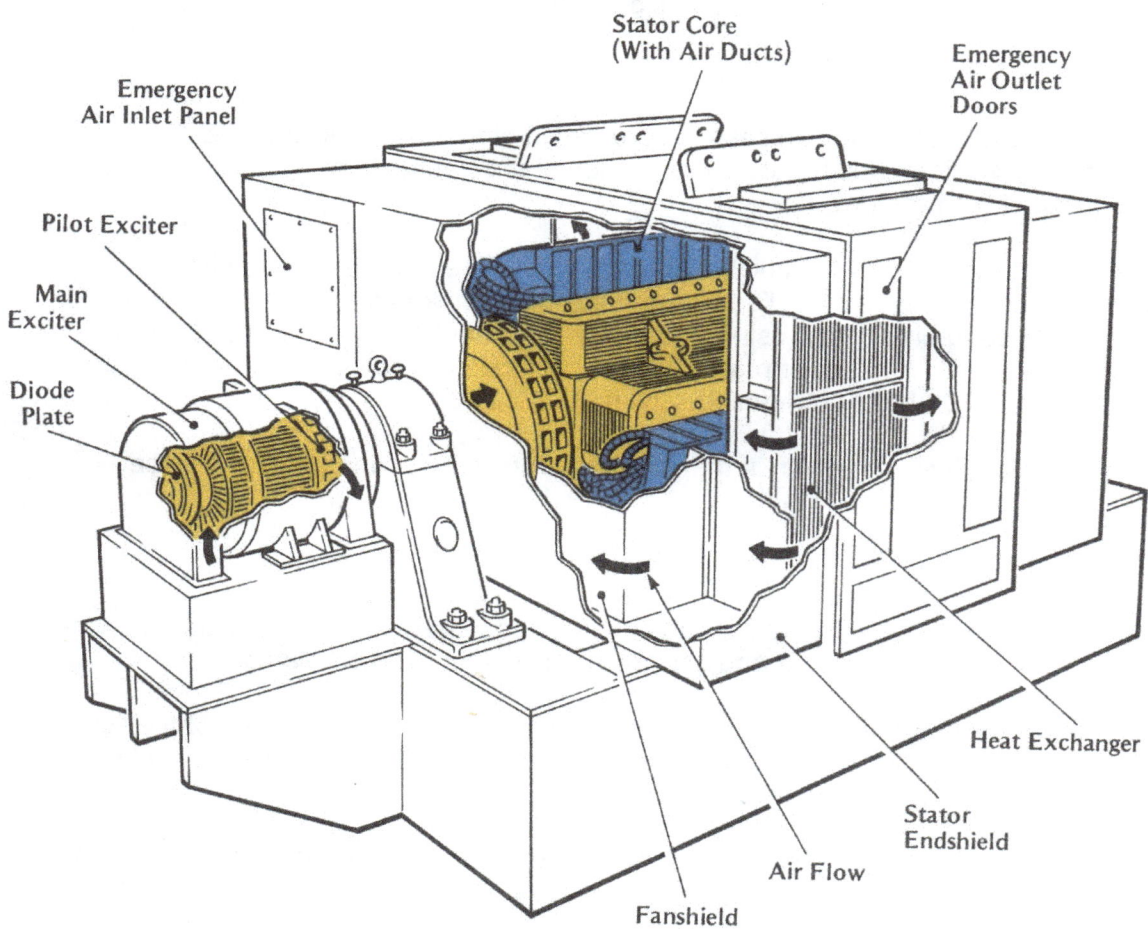

FIGURE 2.1
TYPICAL A.C. GENERATOR

The armature (normally the stator) windings carry the load current, which varies with the loading. These windings have resistance and generate heat at a rate proportional to the square of the current $(W = I^2R)$. The field's exciting winding (normally on the rotor) also carries current. It too has resistance and generates I^2R heat. These two sources of heat, together with iron loss heating, combine to raise the temperature of the machine. All the heat must be taken away by the cooling system if the temperature rise is to be held below the designed limit.

Since the stator heating varies with the square of the load current, doubling the load current gives rise to a four-fold increase in the stator heat generated. It is important therefore that the machine never becomes excessively overloaded. If it does, the cooling system may be unable to handle the heat, and dangerously high temperatures may result.

The generator is cooled by a shaft-driven fan which circulates air in a closed air circuit through all the windings. The air, in circulating, passes through an air/water heat exchanger. Here the heated water is discharged and the cooled air recirculates, as shown by the arrows in the figure. Temperature detectors at various points give warning of overheating; if it is seriously high and continues unchecked, the whole set is usually shut down.

If the cooling system should break down for any reason, panels in the hood can be removed and the machine cooled by natural ventilation through the fan. Under these circumstances however the loading on the generator may have to be curtailed to a value well below its normal rating.

The stator (armature) carries a 3-phase winding consisting of insulated conductors in slots round the inside face. These conductors must be insulated up to the full working voltage of the system. Serious or sustained excess temperature of the winding will cause this insulation to deteriorate or even to break down completely, resulting in an internal flashover and possibly complete write-off of the generator.

The rotor windings, which provide the field, operate at a much lower voltage - of the order of 70V d.c. - so insulation is less of a problem. Nevertheless, if the automatic voltage regulator calls for too much voltage and therefore too much field current, it is still possible to overheat and damage the rotor.

The limitations imposed by overheating the stator and rotor are further discussed in Part 6 System Control, Chapter 1, under the heading 'Capability Diagram'.

2.2   ROTOR CONSTRUCTION

A.C. generators with rotating fields have rotors which fall into two types: salient pole and cylindrical. They are both shown in Figure 2.2.

The salient-pole type is illustrated in Figure 2.2(a). It is by far the most common with offshore generators and also with the smaller sizes onshore. It consists of a solid iron rotor body (square in the case of a 4-pole rotor) onto which pole pieces are bolted. Each pole piece carries one of the field windings as shown in the figure. The poles terminate in pole shoes which spread out the magnetic field in the air gap, but it should be noted that with the salient-pole arrangement the air gap, and so the air gap flux, is far from uniform. Some rotors have damper windings embedded in the pole shoes, but these are not shown in Figure 2.2(a). The salient-pole rotor is commonly used with 4-pole generators. Where there are six or more poles, this is the only type which is practical.

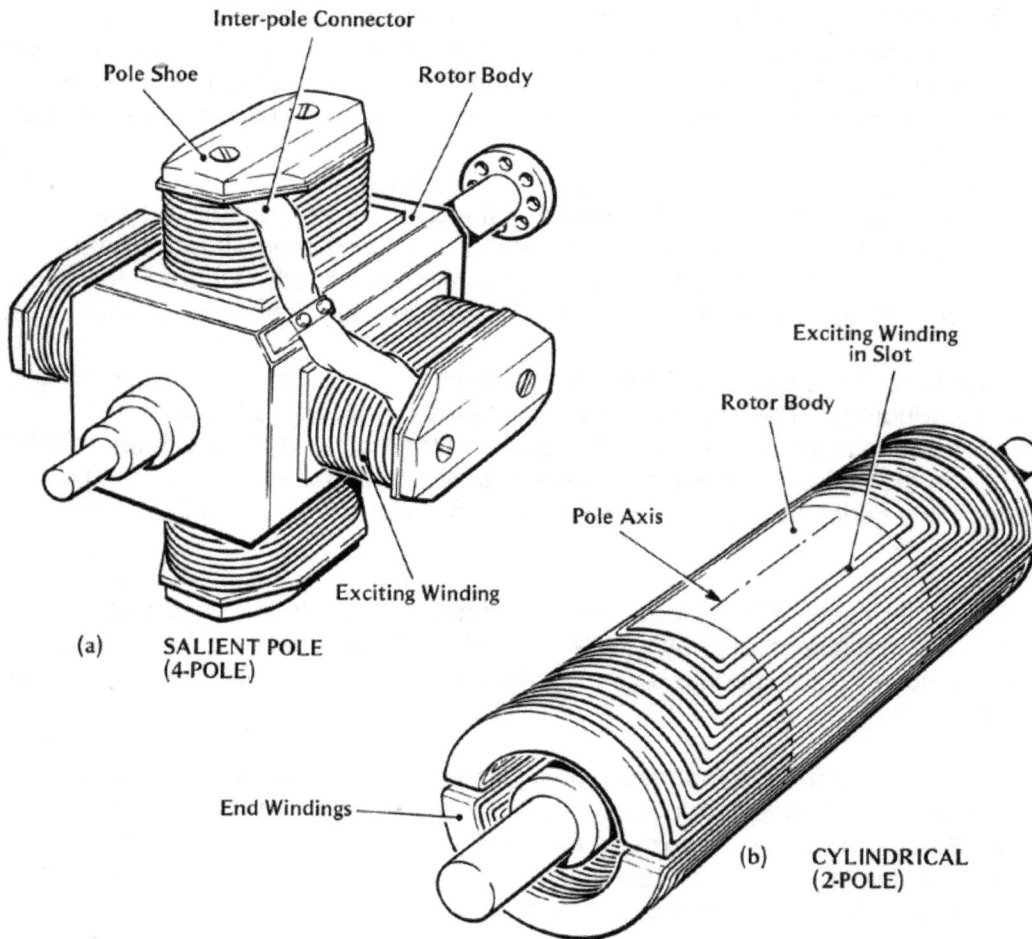

FIGURE 2.2
A.C. GENERATOR ROTORS

The cylindrical rotor (sometimes also called 'turbo type') is, as the name implies, completely cylindrical and has no projections. It is illustrated in Figure 2.2(b). The field windings are embedded and wedged into slots in the rotor surface in a similar way to the stator slots. (The overhang of the end windings has been exaggerated in the figure to make the construction clearer.) The rotor slots cover only part of the surface and are disposed either side of the poles, the whole field winding forming a spiral around each pole centre.

The air gap is uniform, and consequently the air gap flux due to the field winding is almost purely sinusoidal around the gap, being maximum opposite each pole centre. The smooth surface also results in low windage resistance.

Cylindrical rotors are very sound mechanically and are favoured for large, high-speed generators (3 000 or 3 600 rev/mm), where centrifugal forces on a salient-pole rotor would present severe problems. Consequently cylindrical rotors are common with 2-pole generators and are sometimes used with 4-pole types. They are never used with six poles or more, where the rotor construction would become far too difficult.

2.3   HARMONICS

Because the rotor's magnetic field does not have a pure sine-wave shape, the emf which it generates in the armature is not a pure sine-wave either; this is particularly so with a salient-pole rotor.

Although steps are taken in the stator slot arrangements to offset this effect as much as possible and to restore the emf to near sine-wave form, this is only partly achieved, and some impurity remains. It shows up as harmonic voltages in the emf waveform, and it is the odd-numbered harmonics which prevail. In a 3-phase system the third-harmonic voltages (at 150 or 180Hz) are all in phase with each other and cause equal currents through the loads which all return through the neutral conductor if there is one. These third-harmonic currents are sometimes confused with earth-leakage currents since they may, if sufficiently strong, actuate the earth-fault protection in the neutral line. They can be distinguished, however, because their frequency is three times nominal.

In a 3-wire system, where there is no neutral conductor as such, third-harmonic currents cannot flow through the load and generator windings (unless there is an earth fault) because there is otherwise no neutral return path. They can, however, circulate between paralleled generators through their common star-point earths, even without an earth fault, causing additional heating of the stators. The effect of these harmonics increases with the generator loading.

At one time steps used to be taken to restrict the earthing of paralleled generators to one machine only, in order to prevent such circulation. Modern generators, however, produce less harmonics than did the older ones, and they are now usually designed to absorb such circulating currents, so permitting multiple earthing - that is, the individual earthing of each generator.

Technical Specification for Synchronous A.C. Generators requires that the voltage waveform shall be in accordance with the international IEC Publication 60034-1.

This requires that the ratio of the net r.m.s. value of all the harmonic voltages present shall not exceed 5% of the fundamental voltage for small machines up to 1 000kVA, falling to 1.5% for machines greater than 5 000kVA. When calculating the net r.m.s. value, each separate harmonic voltage is 'weighted' by a factor (λ_n, for the nth harmonic) depending on its degree of interference with communications. λ_n varies from about 0.001 for a 100Hz to 1.4 for a 1 000Hz harmonic. Thus, if $E_2, E_3, E_4. \ldots$ are the 2nd, 3rd, 4th .... harmonic r.m.s. voltages and $\lambda_2, \lambda_3, \lambda_4 \ldots$ the weighting factors, then the net r.m.s. harmonic voltage is

$$\sqrt{(E_2\lambda_2)^2 + (E_3\lambda_3)^2 + (E_4\lambda_4)^2 + \cdots}.$$

and this must not exceed the stated percentages of the fundamental r.m.s. voltage.

A further cause of harmonic currents in a generator can be due to the load itself and has nothing to do with the voltage waveform. Loads which include rectifiers are a particularly severe source of harmonic currents. Where an offshore drilling plant is fed direct from the platform's main generating system (as distinct from having separate diesel-driven drilling generators), the SCR units which convert to d.c. for the drilling motors are a considerable source of a wide range of harmonic currents. Because of the action of rectifiers, this range consists entirely of the odd-numbered harmonics. And because the 3rd harmonic currents (and multiples of the 3rd, namely 6th, 9th, 12th ....) are all in phase with one another, those currents cannot flow in a 3-wire system with no neutral return.

Therefore rectifier equipments tend to draw, in addition to the fundamental, the following harmonic currents:

5th	typically	12% of the fundamental
7th	typically	10% of the fundamental
11th	typically	6% of the fundamental
13th	typically	5% of the fundamental
17th	typically	4% of the fundamental
19th etc	typically	3% of the fundamental

These harmonic currents are additional to the fundamental current and cause extra heating in the stator. Their net heating effect is obtained by adding their squares together and taking the square root of the sum (i.e. the net r.m.s. value).

This means that the stator has to carry more current than it would have if its load had been a normal one without harmonics - that is, it must have a higher kVA rating for the same kW active output. Therefore, if the rectified load forms a sizeable part of a generator's capacity, the machine must be under-run in terms of active output if its kVA rating is not to be exceeded, or a special design of generator with a low rated power factor (e.g. 0.6) must be used.

Harmonic currents due to the load, in flowing through the reactance of the generator, cause volt-drops at harmonic frequencies and therefore distortion of the terminal voltage waveform. Due to their higher frequencies passing through the same reactance, these harmonics produce distortion far greater than their magnitudes would suggest. If excessive, this distortion may cause trouble to other consumers, and for this reason Supply Authorities apply rigid limits on the amount of rectified load that may be put onto their systems.

2.4   INSULATION

Generator windings are insulated against the highest voltages to which they may be subjected, and the insulation must withstand a certain specified maximum temperature without deteriorating. There are many insulating materials with different - and often conflicting - properties. They are grouped into a number of classes, depending on the maximum temperature to which they will be exposed and on the insulating material used. The classification is as follows (in accordance with BS 2757 :1986, IEC 60085:1984)

Class	Typical Insulating Material	Ultimate Temperature
Y	Cotton, silk, paper, etc., unimpregnated	90°C
A	Impregnated cotton, silk, etc.; paper; enamel	105°C
E	Paper laminates; epoxies	120°C
B	Glass fibre, asbestos (unimpregnated); mica	130°C
F	Glass fibre, asbestos, epoxy impregnated	155°C
H	Glass fibre, asbestos, silicone impregnated	180°C
C	Mica, ceramics, glass, with inorganic bonding	>180°C

It should be noted that the classification letters do not follow an alphabetical sequence. This is because there were originally only three classes - 'A' 'B' and 'C'. Later intermediate classes were added, and it was decided not to disturb the original well-understood three. Most platform and shore-installed generators are Class 'B' or 'F'.

Certain of the higher-temperature insulation materials may be hygroscopic and therefore not always suitable in any particular environment, particularly where dampness is severe.

It should be particularly noted that the classification depends on the ultimate temperature to which the insulating material may be subjected, for it is this which determines whether or not it will suffer damage when heated. It does not depend on temperature rise alone: if for instance, the ambient temperature is 40°C, a Class 'B' material may be used if the designed temperature rise will not exceed 90°C, so making the ultimate maximum temperature 130°C. Designed temperature rises therefore must take into account the greatest expected ambient temperature in which the machine will operate.

2.5   COOLING

All generators used on platforms and in shore installations are air cooled. The air is circulated past the stator and rotor windings by a fan on the generator shaft. The warmed air itself may be discharged to atmosphere and not used again ('Circulating Air' or CA) or it may be water cooled in a separate cooler with a forced water circulation ('Circulating Air, Forced Water' or 'CAFW'); or in a radiator-type cooler ('Circulating Air, Natural Water or 'CANW'). There are usually alarms if the air or water temperatures exceed certain limits. All the largest gas-turbine generators are CAFW cooled.

The above letter coding was formerly in general use and is well understood. Recently however a new international coding system for cooling methods has been introduced for all rotating machines (BS 4999, Part 21: 1972) and is likely to be met with on modern drawings. It consists of the letters 'IC' followed by two digits. The meanings of these digits are given below for typical platform or shore-installed generators:

First Digit		Second Digit	
0	Free circulation	0	Free convection
1	Inlet duct ventilated	1	Self-circulation
2	Outlet duct ventilated	2	Integral component mounted on separate shaft
3	Inlet and outlet duct ventilated		
4	Frame surface cooled	3	Dependent component mounted on the machine
5	Integral heat exchanger (using surrounding medium)	5	Integral independent component
6	Machine-mounted heat exchanger (using surrounding medium)	6	Independent component mounted on the machine
7	Integral heat exchanger (not using surrounding medium)	7	Independent and separate device or coolant system pressure
8	Machine-mounted heat exchanger (not using surrounding medium)	8	Relative displacement
9	Separately mounted heat exchanger		

Where it is desired to specify the nature of a coolant the following letter code is used in conjunction with the cooling code:

Gases	air	A
	hydrogen	H
	nitrogen	N
	carbon dioxide	C
	helium	L
Liquids	water	W
	oil	U

When nothing but air is used, the letter 'A' may be omitted.

Thus a generator cooled by air with an internal fan and with an air/water heat exchanger using pressurised water from the platform system would be classified IC87, or IC8A/7W, instead of the former CAFW.

The larger generators also have thermocouple-type temperature detectors embedded at various points in the windings. If any one of them exceeds a certain temperature, an alarm is given on the control panel. The panel also has facilities for the operator to scan all the detectors in turn and to read off the actual temperatures.

2.6   BEHAVIOUR UNDER FAULT

Figure 2.3 shows the general construction of a rotating field a.c. generator. For simplicity a 2-pole machine has been chosen. The rotating poles are shown with their exciting field windings and damper windings, and the flux paths are drawn in blue. Outside the field system is the stator carrying the stationary armature winding in slots. The condition depicted is the normal one, with the generator delivering steady load current and with its normal excitation flux.

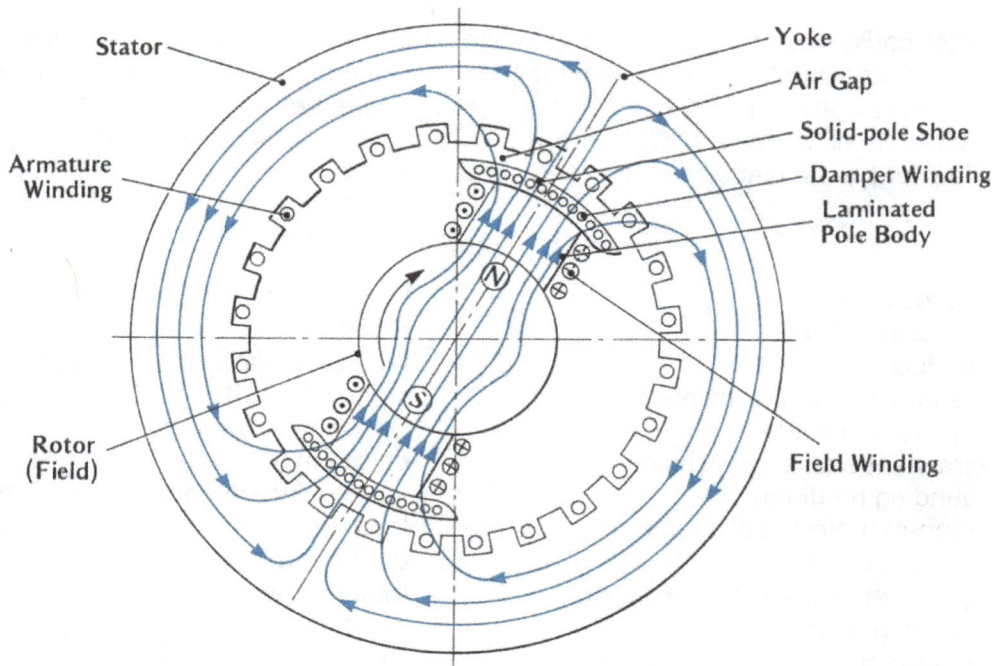

FIGURE 2.3
A.C. GENERATOR - MAGNETIC FLUX PATH

The complete magnetic path is from one field pole (N), through its damper winding, through one air gap, through the armature (stator) coils, through the yoke, back through a second air gap, back through the damper windings and the opposite pole, and finally through the rotor body to the original field pole. The route is indicated in blue in Figure 2.3.

Consider now that a sudden external short circuit is applied. The pattern of current and the flux disturbance which then follows is quite complicated. The sequence of the process is shown in Figure 2.4.

The first step is that the armature current suddenly rises, limited only by the reactance due to the stator iron and air-gap magnetic path. The new air-gap flux due to the sudden increase of armature current (shown in red) tries to penetrate the field poles but is prevented from doing so by the eddy currents set up in the solid-pole shoes, aided by damper windings embedded in the face of the shoe if fitted (see Figure 2.4(a)). The eddy currents induced oppose the flux change (Lenz' Law), and the short-circuit flux from the armature (stator) is deflected along the air gap. Its return path now has a very long air gap, so the reactance due to the magnetic path is very much lower than before. This is called the 'subtransient stage'; the reduced reactance at the beginning of the stage is called the 'subtransient reactance' (symbol X''_d), and the increased current which it allows to pass is called the 'subtransient current', as shown in Figure 2.4(d). It persists for a few cycles. It is during this period that the system protection normally operates to trip the generator breaker, so the breaker must be capable of breaking the highest subtransient current.

After the eddy currents and damper currents have subsided owing to resistance in the pole shoe and dampers, the armature short-circuit flux will have penetrated the outside of the main pole body (usually laminated) - see Figure 2.4(b). Here again it meets opposition, because the changing flux in the pole body induces an emf in the pole winding which causes a current to flow in the closed pole winding/exciter loop to oppose the change (Lenz' Law again). The direction of that opposing current is the same as that of the main exciting current, so the effect of a short circuit is initially to cause a sudden increase in the exciter and main field current. Once again this opposing current is slowly damped out by the resistance of the exciting loop, and the short-circuit flux from the armature gradually penetrates the main poles, so increasing the reactance due to the magnetic path and steadily reducing the short-circuit current. This is called the 'transient stage', and it lasts somewhat longer than the subtransient. The increased reactance at the beginning of the stage (i.e. at the end of the subtransient stage) is called the 'transient reactance' (symbol X'_d), and the reduced short-circuit current which it allows to pass is called the 'transient current'. It is lower and lasts longer than the subtransient, as shown in Figure 2.4(d).

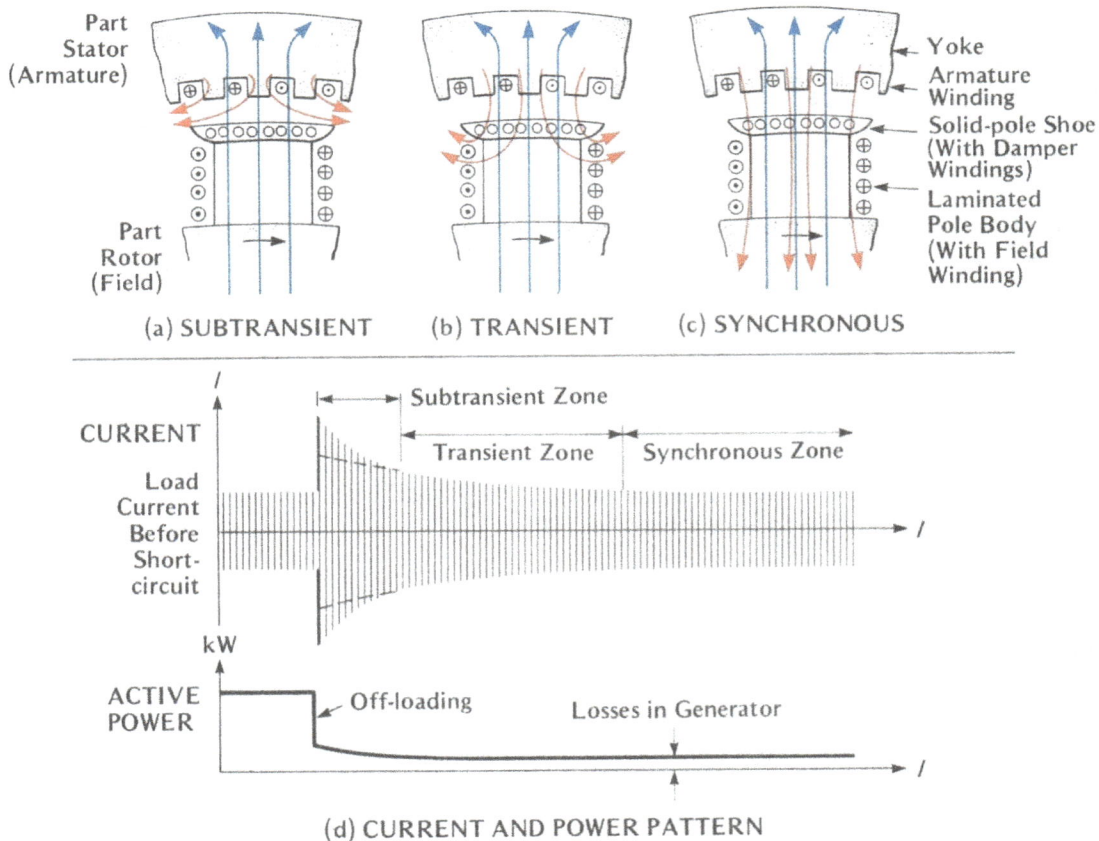

(d) CURRENT AND POWER PATTERN

FIGURE 2.4
A.C. GENERATOR ON SHORT CIRCUIT

Finally, when the armature short-circuit flux has fully penetrated all the pole bodies (see Figure 2.4(c)), all the iron of both stator and rotor is in the magnetic circuit, and the reactance due to its path is at its greatest; the short-circuit current then settles down to its steady value. This is the 'synchronous stage' and may continue indefinitely if allowed to do so. The reactance is called the 'synchronous reactance' (symbol X_d), and the steady-state current the 'synchronous (short-circuit) current', as seen on the right-hand side of Figure 2.4(d).

When this steady state has been reached, the armature short-circuit flux (red), in penetrating the main poles, has partly demagnetised them, since it is in opposition to the main exciting flux (blue). This phenomenon of demagnetisation of the field by the load current is called 'armature reaction'. By weakening the air-gap flux it reduces the generated emf and so reduces the current still further. The steady short-circuit synchronous current may well then be even less than the machine's normal full-load current. This, of course, supposes that the excitation is constant and that no automatic voltage regulation is applied to compensate for the loss of voltage.

In practice Automatic Voltage Regulators (AVRs) are always fitted, but in general they do not act quickly enough to affect the short-lived subtransient stage. As this period gives rise to the fiercest short-circuit currents which the switchgear has to break, the effect of the AVR is not taken into account when calculating fault currents for switchgear. (See also Part 10 'Electrical Protection'.)

To sum up: when a short-circuit is suddenly applied to a generator, the ensuing current goes through three definable stages - subtransient, transient and synchronous - so long as it is allowed to continue. At the beginning of each stage the current is determined by one of three reactances, as follows:

Subtransient reactance X''_d (Current typically 6 times full load)
Transient reactance X'_d (Current typically 3 times full load)
Synchronous reactance X_d (Current typically two-thirds full load)

The above currents assume fixed excitation and no AVR action.

The following points on the three types of reactance should be noted:

Subtransient Reactance determines the initial current peaks following a disturbance and, in the case of a sudden fault, is of importance for selecting the capacity ratings of the associated circuit-breakers.

Transient Reactance covers the behaviour of a generator in the period 0.1 to 3.0 seconds after a disturbance. This generally corresponds to the speed of changes in a system and is usually employed in studies of transient stability.

Synchronous Reactance is a measure of the steady-state stability of the set. The smaller its value, the more stable the machine.

One effect of the heavy short-circuit current on the generator itself is that, if it persisted for more than a few seconds, winding temperatures would rise to a point where insulation could be permanently damaged or may even break down. Automatic protection is therefore provided (see Part 10 'Electrical Protection') to clear the fault as quickly as possible after its onset. Nevertheless, most large generators are designed to carry their short-circuit current for three seconds.

A further, equally important, effect of short-circuit currents is the intense mechanical stresses which they produce by electromagnetic reaction between the current-carrying conductors. These occur from the very first cycle of a fault, and no protection is quick enough to prevent them. The most severe forces occur in the overhang at the ends of the windings. All generators must therefore be constructed to withstand these forces, and the overhangs are specially braced. If movement does occur, this is the most likely place to find it.

2.7   DIRECT AXIS AND QUADRATURE AXIS REACTANCES

All that has been said so far about the generator reactances has assumed that the air gap is uniform. This in turn assumes a generator with a cylindrical rotor.

In practice all offshore, and many onshore, generators are of the salient-pole type which do not have a uniform air gap. This adds a complication which leads to the idea of two different types of reactance: one where the fluxes due to the stator current are opposite the field pole centres - called the 'direct axis' - and the other where the fluxes due to the stator current are opposite the centre of the gap between two adjacent poles - called the 'quadrature axis'. The direct-axis reactances (subtransient, transient and synchronous) are termed X''_d, X'_d and X_d, whereas the corresponding quadrature reactances are X''_q, X'_q and X_q. The two types are combined mathematically in various machine calculations.

In cylindrical rotor machines the quadrature-axis reactances are practically the same as the direct-axis and need not be taken into account separately. This is not so with salient-pole machines; in their case both types of reactance should be used when making a rigorous calculation. However, in practice the error introduced into the calculation of short-circuit currents (see Part 10 'Electrical Protection') by using only direct-axis values is not significant, and indeed it errs on the safe side. For this reason only direct-axis quantities have been used in the previous description of behaviour under fault.

2.8   EXCITATION AND VOLTAGE CONTROL

The different forms of excitation and automatic voltage control are dealt with in Chapter 3.

2.9   NEUTRAL EARTH ING RESISTOR

The star-points of all high-voltage generators on platforms are earthed through a current-limiting 'neutral earthing resistor' (NER). Its purpose is to limit the fault current flowing through the generator if an earth fault develops anywhere on the system (see Part 10 'Electrical Protection').

The NER is separately mounted near the generator and usually consists of a frame containing a heavy grid-type resistance element capable of carrying a large current for a short time. This short-time rating is possible because any heavy fault current will be quickly cleared by the earth-fault protection.

Neutral earthing resistors are therefore given a maximum current rating for a maximum time - for example, '200A for 30 s'. They may also have a continuous current rating - for example '25A cont.' - to cover small earth-leakage and harmonic currents which are not large enough to operate the protection. Their ohmic value goes down to about 10 ohms for the largest offshore generators.

The NER unit sometimes contains also a current transformer to measure the presence of any earth-fault current in order to initiate the protection.

Low-voltage generators are usually solidly earthed without a neutral earthing resistor.

2.10   INSULATED BEARINGS

Bearings of a large machine are often insulated to prevent stray currents from circulating through them. Such currents can arise from emfs being generated in the rotor shaft due to stray magnetic fields. Under fault conditions these stray fields can be very large. Figure 2.5(a) shows how such currents may flow through the bearings.

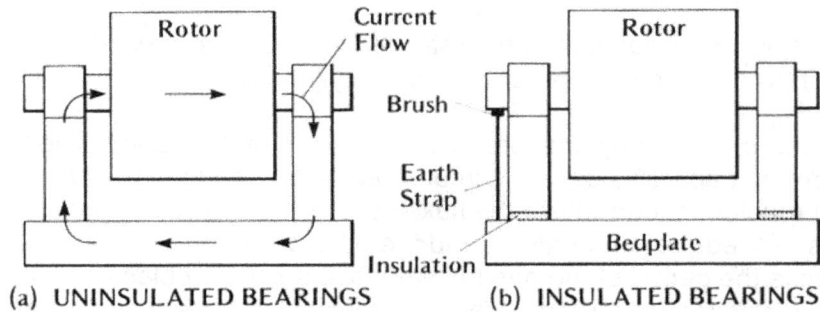

FIGURE 2.5
INSULATION OF BEARINGS

These currents, if allowed to flow, would arc across the bearing surface and cause small craters which would eventually destroy the bearings. Figure 2.5 shows pedestal sleeve bearings, but the same principles apply to ball and roller bearings.

The current path of Figure 2.5(a) can be broken by insulating one or both bearings: the insulation may be at the bearing housing or, more commonly, beneath the pedestal where it seats on the bedplate stool as shown in Figure 2.5(b). The insulation of only one bearing is more usual, but insulating both allows the insulation to be checked.

For reasons of safety the shaft must be at earth potential. Consequently on most machines one bearing (the uninsulated end if only one is insulated) is fitted with an earth strap, one end of which terminates in a brush running on the dry shaft. If the generator is of the 'overhung' type with only one outboard bearing, such as with certain diesel-generator sets, this bearing is insulated and the earthing of the rotor shaft is made through the engine and coupling.

The insulation of the pedestal is carried out by a shim of insulating material between the base of the pedestal and its stool. The holding-down bolts are bushed with insulating material. Sometimes two insulating shims are used with a thin metal sheet between them. This enables the insulation resistance of each part to be measured separately, since the shaft and bed plate are normally both at earth potential.

It is important that, where a bearing pedestal is insulated, no waste material or tools should be allowed to lean against it, as they would short-circuit the insulation.

CHAPTER 3 GENERATOR EXCITATION AND VOLTAGE CONTROL

3.1   GENERAL

The excitation of a generator's field system has already been mentioned in Chapter 2, as it is not possible to describe a.c. generators without referring to their field system and excitation. This chapter discusses the three practical methods of field excitation which may be encountered.

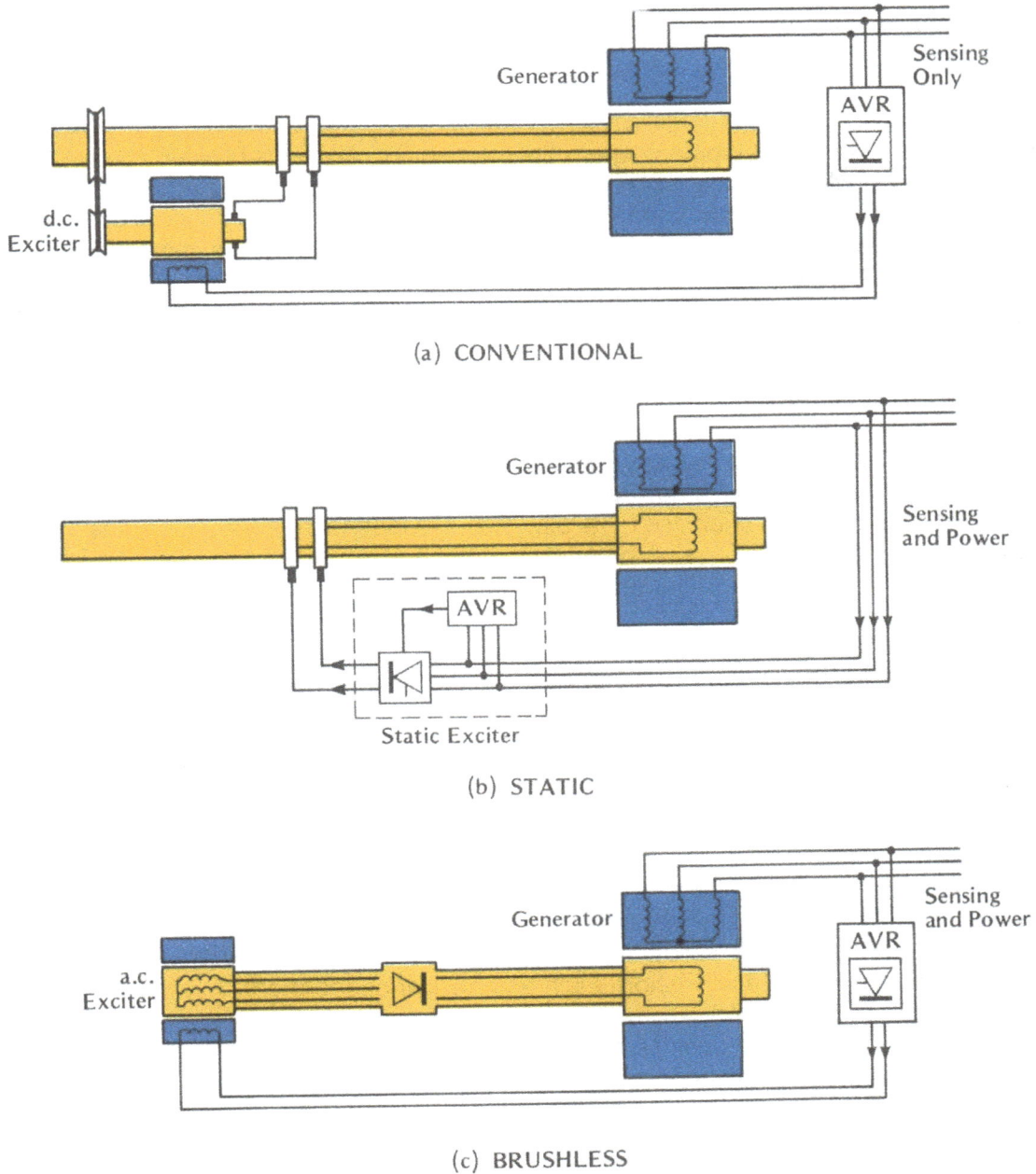

(a) CONVENTIONAL

(b) STATIC

(c) BRUSHLESS

**FIGURE 3.1
A.C. GENERATOR EXCITATION (1)**

3.2 CONVENTIONIAL EXCITATION

Figure 3.1(a) shows the 'conventional' method, where a driven d.c. exciter (in this case belt-driven) feeds its d.c. output through sliprings to the main generator field.

The output voltage is sensed by an automatic voltage regulator (AVR), which regulates the exciter's field so that the exciter output holds the main field at whatever level is necessary to maintain the generator output voltage constant. AVRs are discussed later in this chapter. It will be seen that the control of voltage is a closed loop, and, like any other closed loop servo mechanism, it is subject to certain errors.

3.3 STATIC EXCITATION

Figure 3.1(b) shows a development where the rotating d.c. exciter is replaced by a static electronic exciter, which usually incorporates the AVR. Voltage sensing and excitation power are derived from the main generator output; excitation current is controlled by the AVR, rectified and fed into the main field through sliprings, just as in the 'conventional' case. This is called the 'static exciter' method, and it should be noted that it still requires brushes and sliprings. It is not found on platforms but is widely used onshore, although not to any great extent in oil installations.

3.4 BRUSHLESS EXCITATION (GENERAL CASE)

A further significant development is shown in Figure 3.1(c). Here the shaft-driven rotating exciter has been restored, but it now takes the form of an a.c. generator of the fixed-field type mounted on the main shaft itself. Its a.c. output is taken through connections inside the shaft, through a diode bridge which rotates with the shaft, to the main rotating field of the generator. The field is thus excited by d.c. without the need for brushes and sliprings. It will be seen that this exciter cannot be belt-driven; it must be integral with the main shaft.

As with static excitation, voltage sensing and excitation power are derived from the main generator output. Excitation current is controlled by the AVR, rectified and fed into the fixed field of the a.c. exciter. The a.c. output of the exciter follows the AVR signal, and its output current is rectified by the diodes which rotate with the shaft; the d.c. output from them is in turn passed to the generator's main field. The field current thus follows the AVR signal almost exactly.

It will be seen that the only link between the fixed and moving parts is the magnetic one between the exciter field and its rotating armature: no sliprings and brushes are needed. The method is for this reason called 'brushless excitation', and it will be found, in one form or another, on all platform and onshore main and auxiliary generators.

The principal advantage of brushless excitation over the other two types is that the absence of brushgear and sliprings greatly eases the maintenance problem.

3.5 BEHAVIOUR UNDER SHORT CIRCUIT

In the conventional case (Figure 3.1(a)) excitation power is derived from a separate d.c. generator which is not affected by the voltage on the main generator's output lines. However, with both static excitation and the brushless excitation described above (Figures 3.1(b) and (c)) excitation power (as well as sensing) is derived from the output of the generator itself - true 'shunt excitation'.

Under normal conditions this is quite satisfactory, but under short-circuit conditions the generator's output voltage will drop heavily - it might even vanish. Under this low-voltage output situation the AVR will try to force up the excitation, but, just at the moment it wishes to do so, it has no power available. Under these conditions a collapse of system voltage is possible.

To overcome this a method is employed which makes use of the short-circuit currents themselves to provide the missing excitation.

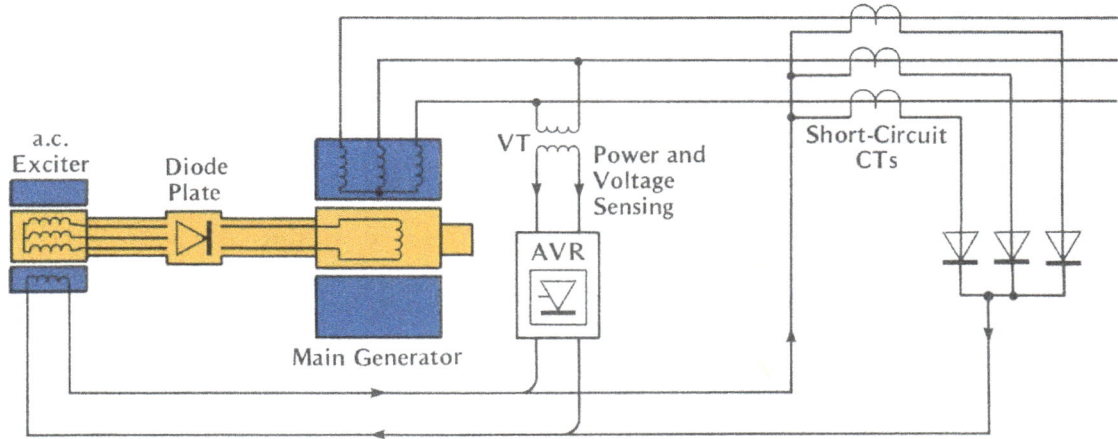

(a) BRUSHLESS WITHOUT PILOT EXCITER

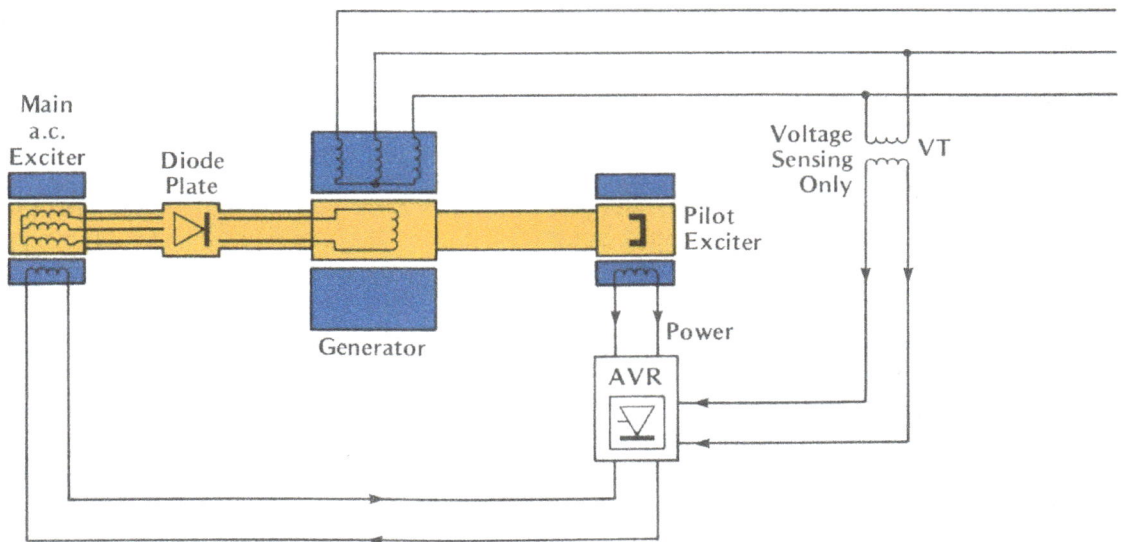

(b) BRUSHLESS WITHOUT PILOT EXCITER

FIGURE 3.2
A.C. GENERATOR EXCITATION (2)

3.6   BRUSHLESS EXCITATION (WITHOUT PILOT EXCITER)

Three heavy current transformers are arranged in the generator output lines as shown in Figure 3.2(a). Their secondary outputs are rectified and passed to the main exciter's field either in parallel with the normal excitation (as shown) or sometimes to a separate field winding in the exciter. Although they take the form of current transformers, these units, when used in this application, are referred to as 'short-circuit CTs'.

Under short-circuit conditions when the generator output voltage is very low, the short-circuit CTs pick up the heavy short-circuit currents and, after they have been rectified, use them to boost the main exciter field, and so the main field. This serves to maintain the generator output voltage under short-circuit conditions - a necessary requirement in network operation so that protection may operate reliably.

Short-circuit CTs are used generally with medium-sized generators with either static or brushless excitation where no 'pilot exciter' is fitted (see below) and where excitation power is drawn from the generator's output. This applies to most basic services generators on platforms and to some main sets.

3.7   BRUSHLESS EXCITATION (WITH PILOT EXCITER)

With large brushless generators a different method is used. Instead of drawing excitation power from the generator output, the AVR has only a voltage-sensing connection. The arrangement is shown in Figure 3.2(b).

The exciter's field is powered independently from a separate high-frequency inductor-type generator called a 'sub-exciter' or 'pilot exciter'. It has permanent magnets as rotating field and is driven by the main shaft. It also provides operating power to the AVR itself. Only the voltage-sensing leads to the AVR are taken from the main generator output. The AVR regulates and rectifies the power from the pilot exciter to the main exciter field. This in turn regulates the a.c. exciter output, and thence the d.c. rectified input to the main field through the diodes, to hold the generator output voltage constant.

The pilot exciter is mounted on the main shaft, usually immediately next to the main exciter (not exactly as in Figure 3.2(b) which is schematic only). It is usually arranged in a single enclosure with the main exciter and the diode plates. Figure 3.3 shows this arrangement.

As in the conventional case, the excitation of the generator is now independent of the generator's output voltage and so is maintained even under short-circuit conditions and without the use of short-circuit CTs. This is the arrangement on almost all platform main generators.

3.8   THE DIODE BRIDGE

In Figures 3.1(c) and 3.2(a) and (b) the diodes are shown for clarity as inside the shaft between the exciter and the main generator. The exciter output is 3-phase, and the diodes are in fact a 3-phase full-wave bridge, requiring six diode elements. Clearly they cannot be buried in the middle of the shaft, and in practice they are mounted on a rotating plate on the extreme end of the shaft at the exciter end, as shown in Figure 3.3 in green. This makes them easily accessible for inspection, testing or replacement.

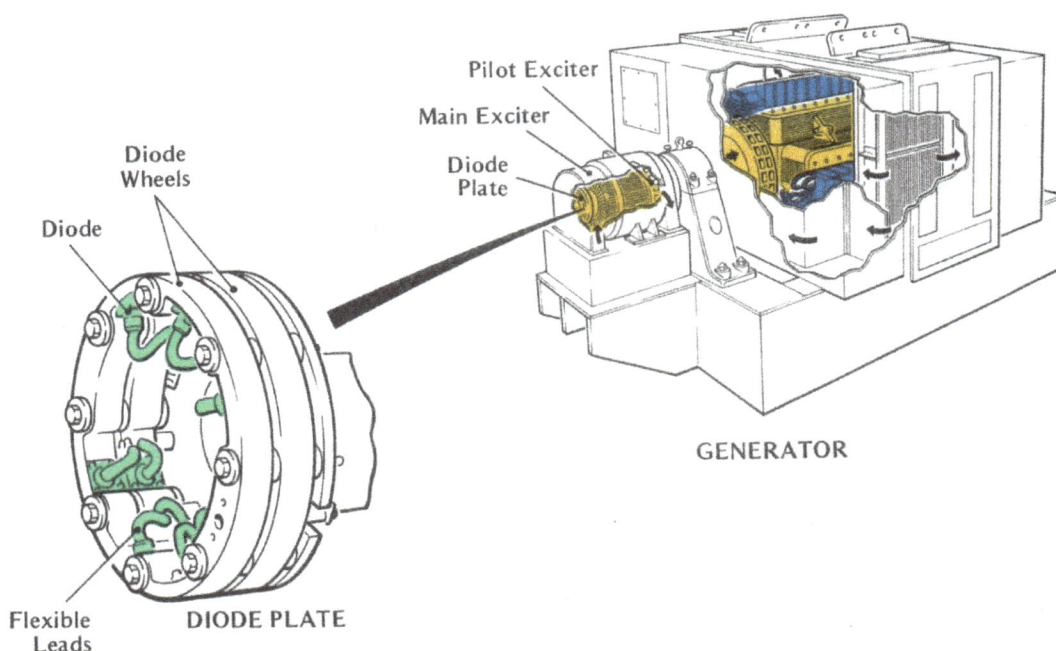

FIGURE 3.3
GENERATOR AND DIODE PLATE

A point on the use of diodes should be noted. If one of the six should fail, either by open-or short-circuiting, harmonic currents flow in the main field circuit. These harmonics are reflected into the field circuit of the main exciter and are detected by a 'diode failure' relay tuned to respond to the principal harmonic frequency; the alarm (or trip) signal from this relay is time-delayed by 10 or 15 seconds to prevent false operation.

A diode failure would have no discernible effect, from the consumer's point of view, on the generator's output voltage. The reduced d.c. output from the diode bridge with one diode faulty would lower the main field's d.c. current slightly, and with it the main generator's output voltage. This would be immediately detected by the AVR, which would increase the excitation until the voltage was restored, and the consumer would not be aware of it. However, the remaining healthy diodes might then be somewhat overloaded, and the situation should be corrected.

With an open-circuited diode the condition would not be serious. The increase of exciter field current would be about 15%, which can be provided by the AVR and carried by the exciter field for some time. Nevertheless, it should be corrected as soon as possible. A short-circuited diode however is more severe, calling for a much greater increase of exciter field current. The AVR and exciter could well be damaged if this condition were allowed to persist.

It is usual for the diode failure relay to give an alarm only, not to trip the breaker and shut down the set. When this alarm appears, generation should be transferred to another machine as soon as opportunity offers; the faulty set should then be stopped and the failed diode replaced. On some sets, however, the diode failure relay actually trips the set.

CAUTION   WHEN MEGGER TESTING A GENERATOR FIELD SYSTEM, ALL DIODES MUST FIRST BE DISCONNECTED OR SHORT-CIRCUITED TO PREVENT THE MEGGER VOLTAGE BEING APPLIED ACROSS THEM AND BREAKING THEM DOWN.

3.9   REGULATION RESPONSE TIME

A further important point resulting from the use of diodes should be noted. When output voltage falls, it is sensed by the AVR and the exciter field is increased. The increased exciter output voltage is passed by the diodes to appear as an increased d.c. voltage across the main field. This causes an increase of main field current at a rate depending on the R/L ratio of the whole field/exciter loop. Therefore the increase is not instantaneous but, because the exciter resistance is appreciable, R is large enough to allow a reasonably quick response.

In order to improve the response time when there is a drop of output voltage, the AVR is made to give the main exciter field a considerable boost, causing a big jump in its a.c. output voltage and so a large rise in the d.c. voltage applied through the diodes to the main field. It helps to overcome the field's natural sluggishness and to build it up more quickly. This is known as 'field forcing'. When the field has reached its new value and the output a.c. voltage is restored, the AVR removes the excess forcing current from the exciter's field.

However, if there is an output voltage rise (for example due to the throwing-off of a large load) it is sensed by the AVR and the exciter field is reduced. The reduced exciter output voltage is now lower than that of the main field, and it is blocked by the diodes. The main field current, flowing in the highly inductive field system, 'flywheels' round the closed circuit formed by the field and the diodes. It decays slowly because it is damped only by the comparatively small resistance of the main field itself (small R/L ratio, and therefore longer time constant L/R).

Thus in a brushless system, whereas the response to a drop in output voltage is reasonably quick, reaction to a rise is appreciably slower. This is particularly significant after a short-circuit has been cleared. During the period of the fault the voltage will have dropped and the AVR will have forced up the excitation, probably to its limit. When the fault is cleared, this overexcitation shows as a large overvoltage on the whole system, which is comparatively slow to recover. This could involve a risk of burn-out of lamps or delicate apparatus.

3.10   AUTOMATIC VOLTAGE REGULATORS (AVR)

AVRs are of many different makes, and various types are found on platforms and onshore installations.

All, however, have certain features in common when used with brushless generators. They are nowadays entirely electronic; they take their operating power from either the main output or the shaft-driven high-frequency sub-exciter (typically at 400Hz), but they sense the voltage to be controlled from the output side of the generator before the circuit-breaker terminals. In high-voltage generators this sensing circuit is taken through a measuring voltage transformer of at least Class 0.5 accuracy (see Chapter 9 of Part 3 'Electrical Power Distribution').

The detailed electronic circuits are not discussed here, but power from the main output or the high-frequency sub-exciter is rectified through thyristors, which are controlled by the voltage-sensing circuits to provide the correct d.c. current to the field of the main a.c. exciter.

3.11   AVR SET-POINT

Like any closed-loop servo, an automatic voltage regulating system holds the voltage constant, within stated errors, at whatever level it has been set. This level is referred to as the 'set-point'.

In an electronic AVR the set-point is adjusted by a variable resistance, or rheostat, in the appropriate part of the circuit. On some generators this rheostat is outside the AVR proper and is mounted on the adjacent generator control panel for manual control; it is usually marked 'Raise Volts/Lower Volts'. On other makes of generator it is arranged for remote control from some distant panel. In such a case the rheostat is motor-driven, the motor being controlled forward or backward by a 2-way-and-off spring-loaded switch marked as above.

When used with a single generator the AVR set-point control does indeed regulate the machine's voltage output, but when used on a generator running in parallel with others, the prime function of the AVR control is not so much to regulate voltage but to adjust the sharing of reactive load between the generators, despite the marking of the control knob or switch. It does, however, have some effect on voltage level, but this is only secondary. The action is fully explained in Part 6 'System Control'.

3.12   A.C. GENERATOR VOLTAGE REGULATION

When a load is applied to the terminals of a generator previously running at no load and without AVR control, the terminal voltage will drop by an amount which depends on the nature of the load. This drop of voltage is called the 'regulation' of the generator at that load. It is usually quoted at full rated load - that is, at the full-load rated current and rated power factor and is expressed as a percentage of the no-load or system voltage. Thus, if V_0 is the no-load voltage and V the generator terminal voltage at full rated load and power factor and with the excitation unaltered, then

$$\frac{V_0 - V}{V_0} \times 100\%$$

is the percentage full-load regulation.

In practice of course the reduced voltage V would be immediately detected by the AVR, which would increase the excitation until the terminal voltage was restored to the system value V_0.

The determination of the generator's internal voltage drops and their effect on regulation is covered fully in Chapter 4.

CHAPTER 4 GENERATOR INTERNAL IMPEDANCE AND REGULATION

4.1   GENERAL

This chapter examines what goes on **inside** a generator. First we will examine the normal 'ohmic' voltage drop due to load current in the main windings. Also we shall consider how the emf which is generated in the armature by the rotating field system undergoes certain changes. They cause it to be less than the emf due to the pure field system and so reduce still further the voltage that appears at the generator terminals.

Take first the simpler case of a d.c. generator.

4.2   D.C. GENERATOR

In a d.c. generator the current which flows from the machine to the external load and back again also flows through the internal resistance of the generator, as shown in Figure 4.1.

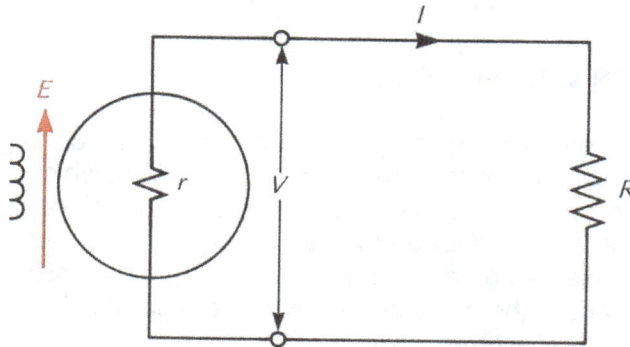

FIGURE 4.1
D.C. GENERATOR - INTERNAL RESISTANCE

If the excitation causes an emf E in the generator armature, the open-circuit voltage V_0 will be equal to E. But if a load current I flows, it flows also through the generator's internal resistance r causing an internal voltage drop equal to $I.r$. When the load is applied therefore the terminal voltage drops from V_0 to a lower value V, where

$$E = V_0 - I.r \qquad \qquad ....(i)$$

The reduction of terminal voltage $(V_0 - V)$ is the 'regulation' of the generator under the stated load conditions. Its value is given by the drop of voltmeter reading when the load is applied (assuming no change in the exciting field).

Since $V_0 = E$, equation (i) can also be written

$$V = E - I.r$$

or $\qquad\qquad E = V + I.r \qquad\qquad ....(ii)$

4.3   A.C. GENERATOR

A similar situation occurs with a.c. generators, but the effect is more complicated, as shown in Figure 4.2.

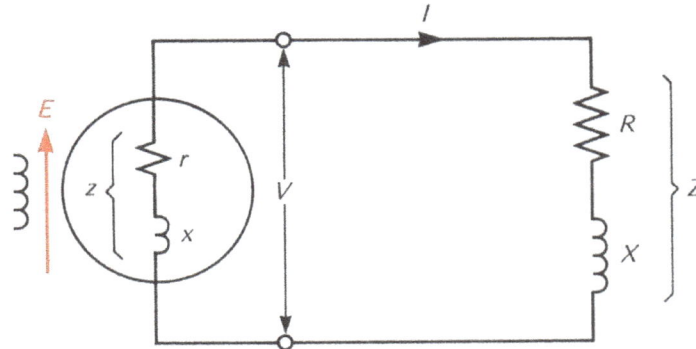

FIGURE 4.2
A.C. GENERATOR - INTERNAL IMPEDANCE

In this case the 'load' may be an impedance which draws both active and reactive current, and the internal resistance of the d.c. generator of Figure 4.1 is replaced by a generator internal impedance z, which consists of both resistance r and reactance x.

The same principles apply as for the d.c. generator - namely that the terminal voltage V_0 is reduced to a lower value V by the internal 'voltage drop' caused by the load current flowing through the internal impedance z. The new terminal voltage is then given by:

$$\overline{V} = \overline{V_0} - \overline{I.z_1}$$

but these are now all vector quantities.

The numerical difference between no-load terminal voltage and the terminal voltage on load, namely $V_0 - V$, is still the regulation of the generator under the stated load condition. Its value is given by the drop of voltmeter reading when the load is applied (assuming no change in the excitation).

When considering what goes on inside a generator it would be logical to take first the generated emf E; then to examine the voltage drop due to the load current flowing through the generator's internal impedance and finally to arrive at the terminal voltage V. However, because the emf of a machine cannot actually be measured (except on open circuit), it would be more convenient to **start** with the terminal voltage V - for that alone is what the external load sees - and then to work backwards through the internal voltage drop, adding it to V, to arrive at the emf E needed to produce the required terminal voltage V. Expressed as an equation (similar to equation (ii)):

$$\overline{E} = \overline{V} + \overline{I.z} \text{ (vectorially)} \qquad \text{....(iii)}$$

This gives a measure of the excitation needed to produce a given voltage at the terminals for any given load current, and, in particular, to find the maximum excitation required from the AVR and exciter to maintain full rated terminal voltage at maximum rated load.

It is known that, if an a.c. voltage V is applied across a resistance R, a current equal to V/R (by Ohm's Law) will flow through it and will be in phase with V. Similarly, if a voltage V is applied across a pure reactance X, a current equal to V/X (by Ohm's Law for a.c.) will flow through it and will lag $90°$ on V.

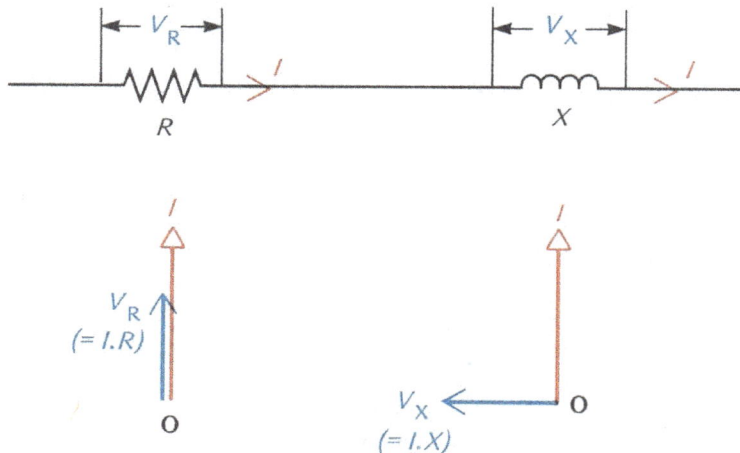

FIGURE 4.3
COMMON CURRENT IN RESISTOR AND REACTOR

If, instead of starting with a common voltage V, a common current I had been flowing through both a resistance R and a pure reactance X, as shown in Figure 4.3, then the voltage V_R equal to $I.R$ and developed across R, will be in phase with I. The voltage V_X, equal to $I.X$ and developed across X, will lead the current I, since the current must lag on the voltage. This is shown in Figure 4.3, where the voltage vectors are in blue, and the currents in red. The voltages V_R and V_X are respectively the voltage drops in R and X due to the common current I flowing through them.

Consider now the voltage drops due to a load current I flowing through the **internal** resistance r and reactance x of a generator armature.

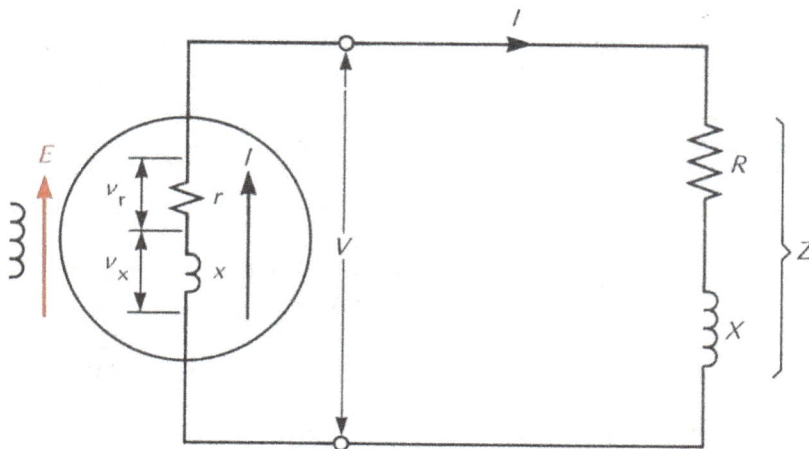

FIGURE 4.4
INTERNAL VOLTAGE DROPS

In Figure 4.4 a generator is supplying an external load Z (resistance R and reactance X) with a current I. The generator is being excited to an emf E, and it has an internal resistance r and reactance x. The terminal voltage while load current is flowing is V.

The external impedance Z causes the load current I to lag by an angle φ on the terminal voltage, cos φ being the power factor of the load. The load current I is then flowing through both r and x (x is assumed typically to be larger than r).

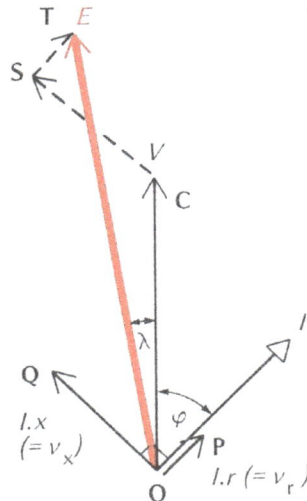

FIGURE 4.5
TERMINAL VOLTAGE AND EMF IN A GENERATOR

As already explained, the voltage drop v_r due to current I flowing through r is $I.r$ and is in phase with I; it is represented by the vector $\overline{OP}$. The voltage drop v_x due to current I flowing through x is $I.x$ and leads I by 90°. It is represented by the vector $\overline{OQ}$.

The total voltage drop $I.z$ in the generator is the vector sum of $\overline{v_x}$ and $\overline{v_r}$ namely $(\overline{I.x} + \overline{I.r}) = \overline{OQ} + \overline{OP}$.

Since $\overline{E} = \overline{V} + \overline{I.z}$ (equation (iii)),

$$\overline{E} = \overline{V} + \overline{OQ} + \overline{OP}$$
$$= \overline{V} + \overline{CS} + \overline{ST} \text{ (Since CS = OQ and ST = OP)}$$

and is so shown in Figure 4.5, the resultant vector $\overline{E}$ being in red.

This shows that, to achieve a terminal voltage V on load, the emf must not only be greater than V numerically but must also in general lead on V in phase. This angle of lead is called the 'power angle' or 'load angle' for that particular load and is given the symbol λ. It represents the advance in the rotor's pole position as compared with its no-load position. The power angle λ should not be confused with the load's phase angle φ.

It should be noted that r and x are the resistance and reactance (in ohms) per phase in a 3-phase system. This assumes a star connection, and V in the formulae is the **phase** voltage (line voltage divided by √3).

Figure 4.5 gives the relationship between the emf E and the terminal voltage V for a given load current. It was evolved by starting with V and evaluating the corresponding E, but equally it shows that, given an emf E, at the stated load current the terminal voltage will be V - that is, it can be used either way.

For example, if the terminal voltage has to be maintained at a steady value V at all loads, then the emf (and so the excitation) can be determined for any stated load current, and in particular for the maximum rated load. From this the size of the excitation system can be deduced.

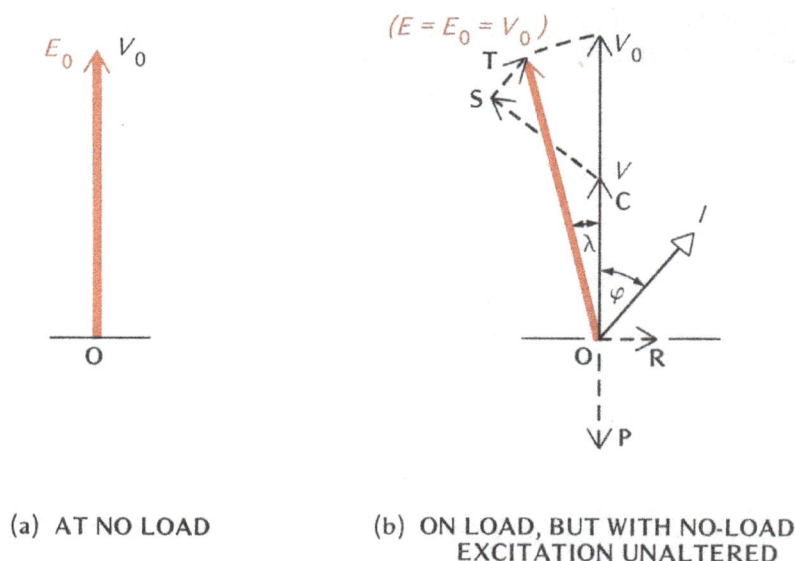

(a) AT NO LOAD

(b) ON LOAD, BUT WITH NO-LOAD
EXCITATION UNALTERED

FIGURE 4.6
TERMINAL VOLTAGE AND EMF IN A GENERATOR
AT CONSTANT NO-LOAD EXCITATION

A special case of this is at no load, where there is no load current I and therefore no internal voltage drops Ir and Ix. The vector E then lies on top of V and is equal to it; that is to say, at no load the terminal voltage is identical with the emf, as shown in Figure 4.6(a). If V_0 is the terminal voltage at no load and E_0 the corresponding emf at no load, then $V_0 = E_0$. This is the only situation where an emf can be actually measured.

As before, V and E are the **phase** voltage and **phase** emf.

4.4   REGULATION

If the no-load terminal voltage V_0 is the system voltage (as it normally will be), then E_0 is also equal to the system voltage. If, as load is applied, the emf were maintained at its no-load value E_0 (that is, if there were no change in the excitation), then the situation would be as shown in Figure 4.6(b), which is a repetition of Figure 4.5 but everywhere reduced in scale. Since the magnitude E is unaltered from its no-load value E_0, the on-load terminal voltage vector V, the current vector and both the volt-drop vectors will be reduced in proportion but will retain their relative sizes and positions.

In Figure 4.6(b), therefore, V_0 is the no-load terminal voltage and V the terminal voltage when a stated load current is applied. The numerical difference between V_0 and V is the drop of terminal voltage, as measured by voltmeter, when the load is applied with no change in the excitation; it is the regulation of the generator at that load.

The regulation of a generator is usually expressed as a percentage of the system voltage when the generator is supplying full rated load current at rated power factor. Then the percentage regulation is

$$\frac{V_0 - V}{V_0} \times 100\%$$

at rated output.

In practice, of course, the situation of Figure 4.6(b) would be immediately detected by the AVR, which senses only the terminal voltage. Finding V reduced below the system voltage, it would raise the excitation and so the emf E. This would increase the scale of Figure 4.6(b) again until it approached that of Figure 4.5 and the terminal voltage V once more reached the system voltage. As will be seen from Figure 4.5, the emf will then be much greater than the system voltage.

4.5   PRACTICAL APPLICATION

The foregoing is the full treatment for determining the emf and the regulation of a generator, taking into account the power factor of the load current and both the internal resistance and reactance of the machine.

In practice the internal resistance of an a.c. generator - particularly of a large one - is negligibly small compared with its reactance. In other words, we can neglect r as compared with x, which means that the internal voltage drop $I.r$ can be neglected as compared with $I.x$. The full Figure 4.5 then reduces to the simpler Figure 4.7 where the vector $\overline{ST}$ $(= I.r)$ has disappeared.

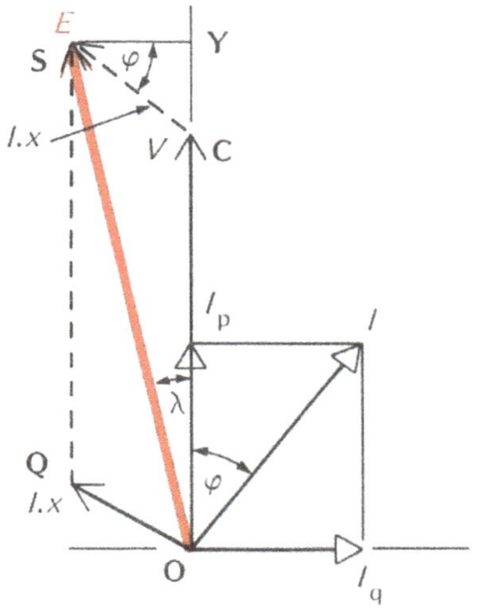

FIGURE 4.7
DIAGRAM NEGLECTING INTERNAL RESISTANCE

The vector $\overline{CS}$, equal to $I.x$, is then the principal element in determining the difference in magnitude between E and V, and so in determining the regulation.

In Figure 4.7 the angle between the vector $\overline{CS}$ and the horizontal is also φ, so that the numerical difference between E and V is approximately CY, or $CS \sin \varphi$. So the difference between the emf and the terminal voltage is approximately $I.x \sin \varphi$. But $I \sin \varphi$ is the quadrature component (I_q) of the load current, so the regulation can be taken as $I_q.X$.

$\overline{YS}$, the other component of $\overline{CS}$, namely $\overline{CS} \cos \varphi (= I.x \cos \varphi)$, has virtually no effect on the voltage difference between E and V, but it does affect the power angle λ. But $I \cos \varphi$ is the in-phase component (I_p) of the load current, so the power angle depends mainly on the in-phase, or active power, component of the load current. The power angle is given approximately by

$$\tan \lambda \ = \frac{SY}{OY}$$

$$= \frac{CS.\cos \varphi}{OC+CY}$$

$$= \frac{I.x \cos \varphi}{V+I.x \sin \varphi}$$

$$= \frac{I_p.x}{V+I_q.x}$$

As before, x is the reactance (in ohms) per phase and V is the **phase** voltage.

To sum up, with the approximations made (that r may be neglected), the regulation or voltage drop within a loaded generator depends principally on the quadrature component of the load current - that is, on the kvar loading only - and on the generator reactance.

The in-phase component of the load current - that is, the kW loading - affects principally the power angle and has little effect on the voltage drop.

A notable case of this occurs when starting large motors. On switching on, not only is the starting current heavy but it is also drawn at a low power factor. In other words a very large reactive load is thrown onto the system - a situation causing the maximum voltage drop in the generators. This will cause a noticeable dip as evidenced by the dimming of lights until the AVR has had time to take charge and restore the voltage.

4.6   SYNCHRONOUS REACTANCE

There is one further matter which has a very marked effect on the value of excitation needed to produce the required emf. It was shown in Figure 2.4 of Chapter 2 that, when a load current - particularly a reactive load current - passes through the armature of a generator, the magnetic field of that current opposes that of field poles and demagnetises them; this is the phenomenon of 'armature reaction'. Consequently, with this effect present, the net field flux is reduced and, for a given exciting current, less emf is generated.

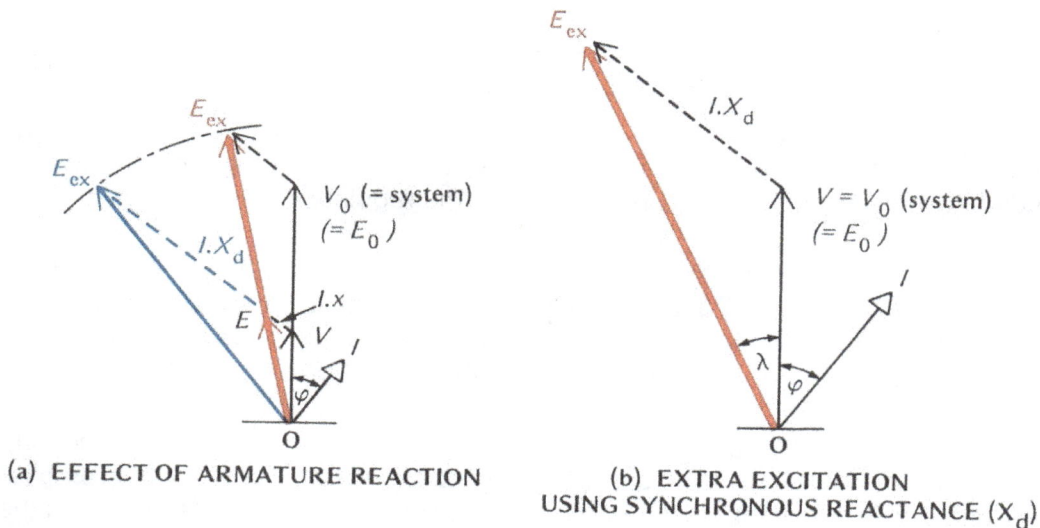

(a) EFFECT OF ARMATURE REACTION   (b) EXTRA EXCITATION
USING SYNCHRONOUS REACTANCE (X_d)

FIGURE 4.8
ARMATURE REACTION AND SYNCHRONOUS REACTANCE

All that has been said regarding the relationship between the emf E and the terminal voltage V still stands. Up until now the emf E has been considered to reflect exactly the field exciting current, but, because of demagnetisation, this is no longer true and the net flux is now less than that produced by the excitation. Consequently the emf E actually generated by the reduced flux is less than would occur with the given exciting current alone. This effect is shown in Figure 4.8(a). E_{ex} represents the emf that would be generated by the exciter field alone, but E is the reduced emf due to the demagnetisation of the field. This results in a smaller V and a smaller I. The heavy reduction in V is as if the exciter emf E_{ex} had remained unaltered but the generator had had a much higher reactance X_d than its real one x. This is indicated by the blue vectors in Figure 4.8(a).

Indeed the effect of armature reaction is taken into account by assuming just this, as shown in Figure 4.8(b). The generator is considered to have a larger, false reactance whose effect is to restore the relationship between a new and much increased excitation emf E_{ex} (not the actual E) and the terminal voltage to that of Figure 4.7 by using a higher value of x. This higher, false reactance is called the 'synchronous reactance' of the generator. Its symbol is X_d. If used instead of the actual reactance x, it gives the same relationship between the excitation emf and terminal voltage as in Figure 4.7.

In fact, taking this artificial synchronous reactance into account, which is normally several times greater than the actual reactance x, many large generators have excitation systems some two to three times that required for the no-load value in order to maintain full terminal voltage at full rated output. Compare the relative sizes of E_{ex} and $V(=E_0)$ in Figure 4.8(b). It also causes a larger power angle λ.

The whole concept of 'synchronous reactance is merely a device for taking into account the armature reaction of a generator when loaded. The voltage drop within a generator is caused partly by the loss of emf due to this armature reaction, and partly by the normal impedance drop due to load current in the windings. By using this wholly fictitious synchronous reactance, the total drop is considered to be due to an increased impedance drop alone. Its use enables currents and voltages to be calculated by ordinary methods as if no demagnetisation were taking place.

It is always used in network 'steady-state' problems such as power flow and load sharing. It is also used in generator design to establish what excitation system will be needed - the sizes of the exciters and main field system, rotor cooling, etc.

4.7 PERCENTAGE REACTANCE

Throughout this chapter reactances per phase have been used directly (x or X_d) and are measured in ohms. In practice an alternative method is generally used, namely to express them as 'percentage reactances'.

Instead of using Ohm's Law $I = V/X$ to determine currents, the generator is considered to be short-circuited and the excitation reduced so that the short-circuit current just equals the rated full-load current. The reduced excitation is expressed as a fraction (or percentage) of the excitation at full rated load; percentage is then the 'percentage reactance' of the generator.

For example, if the rated full-load current of a generator were 1 000A, and on short-circuit this current were achieved with only 25% excitation, the generator would have a 'percentage reactance' of 25%. If the full-load current is divided by the percentage reactance (expressed as a fraction), it gives the current on short-circuit with full excitation.

Thus the short-circuit current in this case $= \dfrac{1\,000}{0.25} = 4\,000$A. If the generator is rated

5 000kVA at full load, then on short-circuit it will deliver $\dfrac{5\ 000}{0.25}$ = 20 000kVA, or 20MVA.

This method of expressing reactances is very useful, for it avoids having to deal with actual voltages and currents for any particular generator. Most drawings give reactances only in percentage form.

In Chapter 9 percentage synchronous, transient and subtransient reactances are given for a typical large generator, together with formulae for converting from percentage to ohmic reactance and from ohmic to percentage. These conversions, however, require knowledge of the rated voltage and current.

The very large synchronous reactances which arise as explained in para. 4.6, result in percentage synchronous reactances being well above 100% - typically nearer 200%. This causes the current of a short-circuited generator to settle (in the 200% case) at only one-half the normal full-load current - that is to say, the generator current is actually reduced by a short circuit (assuming that the excitation does not change) - a result which perhaps may be surprising to some.

A consequence of this is that normal overcurrent protection of the generator may actually fail to work on short circuit. This is investigated further in Part 10 'Electrical Protection'.

4.8   DIRECT AXIS AND QUADRATURE AXIS REACTANCES

In the foregoing discussion it was assumed that the machine concerned had a uniform air gap and was thus of the cylindrical rotor type. All reactances therefore were 'direct axis

It was explained in Chapter 2 that salient-pole generators, which form the majority in offshore and onshore installations, do not have a uniform air gap. In consequence their reactances are of two types - direct axis and quadrature axis.

In practice, omitting the quadrature-axis quantities from the calculation has little effect on the **magnitude** of the excitation E needed to sustain a given load as described in the foregoing paragraphs. The quadrature-axis reactance does, however, have a considerable influence on the power angle λ for that load. Omitting it leads to a calculated angle smaller than actually occurs and may present stability problems.

CHAPTER 5 GENERATOR SPEED CONTROL

5.1   GENERAL

All a.c. generators must run as nearly as possible at constant speed in order that the frequency of the generator's output voltage is held within close limits to the nominal, which on most platforms is 60Hz and on most shore establishments 50Hz. This applies to both gas-turbine and diesel-driven sets. The device which achieves this is called a 'governor

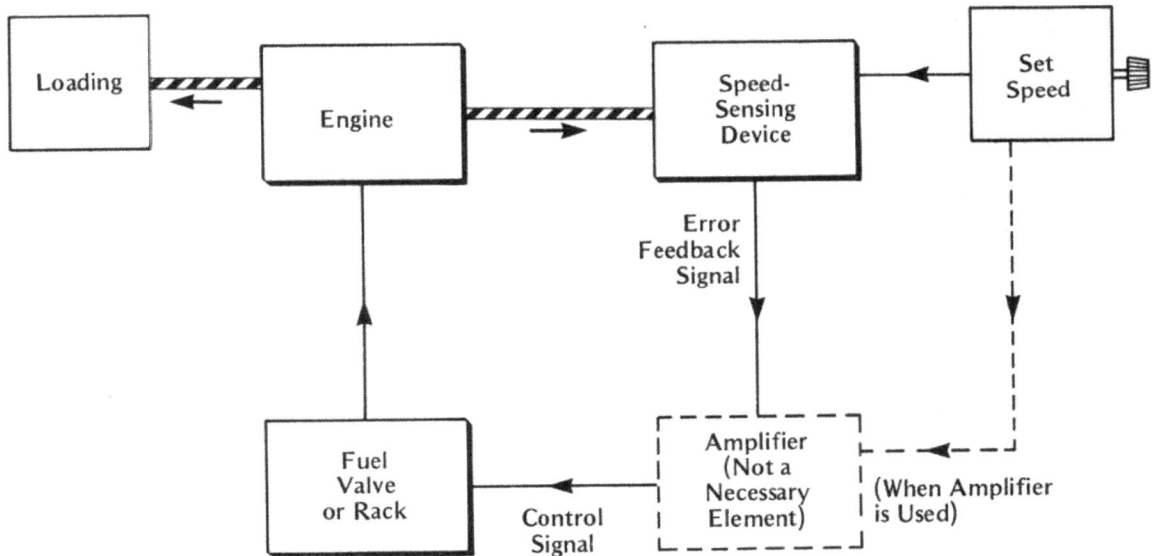

FIGURE 5.1
GOVERNOR CONTROL LOOP

A governor is a speed-sensing device driven by the engine itself or by some mechanical part such as the gearbox, coupled directly to it, or else it derives from the engine a signal which represents the engine speed. It actuates directly, or through an amplifier, the fuel control to the engine. When, for example, load on the engine increases, its speed momentarily drops; this is sensed by the governor, which causes the fuel admitted to the engine to be increased. This in turn increases the engine's torque and raises its speed against the new load until it is restored near to its former level. The control system is automatic and forms a closed loop, like any other automatic servo system, and is illustrated in Figure 5.1.

The above description represents the ideal situation; in practice it is achieved only within certain limits of error which are explained below.

It will be seen that the governor must maintain the engine speed at some arbitrary level. The level can be varied at will by adjusting the governor. This adjustment is called 'speed setting', and the level to which it has been set is the 'set-point'. Once so set, the governor maintains the engine speed at the set-point (within limits of error) at all loads within the engine's rating.

Governors may be mechanical or, more recently, electronic, but the basic control loops for both are the same.

5.2 MECHANICAL GOVERNORS

The mechanical governor is considered first, as it demonstrates the principles more clearly.

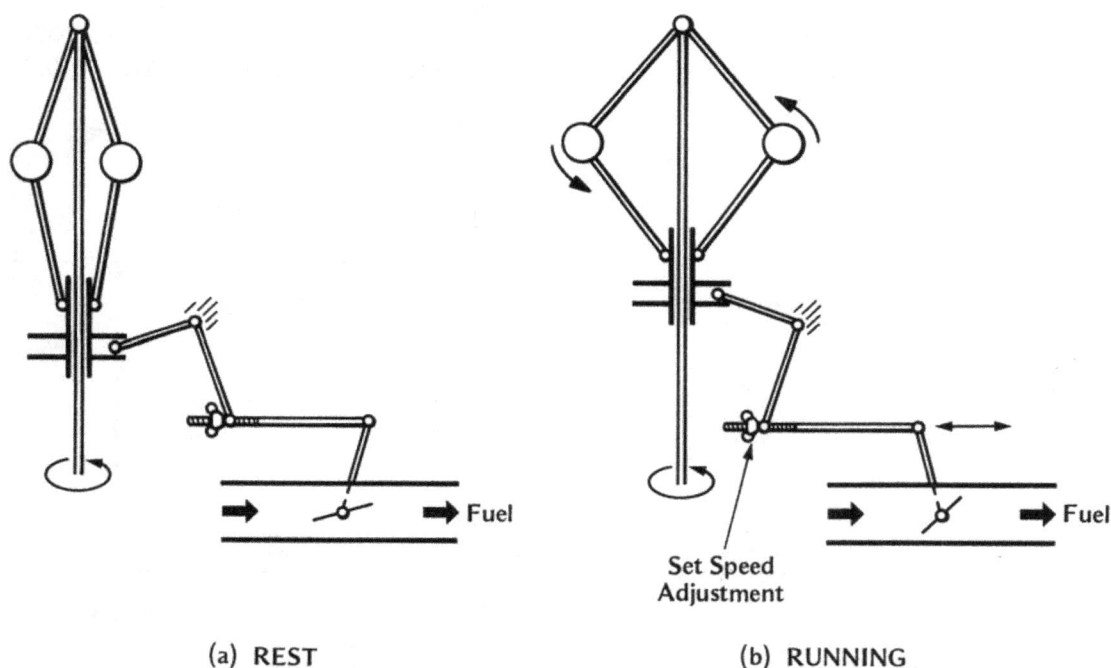

(a) REST   (b) RUNNING

FIGURE 5.2
FLY BALL CENTRIFUGAL GOVERNOR

Figure 5.2 shows probably the oldest form of engine governor, namely the 'flyball' centrifugal type, used originally on steam and gas engines. It consists of a pair of heavy balls held by a link mechanism which is driven by the engine. As the engine rotates, the balls are thrown outwards by centrifugal force against the normal restoring force of gravity. There is no amplifier in this case. As the balls move outwards they raise a sleeve which, by a suitable linkage, operates to reduce the opening of the steam or fuel inlet, shown here for simplicity as a butterfly valve.

When the engine is at rest there is no centrifugal force, and the balls hang in the position shown in Figure 5.2(a); the fuel valve is then wide open. When fuel or steam is admitted the engine starts with a full fuel charge and accelerates. The balls move outwards, raising the sleeve, and gradually close the valve until the steam or fuel charge just balances the engine load, at which point the speed settles down to a steady value, as shown in Figure 5.2(b). The level at which it settles depends on the set-point. This can be adjusted in various ways: in Figure 5.2 it is by adjusting the link between the governor and fuel valve. Lengthening it opens the valve wider and so raises the set speed; shortening it has the opposite effect.

The steady-running condition is shown in Figure 5.3(a), which is a repeat of Figure 5.2(b). Once the speed has settled at its set value, any variations of speed without change of load are closely controlled. An increase causes the balls to move outwards, so closing the valve a little and reducing fuel to check the increase (see Figure 5.3(b)). When the speed has returned to its set value the valve is once again in its former position. A similar effect will occur, but in the opposite direction, for any momentary drop in speed.

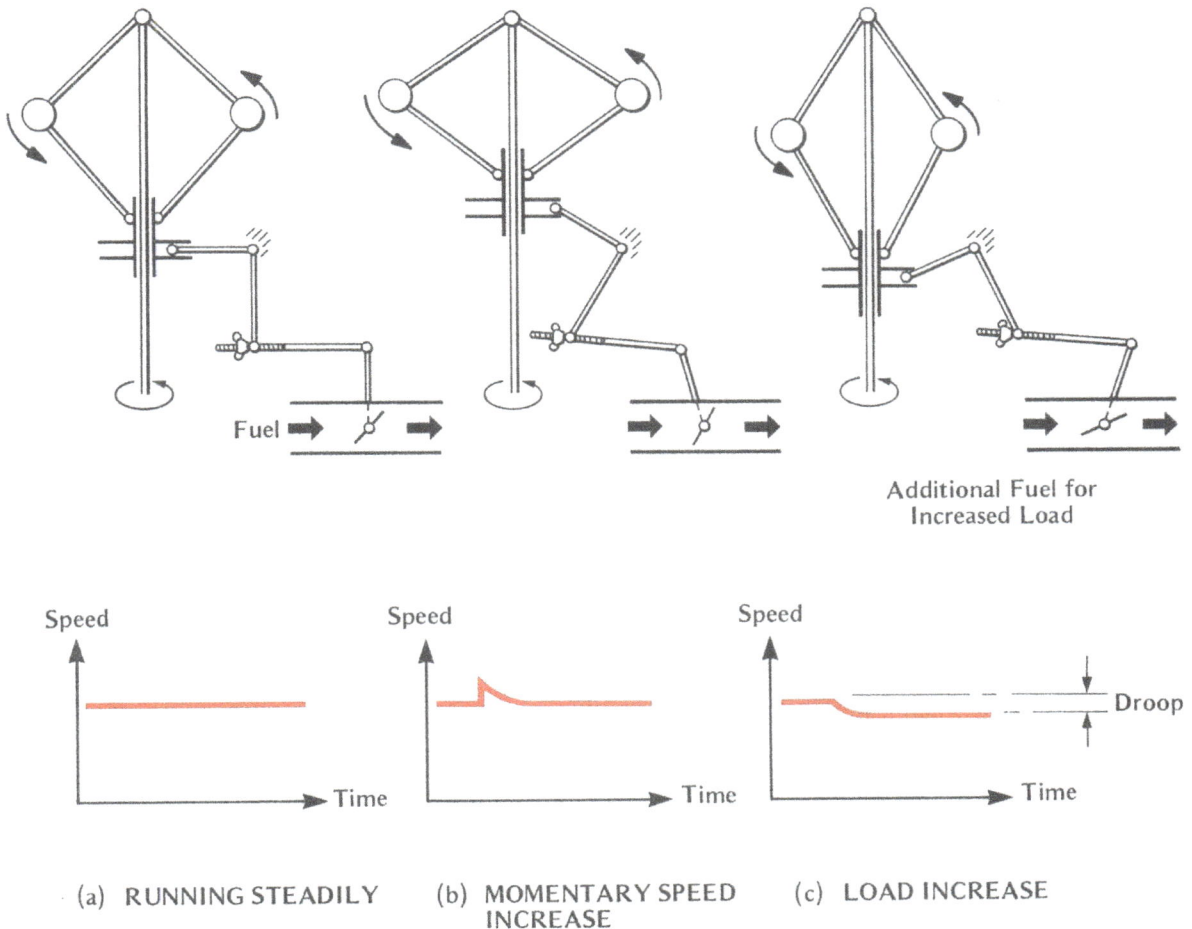

Fuel ➡

Additional Fuel for
Increased Load

(a) RUNNING STEADILY

(b) MOMENTARY SPEED
INCREASE

(c) LOAD INCREASE

FIGURE 5.3
EFFECT OF SPEED AND LOAD CHANGES

The reaction to a change of **load** is different. If the load on the engine increases, the speed at first drops, causing the balls to move inwards. This opens the valve further until the increased fuel produces increased torque to balance the higher load. The deceleration then ceases, and the speed settles at a level somewhat lower than it was before (see Figure 5.3(c)). The amount of speed loss depends on the characteristic of the fuel valve itself - that is, how much extra fuel it admits for a given loss of speed.

From the above it is clear that, although a governor ideally holds the speed constant at its set value, with an increase of engine load it cannot quite achieve this. This is one of the 'errors' referred to in para. 5.1. There is in theory no such thing as a truly 'isochronous' governor, that is one which keeps the speed absolutely constant at all loads, although modern governors do approach this condition. Some degree of error, however small, is necessary for a governor (or indeed for any other closed-loop servo) to work at all.

Figure 5.4 shows, in graphical form, a typical relationship between speed and engine load (in kilowatts mechanical). The fall of speed with increase of load is called 'droop' and is typically about 4% from no load to full load for a mechanical governor. (In Figure 5.4 the slope is exaggerated.) If the set speed at no load is, say, 1 800 rev/mm, then at full load it will be only 1 730 rev/mm, and the frequency of the driven generator will have fallen from 60Hz to 57.6Hz. To offset this the nominal speed to achieve 60Hz could be set at half load, so that the frequency varied from 61.2Hz to 58.8Hz from no load to full load (±2%).

205

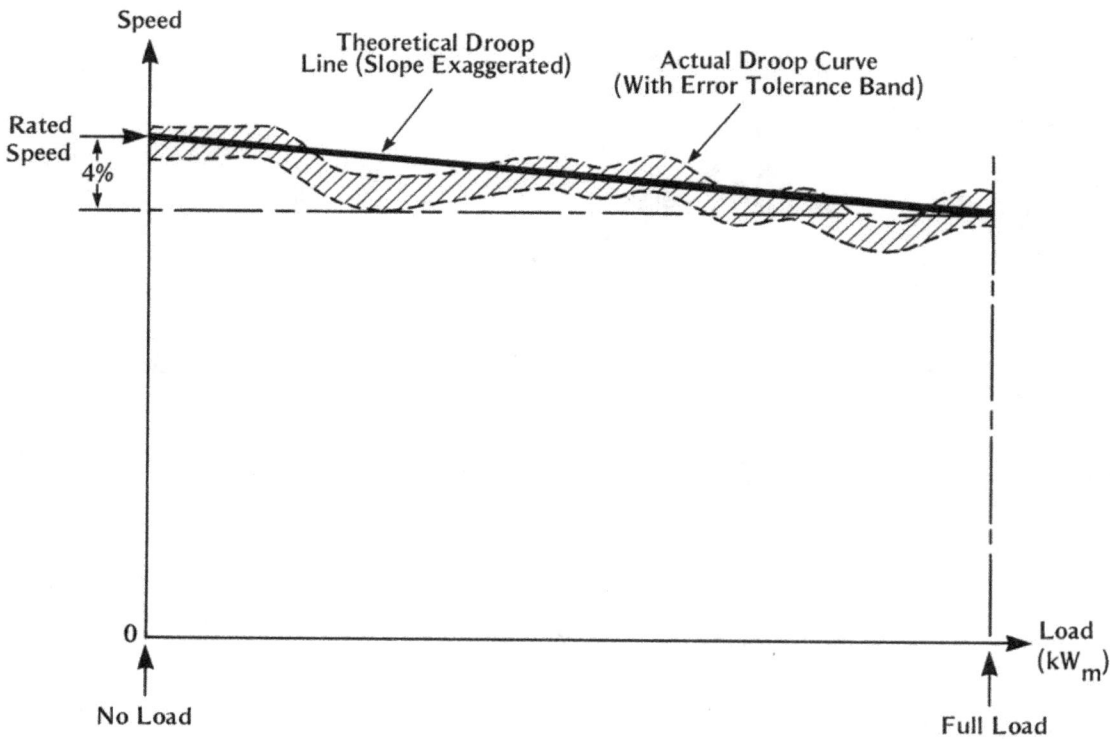

FIGURE 5.4
GOVERNOR DROOP

Other sources of error in a mechanical governor include:

(a) backlash, friction and wear in the flyball and connecting linkages

(b) time-lag in the flyball mechanism (i.e. inertia time to take up new position)

(c) time-lag in the amplifier, if fitted

(d) firing stroke delay (diesel engines only)

(e) non-linearity of the fuel rack or valve

(f) twist in the governor drive

(g) inertia of the rotating parts

All these combine to distort the droop line of Figure 5.4 from its theoretical straight to a somewhat irregular shape. Random errors produce a tolerance band about the mean, as shown shaded.

The effect of most of these errors is self-evident, but three of them need further explanation. Error (a) is likely to become worse as wear takes place with the increased life and usage of the engine. Error (d) occurs only with diesel engines and is due to the next cylinder not necessarily being ready to fire at the moment the governor calls for increased (or decreased) speed. Error (f) may occur if the drive from the engine shaft to the governor is not solid - for example if the drive is taken from the gearbox. This may produce lag or even oscillations.

It will be noted that many of the errors are due to time-lags in various parts of the governor system loop. These all delay the response of the engine to a speed error signal from the governor and, if appreciable, produce the effect of sluggishness.

5.3   MODERN MECHANICAL GOVERNORS

The flyball system of Figure 5.2 is now seldom used. Instead there are rotating weights on the governor shaft, controlled by springs instead of by gravity. This system, however, is still a centrifugal one, and the displacement of the weights still actuates the fuel valve or rack. Instead of the direct linkage of Figure 5.2, most modern mechanical governors use a hydraulic linkage, which is more positive in its action and less liable to backlash and wear. Oil pressure is obtained from a pump driven by the engine or from an auxiliary motor-driven pump, and it fails safe by causing the fuel valve to close if oil pressure fails. The hydraulic system acts as the 'amplifier' of Figure 5.1 between the speed sensor and the fuel valve. It operates the valve by a hydraulic actuator, which converts the governor signal into a hydraulic thrust.

5.4   ELECTRONIC GOVERNORS

Because of the unavoidable errors, including the large inherent droop, of mechanical governors an entirely new type was developed and is now in general use throughout all platforms and most shore installations. This is the 'electronic governor', and those which are found on most platforms are of the 'Woodward', 'Speedtronic' or 'Rustronic' type. It must be emphasised, however, that the governing principles set out in block form in Figure 5.1 apply just as much to an electronic governor as to a mechanical one.

In an electronic governor all linkages, except the final actuator stage, are electrical and therefore not subject to backlash or wear. Consequently a much greater accuracy can be achieved, and a droop of ½% (as compared with 4% for a mechanical governor) is not unusual. Moreover, because of lack of wear, an electronic governor is very consistent in its performance.

One essential difference of detail is that speed is sensed by an inductor-type tacho-generator consisting of an iron toothed wheel rotating past fixed coils. The varying flux as the teeth pass the coils induces in them an emf at a frequency directly proportional to the speed. The other main difference is that the former mechanical or hydraulic linkage is replaced by simple electrical connections (apart from an electrohydraulic actuator referred to below); these have no backlash and are not subject to friction or wear.

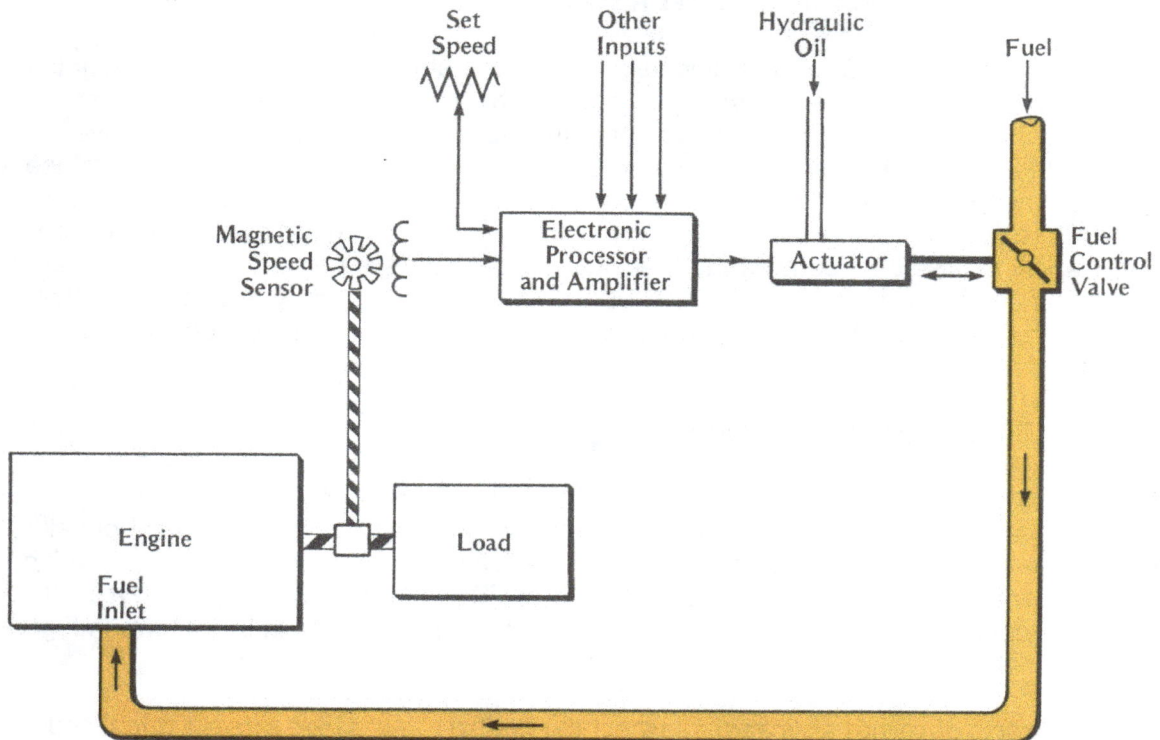

FIGURE 5.5
ESSENTIAL ELEMENTS OF AN ELECTROHYDRAULIC GOVERNOR

The varying-frequency signal is processed and amplified by electronic circuits, and also mixed with certain other signals, to give an electrical output signal representative of the fuel input required. It is converted to a hydraulic signal through a pilot solenoid valve in an electrohydraulic actuator. This is, in effect, a further amplifying stage, and the actuator drives the liquid fuel or fuel-gas valve. The hydraulic oil pressure is derived from an engine-driven pump when the set is running, and from an auxiliary pump when it is at rest or running slowly.

The basic system is shown in Figure 5.5. The closed loop should be compared with the block diagram of Figure 5.1.

5.5   TYPICAL SINGLE SHAFT GAS TURBINE GOVERNOR

The foregoing is a general description of an electronic governor. As applied to various gas turbines and diesel engines the arrangements differ in detail, but the principles remain the same.

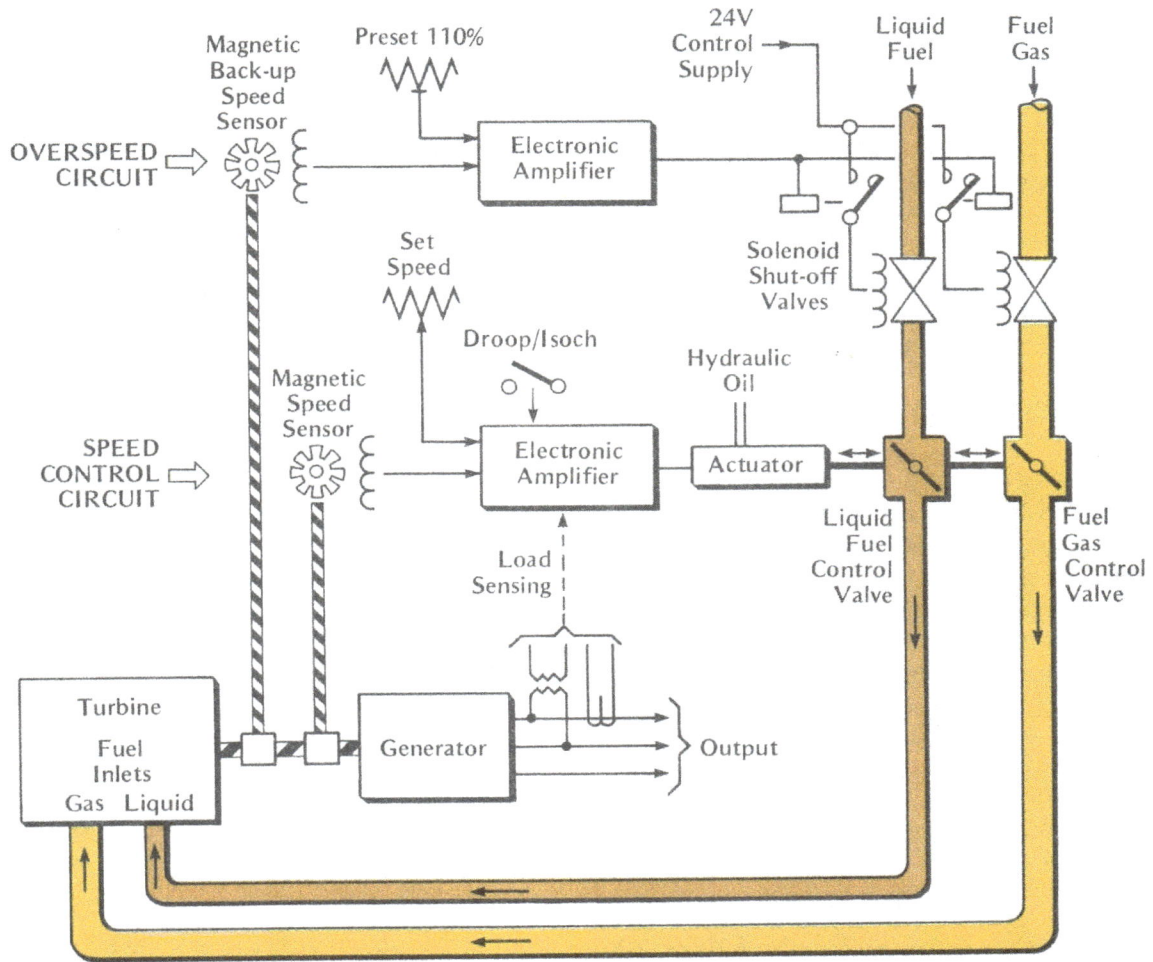

FIGURE 5.6
TYPICAL SINGLE SHAFT TURBINE SPEED CONTROL

The installation for a typical single-shaft gas turbine is shown in simplified form in Figure 5.6. It uses the magnetic speed-sensing pick-up and hydraulic actuator already described. A special feature, however, is the refinement of load sensing.

It has already been explained how various time-lags in the control loop delay the response of the system to a control signal. One of these is the inertia of the rotating parts. In a single-shaft set this is considerable, consisting as it does of the compressor and turbine unit (running at high speed), the gearbox and the generator itself. When there is a sudden increase of load, the rotating mass decelerates relatively slowly because of its high inertia, and there is an appreciable delay before the speed has dropped sufficiently for the sensor to initiate governing action; there are further time-lags in the control loop before the hydraulic actuator admits more fuel. All this amounts to a slow response. By making the governor sense the load change when it occurs, and so anticipate the speed drop, the correcting action can be started earlier and the overall speed of response quickened.

In the system illustrated the electronic processor, in addition to receiving the speed signal from the sensor, receives also a load signal from a wattmetric element on the output side of the generator, as shown in Figure 5.6. These two signals are mixed in the correct proportions - an advantage of electronics - and are balanced against the 'set speed' signal put on manually. The difference, or 'error', causes the actuator to respond to the combined speed/load error signal.

The gas turbine illustrated here is assumed to be dual-fuelled, and the hydraulic actuator controls both the liquid fuel and fuel-gas valves.

5.6   SINGLE SHAFT OVERSPEED PROTECTION

The normal governor control system should prevent excessive speed, but it is covered by a 'back-up overspeed' system in case it should fail. This is similar to the normal speed governor system, except that it operates a shutdown valve instead of the actuator. There is a completely separate magnetic pick-up and electronic amplifier, with a speed setting of about 110%. At this speed the amplifier produces a signal which shuts down the whole engine by allowing the contactors in the circuits of the solenoid-operated shutdown fuel valves to open (a 'fail-safe' arrangement). This is shown in Figure 5.6.

5.7   TYPICAL TWO SHAFT GAS TURBINE GOVERNOR

The installation for a typical two-shaft gas turbine is shown in simplified form in Figures 5.7 and 5.8. Most of such sets installed on platforms operate on single fuel (gas) only, but some have been modified to run on gas or liquid fuel. The governor uses a magnetic speed-sensing pick-up and an electrohydraulic actuator as already described, but as the turbine has two shafts it has certain other refinements added.

The governor control system for a two-shaft turbine was shown basically in Chapter 1, Figure 1.7. Speed is sensed from the output shaft driven by the power turbine, but fuel control is applied to the gas generator only, which runs at a speed different from that of the power turbine, the difference depending on the generator loading. As the gas-generator turbine runs free, it has relatively low inertia and responds quickly to fuel input changes, so load sensing as applied to the single-shaft turbine is not needed.

The main control requirement is to maintain constant speed of the **power turbine**. Other limitations are, however, necessary to ensure that the various limits of the engine's rating are not exceeded. It is arranged that the limitation requiring the least power from the engine, and therefore least fuel, is the one in control.

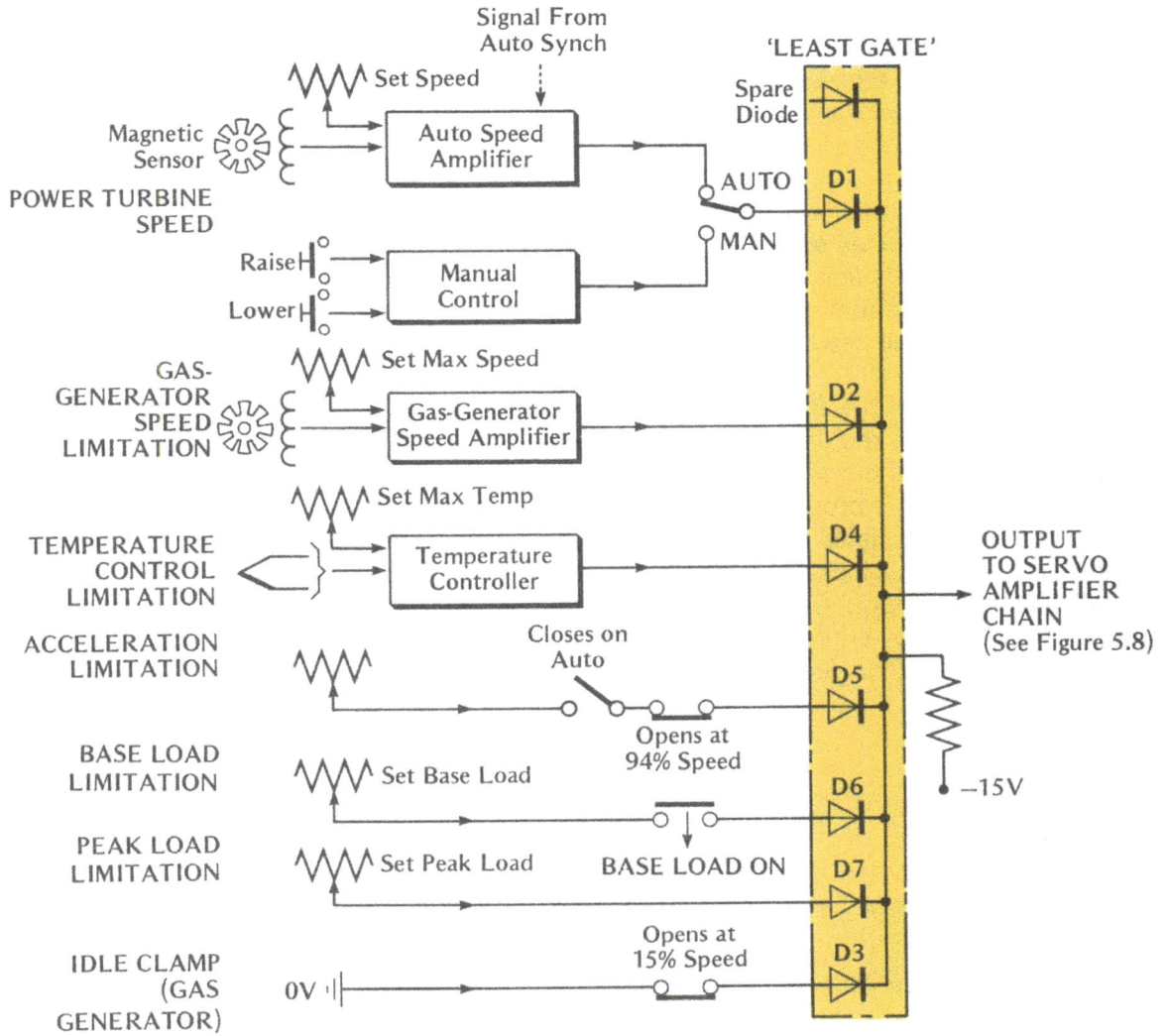

FIGURE 5.7
TYPICAL TWO SHAFT TURBINE SPEED CONTROL (INPUTS)

Figures 5.7 and 5.8 show, in block form, the speed control system and the limitations which are applied to it. They also show how the predominant speed control signal is taken through a chain of electronic and hydraulic units to position the fuel-gas control valve.

At the top of Figure 5.7 is the Power Turbine Speed element. It has the automatic magnetic speed-sensing unit which gives a signal proportional to the power turbine speed. It is compared with the demanded, or 'set-point', speed, and a difference signal is passed on to the output chain.

Next is the Manual Speed Control element. When this is selected the manually set speed signal is passed direct to the output chain.

The third element is the Gas-generator Speed limitation. Its speed is sensed by a magnetic pick-up similar to that on the power turbine shaft, but it is purely an overspeed device. The sensed speed signal is compared with a preset maximum allowable speed setting for the gas generator. If the actual speed exceeds the allowable, the signal to the output chain is reduced to a level which reduces the gas-generator speed to below the allowable limit.

211

The fourth element is the Temperature Control limitation. Thermocouples disposed around the gas-generator exhaust duct detect exhaust temperature. Their signals are averaged and compared with a preset maximum allowable temperature. If the actual temperature exceeds the allowable, the signal to the output chain is reduced to a level which keeps the temperature below the allowable limit.

The fifth element is the Acceleration limitation. If during starting the acceleration is too great, it not only causes excessive temperature but may take other quantities outside their designed limits. This unit, whose level is preset, takes over only during the acceleration period and limits the fuel rate until the speed has reached 94% of maximum. At that point it is taken out of circuit, leaving the other limiting circuits to exercise their individual controls.

The sixth element is the Base Load limitation. If two or more generators are running in parallel and it is desired that one of them (the 'base load' machine) should carry a constant load and that the others should carry all loads above a certain base load maximum, this element is used. It is brought into action only when the 'Base Load' switch is made. The maximum base load level is preset and, as soon as the load reaches this point, the signal to the output chain is locked so that the machine can deliver no further load even though in parallel with the others. Generators on platforms are not usually operated in this manner; it is preferred that they share the load equally or in proportion to their sizes (see Part 6 'System Control'). This feature, though available, is not therefore used.

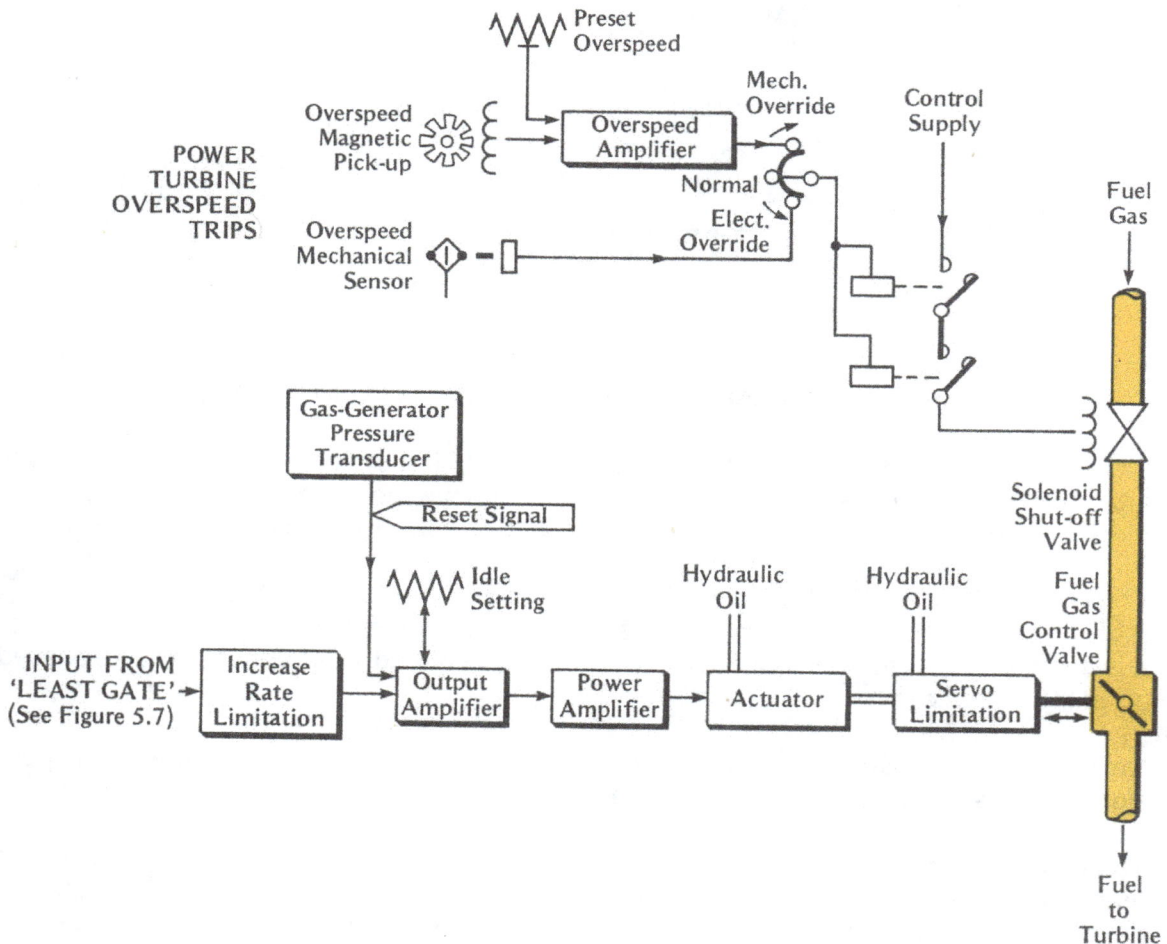

FIGURE 5.8
TYPICAL TWO SHAFT TURBINE SPEED CONTROL (ELECTROHYDRAULIC LOOP)

The seventh element is the Peak Load limitation. It is similar to the Base Load Limit but is preset to the maximum allowable peak load for the turbine. If any attempt is made to exceed this load, the signal is locked, permitting no further increase of fuel.

At the bottom is the element for the Idle Clamp (Gas Generator) limitation. This is in operation while the speed of the gas generator is below idling level - e.g. when starting from rest. At such speeds it prevents all other controls operating. When the gas generator has reached its correct idling speed (15%) the set is ready for acceleration. The Idle Clamp is then automatically disconnected, allowing all the other controls to work normally.

All these individual control signals are fed into a 'Least Gate', shown tinted in Figure 5.7. This is a unit consisting of diodes which select whichever of the eight controls listed above is producing the greatest signal. This is normally the speed signal. That signal alone is automatically selected, and all the others are blocked off by the diodes. (All the control signals are in fact of negative polarity, and the stronger the signal, the less negative it becomes - hence the name 'Least Gate'.) For example, if the set is being controlled automatically, as normal, by the power turbine speed (through diode D1), only that diode is conducting, and all the others are blocked. If now the temperature limit is exceeded, diode D4 conducts and takes over, and all the others, including the speed diode D1, are blocked.

The selected output signal from the Least Gate is passed through an Increase Rate Limit network, which limits the rate at which a power increase can be demanded without interfering with the normal governor action. From here it passes to the main Output Amplifier where it is compared with a feedback or 'reset' signal from the pressure of the gas-generator exhaust, which is proportional to the demanded speed. The difference between these two, or 'error', is fed to a Power Amplifier and thence to the Electrohydraulic Actuator which converts the electrical into a hydraulic pressure signal. This moves a hydraulic Servo Limiter (a hydraulic relay) which in turn moves the fuel valve (or fuel valves in the case of dual-fuelled turbines) to admit more, or less, fuel to the turbine. It continues to move the valve until the gas-generator pressure feedback just balances the governor's demand signal and the error disappears. The fuel valve then stops in its new position.

5.8   TWO SHAFT AUTO AND MANUAL SPEED CONTROL

If the automatic system described above becomes faulty or if for any other reason direct control is desired, the governor can be cut out (but not the other limitations) and the speed controlled by hand. This is done by an 'Auto/Manual' changeover switch and 'Raise' and 'Lower' pushbuttons on the local control panel.

There is a risk, when changing from auto to manual, that the manual setting may not at that moment match that of the automatic, and there would follow a gross change of speed as the fuel valve setting changed abruptly. There is a built-in automatic circuit to prevent this, but normally when changing over from auto to manual the manual setting should first be trimmed to match the auto. Matching is normally indicated by an Auto/Manual Balance Meter on the Local Control Panel.

5.9   TWO SHAFT OVERSPEED PROTECTION

Overspeed protection of this particular gas generator is already covered in para. 5.7 and Figure 5.7, diode D2. For the power turbine there are two separate overspeed devices, one electronic and one mechanical, both completely independent of the main speed control. Both are shown in Figure 5.8.

The electronic overspeed system consists of a separate magnetic pick-up similar to that for normal speed sensing. The signal from this is compared in an amplifier with a preset voltage representing the critical overspeed level; if it exceeds it a signal is passed to an Electronic Overspeed Trip unit which de-energises the solenoid-operated valve by allowing two contactors in series to open and cut off the fuel (a 'fail-safe' arrangement). These contactors and the solenoid are all energised while the set is running.

Separate from this is a mechanical overspeed unit which is basically a centrifugal type. At an overspeed setting slightly in excess of that for the electronic one, a spring-restrained bolt flies out under centrifugal force. It actuates a trip lever mechanically and thence a relay and the main tripping contactors. The mechanical overspeed trip acts as a back-up for the electronic.

Either one can, if required for example for maintenance, be isolated by a '3-position' switch on the local control panel for testing, while leaving the other operative. The switch is marked: 'Mech. Override/Normal/Elec. Override'. The machine is thus never left without overspeed protection.

5.10   LOAD SHARING

Electronic governors have very small inherent droops, typically ½%; they are often referred to as 'isochronous' though of course they cannot be exactly so. This is excellent for single machine running, as speed is held closely controlled, but for load sharing with parallel generators some droop is essential. When so required, droop can be injected artificially into the electronic circuits by operating a 'Droop/Isoch' switch on some makes of governor. This biases them from the load signal which is provided for load sensing (see dotted line of Figure 5.6) and provides an artificial droop. On other types of governor the amount of droop can be varied by direct adjustment of the governor circuits. Load sharing is dealt with more fully in part 6 'System Control'.

CHAPTER 6 DIESEL GENERATOR SETS

6.1   GENERAL

In large onshore installations power is derived from the National Grid. On platforms the main generating sets are always driven by gas turbine, using the platform's own gas as fuel when available, with liquid fuel as an alternative in some cases.

Onshore the grid supply can sometimes fail, and on platforms main generators may also fail, or under certain conditions they may be deliberately shut down. In either case there is loss of main power supply, and it is important that there should be immediately available a quick-starting alternative supply - and this means diesel generation.

All platforms, and most large onshore installations, have one or more diesel-generator sets. In many cases they are arranged to start automatically on loss of mains voltage and to switch themselves onto an emergency switchboard. It is never the intention that such generators should replace the lost main ones, but they should provide limited power for only really essential services such as some degree of lighting, safety, instrumentation, communications, fire and gas detection and so on.

Diesel-driven generators are also required for 'black-start' conditions when no main generators are running but whose auxiliaries must be run in order to start them. Such diesel sets must of course be entirely self-contained, requiring no external assistance to start them.

The construction of a diesel engine is well known and will not be described here. It is usually multi-cylinder, turbo-charged and jacket-cooled through a water/air radiator, some times assisted by a cooling fan. It is usually battery-started, and some sets have an alternative hydraulic starter, hand pumped, for use if the battery becomes discharged, for example after a prolonged shutdown. It is vitally important for diesels which drive emergency generators which are automatically started that the batteries are maintained fully charged ready for an instant start; also that practice starts should be exercised regularly.

6.2   BASIC SERVICES

In all installations the really essential services, which it is vital to keep running even when the normal main power has been lost, are offshore termed Basic Services. The diesel-driven generator is called the 'Basic Services Generator' and its switchboard the 'Basic Services Switchboard', both shown in red in Figure 6.1. The system is usually at low voltage (440V), and positive steps are taken to see that the basic services generator does not feed back into any non-basic low-voltage services or into the high-voltage system. (There are however some exceptions to this practice.)

Under normal conditions on an offshore platform the basic services switchboard is part of the complete 440V distribution system. It is in continuous use and is normally fed through an interconnector from a main 440V board, as shown in Figure 6.1. If power on the main board fails, the basic services board is isolated from it and can be fed direct by its basic services generator, which normally has sufficient capacity for that board and no more. The generator may start automatically on failure of the main 440V power, but quite commonly it must be manually started. The incomer circuit-breaker from the generator is interlocked with the incomer from the main 440V board so that both cannot be closed at the same time; therefore the generator can never feed back into the remainder of the 440V system or, through the transformer, into the HV system (other than with the exceptions mentioned above).

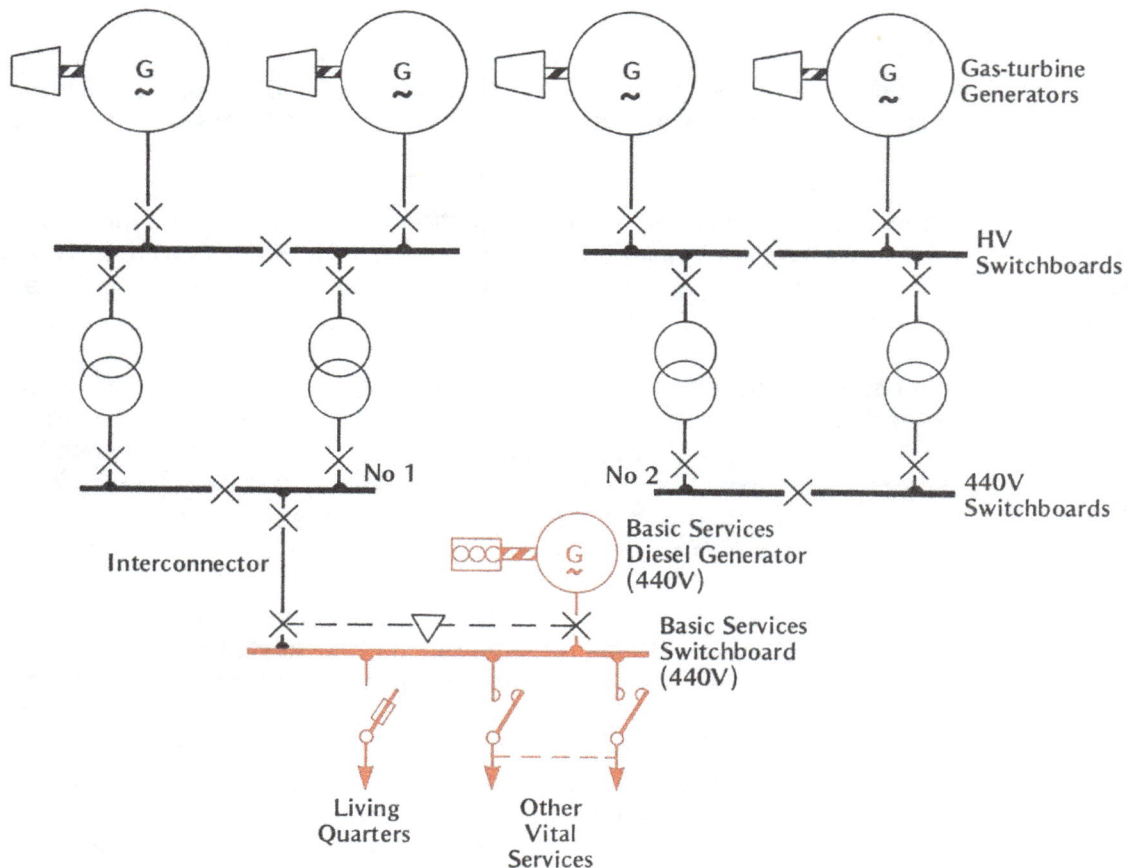

FIGURE 6.1
TYPICAL BASIC SERVICES AND BLACK START GENERATOR ARRANGEMENT

Auto-starting is achieved by providing the basic services busbar with an undervoltage relay which causes the interconnector to open on loss of main supply and the basic services generator to start. When the generator has started and run up, it closes its incomer breaker automatically and in so doing locks out the interconnector. Even when main power is restored, the interconnector breaker cannot be reclosed onto the basic services board until the operator has first opened the generator incomer breaker, so lifting the interlock. The normal interconnector incomer breaker can then be closed, and the system reverts to normal. The basic services generator is afterwards stopped manually and left in a condition to restart whenever needed.

Where the start is manual no undervoltage trip is provided, but instead the act of manually closing the generator incomer breaker also trips and locks out the interconnector incomer breaker. When power is restored, the process is reversed manually.

With regard to the exceptions referred to above, on some of the newer platforms larger diesel-generator sets are fitted which have a capacity appreciably greater than that needed only for the basic services switchboard and its essential loads. In those cases some limited feedback into the system is allowed to power other less essential but still important loads, such as utilities. In that case the interlock between the generator and interconnector breaker is not fitted.

6.3   AVAILABILITY OF BASIC SERVICES GENERATOR

A basic services generator is nearly always needed in a hurry, whether automatically or manually started. It is therefore always left in a 'ready-to-run' state. If automatic, the selector switch is left on 'Auto', even if it had been turned to 'Local' for the previous manual stopping. Ready-use fuel tanks are kept full, oil and water levels correct, battery fully charged and heaters on. These things are checked **daily**, and always after the machine has been run.

Where basic services generating sets are automatically started on loss of main supply, this feature is regularly tested to ensure that it functions correctly. Manual starts on all auxiliary sets are also regularly exercised.

6.4   BASIC SERVICES GENERATOR UTILITIES

Most diesel engines are electrically started from a local battery, usually 24V. When the engine is not in use this battery is kept fully charged by a charger fed from the main a.c. system. An engine-driven d.c. generator charges the battery when the engine is running.

Basic services diesel engines are provided with ready-use fuel tanks with a capacity sufficient for at least 24 hours' full-load running. As main supplies are assumed to have been lost, fuel pumping facilities may not be available, and it may be necessary to refill the tank by hand-pumping from barrels.

Each diesel generator unit is provided with a local control panel on or near the engine mounting, from which the output can be controlled and monitored for speed and voltage. No remote control is exercised from the Electrical Control Room on the generator and interconnector circuit-breakers. All control is local, but there is usually some remote instrumentation in the Electrical Control Room.

CHAPTER 7 GENERATOR PROTECTION

7.1   GENERAL

Electrical plant can be damaged, or destroyed, by operation outside its designed ratings or by fault conditions caused by a breakdown of some part of the system.

The automatic protection of electrical installations, including generators, against such damage is described fully in Part 10 'Electrical Protection'.

In brief, generators are protected against some or all of the following abnormal conditions:

- Overcurrent
- Earth fault
- Differential current
- Reverse power
- Overvoltage
- Undervoltage
- Underfrequency
- Field failure
- Diode failure
- Winding overtemperature
- Turbine trip
- Overspeed.

CHAPTER 8 GENERATOR TESTING

8.1 GENERAL

The general requirements for the testing of all rotating electrical machines, including generators, are laid down in BS 4999 : 1976, Part 60.

As far as generators are concerned, the principal requirements are described below.

8.2 MANUFACTURER'S TESTS

Manufacturer's tests are given three classifications: 'Basic' (formerly called 'Type Tests'), 'Duplicate' and 'Routine Checks'. Basic tests are mainly to prove a new design. They include exhaustive tests to ensure that the design meets the specification and all other performance requirements. They are normally carried out only on the 'first of class' generator, and a Test Certificate is provided on request to confirm the tests. Basic tests may, on special request, be repeated on the first machine of a new, large order, but this is not usual.

Duplicate tests are for performance. They are applied to a generator that is of the same design and construction as one previously made (and in no way altered) and which has already undergone basic tests. The duplicate tests are to ensure that the generator is still in accordance with the original design.

Routine checks are tests to show that each individual generator has been assembled correctly, is able to withstand the appropriate high-voltage tests and is in sound working order both electrically and mechanically.

The three classes of test are listed in Table 8.1.

TABLE 8.1 - MANUFACTURER'S TESTS

Test	Basic	Duplicate	Routine
Resistance of windings (cold)	X	X	-
No-load losses	X	X	X
Temperature rise	X	-	-
Tests for efficiency	X	-	-
Momentary overload	X	-	-
High voltage	X	X	X
Vibration	X	-	-
Short-circuit saturation	X	X	-
Short-circuit losses	X	-	-

Most of these tests are self-evident, but the following additional information is given on the high-voltage test.

For this test, also called a 'withstand' test, a high voltage is applied between the frame and all the generator stator windings, with all other conductors, metal and auxiliary (i.e. heater) circuits bonded to the frame. The actual voltage applied is in accordance with Table 8.2 and is sustained for one minute. It may be at any frequency between 25Hz and 100Hz. It is primarily an insulation test for the generator's windings and is included also in the routine checks to ensure that there has been no fault during assembly of any individual machine.

TABLE 8.2 - HIGH VOLTAGE TESTS

Windings	Test Voltage (rms)
Generator stator windings: <100V, <1kVA >100V, <1kVA 1 – 10 000kVA >10 000kVA and <2 000V 2 000 - 6 000V 6 000 – 17 000V >17 000V	500V + twice rated voltage 1 000V + twice rated voltage 1 000V + twice rated voltage, min 500V 1 000V + twice rated voltage 2.5 times rated voltage 3 000V + twice rated voltage Special agreement
Rotor windings (including related exciters):	10 times the excitation voltage minimum 1 500V maximum 3 500V

8.3 ON-SITE TESTS

Any generator installed on a platform or in an onshore oil installation may be assumed to have undergone its full routine check tests, and its prototype a full basic or duplicate test. On-site tests are therefore only needed to check the original installation and commissioning, and thereafter to ensure that no deterioration has taken place. The remainder of this chapter deals only with tests for the latter purpose.

Deterioration can occur for many reasons: among them are entry of dampness or water leakage in the generator or cable-entry boxes, overheating of the windings due to overloading, or mechanical faults such as vibration or bearing failure.

Both dampness and overheated windings can cause reduced insulation resistance of the windings. After drying out, the generator should be megger tested to ensure that insulation resistance has been restored. Deterioration can be progressive, especially when a machine is little used, and a regular programme of megger testing every generator should be drawn up and the results logged. After temperature correction (see para. 8.4), the resistance levels should be plotted, and, if there is progressive deterioration, this will be immediately apparent.

After repairs to a generator, a megger test should normally be carried out before reconnection if the generator or its connections have in any way been interfered with.

High-voltage 'withstand' tests should never be needed on site unless a major overhaul has been carried out, in which case it would be an engineering or manufacturer's concern.

8.4 MEGGER TESTING

'Megger' instruments are provided which operate at 250V, 500V or 2 500V, and the correct one must be used depending on the rated voltage of the generator to be tested. Normally generators over 415V and all high-voltage generators would require a 2 500V megger.

When the megger is connected and the handle wound up, the voltage should continue to be applied until the needle settles down to a steady value; this might take one minute or more.

When testing the insulation resistance of a winding, all other conductors, metalwork (stator and rotor) and auxiliary circuits such as those for thermistor protection and heaters should be connected to the frame with light wire (fuse wire will do), and the test voltage applied between winding and frame. This is to ensure that not only is the insulation to earth satisfactory for the winding under test, but also that it is adequate to other circuits and elements which are not normally at earth potential - e.g. heater elements and circuits. Where the 3-phase windings are independent and brought out to six terminals, it is advisable to make a test also between pairs of windings by removing the star-point links. However, where generators are star-connected with their star-point internal and permanently made, inter-phase tests are not possible.

Insulation resistance is very dependent on temperature, and, in order to compare one reading with another, it is necessary to reduce the value to a common temperature. This is usually 40°C. Unlike the resistance of a conductor, which rises with temperature, the resistance of insulation **falls** rapidly with increase of temperature.

The graph of Figure 8.1 is used to make this correction by means of a 'temperature coefficient'.

For example, if the observed reading (R_t) is 10 megohms when taken at 70°C, then, using the graph, the temperature coefficient (K_t) is 8.0, and the corrected reading at 40°C (R_{40}) is then 10 x 8.0 = 80 megohms. It can be seen from this example that the correction is considerable when the winding is hot at normal working temperatures.

Typical Temperature Coefficient Chart

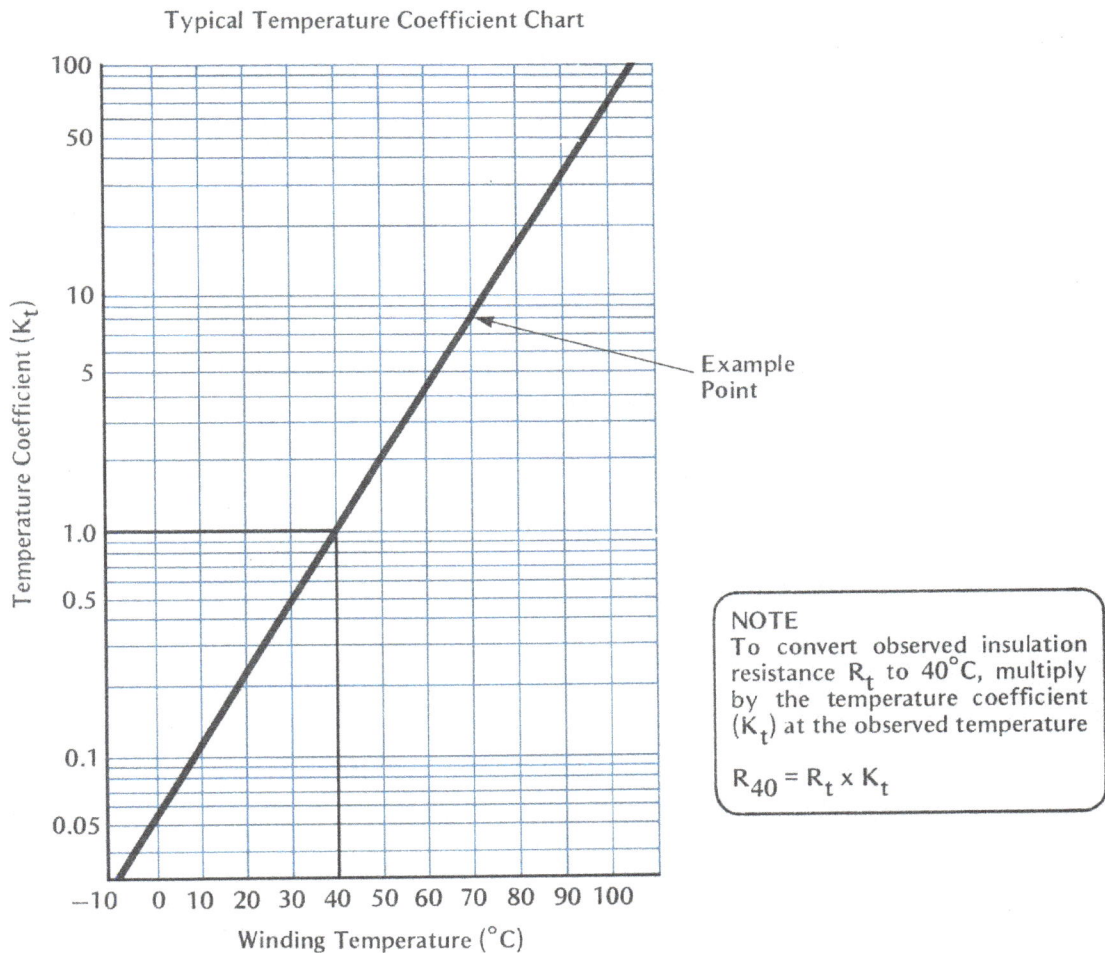

NOTE
To convert observed insulation resistance R_t to 40°C, multiply by the temperature coefficient (K_t) at the observed temperature

$$R_{40} = R_t \times K_t$$

FIGURE 8.1
INSULATION RESISTANCE TEMPERATURE COEFFICIENT

The recommended minimum value of insulation resistance for generators is given in manufacturers' literature. As a guide when precise information is not available, the minimum acceptable value (R_m) for a generator stator winding is given by:

$$R_m = (kV + 1) \text{ megohms when corrected to } 40°C,$$

where kV is the generator's rated voltage in kilovolts. Thus:

for 415V or 440V generators $R_m = 1.4M\Omega$
for 6.6kV generators $R_m = 7.6M\Omega$
for 11kV generators $R_m = 12M\Omega$

CHAPTER 9 USEFUL FORMULAE

POWER:

1 hp	=	0.746 kW$_m$	(= 3/4kW$_m$ approx.)
1 kW$_m$	=	1.34 hp	(= 1 1/3 hp approx.)
1 MW$_m$	=	1 340 hp	

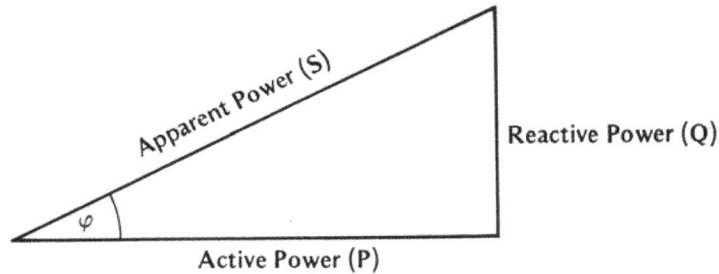

Apparent Power (S)

Reactive Power (Q)

φ

Active Power (P)

POWER TRIANGLE:

$$S = \sqrt{P^2 + Q^2} \quad \text{(volt-amperes)}$$

$$P = S.\cos\varphi \quad \text{(watts)}$$

$$Q = S.\sin\varphi \quad \text{(vars)}$$

EFFICIENCY:

$$\eta = \frac{kW_e\,(\text{output})}{kW_m\,(\text{input})} \quad \text{for a generator}$$

RATING:

If V is rated rms line voltage (3-phase), I the rated rms full-load current and $\cos\varphi$ the rated power factor:

Output (apparent)	=	$\sqrt{3}\,VI$	volt-amperes
Output (active)	=	$\sqrt{3}\,VI\cos\varphi$	watts
Output (reactive)	=	$\sqrt{3}\,VI\sin\varphi$	vars

(**Note**: If kV is used instead of V, the outputs will be in kVA, kW and kvar respectively.)

REACTANCE:

Synchronous reactance (X_d)	- typically 200%
Transient reactance (X'_d)	- typically 30%
Subtransient reactance (X''_d)	- typically 16%

To convert percentage reactance to ohms:

If reactance is $n\%$, V is line voltage and I is rated full-load current,

$$\text{reactance} = \frac{n}{100}\cdot\frac{V}{\sqrt{3}I} \quad \text{ohms per phase}$$

Conversely, if the actual reactance X_d in ohms per phase is known, the percentage reactance n is given by:

$$n = 100\frac{\sqrt{3}IX_d}{V}\%$$

CHAPTER 10 QUESTIONS AND ANSWERS

10.1   QUESTIONS

1.   In a steam turbine what happens to the pressure as the steam passes through?

2.   In a gas turbine why is the exhaust gas temperature monitored?

3.   In a gas turbine why must the air entering the combustion chamber be at high pressure?

4.   How much of the power produced by combustion is needed to drive the air compressor, and why is combustion efficiency so important?

5.   Why does a gas turbine take in far more air than is needed for combustion?

6.   What advantage does a two-shaft gas turbine have over a single-shaft machine?

7.   For speed control, what is the difference between single-shaft and two-shaft machines?

8.   Why can the starter of a two-shaft gas turbine be smaller than the starter of a single-shaft machine?

9.   Why do some large generating sets have ratchet arrangements to turn the turbine rotor for a period after stopping?

10.   How is a catastrophic lubrication failure guarded against?

11.   Why is compressor washing necessary?

12.   Why is it important that a generator should not be overloaded? What is the effect of doubling the load?

13.   To what insulation class are offshore generators specified?

14.   What are the three stages of a generator short-circuit fault, and what level of fault current is passed in each?

15.   Apart from overheating, what effect do heavy short-circuit currents have in generators?

16.   What is meant by 'CAFW' cooling?

17.   What are the three practical methods of generator field excitation?

18.   What is the great advantage of brush less excitation?

19.   What is the disadvantage of static and brushless excitation under fault conditions?

20.   How is this disadvantage overcome?

21.   What precaution must be observed when megger testing a generator field?

22.   What effect do the excitation diodes have upon the AVR response time?

23.   What do you understand by the 'emf' and the terminal voltage of a generator?

24.   A d.c. generator gives 220V on open circuit. If its internal resistance is 2 ohms, what will its terminal voltage be when supplying a load current of 15A?

25.   By what percentage must the excitation of the generator in Q.24 be increased to restore the terminal voltage to 220V?

26.   A single-phase a.c. generator is delivering a current of 20A at 0.8 pf at a terminal voltage of 440V. If the internal resistance is 2 ohms and internal reactance is 10 ohms, make a vector-diagram sketch (to scale) to evaluate the emf E.

27.   What is the power angle under the conditions of Q.26?

28.   A 3-phase a.c. generator is rated 18MVA at 6.6kV. Evaluate the full-load current. If the synchronous reactance is given as 180%, what is its value in ohms per phase?

29.   A single-phase a.c. generator has a no-load voltage of 440V. If its internal reactance is 8 ohms and the internal resistance is neglected, what will be its regulation when delivering full-load current of 40A at 0.8 pf (assuming no change in the excitation)?

30.   Why is speed control of a generator important?

31.   Describe one of the oldest form of speed governor; what is the more modern form?

32.   Can a governor hold generator speed absolutely constant?

33.   What is the basic difference in operation between a flyball mechanical governor and a modern electronic governor?

34.   How does a modern speed governor sense the shaft speed of the generator?

35.   In a single-shaft gas turbine the response to a load increase is slow because of mechanical inertia. How can an electronic governor increase the speed of response?

36.   When is a 'droop' characteristic artificially applied to an electronic governor?

37.   In a two-shaft gas turbine, in addition to the power turbine speed input, what other inputs would the electronic governor typically have?

38.   What overspeed protection is provided for a typical gas-turbine generator?

39.   How is emergency power provided, both onshore and offshore?

40.   What is meant by 'Basic Services'?

41.   How is an emergency generator normally brought into service?

42.   How often should the readiness of the emergency generator be checked?

43.   What on-site generator tests can be carried out by users?

44.   What happens to the insulation resistance of a generator with rising temperature?

10.2   ANSWERS

(Figures in brackets after each answer refer to the relevant paragraphs in the text.)

1.   As energy is used in each stage in turning the turbine, the steam pressure steadily reduces and the steam expands. To compensate for this the turbine wheels increase in diameter from the high-pressure to the low-pressure end.   (1.2)

2.   If it is too high, it indicates a possible combustion fault.   (1.3)

3.   The pressure in the combustion chamber is high, and the air pressure must be higher to enter it.   (1.3)

4.   70% to 80%. A small drop in combustion efficiency therefore causes a greater proportional drop in usable power output.   (1.3)

5.   For cooling.   (1.3)

6.   A proved, aircraft-type jet engine can be used as the 'gas generator'.   (1.4)

7.   In a single-shaft gas turbine the governor controls the shaft speed of the whole machine; in a two-shaft machine the governor controls the speed of the power turbine and driven load only.   (1.6, 1.11 and Chapter 5)

8.   It needs to spin only the compressor/turbine unit of the gas generator. The power turbine, gearbox and generator remain stationary.   (1.7)

9.   To ensure even cooling and prevent 'bowing' of the heavy rotor.   (1.8)

10.   Generator sets normally have an a.c. electric oil pump for pre-start lubrication of bearings, a mechanical pump for normal running use and a d.c. electric pump fed from a battery for emergency use.   (1.9)

11.   Because air pollution can cause encrustation of the compressor blades and reduce their efficiency.   (1.12)

12.   The cooling system will be unable to handle the extra heat generated. At twice the load the heat generated in the stator is quadrupled.   (2.1)

13.   To Class B (ultimate temperature 130°C) or higher (Class F).   (2.4)

14.   Subtransient (typically 6 times full-load current), transient (typically 3 times FLC), synchronous (typically two-thirds FLC).   (2.6)

15.   Mechanical stresses by electromagnetic reaction between current-carrying conductors.   (2.6)

16.   Circulating Air, Forced Water.   (2.5)

17.   Conventional, static and brushless.   (3.2 to 3.7)

18.   The fact that it is brush less, and so requires less maintenance. The only links between the stationary and moving parts are magnetic.   (3.4)

19.   Since both derive excitation power from the generator terminals, under short circuit conditions there may be loss of excitation to maintain the output voltage and thus to operate protective circuits.   (3.5)

20.   By using the short-circuit currents themselves to boost the excitation by means of short-circuit current transformers, or by using a pilot exciter.   (3.6, 3.7)

21.   The diodes must be either disconnected or shorted to prevent breakdown caused by the megger voltage.   (3.8)

22.   They have no effect upon field forcing resulting from an output voltage drop, but delay field reducing, so that response to a voltage rise is appreciably slower.   (3.9)

23.   The emf of a generator is the electromotive force, or voltage, induced in the armature by the relative motion of the field system. It depends on the strength of the field flux and on the speed of relative motion.   (4.1)

        The terminal voltage (V) is the emf due to excitation (E) **less** any internal voltage drops due to the internal impedance of the generator armature or to armature reaction. This difference will in general be vectorial.   (4.2, 4.3)

24.   190V.   (4.2)

25.   By approximately $\dfrac{220}{190}$, i.e. a 15.8% increase.   (4.2)

26.   As Figure.   PF (cos φ) 0.8, therefore φ = 37°

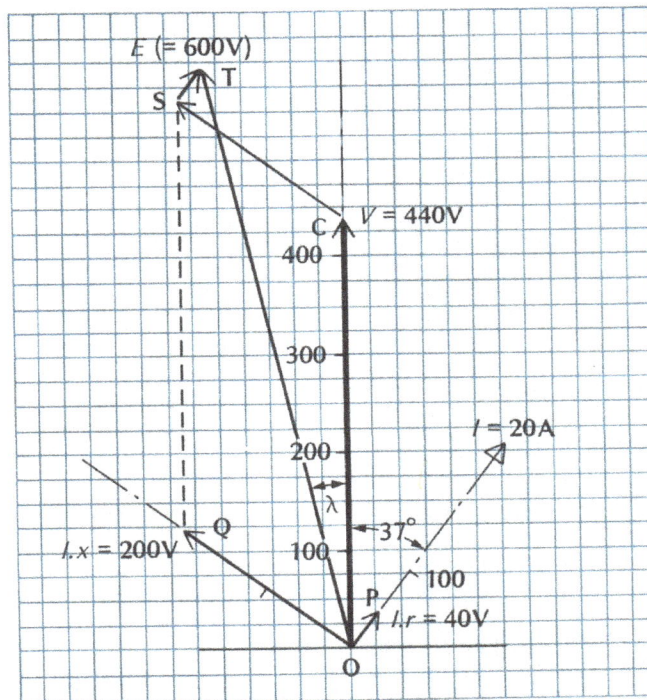

Draw V to scale equal to 440V, and draw I lagging 37° on V.

$$I.r = 20 \times 2 = 40V$$
$$I.x = 20 \times 10 = 200V \text{ (These are the internal voltage drops.)}$$

Scale off $I.r$ and $I.x$ along and leading 90° on I (lines OP and OQ) respectively.

Draw CS and ST respectively equal to each. $\overline{OT}$ is then the emf E under the loading I; it is measured by the same scale as V.

By measurement $E(= \text{OT}) = 600V$.   (4.3)

27. By measurement: $\lambda = 13°$, the power angle. (4.3)

28. Full-load current

$$= \frac{VA}{\sqrt{3}V} = \frac{kVA}{\sqrt{3}kV}$$

$$= \frac{18000}{\sqrt{3}.6.6}$$

$$= 1\,575A$$

   Synchronous reactance

$$= \frac{180}{100} \cdot \frac{6600}{\sqrt{3}.1575}$$

$$= 4.35 \ \text{ohms per phase}$$ (Chapter 9)

29. As Figure.   PF($\cos \varphi$) = 0.8, therefore $\varphi = 37°$.

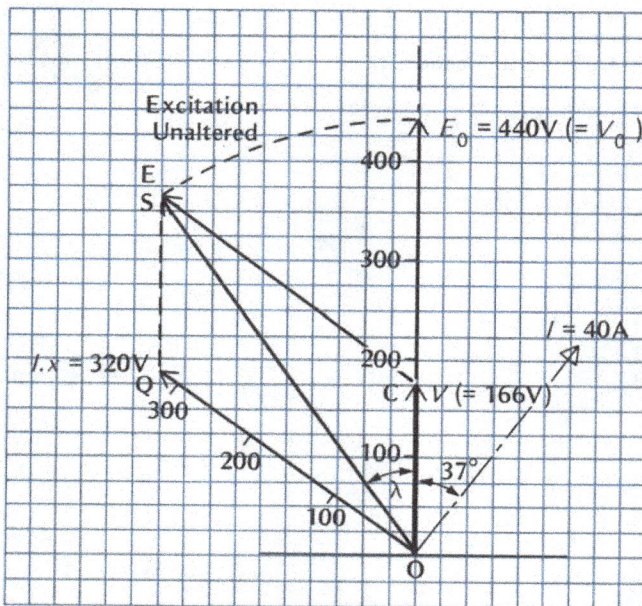

Draw E_0 ($= V_0$) to scale equal to 440V, and draw I lagging 37° on V_0.

$$I.x = 40 \times 8 = 320V \ (I.r \text{ is neglected.})$$

Scale off $I.x$ (line OQ) along line leading 90° on I.
With centre O draw an arc length E_0 to cut the vertical through Q at S.
$\overline{OS}$ is then the emf E under the loading I; it is equal in magnitude to the unaltered excitation E_0 but now leads V.

Draw SC parallel to OQ, cutting the vertical through O at C.
$\overline{OC}$ is then the new terminal voltage V, and SC is the internal voltage drop.

By measurement the terminal voltage $V(= OC) = 166V$.

The regulation $\frac{V_0-V}{V_0} \times 100 = \frac{440-166}{440} \times 100 = 62\%$ (4.4, 4.5)

30.   The generator speed is directly related to the output frequency, which is 60Hz in offshore and 50Hz in most onshore installations. Speed must be held within close limits. (5.1)

31.   Flyball centrifugal governor. Later modernised by the spring-controlled centrifugal governor with hydraulic actuation, and later still by the electronic governor. (5.2, 5.4)

32.   No. Modern governors approach this perfection but never quite reach it. Indeed, in theory, there must always be some degree of error for a servo loop to work at all. (5.2)

33.   None. They are both servo mechanisms, but electronic governors are less susceptible to error and can achieve greater accuracy and consistency. (5.4)

34.   By an inductor-type tacho-generator in which an iron toothed wheel rotates past fixed coils. (5.4)

35.   By sensing the **load increase** as soon as it occurs without waiting for the **speed decrease**. (5.5)

36.   When generators are running in parallel. Load sharing would be unstable without droop. (5.10)

37.   Figure 5.7 shows:

        Gas-generator speed limitation
        Exhaust temperature control limitation
        Acceleration limitation
        Base load limitation
        Peak load limitation
        Idle clamp limitation. (5.7)

38.   An independent magnetic pick-up, similar to that used for speed sensing, but operating an overspeed trip unit (5.6). For some sets this is backed up by a centrifugally operated mechanical cutout and trip. (5.9)

39.   By a diesel-driven generator. (6.1)

40.   Those essential services which it is vital to keep running even when normal main power has been lost. (6.2)

41.   Automatically upon failure of the normal supply. (6.2) (On some platforms however only manual starting is provided.)

42.   Daily. (6.3)

43.   Visual examination and megger tests. (8.3)

44.   It falls. Megger insulation tests should be made periodically between windings and earth and, if possible, between phases. Insulation readings should be corrected to 40°C and plotted. The curve so formed will give a clear indication of any deterioration. (8.4)

PART 3 ELECTRICAL POWER DISTRIBUTION

CHAPTER 1 THE BREAKING OF A.C. CIRCUITS

1.1   GENERAL

The purpose of any switching device is to make and interrupt current. This function applies to any such device from the largest circuit-breaker down to the smallest relay.

The flow of electric current in a circuit is a form of kinetic energy, and it cannot be stopped instantaneously. The energy must be taken out of the system for all current to cease, and this takes a finite time, no matter how small it may be.

1.2   D.C. INTERRUPTION

When the contacts of a switch separate while carrying direct current, that current cannot be stopped there and then but will continue to flow between the opening contacts in the form of an arc. The arc is a column of white-hot gas which is ionised by its heat and provides a conducting path for the current. This path has appreciable resistance which increases as the contacts continue to separate and the arc lengthens. Eventually the resistance becomes so high that the current falls to the point where it can no longer maintain the arc temperature; ionisation ceases, the air ceases to conduct and the arc goes out, finally stopping the current. All energy has been dissipated in the heat of the arc.

1.3   A.C. INTERRUPTION

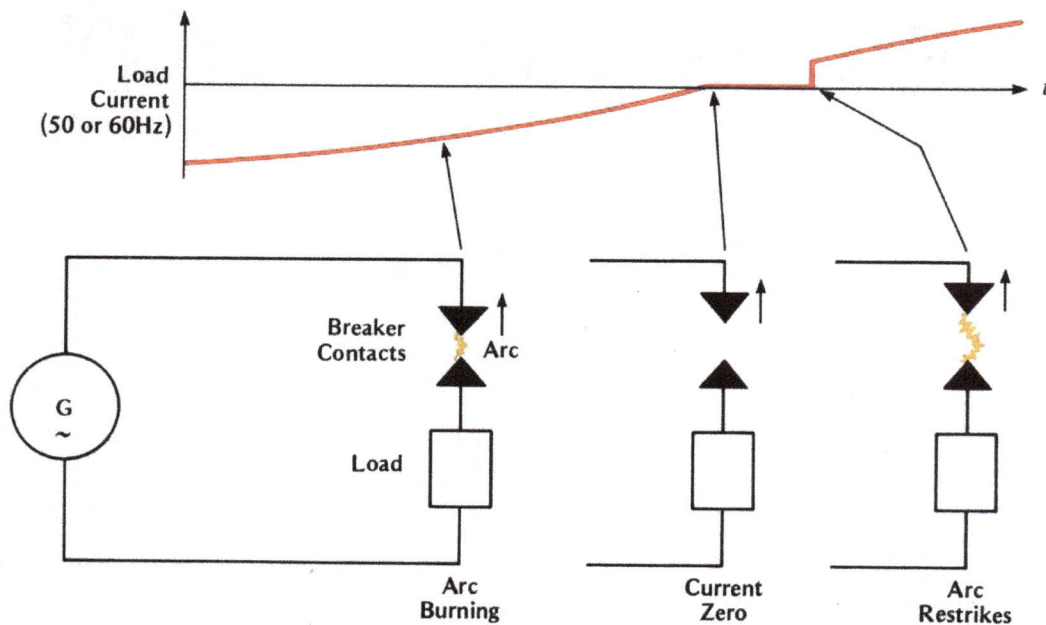

FIGURE 1.1
A.C. CURRENT BREAKING

The mechanism of current interruption in an a.c. switch is totally different from that of d.c. With direct current the arc continues uninterrupted until it can no longer be maintained, after which it goes out and stays out.

With alternating current, however, at each half-cycle the current passes through a natural zero, and the arc which had been carrying that current goes out momentarily for lack of heat. Under the action of the voltage which immediately reappears across the open contacts the gap breaks down, the arc 'restrikes' and will continue to carry current for the next half-cycle until, at the next current zero, it goes out again and the process repeats. This is shown in Figure 1.1.

As the contacts continue to separate, the arc path becomes longer; eventually the gap at a current zero becomes too great for the voltage to break it down, so that the arc does not restrike. The current has then been finally interrupted, and the energy has been dissipated in a series of successive arcs.

It should be particularly noted that, whereas in d.c. the arc will continue unbroken until it is finally suppressed, in a.c. it extinguishes itself naturally twice every cycle of current, restriking each time until it is no longer able to do so.

From here on only a.c. breaking is considered.

1.4   FAULT CURRENTS

When a short-circuit occurs, it may be between two of the three lines of a 3-phase system, or it may involve all three. The fault current may pass between phases as an arc, which has some resistance and so limits the current, or there may be metal-to-metal contact, a

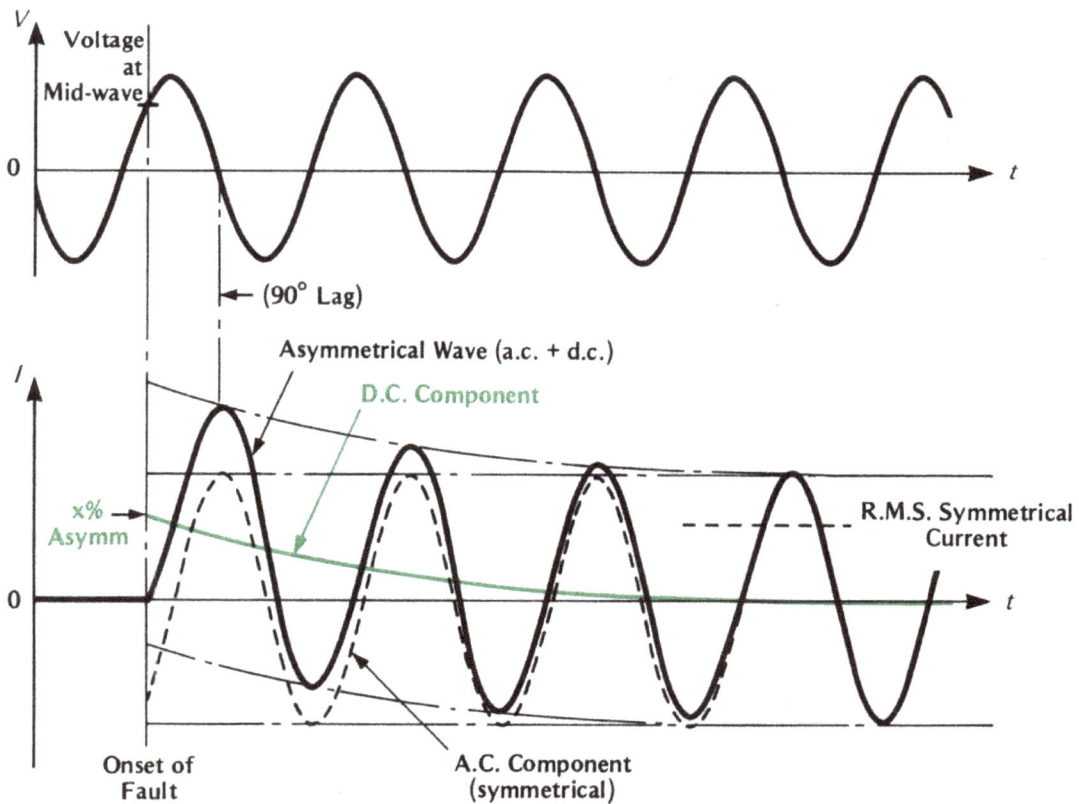

FIGURE 1.2
PARTIAL CURRENT ASYMMETRY AT ONSET OF A FAULT

so-called 'bolted' fault, where the impedance is zero. As an item of switchgear must be able to deal with the most severe possible case it is always assumed that the fault is a 3-phase bolted one, and that the whole circuit is mainly inductive with little resistance.

It was shown in Part 1, Chapter 32, that with an inductive fault the current which immediately follows is in general partially asymmetrical. The asymmetry will be complete (100%) if the fault occurs at the instant of a voltage zero. If it occurs at a voltage peak, positive or negative, the asymmetry is zero (0%) - that is to say, the current wave is then wholly symmetrical.

Figure 1.2 shows the general case where the asymmetry is partial (between 0% and 100%). The point on the voltage wave at which a fault may occur is of course entirely random. So therefore is the degree of asymmetry which will occur in any particular case.

This asymmetrical current wave is regarded as resolved into two parts: a symmetrical a.c. wave plus a steady but decaying 'd.c. component' whose rate of decay is mainly the R/L ratio of the fault circuit; the d.c. component is shown in green in Figure 1.2. The quicker the decay of the d.c. component, the quicker the fault current resumes symmetry.

With complete asymmetry the first peak of the asymmetrical current wave is almost double the amplitude of the a.c. component at that time - that is $2 \times \sqrt{2}(= 2.82)$ times its rms value.

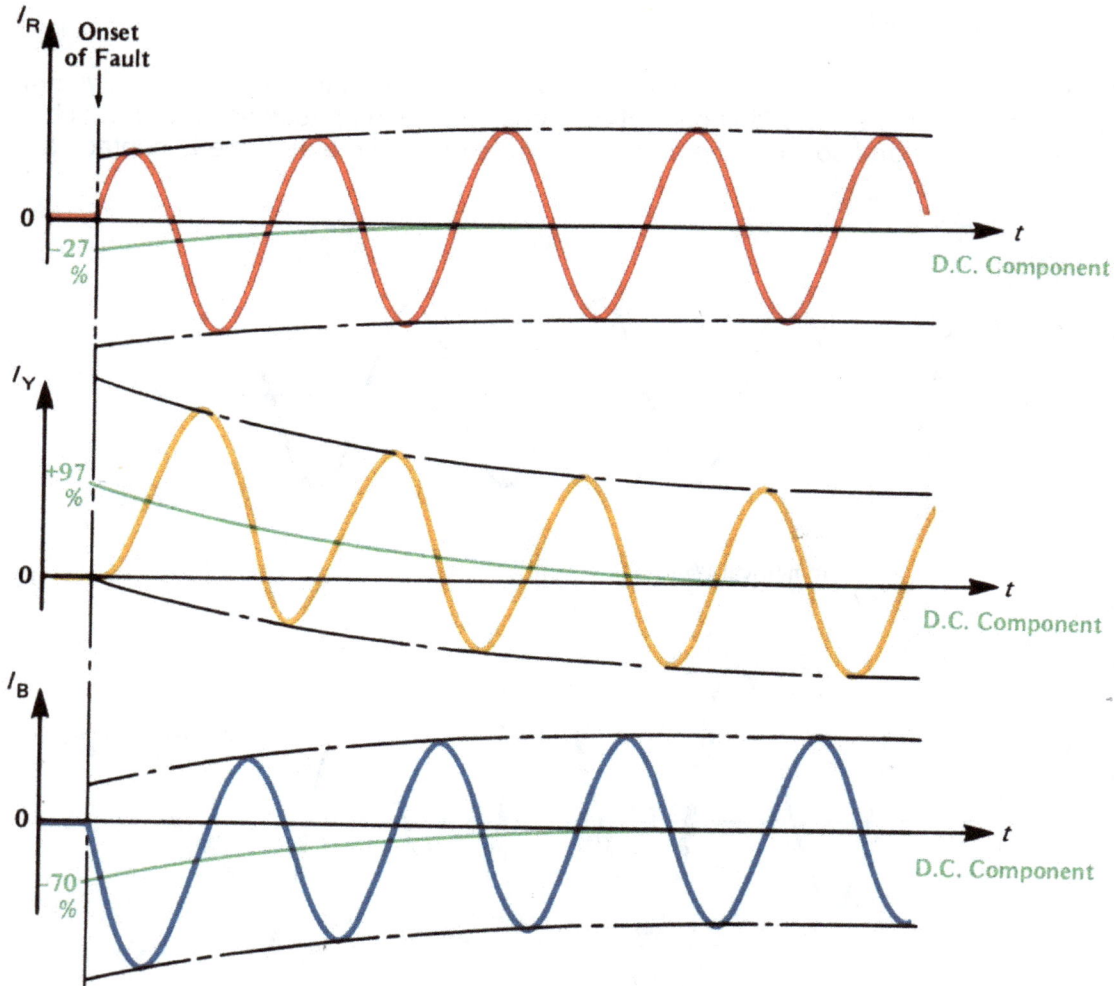

FIGURE 1.3
THREE PHASE CURRENT ASYMMETRY (GENERAL CASE)

However by the time the first current peak is reached there has already been some decay of the d.c. component, and it is usual to take the first current peak as approximately 2.55 times the rms value of the a.c. component. This figure however may differ slightly in special cases.

Although Figure 1.2 shows the a.c. component as having constant amplitude, it does in fact gradually reduce in size as the current moves from its initial subtransient value towards the transient.

For reasons which it is not necessary to go into here, an asymmetrical current is less difficult to break than a symmetrical one. Therefore, in order that the circuit-breaker is able to deal with the most difficult case, it is required, when testing, that the bolted fault shall continue long enough for the d.c. component to decay to a specified level. This level depends on the opening time of the breaker itself (i.e. from instant of trip signal to separation of contacts) and may be of the order of 15%, after which the current is regarded as 'symmetrical'. At Switchgear Testing Stations deliberate delay is introduced between the onset of the fault and the trip signal to the breaker on test to ensure that this is so.

In 3-phase switching the asymmetry will in general be partial in all phases, as shown in Figure 1.3, and the percentages, taking account of sign, will always add up to zero. (In Figure 1.3 they are - 27 + 97 - 70 = 0). If one phase happens to be symmetrical (0% asymmetry) the other two, being displaced 120°, must both be partially asymmetrical.

The breaking capacities of circuit-breakers are always rated in kA (or MVA) **'rms symmetrical'**. The peak asymmetrical current rating in kA may additionally be given. All system fault calculations are made to determine the required rms symmetrical breaking current rating of the switchgear to be installed (see Part 10 'Electrical Protection'). From here on only symmetrical faults will be considered.

1.5   RESTRIKING VOLTAGE

As previously mentioned, the cause of the contact gap breaking down and the arc restriking after a current zero is the voltage which immediately reappears across the gap. It is necessary now to look more closely at the nature of that voltage.

Figure 1.1 shows in simple form the steps in the interruption of an a.c. current supplying a load.

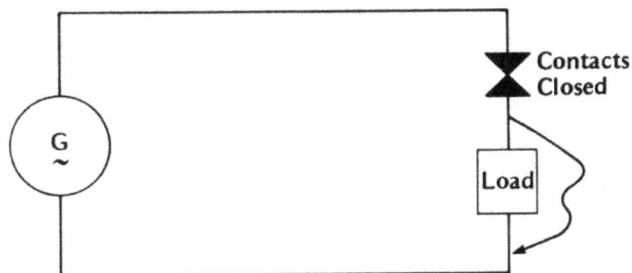

**FIGURE 1.4
LOAD SHORT CIRCUITED**

In the worst case the contacts might have to break the full fault current if the load were short-circuited; the contacts would then be interrupting a dead short-circuit across the generator, as shown in Figure 1.4. This worst case will be considered first; also for clarity, discussion will be kept to single-phase circuits, as in Figure 1.4, although it applies equally to 3-phase circuits.

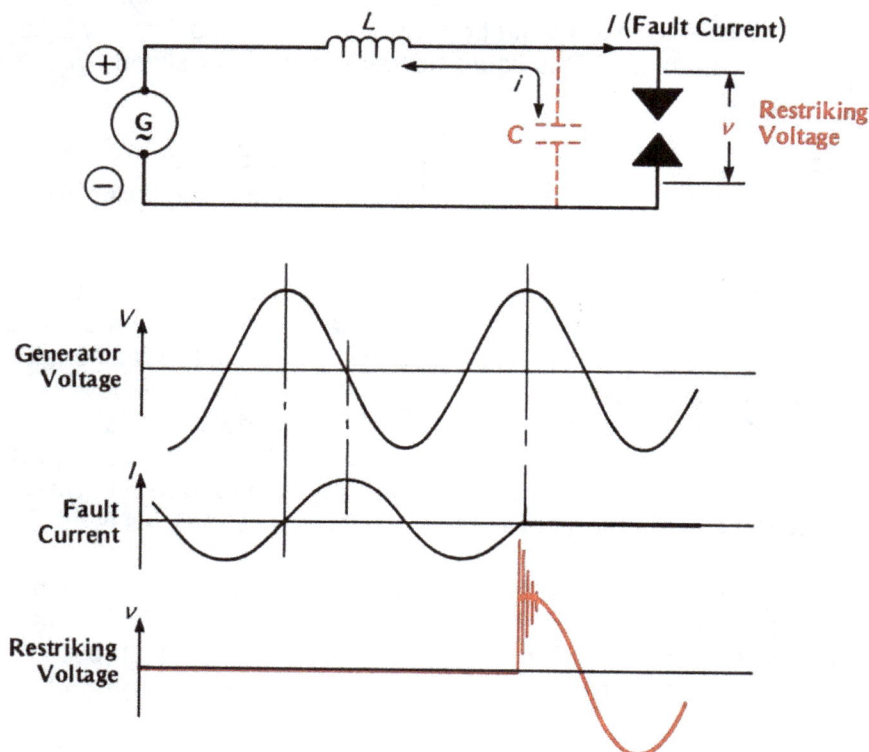

FIGURE 1.5
RESTRIKING VOLTAGE

Consider the idealised circuit at the top of Figure 1.5. Supply is taken from an a.c. generator (50Hz or 60Hz) which is being short-circuited through the circuit-breaker contacts and whose current I the breaker has to interrupt. The generator has internal reactance (and also some resistance, which is neglected here). This reactance is represented by the series inductance L. Figure 1.5 shows the breaker with its contacts separated but with the arc momentarily extinguished at a natural current zero.

The circuit also has a certain amount of self-capacitance, especially in any cable feeder system as well as in the generator itself. Such capacitance is small and is of course distributed, but for simplicity suppose it to be 'lumped' as a capacitor C as shown in red in the figure.

At the onset of the fault the contacts are closed and are carrying the fault current I; they also short-circuit C which is therefore in a discharged state. Even when the arc is established as the contacts begin to separate, C remains effectively short-circuited by the low-resistance arc and therefore discharged.

In Figure 1.5 the upper waveform shows the generator's 50Hz or 60Hz alternating voltage. Before the fault a small current was passing, but with the short-circuit a large current starts to flow, limited only by the generator's reactance. The fault current is therefore lagging almost 90° on the voltage, as shown in the middle waveform of Figure 1.5.

At the first current zero, which therefore occurs at a voltage peak, the arc momentarily goes out but immediately restrikes. This may occur for several zeros, but eventually the gap becomes long enough to prevent a restrike, and the arc stays extinguished.

The short-circuit is now removed from C, which at that instant is still discharged. Suppose the generator's voltage peak at this instant of current zero was such that the positive and negative polarities were as shown in the figure. Then the positive terminal of the generator, which now has C straight across it, will start to charge C, through L, to its own potential, the charging current being i. When C reaches the same potential as that of the generator, the charging current i is flowing strongly through the inductance L and cannot

236

stop. It continues to overcharge C until it reaches double the potential of the generator, by which time i has stopped. But then the overcharged C starts to discharge through L back into the generator. Again i, now reversed, cannot stop, and C discharges again almost to zero. Then the whole cycle repeats.

Thus there is an oscillating charge/discharge current between C and the generator which will continue until it is damped out by such resistance as is present. (Electronics students will recognise this as a 'ringing circuit'.) On each 'swing' the charge voltage of C oscillates between zero and twice the generator peak voltage. But the voltage across C is also the voltage between the open contacts of the breaker and is in fact a 'restriking voltage' appearing across the gap, trying to break it down and re-establish the arc. It succeeds in doing so for several half-cycles, but eventually the gap becomes too long and the arc does not restrike; the full voltage oscillation then takes place. After damping has caused the oscillation to die out, the voltage across C, and so across the open contact gap, settles down to the generator peak voltage. This high-frequency voltage reappearing across the gap is shown in the lowest of the three waveforms of Figure 1.5.

The frequency of the oscillation is very high, chiefly because C is very small. It is given by

$$ f = \frac{1}{2\pi}\sqrt{\frac{1}{LC}} $$

and typically may be of the order of 20kHz. It is so fast that, by comparison, the 50Hz or 60Hz of the generator may be regarded as 'd.c.' and considered not to alter during the very short time of the oscillation. At a frequency of 20kHz the time from current zero to the first peak of restriking voltage is only 25 μs - a very short time indeed.

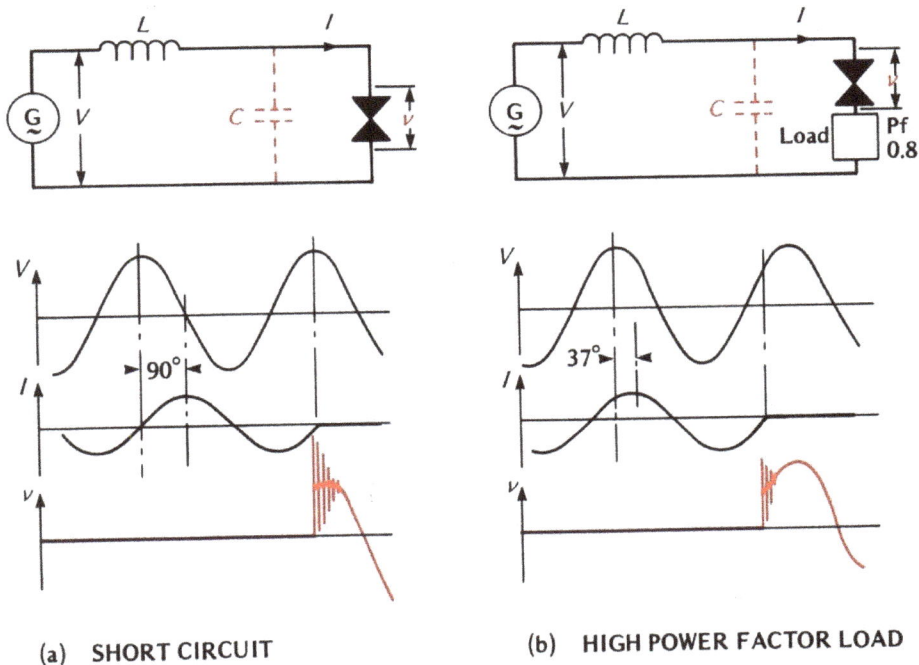

(a) SHORT CIRCUIT (b) HIGH POWER FACTOR LOAD

FIGURE 1.6
EFFECT OF POWER FACTOR ON SWITCHING

This double-height, high-frequency restriking voltage appearing across a breaker when a large a.c. current is broken will travel along all connected conductors as a steep-fronted voltage wave where it could, like a lightning strike, cause insulation problems to connected apparatus with high inductance, such as transformers. This is the well-known 'switching surge' found in all electrical systems both large and small.

The high-frequency oscillating current will also radiate and may give rise to radio interference.

Up to this point only the worst case has been considered - that is, with the circuit-breaker clearing a generator short-circuit. The current is then limited only by the generator reactance and therefore lags nearly 90° on the voltage as in Figure 1.6(a) - a very low power factor.

If the circuit-breaker, instead of clearing a short-circuit, were merely opening a normal load current where the load may have a high power factor, the switching situation would be much eased.

The current is of course lower, which means that there is less energy to dissipate. But also the power factor is much higher (say 0.8), so that the current lags by only a small angle on the voltage (37° if pf = 0.8). Consequently the current zero occurs not opposite a voltage peak but at a point well down the voltage curve, as shown in Figure 1.6(b).

After extinction of the arc at a current zero the discharged capacitance C recharges not from the generator **peak** voltage but from a voltage much less than this - see Figure 1.6(b). The restriking voltage wave, which oscillates to twice the generator voltage **at that moment**, is therefore much smaller than in the low power-factor case.

Breaking a high power-factor current thus results in a considerably lower restriking voltage peak and a generally smaller restriking voltage wave. The lower peak means less liability to restrike, and the breaker will therefore clear in fewer half-cycles of arcing. The travelling restriking wave will also be less intense and is less likely to pose insulation problems on the system.

In short, therefore, the duty on a circuit-breaker is much eased if the power factor of the load to be broken is high. Conversely it is increased with low power factors, that is with highly inductive loads (for example an open-circuited transformer) and especially with short-circuits.

The extra strain with inductive loads is well known in d.c. switching. It occurs also with a.c. switching, but for a totally different reason, as explained above.

It is for this reason that testing authorities lay down that, when a circuit-breaker undergoes short-circuit tests, those tests shall be carried out at a power factor of not more than 0.15.

1.6   OIL CIRCUIT-BREAKERS (OCB)

Oil circuit-breakers are little used on offshore installations but are very common onshore. They are usually confined to high-voltage systems.

An oil circuit-breaker consists of sets of breaking contacts for each phase totally immersed in oil, either in a common tank or in three separate tanks. A mechanism, which may be mechanical, hydraulic or pneumatic in operation, drives all the moving contacts in unison to close the circuit.

It will be recalled from para. 1.5 that the time from a current zero to the first restriking voltage peak is of the order of 25 μs. After an arc has gone out naturally at a current zero, it leaves behind in the gap a mass of vaporised and ionised oil of poor dielectric strength which the reappearing restriking voltage will easily break down.

To prevent a restrike it is necessary to flush out this polluted oil and oil vapour from the gap and to replace it with cool, fresh, clean oil - and all within a matter of microseconds.

Various oil circuit-breaker manufacturers have developed ingenious devices for doing this. They are nearly all, however, variations of a basic concept called the 'cross-jet explosion pot'. This is an insulated, fire-proof pot which surrounds the contacts under the oil and is arranged with ports. It uses the energy of the arc itself to build up internally a great pressure of oil which, at the instant the arc goes out at a current zero, is released and causes a clean, cool supply of unburnt oil to blast away the polluted oil through the ports. The explosion pot, in one form or another, is a well-established arc control device for oil circuit-breakers.

After an oil circuit-breaker has cleared a fault it should be able to continue in service without attention. But it is nevertheless customary to lower the tank at the first opportunity and examine the contacts for wear or burning, and also to replace the oil.

1.7   AIR BREAK CIRCUIT-BREAKERS (ACB)

Air-break circuit-breakers operate with their contacts in free air. Their method of arc control however is entirely different from that of oil circuit-breakers, for it depends on the suppression of the restriking voltage. This gives them, as will be explained, a very different performance characteristic. They are always used for low-voltage interruption and are now tending to replace high-voltage oil breakers up to 11kV and even higher, very largely because of their performance. HV circuit-breakers on most offshore installations are of the air-break type.

Figure 1.7 illustrates the principle of air-break operation. There are differences in detail between various manufacturers, and the method shown is only typical.

The fixed contact is on the left, and the moving contact assembly is hinged. While the breaker is closed, the 'main' contacts (1) which carry the steady load current are held tightly closed by the operating mechanism or latch.

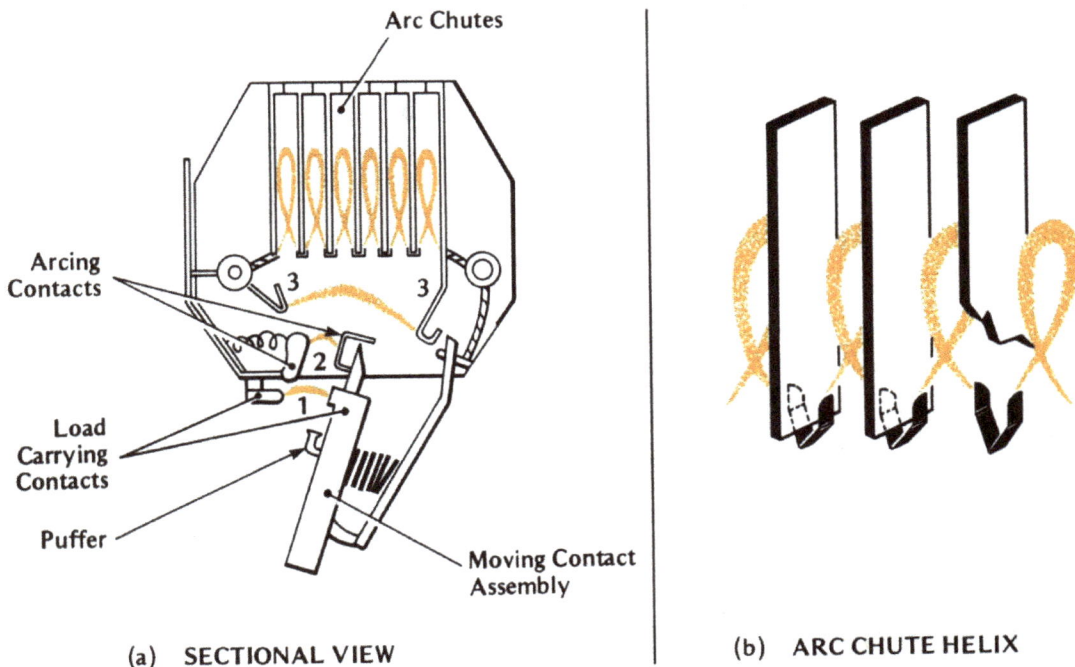

(a) SECTIONAL VIEW   (b) ARC CHUTE HELIX

FIGURE 1.7
AIR BREAK CIRCUIT-BREAKER INTERRUPTION

When released by tripping, the hinged assembly moves to the right, first separating the main contacts across which an arc will form.

Under the electromagnetic forces due to the fixed-contact/arc/moving-contact current loop the arc is driven upwards. During this stage it transfers from the main to separate 'arcing contacts' (2). The main contacts are thereby relieved of any further burning. Still being driven upwards the arc transfers to metal horns (3) on the base of a box called the 'arc chute'.

This box is made of insulating and fire-proof material. It is divided into many sections by barriers of the same material, as shown in Figure 1.7(a). At the bottom of each barrier is a small metal conducting element between one side of the barrier and the other.

When the arc, driven upwards by the electromagnetic forces, enters the bottom of the chute, it is split into many sections by the barriers, but the metal pieces ensure electrical continuity between the arcs in each section; the several arcs are thus in series.

The electromagnetic forces within each section of the chute cause the arc in that section to take up the form of a helix, as shown in Figure 1.7(b). All these helices are in series, so that the total length of the arc has been greatly extended, and its resistance is much increased. This has the effect of reducing the current in the circuit.

Figure 1.7(a) shows the progress of the arc from the time it leaves the main contacts until it is within the arc chute.

When the current next ceases at a current zero, the ionised air in the path of where the arc had been is in parallel with the open contacts and acts as a shunt resistance across both the contacts and the self-capacitance C, shown in Figure 1.8 in red as a high resistance R.

When the oscillation starts between C and L as described for the oil circuit-breaker and shown in Figure 1.5, this resistance damps the oscillation heavily. Indeed it is usually so

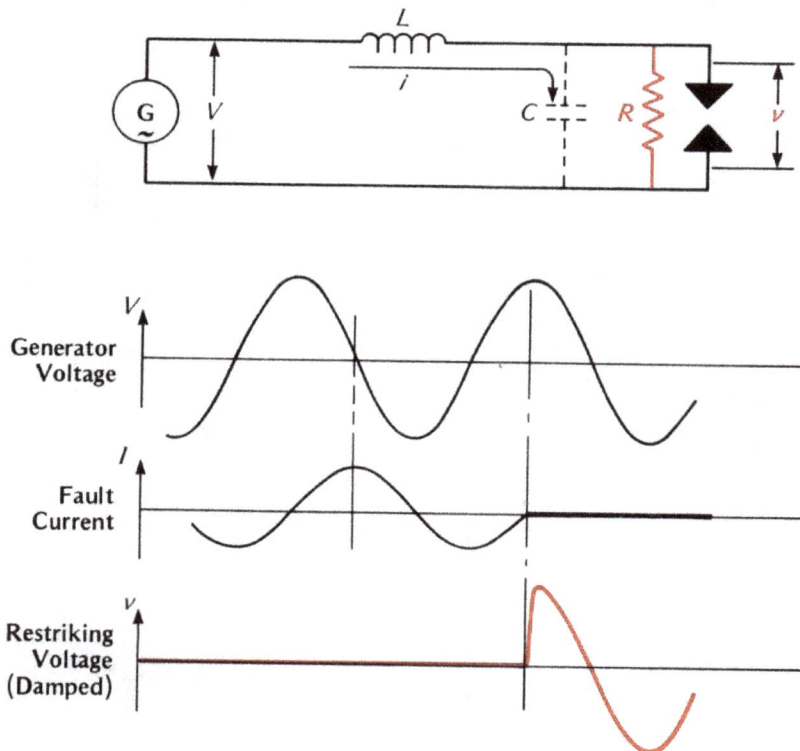

FIGURE 1.8
AIR BREAK CIRCUIT-BREAKER RESTRIKING VOLTAGE

heavy that the damping is 'critical'; the oscillation cannot then take place at all, and the restriking voltage, instead of appearing as a high-frequency oscillation, rises 'dead-beat' to its eventual value of peak generator voltage. This is shown on the lowest waveform of Figure 1.8, which should be compared with that of Figure 1.5.

The effect of this damping is important. Because there is now no restriking voltage peak, and because there is no arc 'residue' across the small main gap, there is less likelihood of the contact gap restriking. The air-break circuit-breaker's performance is consequently good, in that it is able to break circuits in very few half-cycles.

The other great advantage is that, because there is no steep-fronted restriking voltage wave, there is virtually no switching surge to travel down the system and strain the insulation of connected apparatus. There is, of course, practically no radio interference.

It is for these reasons that the air-break circuit-breaker is becoming increasingly attractive to system engineers, and design efforts are being directed to extending it to ever higher voltages and breaking capacities.

There is one feature about the air-break circuit-breaker which however should be mentioned. Its operation depends upon the arc being driven electromagnetically upward from the main contacts into the chute. Whereas with heavy currents these forces are more than adequate, with light currents they are weaker and the drive is less positive. Air-break circuit-breakers are therefore usually fitted with a 'puffer'. This is a small air tube under each arc. Air is compressed in a cylinder operated by the tripping mechanism and blows any reluctant arc upwards into the chute. The puffer can be seen in Figure 1.7(a).

When breakers are tested for performance, tests are always made at light currents as well as with full rated current to prove this feature.

1.8   VACUUM CIRCUIT-BREAKER (VCB)

The vacuum circuit-breaker is becoming increasingly popular, especially in the medium ranges of voltage, because of its good performance and its compactness.

Its method of arc control differs from those of both the oil and the air-break circuit-breaker. Whereas the oil breaker functions by flushing out the 'combustion' products of the arc, and the air-break type by suppressing the restriking voltage wave, the vacuum breaker operates by denying the arc any medium in which to re-form.

Figure 1.9 shows the elements of a vacuum interrupter, which is described in greater detail in Chapter 2. It consists of a glass bottle evacuated to a very high degree; the vacuum area is shaded blue. The contacts inside the vacuum space are of the butt type, and the moving contact, sealed through bellows, travels only a very short distance from the fixed contact when opening. When the contacts first separate, normally in mid-cycle, an arc is drawn and the current continues to flow between them in the arc, supported by vaporised and ionised metal from the contacts.

At a current zero the arc momentarily ceases, and the vaporised metal instantly recondenses onto the contact surfaces, leaving no gas present in the vacuum chamber. When the restriking voltage reappears across the open contacts there is no gas dielectric to break down and therefore no vehicle to support a new arc. The arc consequently does not restrike. It is claimed that vacuum interrupters will break an a.c. circuit in one to two half-cycles - quicker than any other type of breaker.

Because of the recondensing of the vaporised material onto the contact surfaces, there is very little net loss of contact metal over a large number of operations, and therefore very

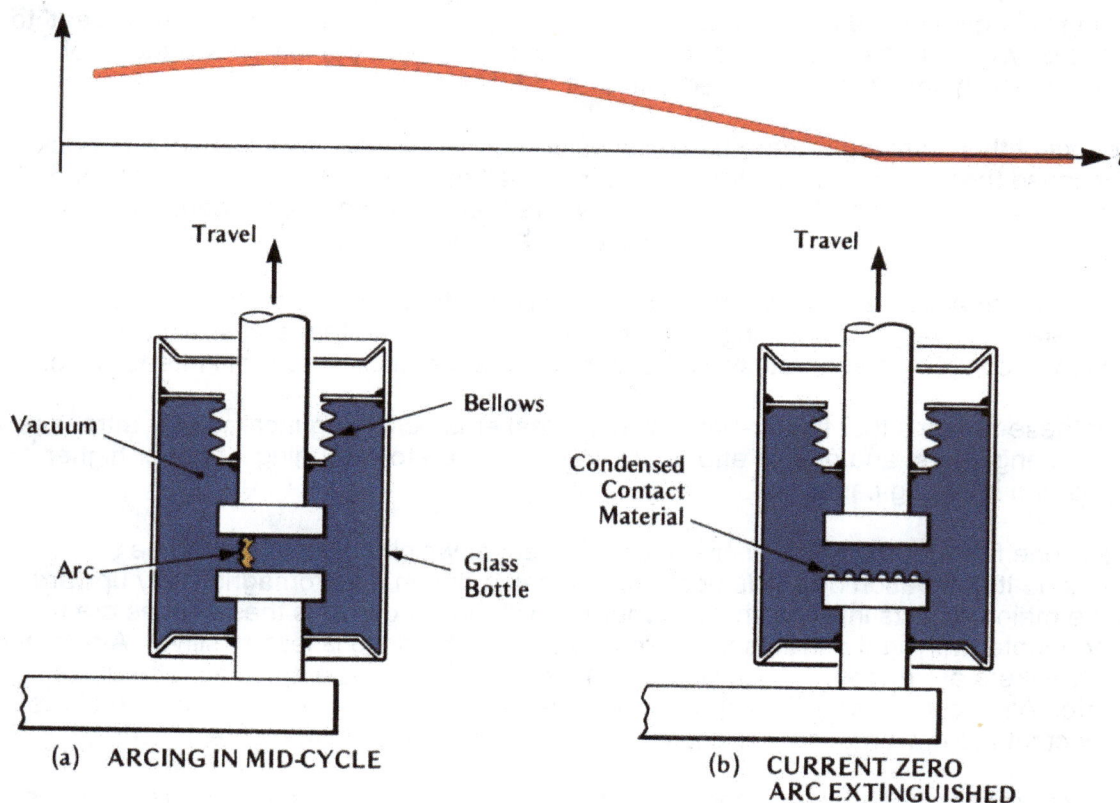

FIGURE 1.9
VACUUM INTERRUPTION

little contact 'wear'. This renders them highly suitable for repeated use as contactors. Such wear as there is is measured by feeler gauge on the operating linkage, and, when it reaches a certain level specified by the manufacturer, the whole vacuum interrupter element must be replaced. Because the contact travel is so small it is recommended that all three bottles should be replaced and the new ones set up and aligned together.

1.9   OTHER TYPES OF CIRCUIT-BREAKER

Circuit interruption by oil, air-break and vacuum circuit-breakers has been dealt with in detail in this chapter, but these three do not exhaust the list.

Air-blast circuit-breakers are widely used onshore, but they are confined to extra high voltage and high breaking capacity transmission systems. They function similarly to oil breakers except that the arc products are forcibly blasted away by the release of compressed air instead of by oil.

Another type is the 'SF$_6$' breaker, which uses sulphur hexafluoride in place of oil. The advantages of using this substance as an insulating and interrupting medium in circuit-breakers arise from its high dielectric strength and outstanding arc-quenching properties. SF$_6$ circuit-breakers are much smaller than air-break circuit-breakers of the same rating, the dielectric strength of SF$_6$ at atmospheric pressure being equal to that of air at 10 atmospheres. The SF$_6$ decomposition products, discharged as a gas following extinction of the very hot arcs, are harmless, but they contain a small amount of fluorine which may react with metallic parts of the breaker.

Neither the air-blast nor the SF$_6$ breaker is at present used in installations either offshore or onshore, and they are not further discussed here.

CHAPTER 2 HIGH VOLTAGE SWITCHGEAR

2.1   GENERAL

The subject of 'Switchgear' is regarded as covering all types of switching devices such as circuit-breakers, contactors and hand-operated switches, as well as fuses and protective devices like relays.

Switchgear is used on both high-voltage and low-voltage systems. It is required to enable generators, feeders, transformers and motors to be connected to and disconnected from the high-voltage or low-voltage system. This switching is necessary both for normal operational purposes and for the rapid disconnection of any circuit that becomes faulty. The switchgear also allows any circuit to be isolated from the live system and for that circuit to be made safe so that work may be carried out on equipment connected to it.

This chapter deals with switching devices as applied to high-voltage systems; low voltage is covered in Chapter 3. Fuses and relays are dealt with in Part 10 'Electrical Protection'. Two types of HV switchgear are considered in this chapter:

(a)   **Circuit-breakers**

Circuit-breakers are used to control generators, transformers, bus-sections, bus-couplers, interconnecting cables between switchboards and the starting and running of very large motors; they are designed to make and break full fault currents. A circuit-breaker may be of the oil-break, air-break or vacuum-break type. Because of the fire hazard only air-break and vacuum-break units are used on most offshore installations, but oil-break circuit-breakers are widely used onshore. They are not designed for continuously repeated operation.

(b)   **Contactors**

Contactors are used to control motor circuits and sometimes transformers. They may be of the air- or vacuum- break type; each type is used both onshore and offshore.

Contactors are designed only to make and carry fault current for a short time, not to break it. Where the system fault level exceeds the limited breaking capacity of the contactor, fuses are inserted in series with the contactor contacts. Contactors are designed to undergo repeated and frequent operation without undue wear.

A switchboard may be made up of a mixture of circuit-breaker and contactor cubicles, depending on the nature of the individual loads and the distribution requirements.

A circuit-breaker or contactor has five ratings:

Voltage. This is the nominal system voltage at which the switch will operate without breakdown.

Normal Current. This is the current which the switch will carry continuously without overheating.

Breaking Capacity. This is the maximum fault current (expressed in kA (rms) or MVA) which the switch will interrupt on all three phases.

Making Current. This is the maximum peak asymmetrical current (expressed in kA) that the switch can carry in any pole during a making operation.

Short-time Rating. This is the maximum time (usually specified as 3 sec or 1 sec) for which the switch will carry, without damage, the full fault current before that current is broken.

The theory and manner in which the various types of circuit-breaker and contactor extinguish the arc and interrupt the current is dealt with fully in Chapter 1. The descriptions which follow are concerned with the actual hardware, its operation and its assembly into switchboards.

2.2   OIL CIRCUIT-BREAKERS (OCB)

In an oil circuit-breaker the contacts operate in a tank of oil (sometimes in three separate but smaller tanks). There are usually two breaks in series per phase, the moving contacts moving together in a downward vertical direction inside the tank. Around each of the six fixed contact tips is placed an arc-control device, usually of the 'cross-jet explosion pot' type. A sectioned view showing one phase of an OCB is seen in Figure 2.1, which shows the circuit-breaker unit in its fixed housing and in the 'Service' position.

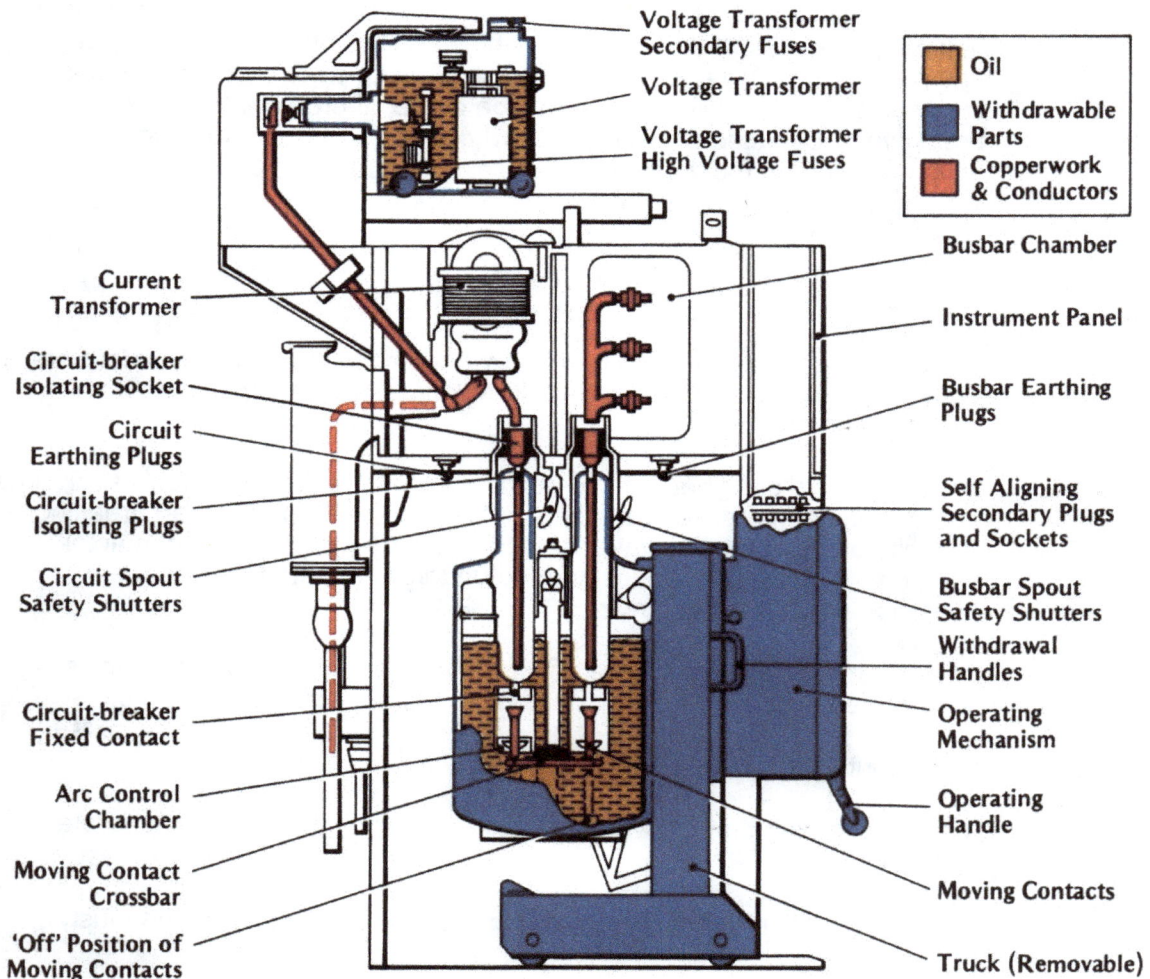

FIGURE 2.1
TYPICAL HIGH VOLTAGE OIL BREAK SWITCHGEAR UNIT (VERTICAL ISOLATION)

Part 3 Electrical Power Distribution

The withdrawable truck is shown in blue, the main copperwork, busbars and conductors red, and the oil is shaded yellow.

The circuit-breaker and its tank, with its six external terminals (two per phase), can be withdrawn vertically downwards clear of the corresponding plug-type connections in the main housing. When so withdrawn the circuit-breaker proper is electrically isolated from the busbars and feeder, and automatic shutters close over the six fixed live connections or 'spouts'. Isolation may be vertical (as in Figure 2.1) or horizontal. After isolation the whole circuit-breaker unit in its tank, together with its operating mechanism, can be drawn out clear of the housing horizontally on its wheels for examination or maintenance.

Mechanical interlocks are provided to ensure that the circuit-breaker unit can never be isolated unless it has first been opened, and that it cannot be reinserted into the housing unless it is already open.

Where a panel is fitted with a voltage transformer (VT), this is mounted in a separate compartment above the cable connection compartment. In Figure 2.1 the VT is oil-immersed in a tank, which can be withdrawn for isolation. A VT may be connected either to the feeder side or to the busbar side of a circuit-breaker, depending on its application, and it is protected by high-voltage fuses mounted inside the VT compartment (also shown in Figure 2.1).

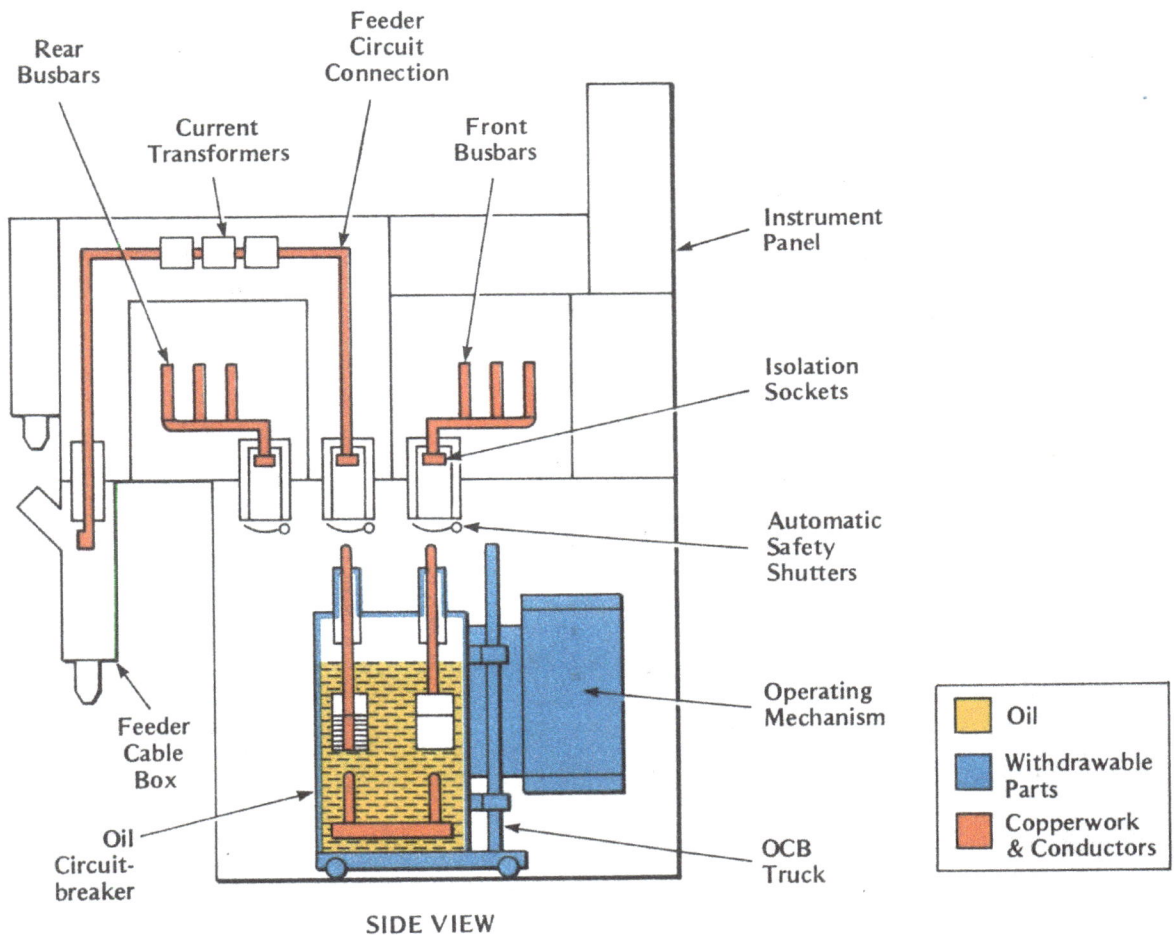

FIGURE 2.2
TYPICAL HIGH VOLTAGE OIL BREAK SWITCHGEAR UNIT
ON DUPLICATE BUSBARS (VERTICAL ISOLATION)

Operating mechanisms to close the breaker may be solenoid, motor/spring, pneumatic or hydraulic and are described more fully in para 2.3.3. The mechanism must be strong enough to close the breaker positively against the 'throw-off' forces of the maximum short-circuit current that can possibly occur at the point in the system where the circuit-breaker is installed (see Part 10 'Electrical Protection'). Circuit-breakers latch-in when closed and are tripped by a separate shunt-trip coil, always energised from an independent d.c. battery-supported source. In a few cases a no-volt coil is used for tripping.

After clearing a full-scale fault an oil circuit-breaker should be able to continue in operation without attention, but it is customary to withdraw it at the first suitable opportunity to examine the contacts and replace the oil.

Oil circuit-breakers, being usually large, are not often made up into switchboards but are assembled into sets with common busbars running through and, where possible, mounted out-of-doors. They are sometimes totally enclosed, with their busbar system, in individual iron enclosures - the so-called 'iron-clad' or 'metal-clad' switchgear. The controlling switchboard would be a separate unit indoors.

On some installations the switchboard contains two independent busbar systems, and the circuit-breaker can be connected to either set. This is the so-called 'duplicate busbar' system.

A typical duplicate busbar arrangement, using vertical isolation, is shown in Figure 2.2. Selection between the rear and front busbars is achieved by moving the circuit-breaker truck backwards or forwards to its correct position under the rear or front pairs of fixed isolation contacts (or 'spouts'). One set of spouts, the feeder connection, is common to both rear and front sets. In the figure the circuit-breaker is shown in the position to engage the front busbars. When in position under the selected spouts, the circuit-breaker is raised to make contact with the chosen busbar system. Interlocks ensure exact positioning before the breaker is raised.

Duplicate busbar systems may also be arranged for horizontal isolation, but the vertical arrangement is more common. They are used more often with oil circuit-breakers than with other types.

Duplicate busbar systems are fairly common in onshore installations. They are also used on some offshore platforms.

2.3   AIR BREAK CIRCUIT-BREAKERS (ACB)

2.3.1 Air Break Circuit-breaker Panel

An air-break circuit-breaker panel is made up of two parts: a fixed portion or 'housing' and a removable circuit-breaker truck. The precise layout varies from one manufacturer to another; a typical arrangement is shown in Figure 2.3.

The unit shown in the figure is horizontally isolated. The circuit-breaker truck (blue) can be clearly seen in the 'Service' position with the breaker's moving contacts open and with the arc chutes above. In this case the busbars run along the bottom of the housing, with the feeder cable box at the back. The busbars and other main copperwork are shown red.

The fixed part is divided into four compartments, each one separated from the others by an earthed sheet-metal barrier. At the front of the panel are two compartments; the upper one is a low-voltage compartment housing low-voltage control equipment, associated fuses and auxiliary cable terminations. The lower one houses the circuit-breaker truck. A door is fitted at the front of each of the compartments. At the rear of the panel are two further compartments each with a bolted cover. The lower compartment houses the busbars and the upper the feeder cable terminations and current transformers, as shown in Figure 2.3.

REAR   FRONT

Voltage Transformer
(in isolated position)   High Voltage
                                Fuse (withdrawn)

Voltage Transformer
(in service position)   Interlocked Access
                                                               Door to Voltage
Feed Shutters   Transformer Compartment

Current Transformers   Low Voltage
                                                               Auxiliary Connection
                                                               Compartment

Circuit Earthing
Switch   Circuit-breaker
                                                               Compartment
Insulating Spouts

Plug and Socket
Isolating Contacts   Air Circuit-
                                                               breaker Truck
Cable Connection   (in disconnected
Compartment   position)

                                                               Moving
                                                               Contacts

Busbars   Air Circuit-
                                                               breaker Truck
Earth Bar   (in service
                                                               position)
Busbar Chamber

Busbar Shutters

Withdrawable
Parts

Copperwork
& Conductors

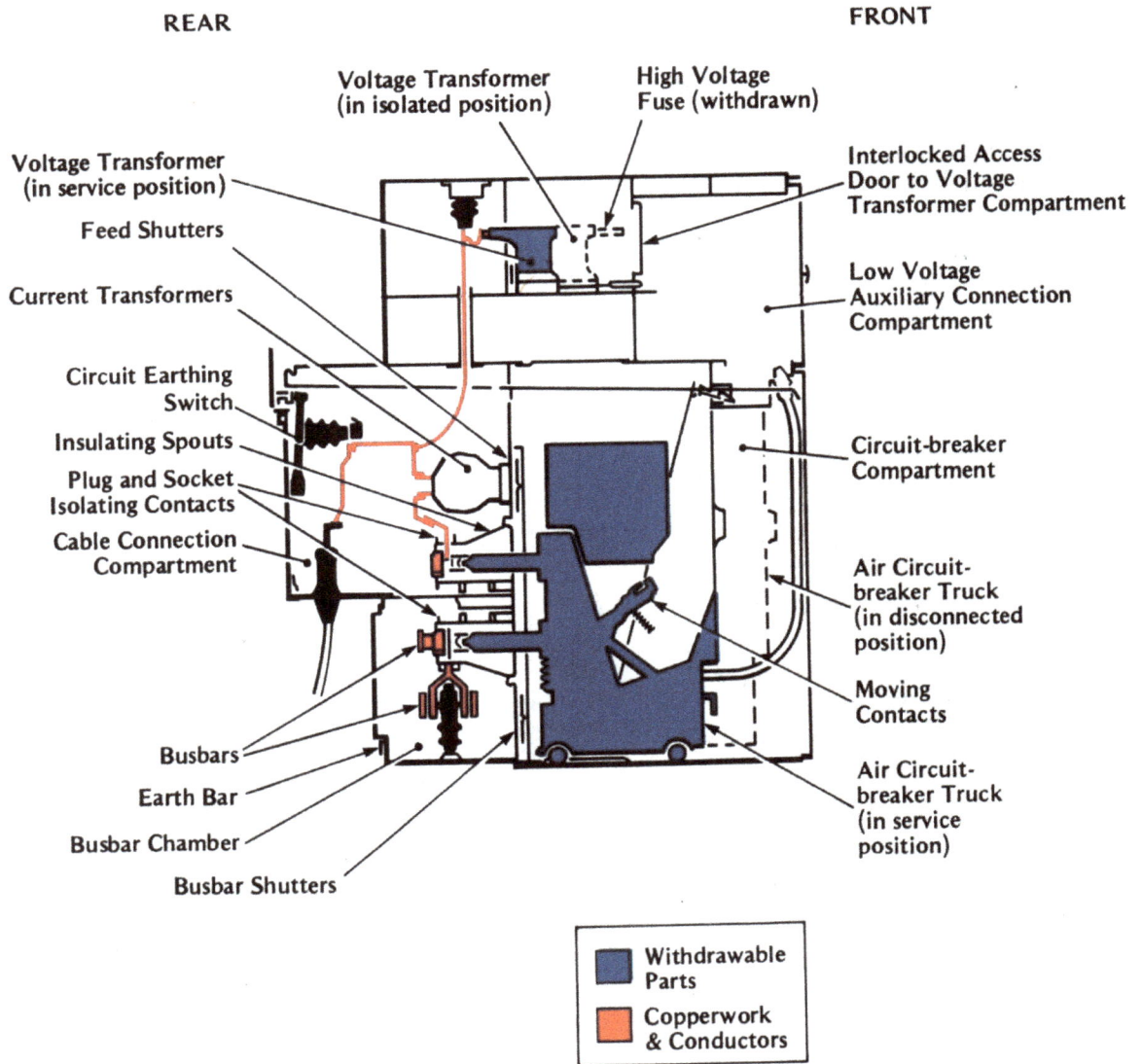

FIGURE 2.3
TYPICAL HIGH VOLTAGE AIR BREAK SWITCHGEAR UNIT
(HORIZONTAL ISOLATION)

The circuit-breaker is mounted on a truck supported on rollers which can be moved in a fore-and-aft direction within the circuit-breaker compartment, or removed from it. At the back of the circuit-breaker compartment are two steel shutters, one to cover the fixed feeder contacts and the other the busbar contacts. When the circuit-breaker is moved into the 'Service' position, the shutters open to allow plugging contacts at the rear of the circuit-breaker to enter the insulated spouts which house the fixed feeder and busbar contacts. The shutters provide a safety barrier to prevent human contact with live metal when the circuit-breaker is disconnected or removed from the panel.

Where a panel is fitted with a VT, this is mounted in a compartment above the cable connection compartment. A voltage transformer may be connected either to the feeder side or to the busbar side of a circuit-breaker, depending on its application, and it is protected by high-voltage fuses mounted inside the VT compartment (also shown in Figure 2.3). The cover of this compartment can only be removed when the VT has been isolated.

On the dead-front panel in front of the VT compartment may be mounted indicating instruments, controls and the protective relays concerned with that item of switchgear.

2.3.2 Air Break Circuit-breaker Truck

A typical circuit-breaker on its truck is shown in Figure 2.4. The switching contacts are shown open, and above the fixed contacts are the arc chutes. On the left are the two sets of three isolating contacts, with their heavy contact tips, which engage with the fixed busbar and feeder sets of isolating contacts in the housing. When the truck is withdrawn, safety shutters drop to cover the live fixed contact spouts; these shutters are automatically lifted as the truck is reinserted into the 'Service' position.

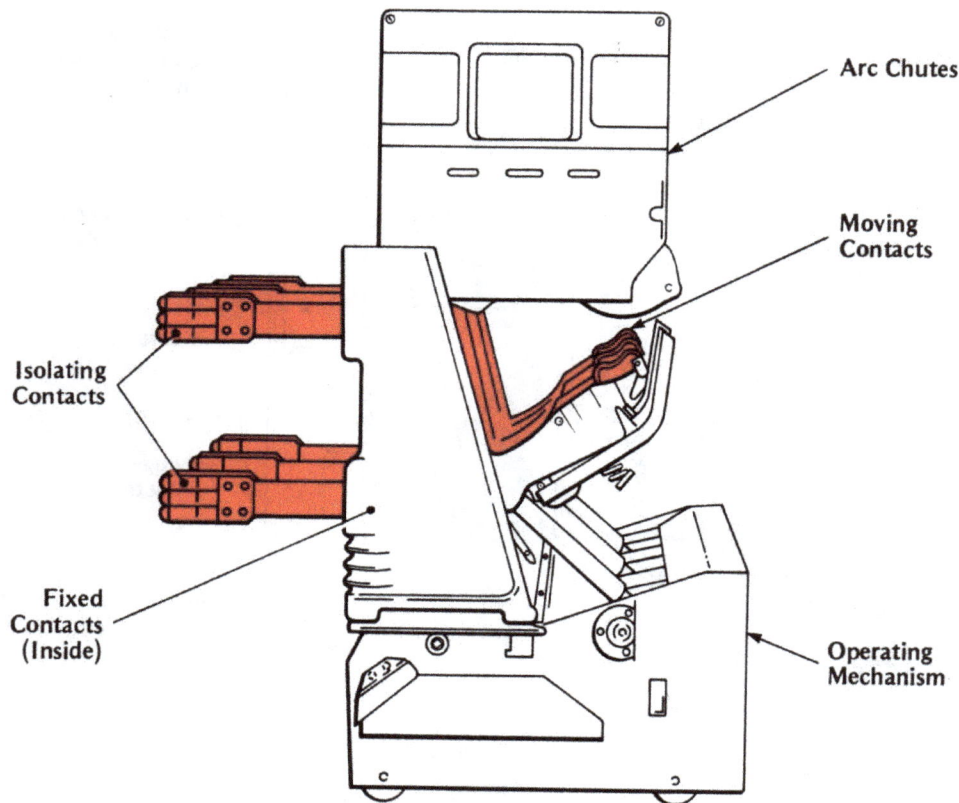

FIGURE 2.4
TYPICAL AIR BREAK CIRCUIT-BREAKER AND TRUCK

The motor/spring operating mechanism is housed in the base. On the face on the right-hand side (not seen in the figure) are the handle for manually charging the spring, mechanical indicators for showing whether the spring is charged or discharged, indicators for showing whether the breaker is open or closed, and mechanical pushbuttons for releasing the spring (i.e. closing) and for tripping the breaker; these are marked 'I' and 'O' respectively.

2.3.3 Operating Mechanisms

The circuit-breaker closing mechanism for either an oil-break or air-break circuit-breaker is usually solenoid or motor/spring powered. The solenoid uses d.c., usually at 110V, and operates on the breaker linkage. Alternative hand operation is provided.

The closing mechanism operates through a toggle link which, when closed, is 'over-toggled' and therefore solid. When a trip signal is given, a trip coil breaks the toggle and allows the circuit-breaker to open under its trip springs. If the breaker should be closed onto a fault, the toggle breaks immediately, so allowing the breaker to re-open freely. This arrangement, known as a 'trip-free' mechanism, protects the operator against injury by the breaker's throw-off forces if he unwittingly attempts to close it by hand onto a fault.

A motor/spring mechanism charges a powerful helical spring by motor. When charged, the spring latches and remains so until released. On receipt of a closing signal, the latch is released and the spring closes the breaker through its linkage. Immediately the breaker has closed, a limit switch causes the motor, usually d.c. operated, to recharge the spring ready for the next closing operation. In the event of motor power failing, the spring can be recharged manually by a detachable handle. The full operating circuit is explained in para. 2.7.

An essential difference between the solenoid and the motor/spring closing systems is that the solenoid takes a very large current during its quick stroke, necessitating heavy cable leads, whereas the motor takes about five seconds to recharge the spring - that is, to deliver the same energy - so that the motor draws far less current. The motors can, if the system requires it, operate on a.c.

Both systems require a separate trip signal to actuate the shunt-trip coil; this releases the closed mechanism and causes the breaker to open under its own separate trip springs. It can also be tripped mechanically by hand.

A number of complete housings, together with their withdrawable circuit-breakers, can be assembled into a single switchboard, as shown typically in Figure 2.5.

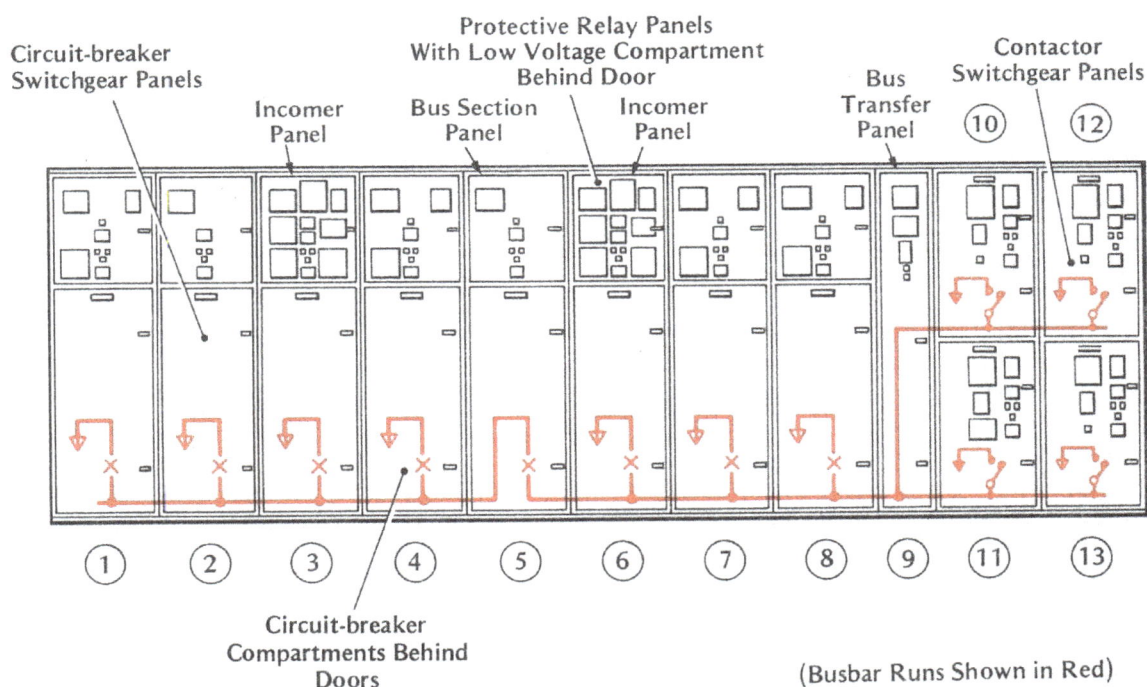

FIGURE 2.5
TYPICAL HIGH VOLTAGE SWITCHBOARD

The panels, whose upper parts carry the local controls, protective relays and indicating instruments, form a continuous switchboard. The circuit-breakers are at the bottom behind doors which can be opened for mechanical control of the breaker or for withdrawing it.

The switchboard acts as an electrical 'manifold', with a common set of busbars running the length of the board. In most offshore boards the busbars run along the bottom, sometimes encapsulated in epoxy resin, and tee-offs are made from them to the lower contacts of each circuit-breaker (Panels 1-8) and contactor (Panels 10-13). This is shown in red in Figure 2.5 for a typical high-voltage switchboard.

At a busbar transfer panel, such as Panel 9, the copperwork run is changed from the bottom level (for the circuit-breakers) to two levels for the tiers of contactors. This arrangement is also shown in Figure 2.5. Sometimes (as here) use is made of the front of a narrow transfer panel to accommodate extra relays.

At a bus-section panel, Panel 5, the bottom run on one side passes into the bus-section circuit-breaker, and from its top contacts back to the bottom level to feed the circuit-breakers on the other side. By studying Figure 2.5 it will be seen that, even when the busbars on one side of a section breaker have been made safe for maintenance by complete isolation and earthing down, the bus-section cubicle may still have live copperwork on the other side of the circuit-breaker. Particular care is therefore necessary when working on, or near to, bus-section cubicles.

Switchboards other than the one described and made by other manufacturers, will differ in the detail of their busbar arrangements. Nevertheless the principles explained above will apply in all cases.

2.4 AIR BREAK CONTACTORS

Contactors (sometimes called Motor Switching Devices (MSD) when used with motors) are generally smaller and lighter than circuit-breakers operating on the same system because they carry less load current and, more important, they do not have to break full system fault current. However they must be able to **carry** it for the specified time (3 sec or 1 sec), to close onto it positively without bounce, arcing, or welding, and to operate repeatedly.

For these reasons high-voltage contactors only occupy half the space taken by the circuit-breaker, and they are usually mounted in pairs, one above the other, in a panel the same size as for one circuit-breaker. Four such contactor panels can be seen on the right of the switchboard in Figure 2.5.

The contactor itself operates on the same principle as the air-break circuit-breaker, with contacts in air and an arc chute. The closing mechanism is always solenoid-operated, and particular attention is paid to the linkage to make it robust and suitable for repeated operation, such as for motor starting.

Contactors may be 'latched' or 'unlatched'. In the former case they hold-in by latch after closing, and the closing solenoid is then de-energised. They require a separate trip signal to open them by shunt-trip coil, as with the circuit-breaker.

Much more common, however, is the unlatched contactor. It is held closed by its closing solenoid, usually with reduced current through an economy resistor. The solenoid remains energised throughout. When it is desired to trip the contactor, the solenoid is simply de-energised and the contactor 'falls off'. It also falls off if the supply to the solenoid fails. Such contactors are said to have an 'inherent undervoltage' feature, which means that, unlike the latched type, they will open automatically if the operating voltage fails.

Because a contactor breaks normal currents but cannot break a heavy fault current, it is usually backed up by fuses (see Part 10 'Electrical Protection'). These are a set of high rupturing capacity (HRC) fuses in series. They are so chosen that, with currents in excess of those which can be safely broken by the contactor, the fuses will blow first, so interrupting the current and leaving the contactor to open on a dead circuit. These fuses, which are quite large, are usually mounted at the bottom of the contactor unit, which must be withdrawn in order to get at them.

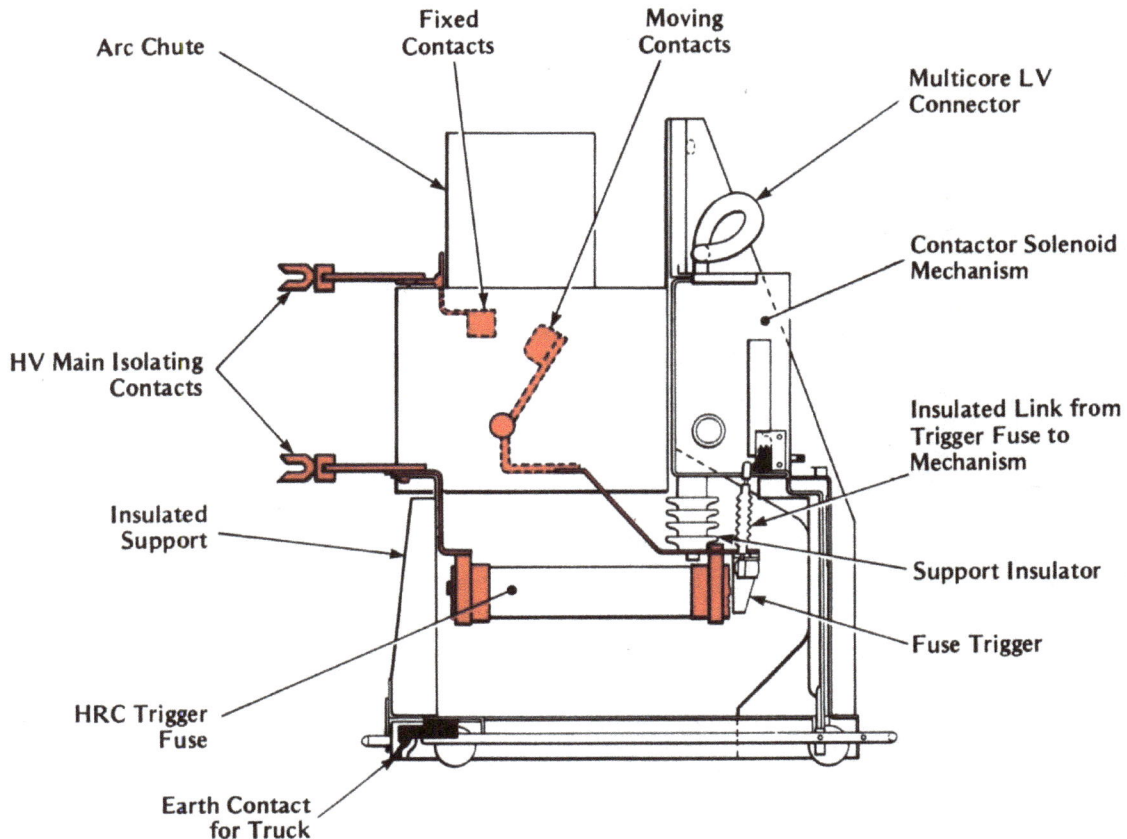

FIGURE 2.6
HIGH VOLTAGE CONTACTOR UNIT

Figure 2.6 shows a typical half-height HV contactor unit. The fuses can be seen at the bottom. The contactor unit, with fuses, is horizontally isolated and can be withdrawn on rails for servicing, maintenance or fuse changing. A special carriage with rails is required to withdraw an upper unit. The whole unit shown in Figure 2.6 is withdrawable clear of the switchboard.

The HRC fuses are of the 'trigger' type. The blowing of any one fuse will mechanically release the contactor, so ensuring a break in all three phases.

2.5   VACUUM CIRCUIT-BREAKERS AND CONTACTORS

Circuit-breakers and contactors which use vacuum interrupters form another class of switching device. The principle upon which vacuum interrupters control the arc and break heavy currents is described fully in Chapter 1. The paragraphs which follow describe how the interrupters are actuated and how the circuit-breaker or contactor is integrated into a switchboard panel.

The only important differences between a vacuum circuit-breaker and a vacuum contactor are that the circuit-breaker has a much larger rated breaking capacity and is latched, whereas the contactor is usually unlatched. Apart from these features and, of course, size, the description which follows applies to both.

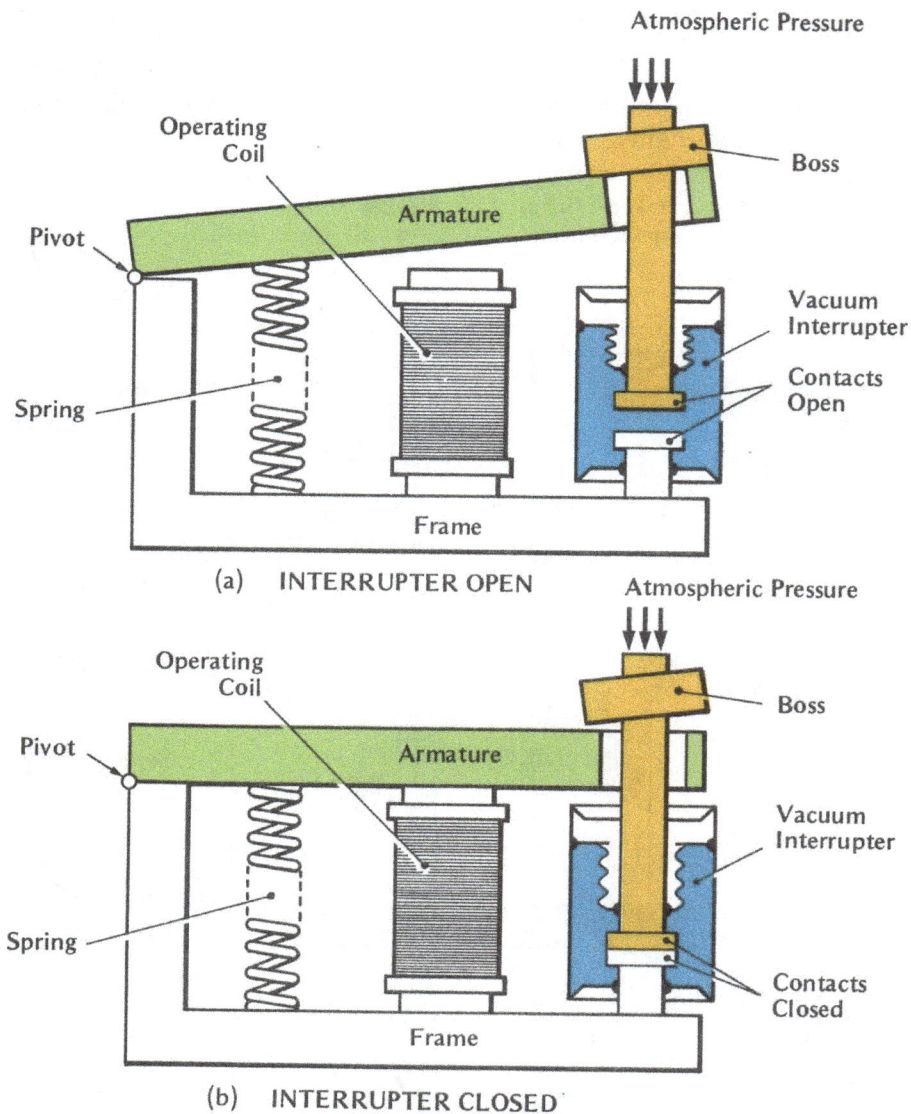

Atmospheric Pressure

Operating Coil

Pivot

Armature

Boss

Spring

Vacuum Interrupter

Contacts Open

Frame

(a)   INTERRUPTER OPEN

Atmospheric Pressure

Operating Coil

Pivot

Armature

Boss

Spring

Vacuum Interrupter

Contacts Closed

Frame

(b)   INTERRUPTER CLOSED

FIGURE 2.7
VACUUM CONTACTOR

A typical vacuum interrupter element, together with its operating mechanism, is shown in Figure 2.7. Figure 2.7(a) shows the interrupter in the open position. The armature arm (green) is held open by a compression spring, and the moving contact (yellow) is lifted by a boss on it into the open position and held there. The vacuum area is coloured blue; it is contained on its upper side by a bellows to take up the movement of the upper contact.

When the closing coil is energised it attracts the armature downwards away from the boss (Figure 2.7(b)). The moving contact is then free to move and is forced downwards by atmospheric pressure, opposed only by the vacuum. It should be particularly noted that, with this design, contact pressure is maintained by atmospheric pressure only.

To open the contacts the closing coil is de-energised, the spring takes over and drives the armature upwards. It strikes the boss and takes the moving contact sharply with it against the atmospheric pressure, so returning to the position of Figure 2.7(a). This mechanism will be seen to be of the normal unlatched type, used when operating as a contractor, which opens on the de-energising of the operating coil. When used as a circuit-breaker the motion would be latched and a shunt-trip coil used to release it.

Other vacuum switches, particularly vacuum circuit-breakers, have different types of mechanism and do not always use atmospheric pressure. Those which latch require a separate tripping coil which usually operates by breaking the latching toggle.

Because vacuum interrupters are sealed for life, there is no question of contact replacement. Since the vaporised metal is always redeposited on the contacts, wear is in any case small. It can be detected by indicator or the use of feeler gauges. If the wear of any one of the set of three exceeds the manufacturer's recommendation, all three interrupters must be replaced since the small contact travel makes the adjustment critical.

The 3-phase vacuum interrupter unit, together with its operating mechanism and back-up fuses, is vertically isolated within its own panel. The whole unit is drawn downwards by operation of an external handle, as shown in Figure 2.8.

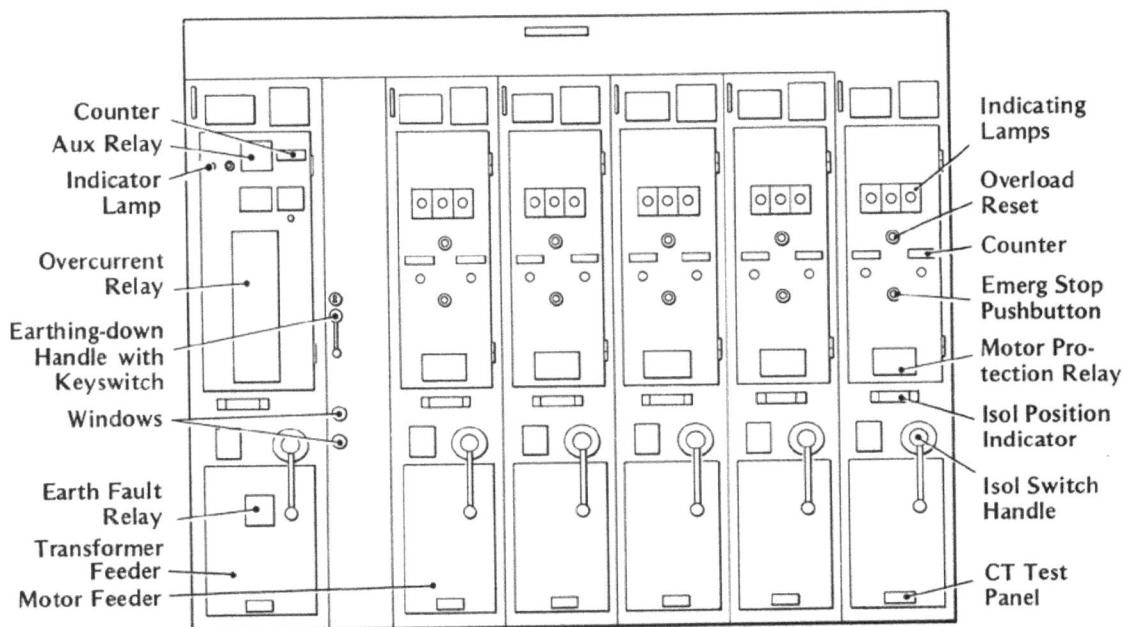

FIGURE 2.8
TYPICAL VACUUM CONTACTOR SWITCHBOARD

Interlocks are provided to prevent isolation unless the interrupter is open. In those contactor panels which feed motor circuits the act of downward isolation also automatically puts an earth on the isolated feeder cable. In contactor panels which feed transformers the feeder earth is separately applied by an external earthing handle which cannot be moved until the interrupter unit is fully isolated.

A set of vacuum contactor panels, sometimes also with vacuum circuit-breaker panels, can be assembled to form a complete switchboard, as shown in Figure 2.8. Vacuum switch panels are very narrow compared to the equivalent air-break circuit-breaker panels and lend themselves well to very compact, space-saving switchboards.

2.6   SWITCHBOARD DISTRIBUTION

A high-voltage switchboard is an assembly point which receives power from the HV generators or other sources, controls it and distributes it. A common busbar system runs through the board to which the power sources are connected through switchgear. The busbars act as a 'manifold', and feeders are taken from it, through circuit-breakers or contactors, to all power-consuming services such as transformers, motors or interconnectors.

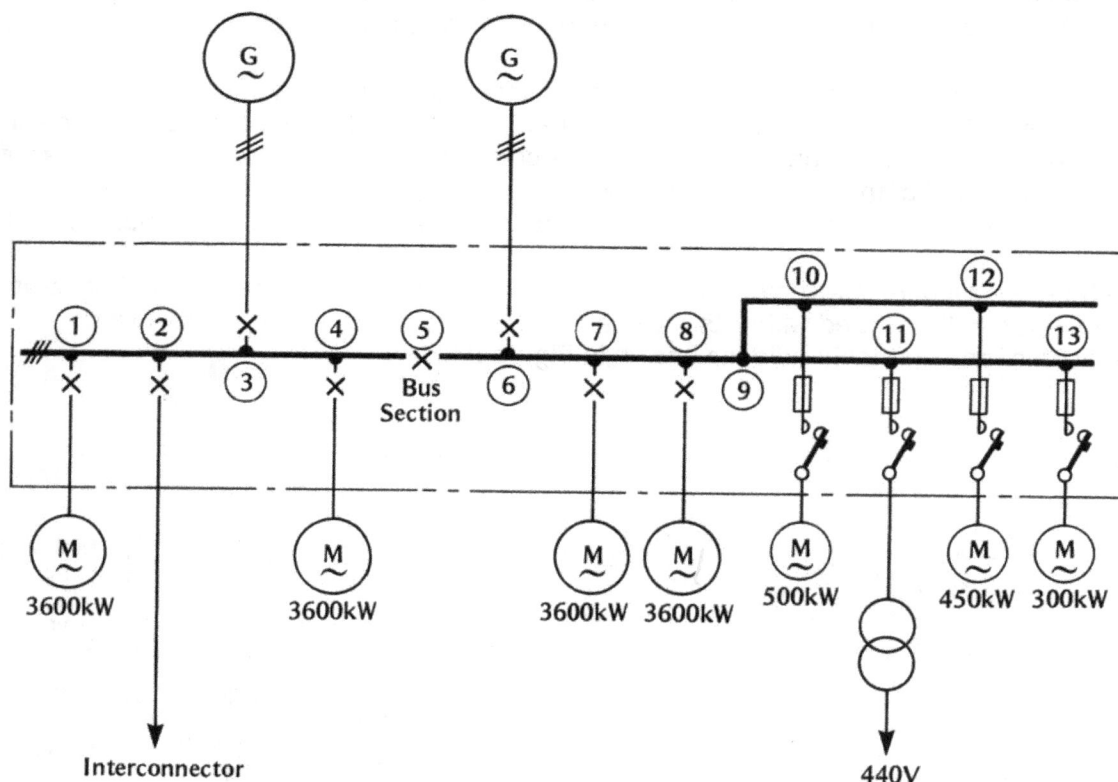

FIGURE 2.9
TYPICAL OFFSHORE HIGH VOLTAGE SYSTEM

Figure 2.9 shows, in diagram form, a typical air-break HV switchboard. It is in fact the diagram of the HV switchboard shown pictorially in Figure 2.5. It shows the two generator incomer breakers (3 and 6), the bus-section breaker (5), four feeder breakers (1, 4, 7 and 8) supplying very large motors which are too big for contactors, and one interconnector breaker (2). On the right are the four contactor panels. In Panel 9 (a bus transfer panel) the busbar splits into two - one for the high-level and one for the low-level contactors. There are three contactors (10, 12 and 13) feeding medium-sized motors and one (11) feeding a transformer; all four contactors have back-up fuses.

2.7   CIRCUIT-BREAKER CONTROL

The closing mechanisms of circuit-breakers met with on offshore and onshore installations are almost always solenoid or motor/spring operated, as described in para. 2.3.3, and all breakers have separate shunt-trip coils.

Solenoids are operated through their own contactors, usually from a 110V d.c. supply. The control circuits are simple and need no explanation here.

Motor/spring mechanisms are more complicated. Their circuits require switches and contacts to control the close and trip initiation and to ensure the correct sequence of operation. Circuits are also necessary for indication, for charging the closing spring and for operating the closing release. Auxiliary switches are therefore provided, mechanically linked to the circuit-breaker mechanism. Also fitted are carriage switches or isolating contacts which are operated by the movement of the truck to the 'Service' or 'Isolated' positions.

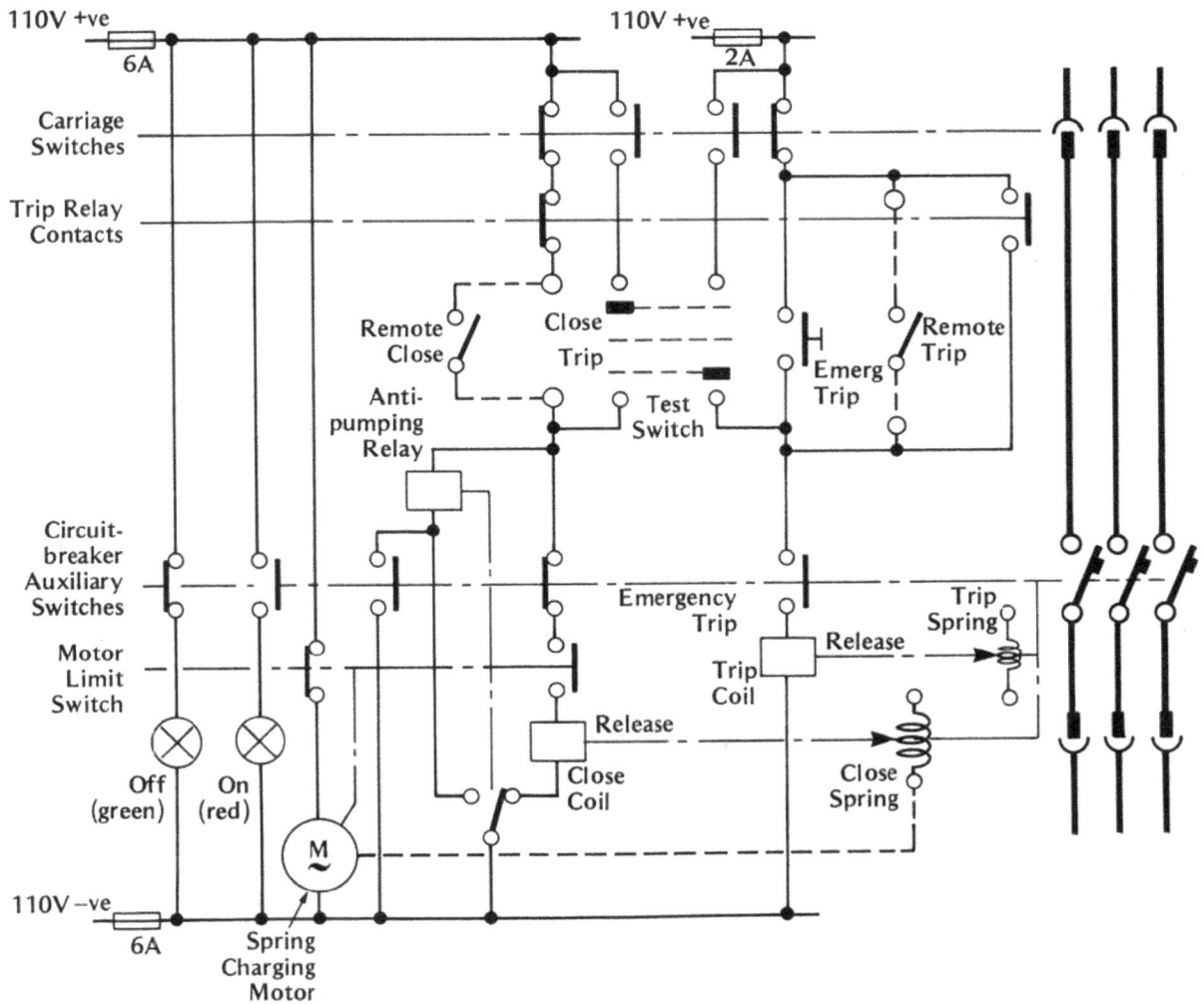

FIGURE 2.10
TYPICAL CIRCUIT-BREAKER CONTROL CIRCUIT (MOTOR/SPRING OPERATED)

A typical circuit-breaker motor/spring control arrangement is shown, in diagram form, in Figure 2.10. Contacts are shown with the circuit-breaker open and in the 'Service' position, all relays de-energised and the closing spring discharged.

When the 110V d.c. supply is connected, the spring charging motor runs; after the closing spring has been charged and automatically latched, auxiliary limit contacts operate to stop the motor and prepare the closing circuit.

Some circuit-breakers may only be closed remotely by the manual operation of a switch at a remote electrical control panel; others may be closed either remotely or by a local switch on the switchgear panel, in which case a 'Local/Remote' selector switch is fitted. The circuit illustrated in Figure 2.10 is arranged for remote closing only. Once the spring is charged, operation of the remote switch to the CLOSE position completes the closing circuit and energises the closing coil; this releases the latch and the spring closes the breaker. When the circuit-breaker has closed, an auxiliary switch opens, disconnecting the closing coil, and at the same time the motor, reconnected by the limit switch, runs to recharge and latch the spring.

Normally the circuit-breaker is tripped by operation of the remote switch; in addition a local 'Emergency Trip' pushbutton is fitted to the switchgear panel. If a system fault causes the protection to operate, the trip relay tripping contact closes. Closing any of these tripping contacts energises the trip coil, releases the holding latch and allows the breaker to open.

255

If the circuit-breaker closes onto a faulty circuit it receives an immediate trip signal from the protection, which causes it to trip; if the operator continued to hold the operating switch to CLOSE, the circuit-breaker would 'pump' in and out until the control switch was released. To prevent this, an anti-pumping circuit is provided. When the circuit-breaker closes, and while the control switch is still held to CLOSE, an auxiliary switch completes a circuit to energise the anti-pumping relay. The contacts of this relay change over to break the closing coil circuit and complete a hold circuit for the anti-pumping relay. Thus the breaker will open if tripped by an immediate fault condition, but no further closing operations can take place until the control switch has been released. This provides a 'one-shot' closing facility.

For maintenance it may be necessary to close and trip the circuit-breaker by local control while it is disconnected and isolated from the busbars. When the truck is in the 'Isolated' position, carriage switch contacts operate to change over the closing and tripping circuits from their normal connections to a local test switch. This may now be used to operate the circuit-breaker.

Auxiliary contacts operate lamps to indicate when the circuit-breaker is open or closed. Additional contacts may provide control features special to a particular circuit, such as the automatic start of standby motors, load shedding, sequence starting or other special requirements.

Where solenoid closing is used alone, its high d.c. current requires the closing solenoid itself to be switched by a separate contactor.

Separate fuses are provided in each switchgear cubicle for the closing and tripping supplies so that each may be isolated without affecting the other, and to provide protection to each auxiliary circuit appropriate to its loading.

The circuit-breaker motor/spring control system shown in Figure 2.10 and described in this section is typical of what may be found in many installations, but variations exist.

2.8   TRIP CIRCUIT SUPERVISION

There may be long periods of time when circuit-breakers are not called upon to trip. It is vital however that, when the occasion arises such as the onset of a fault, the trip circuit operates and the breaker trips successfully. A 'Trip Circuit Supervision' relay is provided on each circuit-breaker unit. It monitors the trip circuit and its power supply continuously, both when the circuit-breaker is closed and also when it is open.

A trip circuit supervision schematic diagram is shown in Figure 2.11. The trip circuit supervision relay consists of three elements: 'a', 'b' and 'c'. The trip circuit is monitored by passing a small current derived from the tripping supply through the trip coil and all associated trip circuit wiring in series with the coils of relay 'a' or 'b' or both. This current is sufficient to operate relays 'a' and 'b', even when they are connected in series, but, being limited by resistors, is not high enough to trip the breaker. Relays 'a' and 'b' are energised individually when the contacts across which they are connected are **not** closed. Relay 'c' is a hand-reset flag relay with alarm contacts which are closed if the trip circuit is healthy, but they open to give an alarm if it is not.

If the circuit-breaker is open, both relays 'a' and 'b' are energised in series. If the circuit-breaker is open but a trip initiation signal is present, say from a hand-reset tripping relay, then relay 'b' alone is energised. When the circuit-breaker is closed, relay 'b' is de-energised and only relay 'a' monitors the trip circuit.

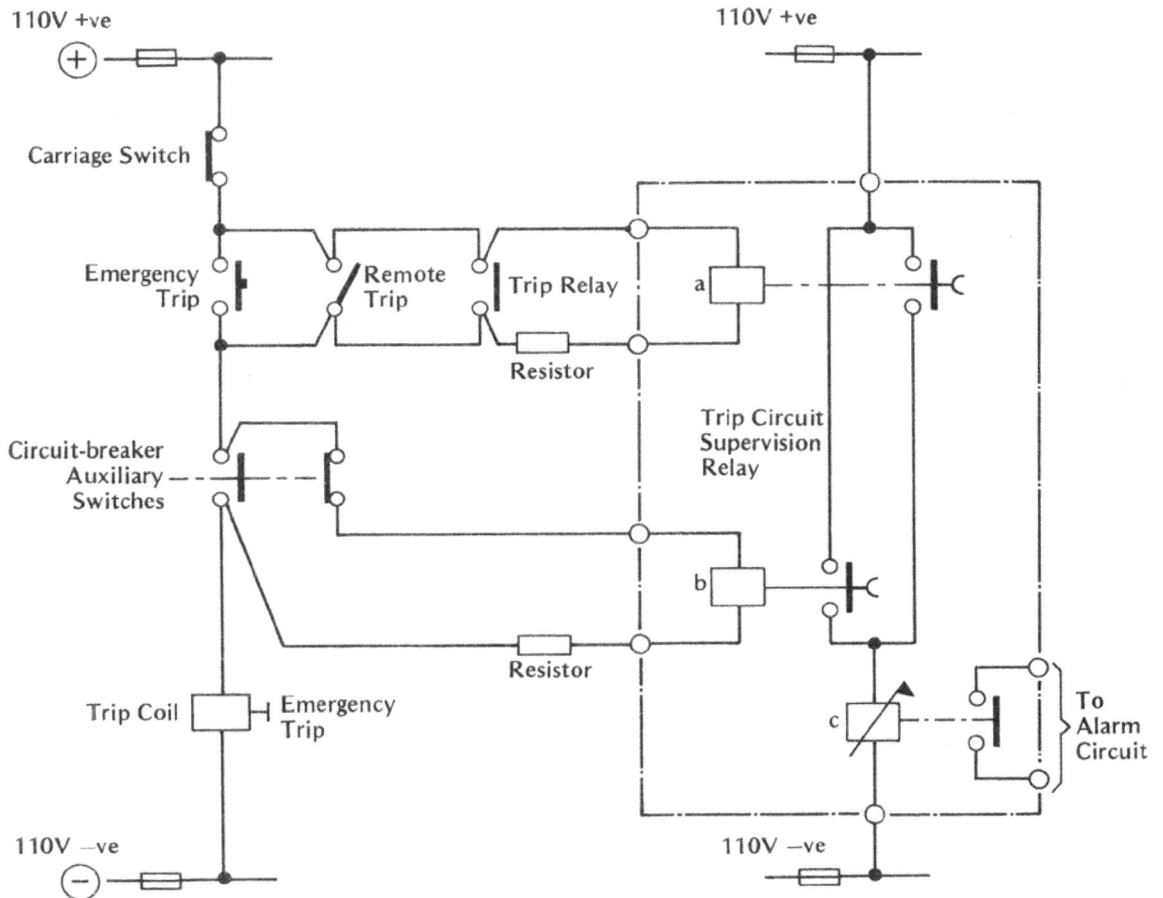

FIGURE 2.11
TRIP CIRCUIT SUPERVISION

If the trip supply fails or the breaker fails to open when a trip signal is initiated, then both relays 'a' and 'b' are de-energised and an alarm is initiated when relay 'c' releases. This happens when:

- the trip supply fails;

- any part of the trip circuit is open-circuited;

- the breaker fails to open;

- the trip circuit supervision relay auxiliary supply fails.

The opening of the contacts of relays 'a' and 'b' is delayed by about 400ms to allow for transient dips in the 110V d.c. supply and to allow time for the circuit-breaker to open without initiating an alarm.

2.9   FUSES

The only high-voltage fuses fitted in an HV switchboard are those which form a back-up to a contactor (air-break or vacuum) and those on the HV side of a voltage transformer. All are of the high rupturing capacity (HRC) type.

The contactor fuses are of the open type but are embodied in the contactor unit itself. This forms adequate protection since it is necessary to isolate the unit, and in the case of the air-break type to withdraw it, in order to gain access to the fuses. They can be seen in Figure 2.6.

Where fuses, whether HV or LV, are used in series with a contactor, their purpose is to protect the contactor itself against having to open on a fault current which is in excess of its rating. Fuses used in this manner are termed 'back-up fuses' and are selected with reference to the contactor's own inverse-time characteristic. This process is fully described in Part 10 'Electrical Protection'.

Voltage transformer HV fuses form part of the VT itself. The VT compartment can only be opened after isolation, after which the fuses are accessible. Although the VT fuses have a very small current rating, they still have to be able to break a full-scale short-circuit current if a fault should develop on the VT itself.

There are many low-voltage fuses in the control and instrumentation circuits. These are of the type described in Chapter 3 for LV switchgear.

2.10   CIRCUIT AND BUSBAR EARTHING DOWN

To ensure that a high-voltage circuit is safe to work on, it must be disconnected (isolated) from all sources of supply, and it must also be solidly earthed down. Earthing is carried out at the isolating switchgear.

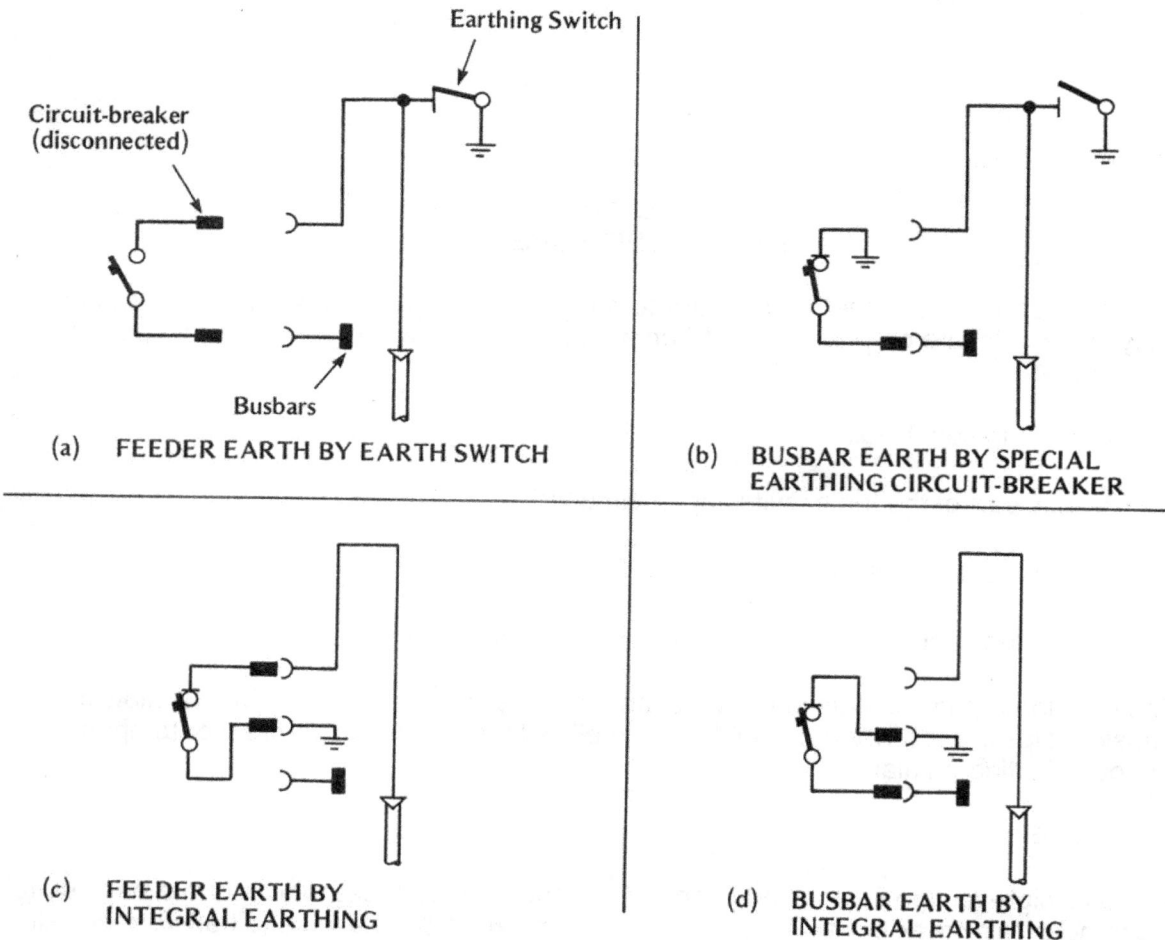

(a)   FEEDER EARTH BY EARTH SWITCH

(b)   BUSBAR EARTH BY SPECIAL EARTHING CIRCUIT-BREAKER

(c)   FEEDER EARTH BY INTEGRAL EARTHING

(d)   BUSBAR EARTH BY INTEGRAL EARTHING

FIGURE 2.12
FEEDER AND BUSBAR EARTHING

Earthing-down procedures vary from one type of switchgear to another; the manufacturer's operating instructions must be consulted in each case.

Four examples of the use of switchgear for safety earthing are shown in Figure 2.12. In each case the circuit-breaker or contactor truck must be moved out of the 'Service' position before a permissive interlock key can be released which permits earthing to take place.

Figures 2.12(a) and (b) show one type of earthing system. In Figure 2.12(a) the feeder is earthed through an earthing switch that can only be closed after the permissive interlock key has been inserted and operated. The permissive interlock key is released only when the switchgear truck is either locked in the 'Isolated' position or completely removed from the panel. For busbar earthing the same switchgear uses a special earthing circuit-breaker truck shown in Figure 2.12(b) which temporarily replaces the normal truck. Interlocks described below ensure that the busbar cannot be earthed until it is disconnected from any possible source of supply.

Figures 2.12(c) and (d) show a different type of switchgear with 'integral earthing', where the busbar or the feeder contacts on the circuit-breaker are moved mechanically to connect with a set of earthed contacts mounted between the busbar and the feeder spouts. The busbar or feeder contacts can only be moved after the appropriate permissive interlock key has been inserted into the front panel of the switch-truck and operated. This key, either busbar or feeder, is captive on the shutter operating mechanism inside the switchgear panel and is only accessible when the truck is removed. When either key is withdrawn, the appropriate shutter will not open when the breaker truck is pushed in, so preventing accidental access to live metal. To earth a feeder, the busbar shutters must be immobilised, and vice versa.

Where any earth is applied, a key is released from the switchgear unit concerned, to be placed in a Lockout Box. The earth cannot be removed until the key is returned.

Before a section of **busbars** is earthed down, all possible in-feeds must be disconnected. For example, one-half of a busbar system must be isolated not only from its generator incomers but also from the other half through the section breaker; also from any interconnectors and from any transformer that could feed back into the busbar. To ensure this, the individual keys released when all the possible in-feed breakers have been isolated are inserted into a key-box. When all the keys are home, a master key is released which permits the busbar earthing connection to be made or the earthing switch to be inserted.

Full and detailed instructions for isolating and earthing down are contained in 'Standing Instructions Electrical'.

2.11 PANEL HEATERS

Each switchboard panel is fitted with an anti-condensation heater; this is usually energised at 240V or 250V a.c. and controlled either by a hand switch or by an auxiliary switch that connects the heater when the circuit-breaker is open.

CHAPTER 3 LOW VOLTAGE SWITCHGEAR

3.1   GENERAL

Switchgear is required to enable power sources to be connected to and disconnected from the low-voltage distribution system. This switching is necessary both for normal operational purposes and also for the rapid and automatic disconnection of any circuit that becomes faulty. The switchgear also allows any circuit to be isolated from the live system and for that circuit to be made safe so that work may be carried out on the equipment connected to it.

This chapter deals with switching devices as applied to low voltage (typically 415V or 440V). Three types of LV switchgear are considered:

(a)   Circuit-breakers

Circuit-breakers are used to control inputs from transformers, section breakers on switchboard sections, interconnectors between LV switchboards and inputs from auxiliary LV generators. All LV circuit-breakers are of the air-break type.

(b)   Contactors

Contactors are used to control mainly motor circuits. They are always of the air-break type and are usually enclosed in the individual cubicles which form that part of a low-voltage switchboard referred to as a 'Motor Control Centre' (MCC).

Contactors are designed only to make and carry fault current for a short time, not to break it. Where the system fault level exceeds the limited breaking capacity of the contactor, fuses are inserted in series with the contactor contacts. Contactors are designed for remote operation and to undergo repeated and frequent operation without undue wear.

(c)   Moulded Case and Miniature Circuit-breakers

These form a special class of lightweight compact circuit-breakers for mounting onto or behind panels. They are designed for hand operation only but have built-in protective tripping arrangements.

3.2   CONSTRUCTION

3.2.1 Main Air Break Circuit-breakers (ACB)

The main low-voltage circuit-breakers are always of the air-break type whose construction and operation are similar to those for high-voltage ACBs described in Chapter 2 and shown in Figure 2.2. Being designed for low-voltage systems their insulation levels are of course lower, but, by the same token, their normal rated currents and their short-circuit current ratings are considerably higher. This leads to generally heavier copperwork, to large arc chutes and especially to heavy switching contacts and isolating contacts.

Like their HV equivalents, LV circuit-breakers are horizontally isolated, with similar interlocks to ensure the correct sequence of operations when being withdrawn or reinserted.

Being smaller in size they are usually mounted in pairs, one above the other in an LV switchboard, presenting a dead-front panel face.

Most LV systems are 4-wire. Some main circuit-breakers are 4-pole, but most are 3-pole with an unswitched neutral connecting link.

LV circuit-breakers are rated from 800A to 4 000A normal current. They come in standard ranges of breaking capacity, which in British Standards is 35kA, 43kA and 70kA rms symmetrical. These currents at 415V are equivalent to 25MVA, 31MVA and 50MVA respectively, or at 440V are equivalent respectively to 27MVA, 33MVA and 53MVA. Because of the heavier normal and short-circuit currents found in LV systems, the circuit-breakers usually have much heavier breaking and isolating contacts than those of the HV types.

The circuit-breaker closing mechanism may be operated by solenoid or motor/spring. Descriptions of both these methods will be found in Chapter 2, para. 2.3. Tripping is by a separate shunt-trip coil, always powered from an independent battery-supported d.c. supply.

The basic circuit-breaker control circuits are essentially the same as those described in Chapter 2 for high-voltage switchgear. Some circuit-breakers are equipped with an additional release device in series with the main circuit which trips the circuit-breaker instantaneously if it is closed onto a fault; it does not operate under any other circumstance. This series tripping release is part of the circuit-breaker, not a protective relay, and requires to be reset by hand after operation. Where this device is fitted, the anti-pumping circuit is unnecessary and is omitted.

Closing and tripping of low-voltage circuit-breakers is usually carried out by operating switches on a remote electrical control panel. In some cases this remote control facility is not provided, and switching is done locally at the switchboard.

The procedure for disconnecting a circuit-breaker and removing it from its panel depends on the type of switchgear; the maker's instructions must be consulted in each case. To avoid personal injury, the closing spring must always be discharged before removing a spring-closed circuit-breaker from its switchgear panel.

3.2.2 Contactors

All contactors form a part of the individual distribution cubicles which make up an MCC. Almost all are unlatched.

Each contactor is rated according to the service which it feeds, which may vary from a fractional horsepower motor drawing one ampere to a large 250kW motor drawing over 400A. Consequently the contactor cubicle may vary in size from 'one tier' deep up to 'seven tiers' deep (see para. 3.3.2).

Each contactor operating coil is supplied, through its control circuits, from the switchboard busbar either direct at 415V or 440V, or through a small step-down transformer. This ensures that, if busbar power fails, all connected contactors 'drop off' and keep their motors disconnected until each can be individually restarted.

Every contactor is backed up by a set of high rupturing capacity (HRC) fuses housed in the same cubicle. These are rated according to the fault level at the switchboard (typically 31MVA or 50MVA, equivalent to 43kA or 70kA at 415V). The correct size of fuse is chosen so that, in the event of a fault in the feeder circuit which exceeds the ability of the contactor to clear it, the fuse will blow first, leaving the contactor to open on a dead circuit. The choice of back-up fuse is further discussed in Part 10 'Electrical Protection'.

Facilities are provided for testing the contactor while isolated from the busbar. While so isolated a separate test supply can be applied to the contactor coil to check its operation without actually starting the motor.

3.2.3 Moulded Case Circuit-breakers

A type of low-voltage circuit-breaker widely used in most installations is the Moulded Case Circuit-breaker, or MCCB for short.

Shown in Figure 3.1, it consists of a moulded plastic case containing a switching element which is operated manually by an external handle or 'dolly'. Because the original design was American, the dolly position is down for 'Off' and up for 'On'. MCCBs can be used for switching either a.c. or d.c. circuits. They are usually mounted, when used on distribution panels, behind the panel, and only the dolly shows. Other arrangements however, such as surface mounting, are also found.

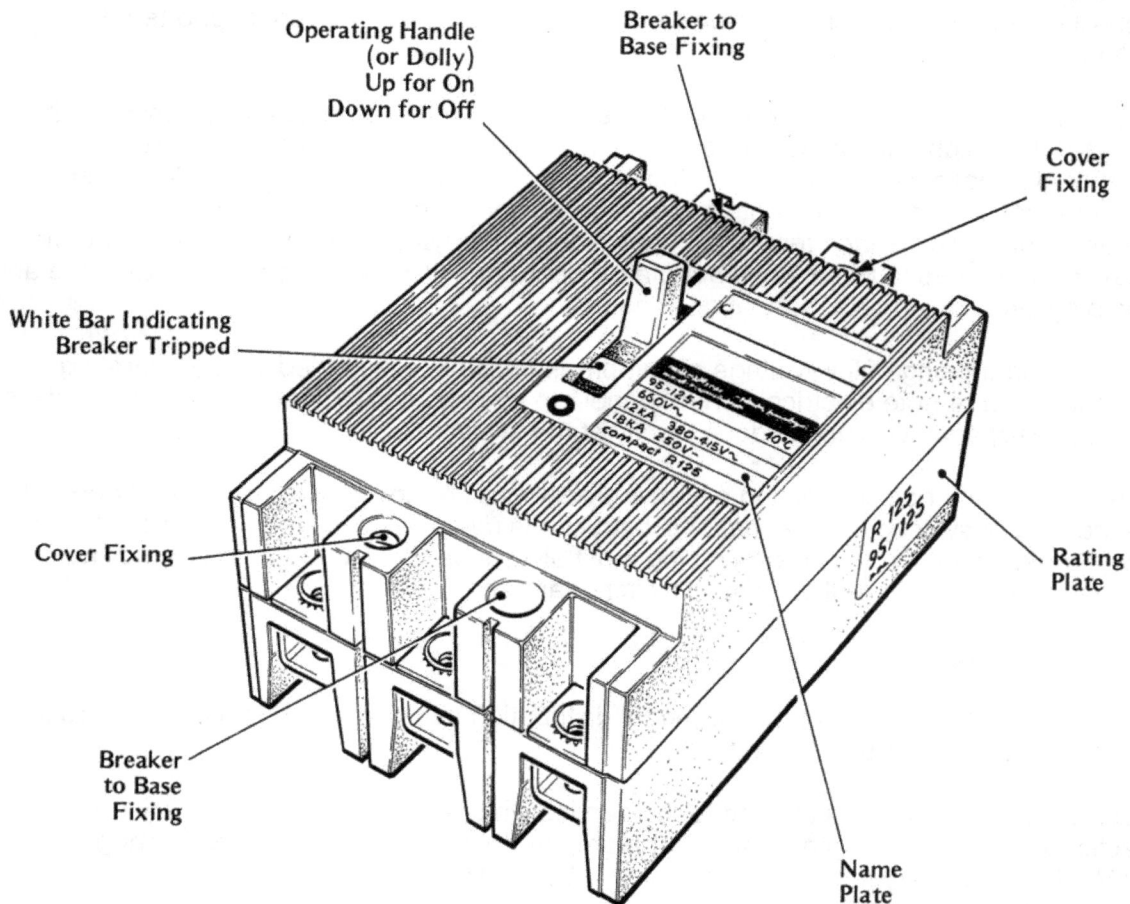

FIGURE 3.1
MOULDED CASE CIRCUIT-BREAKER

Most of the MCCBs used onshore and offshore are 3-pole, but very occasionally a 4-pole version is fitted. They are also supplied as 2-pole (for example for d.c. switching), but this is usually a 3-pole type with one pole omitted.

MCCBs are very compact and have a high breaking capacity for their small size. Where the system fault level at the point where an MCCB is used exceeds its fault-breaking capacity, separate HRC back-up fuses must be used in series, as described above for contactors. MCCBs are made by a number of manufacturers, and different makes and sizes are used in installations. The following description, therefore, can be no more than very general.

A MCCB as used in onshore and offshore installations is normally fitted with two separate overcurrent devices. One is a thermal element in each pole having an inverse-time characteristic, and the other an instantaneous 'high-set' electromagnetic element in two of the three poles; this operates instantaneously but only on the highest fault currents and then overrides the thermal element. Both trip the circuit-breaker when the current reaches the set operating level in any of the poles.

Typical MCCBs used in installations have normal current ratings of either 125A or 250A, according to the circuits they control. The breaking capacities of these two sizes are given below.

Normal Current	Breaking Capacity to BS Rules		
	MVA (3-phase, 440V)	Equiv. kA (a.c.) (rms symmetrical)	kA (d.c. 250V)
125A	7.5 MVA	10kA	18kA
250A	15 MVA	20kA	25kA

The maximum currents that can be handled by these two sizes of MCCB are therefore 125A and 250A, but they can be arranged to trip at lower currents. This is achieved by fitting a separate 'trip unit' to the breaker. In the 250A size the trip units are interchangeable, but in the 125A size they must be fitted at the time of ordering and cannot thereafter be changed; if a different trip setting is needed, the whole 125A MCCB must be replaced by one with the new setting. Sizes larger than 250A are manufactured but are not installed in platforms.

The following trip units are available:

125A size	250A size
125A	250A
100A	200A
75A	160A
50A	125A
25A	100A
15A	60A

There is normally no adjustment of current setting in any of the thermal units, but the electromagnetic elements have settings adjustable in five steps, numbered 1 to 5, by means of a control knob on the trip unit. The range of adjustment of the high-set instantaneous overcurrent trip element is typically from 6 to 13 times the full-load rating of the **trip unit** (note, not of the MCCB itself).

When a closed MCCB self-trips due to operation of either of its overcurrent trip elements, the main contacts open fully, but the dolly goes to a mid, 'half-cock' position where it shows a white line in the dolly window (see Figure 3.1). This indicates which of a number of MCCBs has tripped and enables it to be distinguished from those which were already open. Before the MCCB can be reclosed by hand, the dolly must first be moved to the 'Off' position.

MCCBs are 'latched' breakers in that they do not, like contactors, fall out automatically if the service voltage disappears; they must be individually opened by hand. However the 250A size (only) can be fitted with an undervoltage release, but none is so fitted on offshore installations.

Although MCCBs are self-tripping, they are not normally remote controlled. However, the 250A size (only) can be provided with a shunt-trip release which enables it to be electrically tripped from a remote point. It cannot be remotely closed without an additional operating mechanism. Neither of these features is employed in most installations except for the shunt trip on certain offshore d.c. distribution boards.

Many MCCBs are used in onshore and offshore installations as incomer isolators for sub-distribution boards such as lighting or sundries panels. When used in this way they are pure isolators and do not have a protective function. Such MCCBs usually have their trip units removed and can only be opened by hand.

3.2.4 Miniature Circuit-breakers

The range of MCCBs already described extends down to a series of smaller breakers of ratings up to 70A maximum. These are known as Miniature Circuit-breakers (MCBs). (*The term MCB should not be confused, as it often is, with MCCB for the moulded-case type.*)

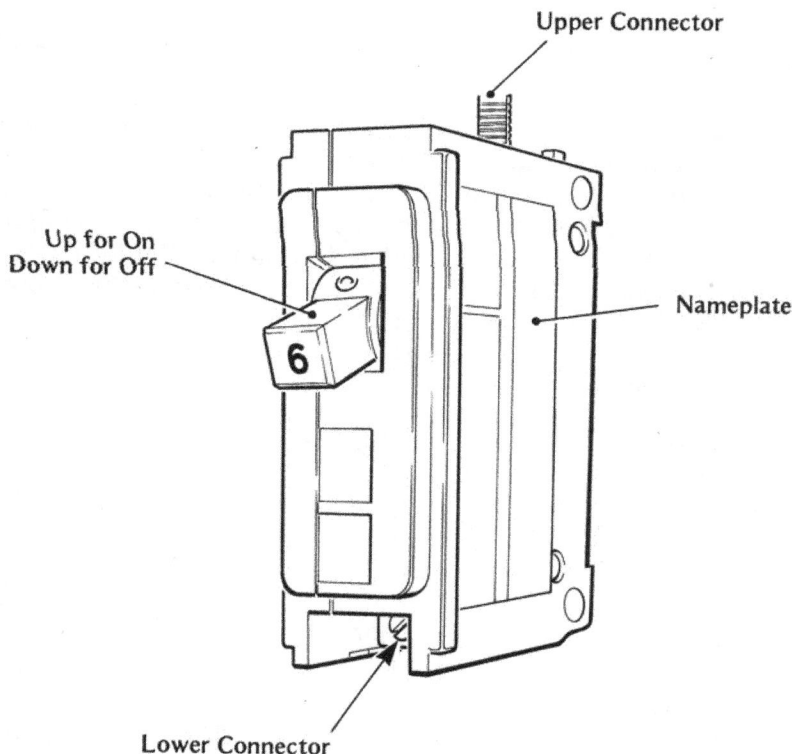

FIGURE 3.2
MINIATURE CIRCUIT-BREAKER

A typical single-pole MCB is shown in Figure 3.2. In operation it is generally similar to, though physically smaller than, the moulded-case design. The MCB has a moulded plastic case and is manually operated by an external dolly. It is manufactured as a 1-pole, 2-pole, 3-pole or 4-pole unit, but the commonest types in most installations are the 1-pole and

2-pole models which are used in single-phase-and-neutral sub-distribution panels. Such panels usually mount the MCBs horizontally in blocks of 6 or 12. Other forms of mounting are also found.

Sometimes where the 3-phase switching is desired, instead of using a 3-pole MCB, three 1-pole units are physically ganged together by a bar joining the dollies.

Each pole of an MCB is protected by two separate overcurrent devices. One is a thermal element having an inverse-time characteristic, and the other an instantaneous 'high-set' electromagnetic element; this operates only on the highest fault currents and then overrides the thermal element. Both trip the circuit-breaker when the current reaches the set operating level in any of the poles.

The size ranges of MCBs vary from one manufacturer to another, but only a limited number are used in most installations. Two typical sizes have normal current ratings of 70A and 32A, but they may be fitted with one of several different trip units, as follows:

70A size		32A size	
3A	25A	1A	10A
6A	32A	2A	15A
10A	40A	3A	20A
16A	50A	5A	25A
20A	70A		32A

The trip unit fitted in any given MCB is usually denoted by a figure on the dolly. The unit is fitted during manufacture and cannot be changed. (Figure 3.2 shows a 6A trip unit fitted.)

The fault current breaking capacity of an MCB differs according to the number of poles, the working voltage and, in some cases, the trip unit fitted. The maximum breaking currents are as follows:

70A size

Trip Unit	Breaking Current (kA)			
	250V a.c.		440V a.c 3-pole	125V d.c. 2-pole
	1-pole	2-pole		
3A	1kA	1kA	1kA rms symm (=0.75MVA)	1kA
6A	1kA	1kA	1kA rms symm (=0.75MVA)	1kA
10A	5kA	8kA	5kA rms symm (=3.8MVA)	10kA
16 to 70A	5kA	8kA	3kA rms symm (=2.3MVA)	10kA

32A size

Trip Unit	Breaking Current (kA)			
	250V a.c.		440V a.c 3-pole	125V d.c. 2-pole
	1-pole	2-pole		
1 to 5kA	6kA	10kA	3kA rms symm (=2.3MVA)	5kA
10 to 32A	6kA	8kA	3kA rms symm (=2.3MVA)	5kA

The electromagnetic trips in the 70A size operate at 7 to 12 times the current rating of the thermal trip unit fitted (note, not of the MCB itself). In the case of the 32A size the range is 6 to 9 times.

Both sizes of MCB are also manufactured in an alternative 'high breaking capacity' version giving approximately 50% higher breaking current, but these are not used on offshore or onshore installations.

If the system fault level at the point where the MCB is used exceeds its fault-breaking capacity given in the tables opposite, a properly selected back-up fuse of correct characteristic must be placed in series, as described in Part 10 'Electrical Protection'.

MCBs are 'latching' devices and do not, like contactors, fall out automatically if the service voltage disappears; they must be individually opened by hand. They cannot be fitted with an undervoltage release, nor can they be remote-controlled.

3.2.5 Earth Leakage Circuit-breakers

A special type of miniature circuit-breaker, very sensitive to small earth-leakage currents, is described in Part 10 'Electrical Protection', Chapter 5.

3.3   LOW VOLTAGE SWITCHBOARDS

3.3.1 Centre (Incomer) Section

An LV switchboard is usually supplied from one or two step-down transformers fed from the HV system. LV operating voltages are normally 415V onshore and 440V offshore.

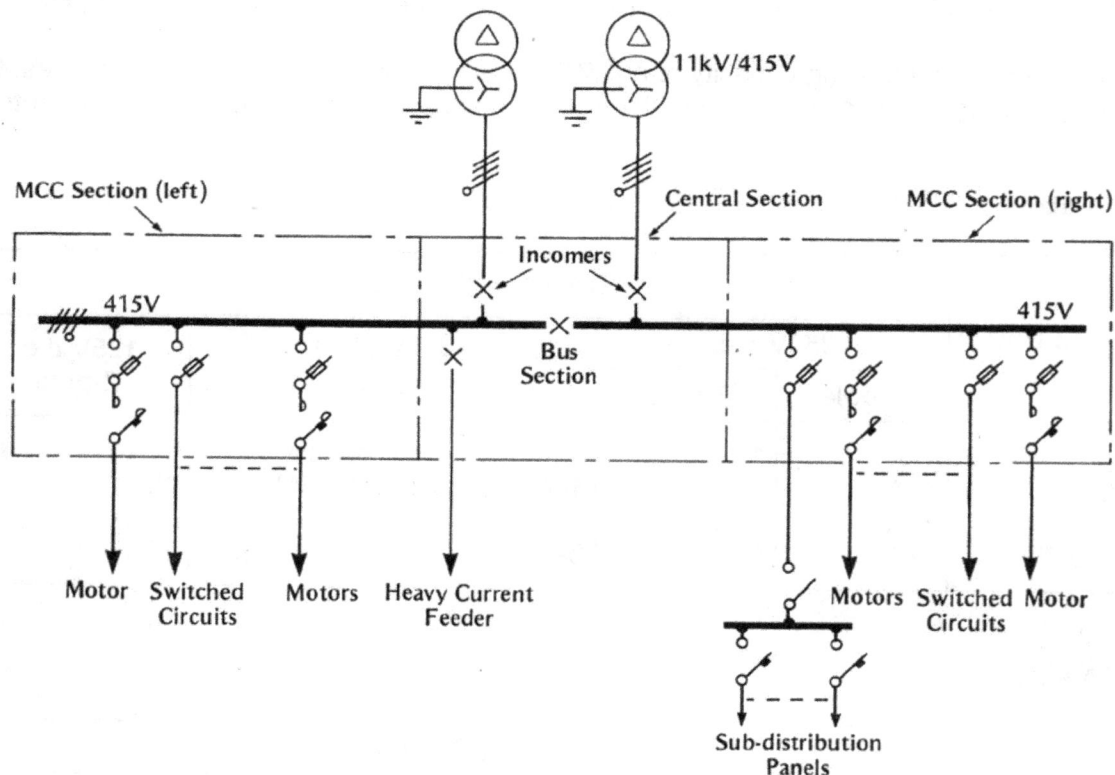

FIGURE 3.3
TYPICAL ONSHORE 415V SYSTEM

Figure 3.3 shows diagrammatically such an onshore LV system with two transformer incomers, a bus-section breaker, a heavy feeder and two grouped distributing sections, left and right, also called Motor Control Centres (MCCs). (Note: This is a somewhat misleading term, as by no means all the feeder circuits supply power to motors.)

The circuit-breakers usually form the centre section, with power being passed to left and right by the busbars. Heavy-current feeders and the larger interconnectors feeding power to or from other LV switchboards sometimes require circuit-breaker protection and are then brought into the centre section.

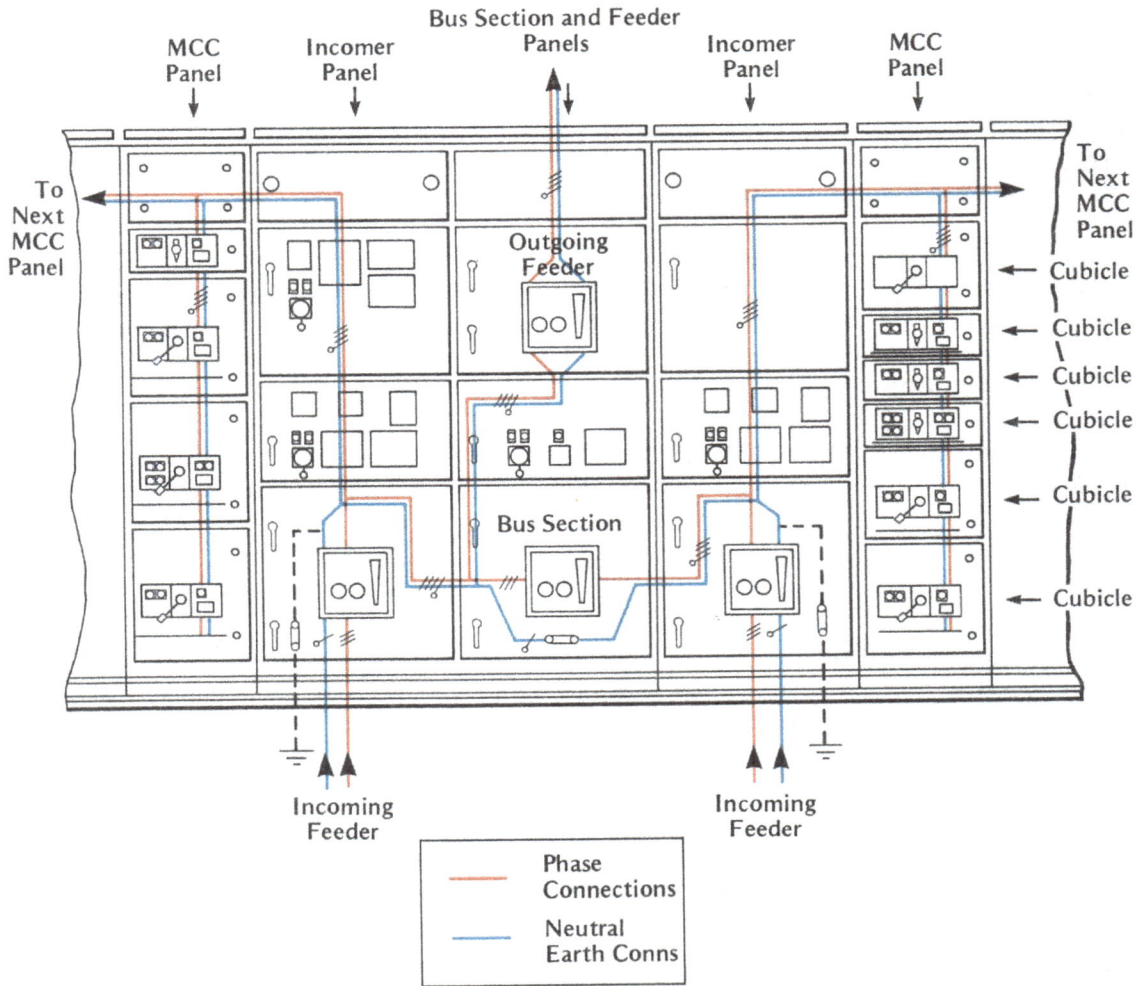

FIGURE 3.4
PART OF TYPICAL OFFSHORE 440VSWITCHBOARD

Part of a typical offshore 440V switchboard is shown in Figure 3.4. The part shown consists of five panels mounted side by side; the centre three panels contain cubicles for the incoming feeder, bus-section and heavy feeder circuit-breakers; the associated protective relays, control switches and indication equipment are mounted on the fronts of each panel.

On each side of the circuit-breaker panels are MCC panels, one of which is shown in Figure 3.4 on each side of the centre panels. Further MCC panels are added as required. Each contains a number of motor control contactor cubicles and fuse-switch cubicles mounted one above the other to control the outgoing circuits. The fuse-switch cubicles control those circuits not associated with motors such as sub-distribution boards or welding sockets.

Additional MCC panels are mounted on each side of the centre section to house the feeder cubicles necessary to meet the requirements of the system concerned. A large switchboard may include as many as 30 or more MCC panels.

The arrangement of the busbars and circuit connections is shown diagrammatically in colour on Figure 3.4; the main busbars are shown in red for the phases and blue for neutral and run through busbar chambers at the top and to the rear of the panels; they are connected through the length of each section of switchboard. Power is supplied to each outgoing feeder cubicle by a set of dropping busbars (also shown in red and blue) housed in a vertical enclosure at the rear of each MCC panel.

On many switchboards each incoming switchgear panel has provision for earthing the neutral busbar through a bolted link, shown in black. It is not switched with the breaker. When the incoming supply is from a transformer, this link is closed, earthing both the neutral busbar and the star-point of the transformer; this provides an earth for that particular part of the LV system. On systems where the transformer star-point is earthed direct, this feature is not provided at the switchboard.

On some installations where dry-type encapsulated transformers are used, the transformers themselves form part of the LV switchboard, installed behind a panel and with their LV terminals connected directly onto the copperwork of the incomer panel.

3.3.2 MCC Section

Each outgoing circuit (other than interconnectors) is controlled by a feeder cubicle on one of the MCC 'wings' of the switchboard. These cubicles are of different types, depending on

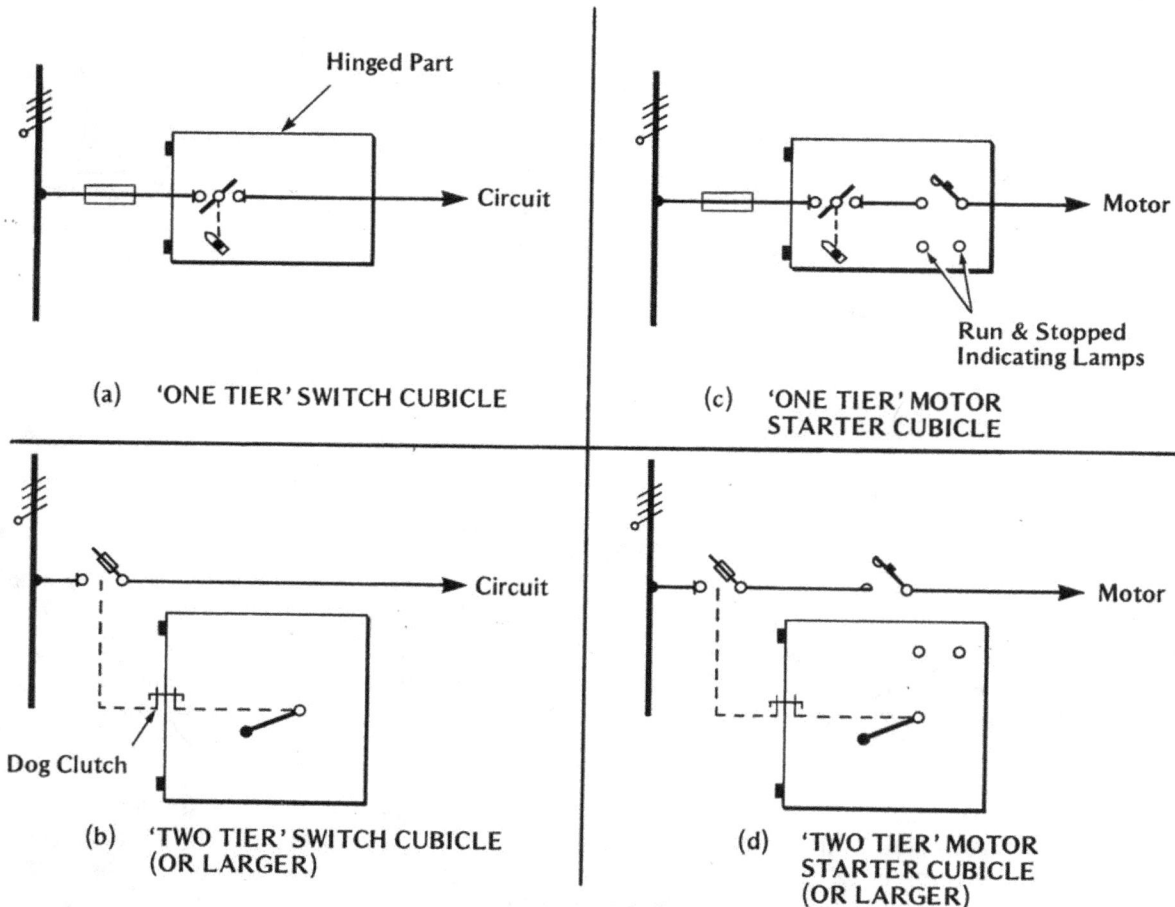

(a) 'ONE TIER' SWITCH CUBICLE

(c) 'ONE TIER' MOTOR STARTER CUBICLE

(b) 'TWO TIER' SWITCH CUBICLE (OR LARGER)

(d) 'TWO TIER' MOTOR STARTER CUBICLE (OR LARGER)

FIGURE 3.5
TYPICAL MCC CUBICLES

the manufacturer. The following description is typical and is widely used in both onshore and offshore installations.

The MCC feeder cubicles occupy the full width of an MCC panel, but their vertical height depends upon the rating and function of the unit. The smallest unit occupies one module of height - a 'one-tier' cubicle - and there is space for ten of them. The largest is seven tiers high. In practice a panel usually contains a mixture of cubicles of different heights to suit the particular distribution requirement.

In the smallest (one-tier) feeder cubicles the HRC fuse bases are permanently fixed to the busbar droppers and, although they are shrouded, care is needed when withdrawing or replacing them. Such feeder cubicles have a simple rotary isolating switch on the hinged panel-front door, as shown in Figure 3.5(a).

Larger sizes of switch panel are provided with a fuse-switch for isolation. This is operated by an isolating handle on the door of the unit with a mechanical drive to the fixed fuse-switch through a dog-clutch which is engaged only when the door is closed. The fuses are dead when the fuse-switch is off. This is shown in Figure 3.5(b).

Both isolating switches and fuse-switches are interlocked with the doors of their associated switch unit so that the door cannot be opened unless the switch is off.

All MCC feeder cubicles used for motor control have a contactor in the circuit following the isolator switch or fuse-switch; the two types are shown in Figures 3.5(c) and 3.5(d).

It is possible to test the contactors without actually starting the motor. When the cubicle door is opened, the isolator having first been opened, a switch inside can be closed to provide an alternative supply to the contactor coil. The contactor can then be operated while its main contacts are isolated from the mains. A small cubicle at the top of the MCC panel provides the test supply through small distribution fuses; it is labelled 'TEST' or 'CONTROL'.

Motors are normally started and stopped by remote control from the control room or the motor site. Starting pushbuttons or switches at those points cause the contactor at the MCC to close. It is very rare for provision to be made to start a motor at the MCC cubicle itself (some ventilation fans are exceptions). However each motor cubicle at the MCC has an emergency stop pushbutton.

3.4 FUSES

Fuses are used with low-voltage switchgear:

(a) as back-up for distribution contactors, or

(b) for various control and instrumentation circuits.

In all cases they are of the HRC type.

When used as back-up the fuses are inside the individual distribution cubicles on the MCC section of the switchboard. In one design they are either direct on the busbars (for one-tier units) or embodied in the isolating switch as a 'fuse-switch' in larger units - see Figure 3.5. In this case protection against accidental contact is afforded by the cubicle enclosure itself. Access to a fuse-switch is only possible after the fuse-carrying blades of the switch have been put in the isolated (open) position and the door opened.

Great care is needed with the busbar fuses in one-tier cubicles. Although the door cannot be opened until the isolating switch has been opened, the fuses themselves are still connected to the live busbar, though not carrying current - see Figure 3.5(a) and (c). Although the fuse links are well shrouded, caution should be shown when removing or replacing them.

Low-voltage control and instrument fuses are usually panel-mounted in their own carriers. Their physical size is determined by their normal current rating. Their breaking capacity is determined by the fault level current of the circuit in which they are connected

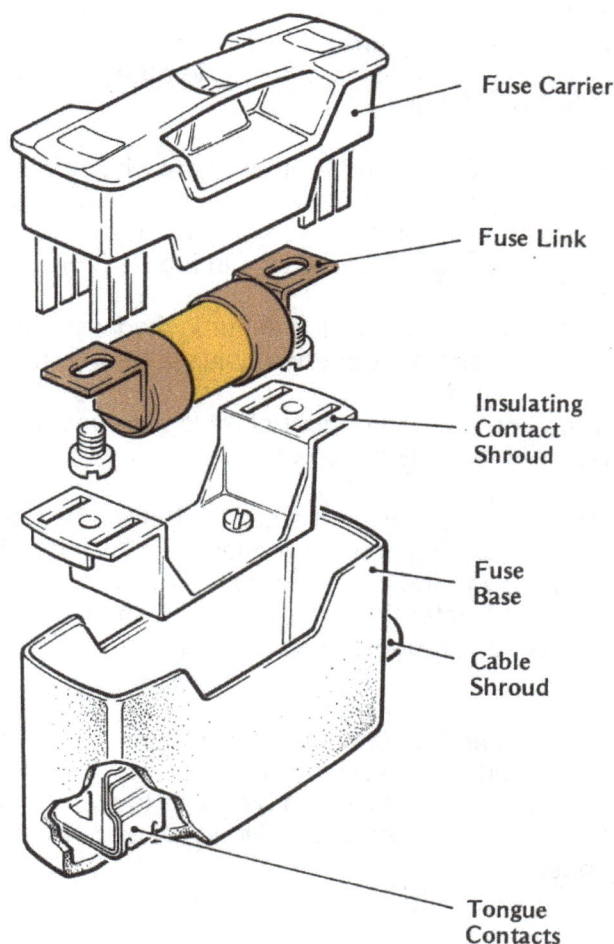

FIGURE 3.6
COMPLETE LV FUSE UNIT (TYPICAL)

A typical low-voltage fuse assembly is shown in Figure 3.6. The replaceable ceramic cartridge with its metal terminal caps is known as the 'fuse link'. It is held in an insulated 'fuse carrier' which completely shrouds all live metal. The carrier is supported on an insulated 'fuse base', where it is firmly fixed by various mechanical means, among them tongue contacts, butt contacts held by insulated screw pressure, or wedge contacts pressed in by insulated screws. A tongue-contact type is shown in Figure 3.6.

A full description of how an HRC fuse operates to interrupt current is given in Part 10 'Electrical Protection'.

3.5   BUSBAR BRACING – SHORT-CIRCUIT FORCES

Figure 3.7 shows two busbars, A and B, installed side by side, supported on post insulators and carrying a single-phase current.

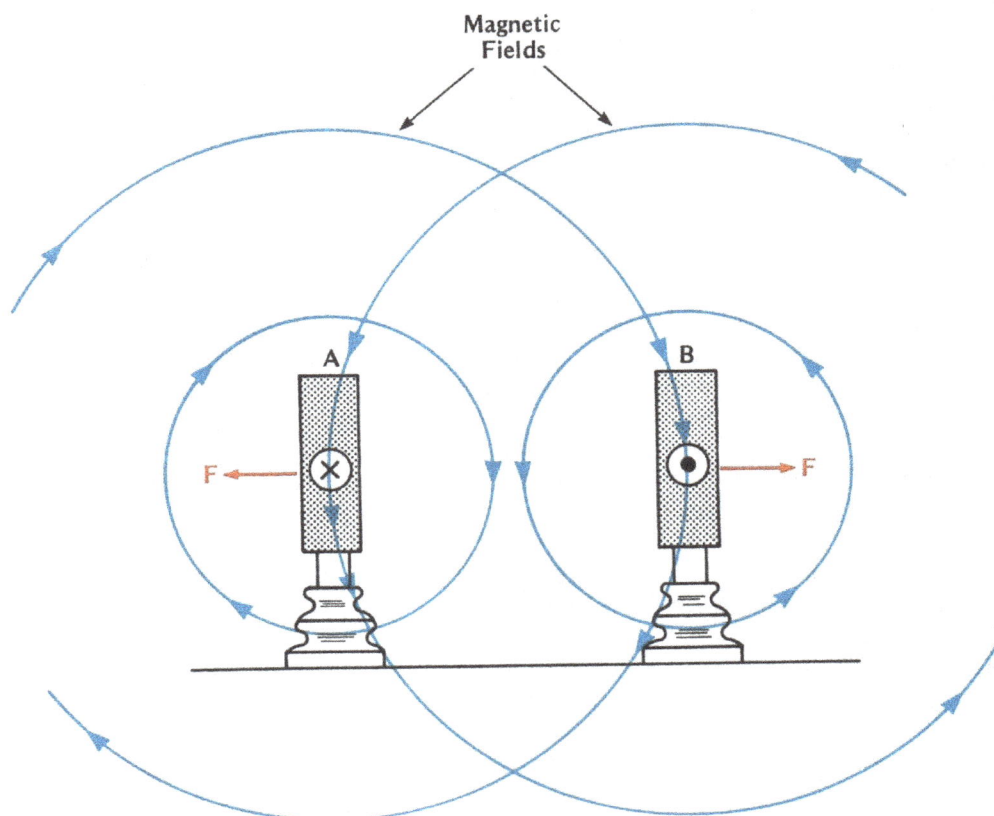

FIGURE 3.7
FORCES ON A BUSBAR

This current, at the instant shown, is assumed to be flowing down (into the paper) in the left-hand bar A, and up (from the paper) in the right-hand bar B, as indicated in the figure.

By Oersted's rule the current in bar A will give rise to a circular magnetic field around it in a clockwise direction (corkscrew motion), whereas the current in bar B will give rise to a circular magnetic field in the anti-clockwise direction. These fields are shown in the figure, and each field system cuts the other busbar.

By Fleming's Left-hand Rule (see Part 1 Electrical Theory) the interaction between the current in A and the magnetic field from B which cuts it is to produce on bar A a mechanical force F to the left. Similarly the interaction between the current in B and the magnetic field from A produces on bar B an equal mechanical force F to the right.

Half a cycle later both the currents and both the magnetic fields are reversed, so that the mechanical force on each, produced according to Fleming's Left-hand Rule, is unaltered in direction.

Thus both bars are subjected to outward mechanical forces trying to push them apart. They are resisted only by the post insulators which undergo a shearing and cantilever stress tending to break them. This force alternates in magnitude as the current cycles, but not in direction. It reaches a peak value twice each cycle: once when the currents peak as shown in the figure, and again half a cycle later when they peak in the opposite direction.

The magnitude of the force on a bar depends on the strength of the current in the bar and also on the strength of the magnetic field cutting it, which itself depends on the strength of the equal and opposite current in the other bar. The force is thus proportional to the current squared. It also depends on the distance between the bars, becoming less as the spacing is increased.

The following calculation gives some idea of the scale of these forces, particularly under conditions of short-circuit when the currents may be very large indeed. If the currents were, say, 50 000A d.c. in each bar, and if the bars were spaced 3 inches (7.5cm) apart, the outward force on each bar would amount to no less than 670 kgf per metre run of bars, or nearly ¼ ton-force per foot run.

If the current were 50 000A **alternating** (rms), the peak current would be √2 times this, namely over 70 000A, and the peak forces would be doubled to nearly ½ ton-force per foot run. It does not end there: on switching on, or at the onset of a fault, the current might be 100% asymmetrical (see Chapter 32 of Part 1), in which case the current could peak to 2.55 times the rms value, namely to nearly 130 000A, which would give momentary forces of over 1½ tons-force per foot run.

It can easily be seen that such enormous forces would instantly break up the busbar system unless the bars were firmly braced at close intervals against lateral movement. The forces due to one single asymmetrical peak could cause the damage which would initiate a catastrophic break-up. Short-circuit currents of the order of 50 000A rms are not uncommon in LV systems.

Care must be taken in switchboards to brace not only the busbars themselves, but also any 'droppers' or tee-offs through which a short-circuit current may pass.

In 3-phase switchboards, where three busbars are usually arranged side by side, the currents, being 120° apart in time, are not equal and opposite as described above for single phase. In different parts of the cycle the currents in adjacent bars will sometimes be in opposite directions and sometimes in the same direction. Therefore the mechanical force on any bar is at one instant acting to blow it outwards, and at the next to draw it inwards. If the spacing is not sufficient (bolt-heads can be a source of trouble) or if the bar or bracing is not stiff enough, a bar might even touch its neighbour, so transferring the original external short-circuit into the switchboard itself, with a high risk of fire. Possible mechanical resonance of the bars or of any droppers to the 50Hz or 60Hz supply must also not be overlooked.

Although busbar forces have been described here in relation to low-voltage switchgear, they occur equally in high-voltage switchgear. However the current levels there, even short-circuit levels, are generally much lower and the busbar spacing greater; consequently the problem is most acute on low-voltage switchboards.

3.6   HOLLOW BUSBARS - SKIN EFFECT

If a conductor, such as the one shown in Figure 3.8, is carrying a current, that current will normally make use of the conductor's whole cross-section area. If the cross-section is regarded as made up of a number of equal, thin elements each carrying some of the total current, then all these elements are in parallel and have the same resistance for a given length of conductor. The current will divide equally between them - that is to say, it will distribute itself uniformly over the whole cross-section.

This is certainly true of direct currents and also of alternating currents at power frequencies if they are not too large.

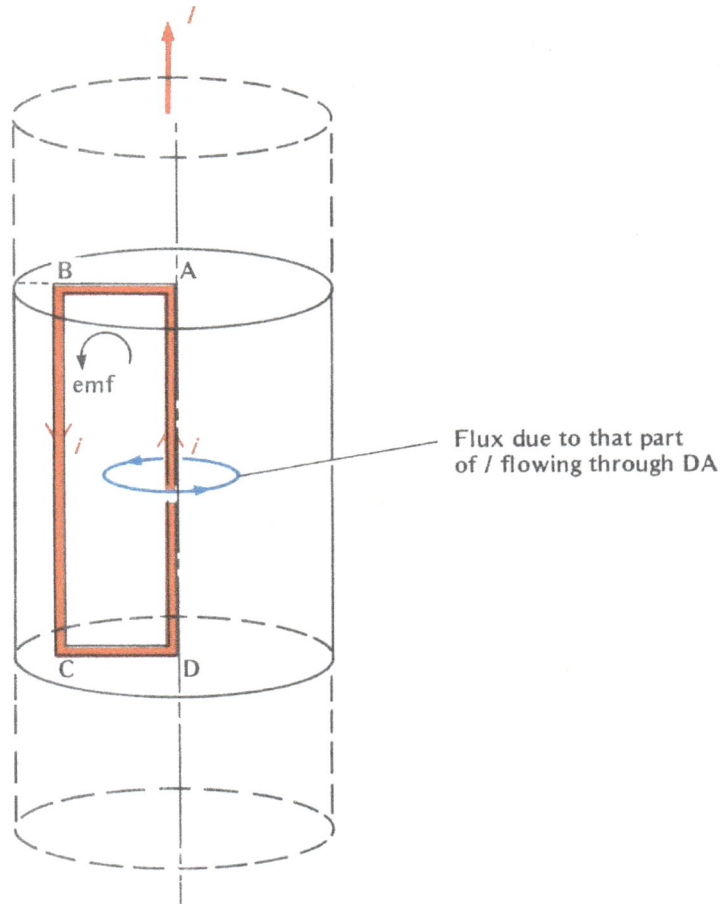

FIGURE 3.8
SKIN EFFECT - ELEMENT OF CONDUCTOR

To examine how the conductor behaves when carrying an alternating current, consider a solid conductor of circular section as shown in Figure 3.8. Take a short length of this conductor, for example between points A and D. Assume the current (I) to be flowing upwards as indicated, the direction being taken at a certain instant of time during the alternating cycle when the current is rising.

Consider now a rectangular element ABCD within the conductor, the side DA along the axis, AB and DC in a radial direction, and the side BC **inside** the conductor and parallel with AD. The thin strip of conducting material around the sides of the rectangle forms a closed conducting loop.

That part of the main current I flowing through the centre part DA of the conductor causes a circular magnetic field around it, anti-clockwise as seen from above and as shown in the figure, and in phase with the current causing it. This alternating flux passes through the conducting loop ABCD, causing an emf to be generated in it by Faraday's Law, that emf being anti-clockwise at the instant shown in the figure, but lagging 90° in phase on the flux, and therefore on the conductor's main current.

The emf induced in the closed rectangular element ABCD causes a current i to flow round it, again anti-clockwise at the instant shown. Because the loop is mainly inductive this current will lag almost 90° on the emf causing it, which, as shown above, itself lags 90° on the flux and so on the main current. Therefore the loop current lags a total of 180° on the main current - that is to say, it is anti-phase with it.

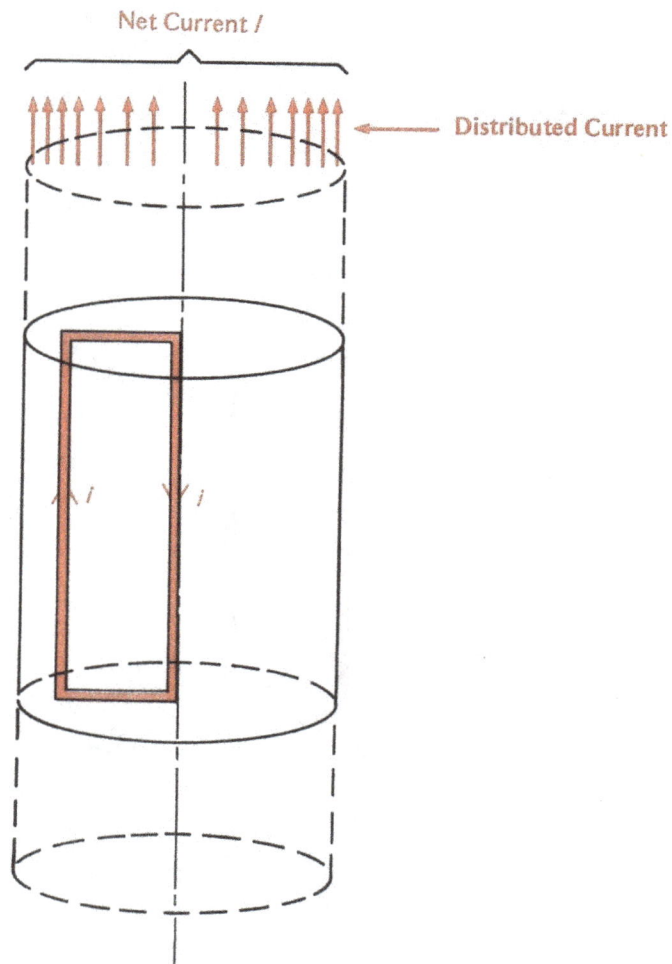

FIGURE 3.9
SKIN EFFECT - CURRENT IN A CONDUCTOR

Although Figure 3.8 suggests that the current flow *i* in the loop is anti-clockwise at the instant shown, in fact, because of its 180° phase lag, it is opposite in sign, and the actual directions at that instant are as shown in Figure 3.9.

This figure shows clearly that the loop current opposes the main current at the centre but adds to it towards the outside. The current density at the centre is therefore reduced, but it is increased as one goes outwards. The effect is dependent on the magnitude and on the rate of change of the flux. The former is directly proportional to the conductor's main current, and the latter to its frequency. The effect is also greater with large diameter conductors.

The result is an increasing concentration of the main current towards the outer layers and a decrease at the centre. This phenomenon is known as 'skin effect'. It is not present at all with d.c., since the circular magnetic field does not alternate and so produces no emf. With normal levels of a.c. power currents the effect, though present, is negligible. But when currents of several thousand amperes are flowing, the internal magnetic field is large and the skin effect becomes pronounced.

By the time currents of the order of 3 000 to 4 000A are reached, the effect is so strong that the bulk of the current is flowing in the outer parts of the conductor and very little in the centre. The centre, in fact, is hardly being used and is a waste of expensive copper. With currents above these levels it is not uncommon to install hollow busbars, which are in effect a continuous skin having no centre. A further advantage of using hollow busbars is that cooling air or other medium may be passed along their full length. There is no need for the hollow busbar to be circular; it may be a rectangular extrusion or made up of plates.

Although LV switchboards with busbar ratings up to 4 000A are found on some offshore platforms, hollow busbars have so far not been used. They may however be found on onshore installations of very large current capacity.

The phenomenon of skin effect is well known in the radio world where high frequencies up to many megahertz are used, but it is not generally realised that it also occurs at power frequencies.

CHAPTER 4 SWITCHGEAR TESTING

4.1 GENERAL

All high- and low-voltage switchgear is subjected to extensive tests by the manufacturer before delivery to the customer. While it is not the job responsibility of operators and maintenance staff to carry out these tests, it is obviously an advantage to have some knowledge of them. They are summarised below.

4.2 MANUFACTURERS' TESTS

Switchgear testing, which is normally taken as meaning circuit-breaker testing, is a very complex subject, of interest chiefly to switchgear designers and manufacturers. Tests fall into two categories, as follows:

(a) **Routine Tests**

These are tests to which each individual circuit-breaker is subjected; they comprise the following:

- Check of mechanical operation, including measurement of tripping time.

- Overvoltage withstand: a test of insulation soundness using a d.c. voltage proportional to the rated voltage of the circuit-breaker, usually of one-minute duration.

- Heat run: a continuous run at full rated current with contacts closed, to check that, when temperature has stabilised, it does not exceed the specified level.

(b) **Type Tests**

These are tests made on a prototype circuit-breaker which is representative of others to demonstrate that the design complies with the stringent requirements of the system in which it is to operate; these tests may include the foil owing:

- Closing and trip tests at 10%, 30%, 60% and 100% of the rated breaking rms symmetrical fault current at not more than 0.15 power factor. The breaker must not show 'distress', and there should not be undue contact burning.

- A trip test at 100% of the rated breaking peak asymmetrical fault current in one pole and at more than 0.15 power factor.

- An impulse test to simulate the effect of a lightning strike on the system. A steep-fronted, high-voltage pulse rising to its maximum in 1.2 µs and falling to 50% in not more than 50 µs, is applied. The circuit-breaker must not trip or flashover.

Trip and impulse tests are carried out at full voltage and with full fault current. They call for the use of a large test generator capable of giving out the full rated breaking MVA without significant drop of voltage. Such test generators are of very special design and are capable of undergoing repeated short-circuits. There are only a few in the country, located at special Short Circuit Test Stations. The largest station in the UK is capable of delivering 6 000MVA. There are also stations on the Continent of which KEMA in Holland is one of the largest.

All the UK stations are administered by the Association of Short-circuit Testing Authorities (ASTA). On successfully completing a full type test they issue an 'ASTA Certificate' of rating, which is complete with oscillograms taken during the test.

Any purchaser of a circuit-breaker of identical design can obtain a copy of its ASTA Certificate. He would not normally have his own purchased circuit-breakers undergo repeat short-circuit tests. Only if he had called for any change in the design which might affect the breaker's performance and so the validity of the certificate would new tests and a new certificate be necessary. Such special tests would be extremely expensive.

The theory of a.c. circuit interruption is discussed in Chapter 1. The oscillograms supplied with the ASTA Certificate serve to confirm that the restriking and other waveforms of the circuit-breaker are in accordance with the design.

4.3   USER CHECKS

Operators and maintenance staff, while it is not their responsibility to consider the tests referred to above except as a matter of interest, are required to apply certain routine checks and tests to circuit-breakers and switchboards at the intervals laid down in the appropriate maintenance schedules. These tests include:

- Visual examination of the whole switchboard, inside and out, for cleanliness, mechanical damage, corrosion or signs of overheating or leaking where applicable. Also checking of busbar and copperwork for bolt tightness and cleanliness.

- Megger testing each part of the switchboard busbar system with all circuit-breakers open. Tests between each phase and earth (with the other phases earthed), and between phases.

- Megger testing each circuit-breaker in turn (while isolated) between each phase and earth (with the other two phases earthed), and between phases.

- Simulation of overcurrent protection on each circuit-breaker by current injection. This may be of two types - secondary injection' or 'primary injection' (see para. 4.4).

- Simulation of earth-fault, undervoltage or other protection on each circuit-breaker by manual operation of the protection devices, and checking the alarm indications and that the circuit-breaker trips.

- Visual and electrical checks of the local trip-and-close battery (where fitted) and its charger(s).

- Simulation of trip circuit failure and checking correct operation of Trip Circuit Supervision with each mode of failure.

- Taking an oil sample from each oil circuit-breaker (where applicable) for laboratory testing at specified intervals.

- Examination of the circuit-breaker contacts after a stated period or a stated number of normal operations, or after a fault clearance. This includes oil draining where applicable and renewal of oil and of contacts as necessary.

4.4 CURRENT INJECTION TESTING

Current injection tests are made on switchgear to check that the various protective systems operate properly and at the correct preset current levels. Current injection is of two distinct types, 'secondary' and 'primary'.

4.4.1 Secondary Injection

Secondary injection consists of introducing a variable controlled current from a separate supply source into the circuit which is normally fed by the secondary of each current transformer. The CT secondary itself is at the same time disconnected and short-circuited. Most relays have a test block with links which enable this to be done without disturbing the wiring. Varying the injected current enables the operating settings of the connected relays to be set up or checked and the continuity of the relay circuit to be verified. Secondary injection does not test the CT itself.

4.4.2 Primary Injection

Primary current injection achieves a similar object but consists of injecting a variable but heavy current into the **primary** of the current transformer; this avoids disturbing the secondary circuit in any way. As the primary usually consists of a bare copper conductor bar, connection must be made direct to the bar on both sides of the CT. In some makes of circuit-breaker heavy lugs are provided on the bars for this purpose.

The heavy current for primary injection is usually obtained from a step-down portable transformer fed from the 240V single-phase station supply. The transformer is fairly large, about 10kVA, and can supply up to 1 000A of current. Although primary injection has the advantage that it tests the complete installation, including the CT itself, it is heavy and cumbersome and is consequently less used in the field than secondary injection.

Some current transformers are provided with a separate test winding which is used for primary injection. In that case, instead of injecting a very large current into the primary bar, a much smaller current can be injected into the test winding, and the CT itself would still be included in the test. This method, where it can be used, obviates the need for cumbersome primary injection equipment.

4.5 BUSBAR DUCTER TESTS

Continuity resistance of busbar copperwork, especially across joints, needs to be regularly checked to prevent overheating. It is measured by a specially-sensitive continuity tester called the 'Ducter' ohmmeter. This is a portable device, similar to a megger in appearance, but with an ohmmeter scaled to read down to 1 $\mu\Omega$ or up to 10 Ω in six ranges.

This instrument is designed to measure very low values of resistance while a d.c. test current is flowing. Although principally used for measuring continuity resistance across busbar joints, it can also be used for measuring earth bonding and switch and circuit-breaker contact resistance.

The tester is entirely self-contained and incorporates its own rechargeable battery. The ohmmeter is scaled 0 - 100 $\mu\Omega$, but a 6-way range selector switch extends this up to 10 Ω. The meter is of the cross-coil type (similar to that of a megger), which gives a reading independent of the state of the battery voltage.

The instrument has four terminals: C1 and C2 which apply test current through the joint to be tested, and two potential terminals P1 and P2 connected to test prods which are applied either side of the resistance to be tested. The voltage detected between P1 and P2 is, by Ohm's law applied to the known test current, proportional to the resistance to be measured. This voltage is amplified within the tester and applied to one coil of the ohmmeter, the other being fed from the test current. The meter reads the continuity resistance in microhms direct.

The instrument incorporates a battery tester. If it indicates a low charge state, the battery must be recharged before use. A built-in charger enables this to be done from the local a.c. supply voltage. A completely discharged battery requires 16 hours to recharge.

In some installations the tests across busbar joints and joints on other copperwork such as busbar droppers, cable joints and on the primaries of any wound-type CTs must be carried out with a test current of at least 20A d.c. The voltage drops so measured are interpreted on a comparative basis, and for identical types of connection or circuit the measured values must not differ by more than 20% from each other.

CHAPTER 5 QUESTIONS AND ANSWERS: CHAPTERS 1 - 4

5.1   QUESTIONS

1.   What is the fundamental difference between breaking a d.c. and an a.c. circuit?

2.   If an a.c. circuit undergoes a sudden short-circuit, what is the nature of the current waveform which immediately follows?

3.   Make a sketch of the current waves which immediately follow a short-circuit on a 3-phase system.

4.   What do you understand by the term 'bolted fault'? How does such a fault differ from the more likely arcing fault?

5.   What is the approximate ratio of the peak value of a short-circuit current to the rms value of the symmetrical current at that time?

6.   What effect will there be on the current in a purely inductive circuit following a short-circuit which occurs at:

    (a)   a voltage peak (positive or negative)
    (b)   a voltage zero
    (c)   a voltage between zero and peak?

7.   If a 'bolted' short-circuit occurs across all the phases of a 3-phase inductive circuit, what effect will it have on the three current waveforms?

8.   What do you understand by the term 'a.c. component' and 'd.c. component' of a current waveform?

9.   How are the breaking capacities of circuit-breakers normally given?

10.   What do you understand by 'restriking voltage' in a circuit-breaker?

11.   What are the three main types of high-voltage switchgear met with on onshore and on offshore installations?

12.   What are the essential differences in arc control methods used by the three types of Q.11?

13.   How does the nature of the restriking voltage differ between an oil circuit-breaker and an air-break circuit-breaker?

14.   Why is a 'puffer' often employed with an air-break circuit-breaker?

15.   What is the principal difference between a circuit-breaker and a contactor?

16.   Why is a contactor usually fitted with back-up fuses?

17.   What precautions are necessary before withdrawing a circuit-breaker truck from, and reinserting it into, its switchboard housing?

18. How is safety ensured when withdrawing a voltage transformer from a high-voltage switchgear unit?

19. What types of circuit-breaker operating mechanisms are used? Describe each briefly.

20. What do you understand by 'Trip Circuit Supervision'? Why is it necessary?

21. Why is it necessary to earth down a feeder or a busbar when work is to be done on it? Describe two methods of doing this.

22. What type of main (incomer and bus-section) switchgear would you expect to find on a low-voltage switchboard?

23. How is general distribution arranged from an LV switchboard?

24. Describe briefly (a) a moulded-case circuit-breaker (MCCB) and (b) a miniature circuit-breaker (MCB).

25. How is access gained to the inside of an MCC distribution cubicle of an LV switchboard?

26. If you wish to test the operation of an LV motor starting contactor without operating the motor, how would you do it?

27. Can you start and stop a motor from its MCC cubicle?

28. Why are busbars strongly braced in switchboards?

29. Why are hollow busbars sometimes used for very heavy currents?

30. What tests would you make on a circuit-breaker to establish (or check) the over-current protective relay settings? What are the two main methods of testing?

31. What are the advantages and disadvantages of these two types of test?

32. How would you test the soundness of busbar and similar heavy-current joints in a switchboard? How are the readings compared?

5.2 ANSWERS

1. With d.c. interruption the arc continues to burn until it collapses by lengthening either naturally or under the action of electromagnetic forces, so increasing its resistance and reducing the current. With a.c. interruption the arc goes out naturally twice every cycle at each current zero, restriking each time. The aim of the switchgear is to allow it to go out naturally and then to keep it from restriking.

(1.2, 1.3)

2. In general the current waveform immediately after onset of a short-circuit is asymmetrical, the degree of asymmetry being random and depending on the instant of short-circuit. If this instant occurs at a voltage peak (positive or negative), the asymmetry is zero, but if it occurs at a voltage zero the asymmetry is complete (100%).

3. See Figure 1.3.

(1.4)

4. In order that a fault current may be calculated without having to consider the unknown effect of arc resistance, faults are assumed to have zero impedance at the fault point - that is, it is assumed that a solid conductor is bolted across all phases. An arcing fault will in practice introduce some resistance and reduce the calculated fault current.

(1.4)

5. Approximately 2.55 times.

(1.4)

6. (a) Current wave will be symmetrical.
   (b) Current wave will be completely (100%) asymmetrical.
   (c) Current wave will be partially asymmetrical.

(1.4)

7. In general all three current waves will be partially asymmetrical, the sum of their asymmetries (taking sign into account) being always zero. Even if one of them is symmetrical (0%) or wholly asymmetrical (100%), the other two, being displaced 120°, will be partially asymmetrical.

(1.4)

8. An asymmetrical wave may be considered as resolved into two parts: a symmetrical alternating current wave called the 'a.c. components', and a steady, but decaying current called the 'd.c. component'. The rate of decay of the d.c. component depends mainly on the R/L ratio of the faulted circuit. When the d.c. component has decayed to not more than 15%, the current is deemed to be 'symmetrical'.

(1.4)

9. Breaking capacities are given either in 'kA' or 'MVA'. In both cases the current is assumed to have recovered symmetry down to at least 15%. The kA (or MVA) rating is therefore the 'rms symmetrical' value. The rated peak asymmetrical current may also be given.

(1.4)

10. Restriking voltage is that voltage which reappears across the circuit-breaker contacts immediately after the arc has extinguished.

(1.5)

11. Oil-break (OCB), air-break (ACB) or vacuum (VCB).

(1.6 - 1.8)

12. Oil-break controls the arc by allowing it to extinguish naturally at a current zero and attempts to prevent its restriking by introducing clean, cool oil into the former arc path. (1.6)
Air-break introduces resistance to the arc path such that the restriking wave is critically damped. (1.7)
Vacuum interrupter operates by condensing the vapour immediately the arc goes out at current zero and denying it a material in which to restrike. (1.8)

13. In an oil circuit-breaker the restriking voltage usually takes the form of a high-frequency surge, whereas in an air-break circuit-breaker it is a critically damped wave. (1.5, 1.7)

14. Because with low-current breaks there may not be enough electromagnetic force to drive the arc up into the arc-chutes for splitting, this movement is encouraged by a small blast of air to drive it up mechanically. (1.7)

15. A circuit-breaker's purpose is primarily to interrupt fault currents. It may be used as an operational switch, but only occasionally. A contactor on the other hand is designed for repeated operation and not, in general, for interrupting faults. (2.1)

16. Because a contactor has only limited breaking capacity, it must usually be protected by back-up fuses. These are selected to blow before the contactor can open if the fault current is beyond the contactor's breaking capacity. (2.1)

17. Before a circuit-breaker truck is withdrawn, the breaker must be open and its closing mechanism disconnected. If spring-operated, its spring must be discharged.
Before a circuit-breaker truck is reinserted, the breaker must be open, its spring discharged and its closing mechanism disconnected. (2.2)

18. A voltage transformer in an HV switch unit is behind a barrier. The barrier door cannot be opened until the VT has been withdrawn and isolated on its HV side. (Figs. 2.1 & 2.2)

19. Mainly solenoid and motor/spring. A solenoid, usually d.c.-operated, closes the breaker by direct trip-free linkage. A motor/spring mechanism closes the breaker by discharge of a closing spring, which is then immediately recharged by the motor (using much less current than the solenoid) ready for the next closing. The spring can also be charged by hand and released manually. (2.3.3)

20. When a circuit-breaker has been standing for long periods without operation, there is no certainty that its trip coil or the supplies to it are still in working order. A Trip Circuit Supervision relay monitors the continuity of the trip coil itself and of its connections by passing a small current continuously through it, giving an alarm if it becomes interrupted for any reason. It also gives an alarm if the trip supply fails. (2.8)

21. When work is to be carried out on any part of an electrical system, that part must be isolated from all power sources and locked out. As an added precaution all parts being worked on must be earthed down, so that all cables are discharged and no voltage can appear at any point.

On HV systems this isolation and earthing-down, whether of feeder cables or busbars, is carried out at the switchboard itself. On one type of breaker the isolated contacts are 'swung' to apply the earth where required; on another type feeder cables are earthed down by an interlocked earthing switch. On the latter type for earthing down the busbars the normal circuit-breaker is replaced by a separate 'earthing breaker' - see Figure 2.11. (2.10)

22.   Usually air-break circuit-breakers.   (3.1)

23.   General distribution is through Motor Control Centres associated with most LV switchboards. The MCC consists of many panels; in each panel are a number of cubicles each containing an isolating fuse-switch or switch-fuse and, in the case of motor feeders, also a contactor.   (3.3)

24.   (a)   An MCCB is a very compact, usually 3-phase, circuit-breaker which can be closed and opened only by hand through a front dolly. It also includes a thermal overcurrent trip unit as well as a 'high-set' electromagnetic instantaneous overcurrent trip of a setting suited to the circuit in which it is used.

      (b)   An MCB is a smaller version of the MCCB and may be 4-pole, 3-pole, 2-pole or 1-pole. It has similar types of thermal and electromagnetic overcurrent trips; these cannot be interchanged.   (3.2.3)

25.   The door cannot be opened until the isolating switch handle is moved to the 'off' position.   (3.3.2)

26.   Inside contactor cubicles is a switch which applies a separate test voltage to the contactor coil. Since the circuit is already isolated by the fuse-switch, test closing of the contactor cannot start the motor.   (3.3.2)

27.   Except in very rare cases motors cannot be **started** from the MCC (they can only be started locally or by remote control). All can however be **stopped** by an emergency stop pushbutton at the MCC.   (3.3.2)

28.   With heavy short-circuit currents in adjacent busbars, intense electromagnetic forces are generated which can physically move the busbars and connected copper-work unless they are strongly braced.   (3.5)

29.   Due to 'skin effect' when carrying large currents (4 000A and above), only the outer parts of a conductor effectively carry the current, and the inner parts are wasted. Busbars are therefore sometimes made hollow; this also assists in their cooling.   (3.6)

30.   Overcurrent relay settings can be made, or checked, by injecting a variable current into their circuits. This is done by either secondary injection or primary injection . In the former the CT secondary is disconnected and shorted, and test current is injected into the remainder of the secondary circuit, including the relays. With primary injection a heavier current is injected into the primary of the CT, leaving the secondary circuit undisturbed.   (4.4)

31.   The advantage of secondary injection is that the test equipment is relatively light; its disadvantages are that the CT itself is not tested and that the CT secondary circuit must be disturbed. The advantages of primary injection are that the CT itself is subjected to test and that it is not necessary to disturb any of the secondary protective circuits. Its disadvantage is that primary injection equipment is heavy and cumbersome.   (4.4)

32.   By the use of 'Ducter' equipment. This injects a known d.c. test current through the joint and measures the voltage drop across it. It displays the resistance across the joint in $\mu\Omega$ or mΩ. The resitance readings across similar joints should not vary by more than 20%.   (4.5)

CHAPTER 6 POWER TRANSFORMERS

6.1   GENERAL

In onshore installations main electric power is taken in bulk from the National Grid through Area Boards at voltages up to 33kV. In offshore installations however main power must be generated locally, and this is usually done at 6.6kV.

Note that 6.6kV is a **nominal** rating; in practice generation is more usually at 6.8kV to allow for voltage drop in the network.

There are instances of generation at other levels, but these are always 'high voltages'. Some large loads are fed directly from the HV system, but for most purposes the supplies are needed at low voltage, typically 440/250V offshore and 415/240V onshore. These are provided through 3-phase power transformers.

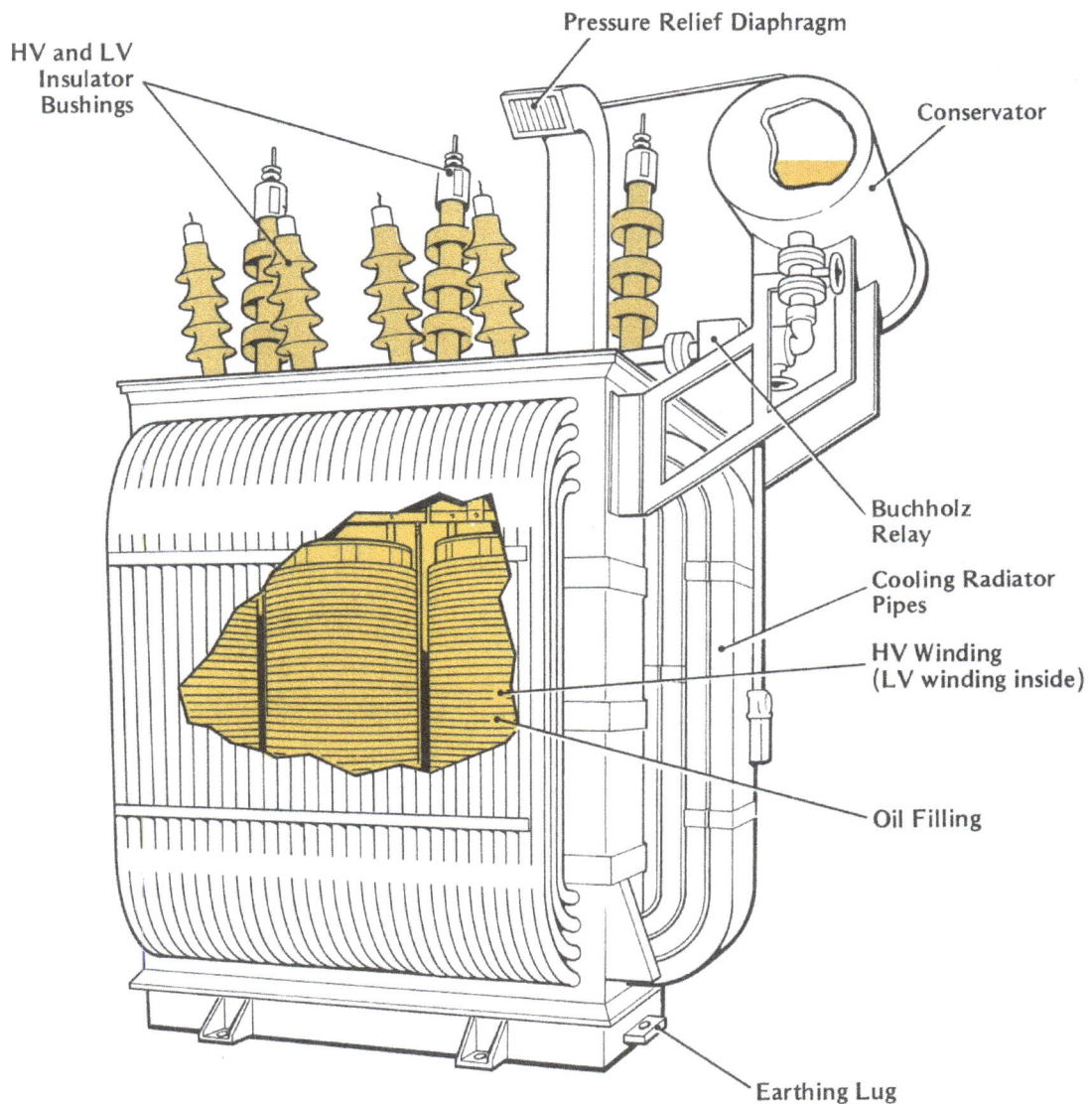

FIGURE 6.1
THREE PHASE OIL FILLED TRANSFORMER

6.2   CONSTRUCTION

6.2.1 General

Power transformers are always enclosed in a tank or similar protection. They may be liquid cooled, air cooled or dry type (encapsulated or open). If liquid cooled, the coolant may be mineral (hydrocarbon) oil, silicone oil or some artificial liquid such as 'Askarel'.

The internal construction of all power transformers is similar. The windings are stacked around a 3-limbed laminated iron magnetic core, the low-voltage windings innermost and the high-voltage windings outside them - the best arrangement for insulation. It can be seen in the cut-away portion of Figure 6.1. Ducts are arranged through both windings on each limb to assist cooling. The terminations of the windings are brought out to cable boxes for external connections (see para. 6.8) or, for large outdoor transformers, to terminal bushings.

6.2.2 Liquid (Oil) Filled Transformers

The largest bulk-power transformers are usually in a single tank, completely filled with oil and with a header tank called a 'conservator' on the roof. This maintains a static head of pressure on the oil and also allows free expansion and contraction. The transformer of 6.1 is of this type.

In the pipe connecting the main tank to the conservator there is often inserted a device called a 'Buchholz Relay'. It has two elements: one traps and collects any small gas bubbles evolved at a winding due to the early stages of a possible breakdown of insulation. If sufficient gas has accumulated, a float switch gives an alarm. The other element is a pivoted vane. If a major fault occurs inside the tank, the displaced oil surges past the vane, causing it to swing, make a contact and trip the supply breaker. The Buchholz relay is further described in Part 7 Control Devices.

The oil coolant is heated by the I^2R losses of the currents in all the windings and also by the iron losses. It circulates through a closed cooling system by thermosyphon action (in a few cases by pumping). Heat is extracted from the coolant through radiating tubes or fins either by natural convection, by forced cooling from fans or, more rarely, by water cooling. The cooling of transformers is dealt with more fully in para. 6.6. The whole arrangement can be seen in Figure 6.1.

Facilities are provided for oil filling, draining and sampling for test. Oil samples are taken periodically for insulation testing in the laboratory, where they are examined for deterioration or water pollution.

A tapping switch, normally for off-load use only, is usually fitted for changing the transformer taps. The larger system network transformers may have on-load tap-changing gear - see para. 6.9 for details.

Smaller oil-filled transformers are usually sealed, with an air space above the oil instead of a conservator, to allow expansion when hot. In appearance they would look like the Askarel-filled type shown in Figure 6.2.

Silicone oil is increasingly used instead of mineral oil because it is non-flammable. The construction is however similar.

6.2.3 Liquid (Askarel) Filled Transformers

In onshore installations where transformers can be installed at a distance from other plant, the fire risk is no greater than normal, and mineral oil-filled transformers are generally used. Offshore however the fire risk is crucial, and other designs of transformer are necessary.

For this reason power transformers on offshore installations do not use mineral oil as the insulating and cooling medium. Instead, a non-flammable silicone oil or a non-flammable liquid such as Askarel is used which is sometimes described by a trade name such as 'Pyroclor' or 'Pyralene'.

The basic transformer design using silicone instead of mineral oil is no different from that described for oil-filled transformers. But Askarel, apart from its fire-resistant properties, has other characteristics, namely:

- it is expensive,
- it evaporates readily when in contact with air,
- it is very penetrating,
- it is heavy,
- it is toxic and must not be allowed to come into contact with the skin or eyes,
- it must on no account be discharged into the sea because it is non-biodegradable.

The external appearance of a typical Askarel-filled transformer is shown in Figure 6.2. Its internal construction is as described in para. 6.2.1, but, additionally, the transformer is hermetically sealed to prevent loss of liquid by evaporation. In some designs the main cover flange joint is welded up, as is also the filler plug after filling with liquid, and sometimes even the drain plug. The expansion space above the liquid is filled with dry air or inert gas and has sufficient volume to ensure that the pressure inside the tank is limited to a safe value even at maximum temperature and expansion. A pressure/vacuum gauge is fitted. It indicates zero at normal temperature, a positive pressure at high temperature and a partial vacuum at low temperature. This gauge is a constant monitor on the state of the sealing. A sight-glass also gives a direct measure of the liquid level at all temperatures.

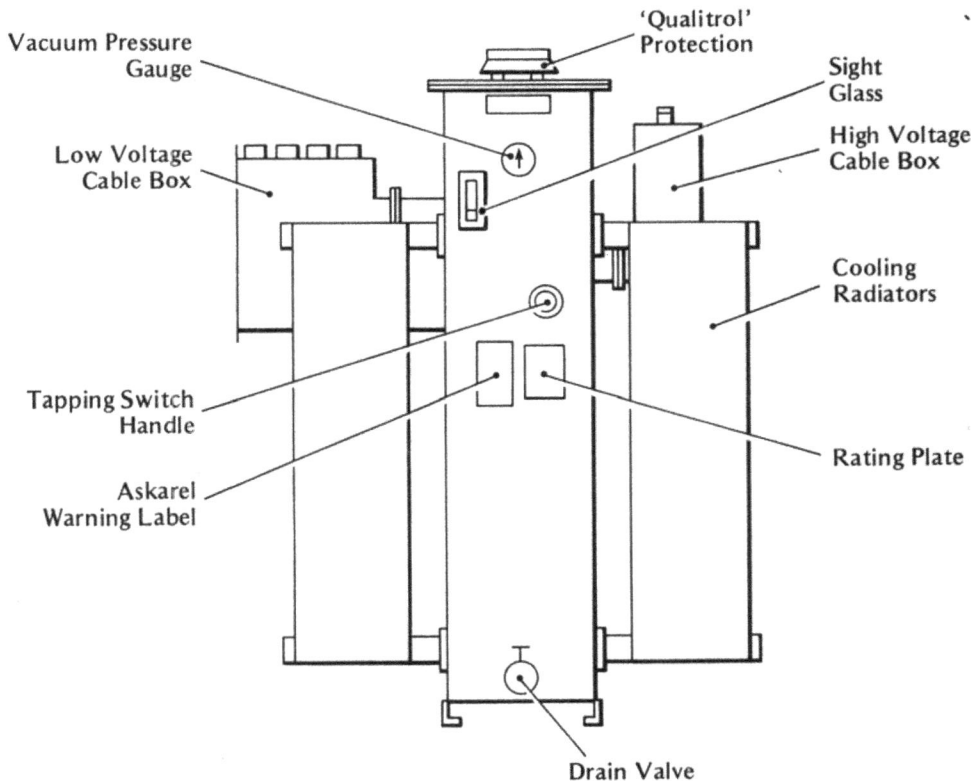

FIGURE 6.2
TYPICAL 2 000kVA SEALED POWER TRANSFORMER (ASKAREL FILLED)

The off-load tapping switch spindle has five positions. Because Askarel is a penetrating liquid it is not easy to provide an adequate seal where the operating shaft of the tapping switch comes through the tank-wall of the transformer. Some designs rely on a simple packing gland whereas others back this up by enclosing the whole tapping switch handle in an auxiliary box with a bolted cover. This box is filled with Askarel and provides a second barrier to prevent liquid from leaking out of the main tank. With the latter design it is necessary to drain and open up the auxiliary box to get at the tapping switch handle.

If a fault develops inside the transformer, the Askarel breaks down and forms a gas leading to a gradual or sudden rise in pressure, depending on the severity of the fault. To prevent the transformer tank from splitting, a spring-loaded 'Qualitrol' pressure relief device on the top operates to release the excess pressure; at the same time contacts within the device close to trip the incoming supply and give an alarm. Pressures encountered in normal service are not high enough to operate the device, which is essentially an emergency relief valve. The Qualitrol device is further described in Part 7 Control Devices.

In case the tank should ever crack or split and so spill the Askarel, these transformers are always erected with a sill around their mounting places. The sill is high enough to contain the entire filling of its transformer and to prevent the toxic liquid spreading. After a spill the liquid must be collected and put into containers, along with mopping-up rags, and sent ashore for disposal. **ON NO ACCOUNT MAY THEY BE DROPPED INTO THE SEA.** Personnel cleaning up must wear protective clothing, gloves and goggles.

Though widely used offshore at first, Askarel-filled transformers are gradually being phased out in favour of silicone oil-filled types because of the toxic nature of Askarel.

6.2.4 Dry Type Transformers

On some offshore installations transformers are used where the windings are encapsulated in epoxy resin and the whole block is air cooled. To assist the cooling, air ducts are arranged through the solid encapsulation.

Such transformers are often given a dual rating (e.g. 2 000/2 500k VA): the lower one where the cooling air circulates naturally, and the higher one where it is assisted by fans. It is arranged that the fans start automatically when the loading of the transformer exceeds the lower rating.

Dry-type power transformers can readily be built into their own LV distribution switchboards to form a single unit, thereby saving LV cable boxes and cables and bringing the incoming feeder copperwork right up to the transformer's LV terminals.

Small low-voltage transformers used as part of other equipment - for example, in battery chargers or inverters - are usually open type, air cooled without any enclosure. They may be 3-phase or single-phase. Such open-type transformers are protected by being housed within the parent equipment's main enclosure.

6.3   RATINGS

The capacity of transformers is always given in kVA or MVA, because the heating depends only on the actual current and is not affected by the power factor of that current.

The ratings of most offshore power transformers extend over a range of about 400kVA to 2 500kVA, depending on their duty. In onshore installations the ratings may go up to 30MVA or more; National Grid sizes may go up to 750MVA.

A transformer is designed to give a nominal secondary voltage from a nominal primary voltage - for example 11 000/415V or 6 600/440V. Due to voltage drop within the transformer itself, the actual turns ratio must be somewhat lower than this if the nominal secondary voltage is still to be achieved at full load. In the two examples cited above the turns ratio (that is, the no-load ratio) would need to be about 11 000/435V and 6 600/460V respectively.

Alone among electrical plant, transformers are required by British Standards to have their no-load rating displayed on their nameplates (generators and motors have their full-load ratings). The nameplate figure is therefore sometimes misleading in that it suggests a 435V or 460V system, whereas the nominal system voltage is still 415V or 440V. In some documentation only **nominal** voltages are normally used, notwithstanding any transformer name-plate figures. Errors due to this misunderstanding may often be found on other drawings.

6.4   IMPEDANCE VOLTAGE AND REGULATION

In a manner similar to generators, transformers present impedance to the flow of through-currents. This impedance is measured by the percentage voltage applied at rated frequency to the primary winding necessary to circulate full rated current in the secondary when short-circuited.

The effect of transformer impedance is to cause an internal voltage drop when load current is passed through the transformer. Exactly as with a generator, the greatest drop is caused when a reactive current passes through the **reactance** of the transformer. (The vectorial treatment of impedance loading on a generator is fully covered in Part 2, Chapter 4.)

The internal drop due to load current causes a reduction of the secondary terminal voltage below its open-circuit level. This reduction, usually expressed as a percentage of the no-load voltage, is termed the 'regulation' of the transformer under the stated load conditions. Since the impedance of a transformer is almost wholly reactive, it follows that the greatest regulation occurs when a highly reactive load is applied.

If E is the nominal rated line voltage applied to the primary winding, and if E_{sc} is the line voltage which, when applied to the primary, will circulate full rated current in the short-circuited secondary, then

$$Z = \frac{E_{SC}}{E}$$

E_{sc} is called the 'impedance voltage' of the transformer and is usually expressed as a percentage of E. This same percentage gives the value of Z, which is the 'percentage impedance of the transformer'.

This impedance is almost pure inductive reactance and ranges in value from about 5% to 10% for the sizes of transformers in use. The measured percentage impedance is marked on each transformer nameplate and is used, together with other circuit impedances, to calculate the symmetrical short-circuit level on the low-voltage system. (See Part 10 'Electrical Protection'.)

At the instant of switching on a transformer, while the core is unfluxed and therefore offers no reactance, a large 'inrush current' will flow which, although transient, may achieve a value of up to five times full-load primary current. This disappears quickly after switch-on.

Once the core is magnetised, the impedance of a transformer to fault currents is constant; this contrasts with a generator whose reactance changes from subtransient through transient to synchronous as a fault progresses.

6.5   INSULATION

6.5.1 Dry Type Transformers

The maximum temperature to which the windings of dry-type transformers may be allowed to rise depends on the type of insulating material round the conductors. These transformers are classified according to the insulating material used, and to each class is allotted a maximum ultimate temperature. The classification is as follows (according to BS 171-3 : 1987 and BS 2757 : 1986):

Class	Typical Insulating Material	Ultimate Temperature
A	Impregnated cotton, silk, etc.; paper; enamel	$105^{\circ}C$
E	Paper laminates; epoxies	$120^{\circ}C$
B	Glass fibre, asbestos (unimpregnated); mica	$130^{\circ}C$
F	Glass fibre, asbestos, epoxy impregnated	$155^{\circ}C$
H	Glass fibre, asbestos, silicone impregnated	$180^{\circ}C$
C	Mica, ceramics, glass, with inorganic binders	$> 180^{\circ}C$

It should be noted that the classification letters do not follow an alphabetical sequence. This is because there were originally only three classes - 'A', 'B' and 'C'. Later intermediate classes were added, and it was decided not to disturb the original well-understood three.

Certain of the higher-temperature materials may be hygroscopic and therefore not always suitable in any particular environment, particularly where dampness is severe.

It should be particularly noted that the classification depends on the **ultimate** temperature to which the insulating material may be subjected, for it is this which determines whether or not it will suffer damage when heated. It does not depend on temperature **rise** alone. If, for instance, the ambient temperature is $40^{\circ}C$, a Class 'B' material may be used if the designed temperature rise will not exceed $90^{\circ}C$, so making the ultimate maximum temperature $130^{\circ}C$. Designed temperature rises must therefore take into account the greatest expected ambient temperature in which the transformer will operate.

6.5.2 Liquid Filled Transformers

Liquid-filled transformers are not classified for insulation as are the dry type. There is an overall requirement that the temperature **rise** of the windings shall not exceed $65^{\circ}C$, and that the temperature rise at the top of the liquid shall not exceed $60^{\circ}C$ if the transformer is sealed or has a conservator.

6.6   COOLING

The cooling system of a given transformer is identified by a 4-letter code, as follows:

> 1st and 3rd letter: kind of cooling medium
> 2nd and 4th letter: kind of circulation

The code symbols for the first and third letters are:

Mineral oil	O
Synthetic insulating liquid	L
Gas	G
Water	W
Air	A
Solid insulant	S

The code symbols for the second and fourth letters are:

Natural circulation   N
Forced circulation   F

Examples of the use of this code are:

Oil-filled, thermosyphon circulation, natural ventilation	ONAN
Askarel-filled, thermosyphon circulation, natural ventilation	LNAN
Dry-type encapsulated, fan cooled	SNAF
Oil-filled, pumped circulation, water cooled by pump	OFWF

6.7   THREE PHASE TRANSFORMER CONNECTIONS

A 3-phase transformer has a 3-limb core. For transformers designed to BS 171 the terminals of the windings mounted on each limb are identified by a letter as shown in Table 1 overleaf.

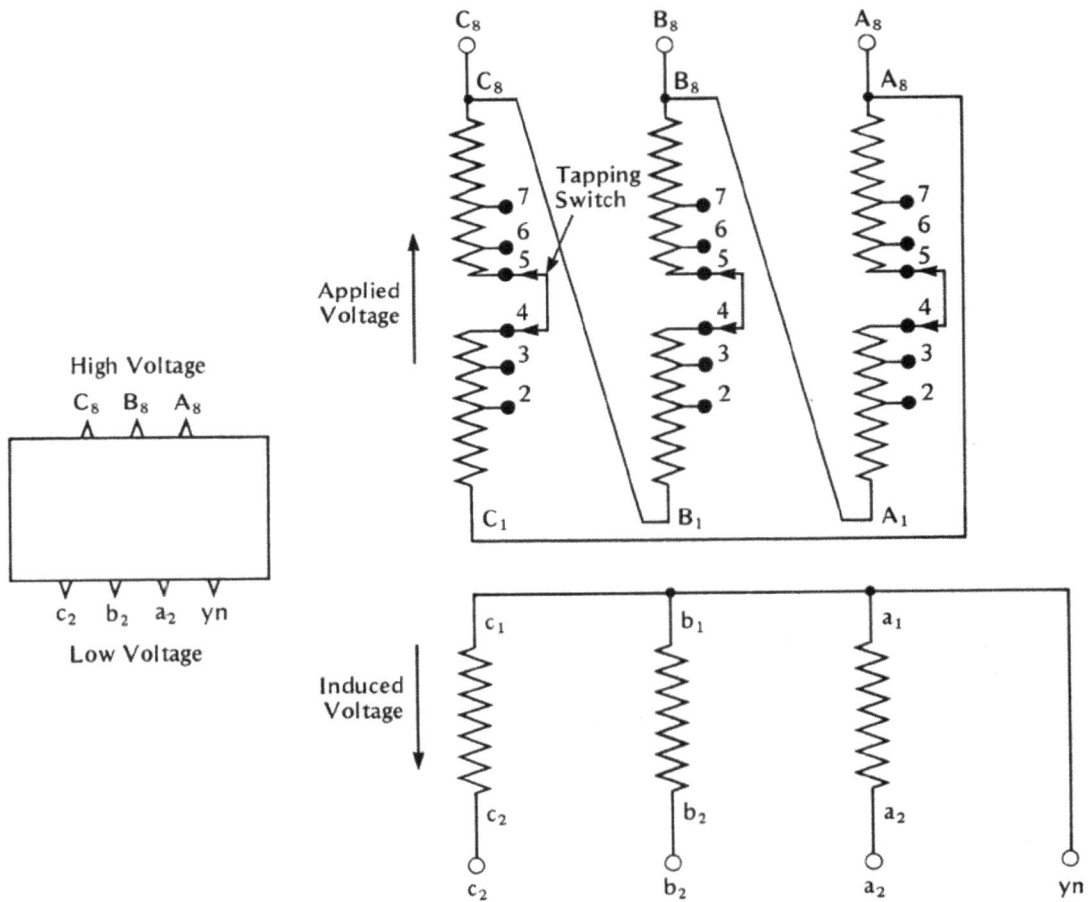

FIGURE 6.3
WINDING AND TERMINAL MARKINGS

TABLE 6.1 DESIGNATING LETTERS FOR 3-PHASE TRANSFORMERS

Winding	Designating Letter		
	Limb 1	Limb 2	Limb 3
High Voltage	A	B	C
Low Voltage	a	b	c
Tertiary (if fitted)	3A	3B	3C

The external connections to the high- and low-voltage windings are brought out of the tank through bushings. These terminals are labelled using letters appropriate to the winding concerned as shown in Figure 6.3. When viewed from a position facing the high-voltage side of the transformer, the phase sequence is A-B-C from left to right. The subscript numbers identify the winding terminations, including tappings, numbered in the direction of the applied or induced voltage at a given instant.

Three-phase windings can be connected in delta, star or zig-zag (not very common); the star or zig-zag connection must be chosen if a star-point is required to provide a neutral for a 4-wire system or for earthing. A common arrangement for 3-phase power transformers in both onshore and offshore installations is for delta-connected high-voltage windings and star-connected low-voltage windings, with the star-point brought out to provide a neutral and earth for the low-voltage system.

A delta-connected winding is designated by the letter 'D', a star-connected winding by 'Y' and a zig-zag winding by 'Z'. Capital letters are used for the high-voltage windings and lowercase for the low-voltage. Thus 'Dy' stands for delta HV/star LV; Yy for star HV/star LV, and so on. When the star-point of a star-connected winding is brought out it is designated 'YN' for a high-voltage or 'yn' for a low-voltage winding.

The winding connections for a delta/star transformer having a delta-connected high-voltage winding are shown in Figure 6.4, which also shows the vector relationship between the voltage applied to each high-voltage winding and the induced voltage in each corresponding low-voltage winding, the reversal between secondary and primary being ignored.

Taking the **phase-to-neutral** vector of 'A' phase high-voltage as reference vector at 12 o'clock, the corresponding 'a' phase low-voltage vector leads by 30° and is therefore at 11 o'clock. Thus the vector symbol in this particular connection arrangement is 'Dy11', which describes the high- and low-voltage winding connections and the angular displacement between primary and secondary voltages. Other winding arrangements are sometimes used, and for full particulars of these, together with their vector symbols, reference should be made to BS 171 - Specification for Power Transformers.

In the case shown above the vector symbol is sometimes written 'Dyn11' to draw attention to the neutral's being brought out on the secondary (low-voltage) side.

Transformers of different vector groups must not in general be paralleled. If all the primaries are supplied from a common source, the secondaries of differing groups such as Dy11, Dy1, Yy0 will have different phase relationships. For example, there will be 60° difference between Dy1 and Dy11 (which leads on it), or 30° difference between Yy0 and Dy1 (which lags on it). Such out-of-phase secondaries must never be paralleled, even though their primaries may be in parallel.

The exception is that groups with the same clock numbers, such as Dy11, Yd11, Yz11, may be paralleled, provided that there is no other objection, since the secondaries are all in phase.

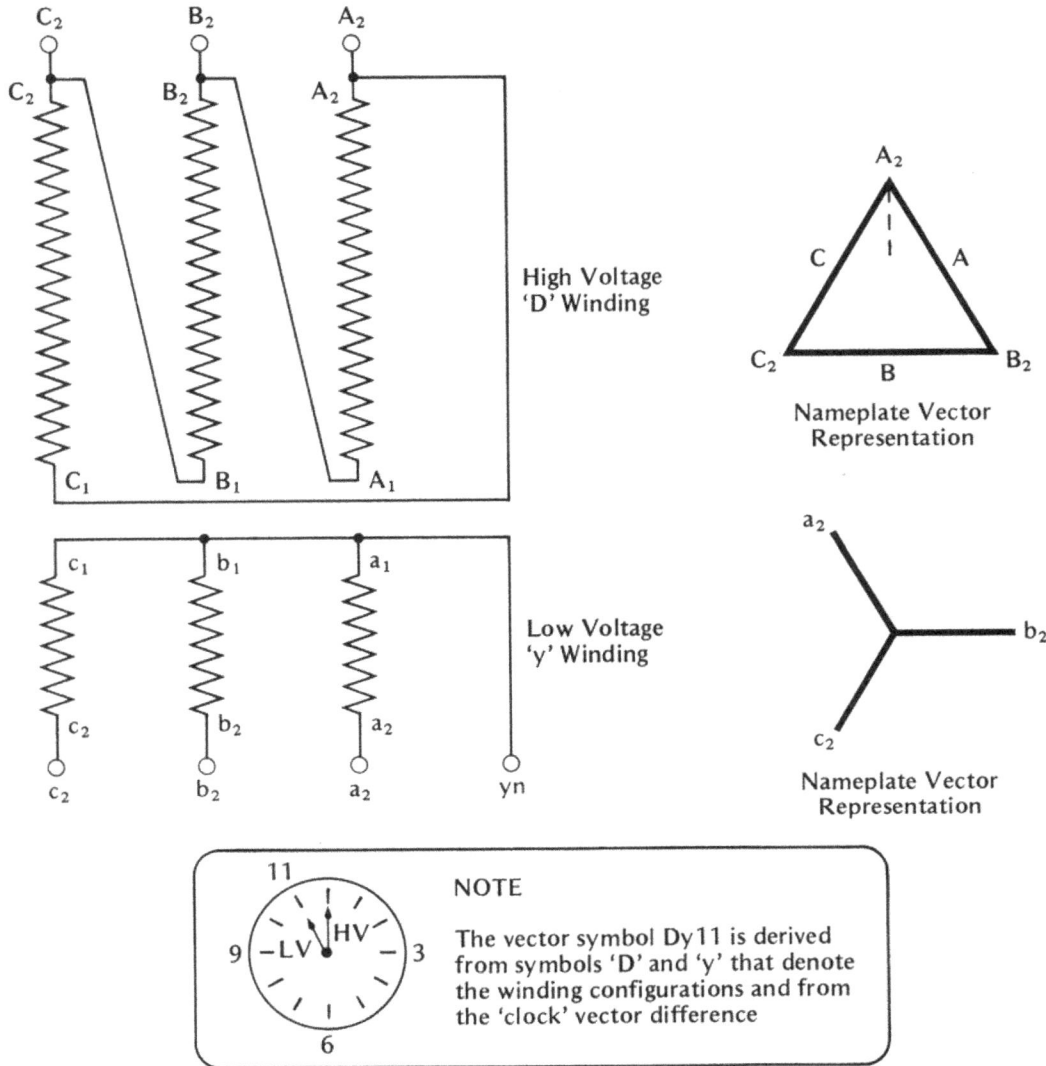

High Voltage 'D' Winding

Low Voltage 'y' Winding

Nameplate Vector Representation

Nameplate Vector Representation

NOTE

The vector symbol Dy11 is derived from symbols 'D' and 'y' that denote the winding configurations and from the 'clock' vector difference

FIGURE 6.4
VOLTAGE VECTOR SYMBOLS

6.8 CABLE BOXES

The terminals of large transformers which are connected to external lines are brought out through ceramic bushings in the cover (shown in Figure 6.1). The terminals of other transformers have to be connected to cables. This applies particularly to transformers on offshore installations and to most transformers in onshore oil installations.

The windings are connected to cables through cable boxes fixed to the transformer tank. If cables are used on both HV and LV sides, the cable boxes would be on opposite sides of the tank, as seen in Figure 6.2.

On most transformers the current on the HV side is low enough to be carried by a single, 3-core cable which enters the HV box through a sealing gland and divides inside. Sometimes small current transformers are also fitted inside the cable box.

On the LV side currents are much heavier and often exceed 3 000A. The LV cable box is therefore much larger. In order to carry such currents, three, or sometimes four, single-core cables are required for each phase. This results in the cable runs in the area of a transformer being very heavy and often difficult to accommodate. In addition there may be two similar-sized cables to carry the neutral, or 4th-wire, current.

On some installations the LV cable box is dispensed with and the windings are connected directly to the switchboard by busbar-type copperwork in an enclosed duct which is brought right up to the side of the transformer.

6.9   TAP CHANGING

6.9.1 General

Tappings are usually provided to vary the transformer's turns ratio by up to ±5%. The correct tap is set when the installation is first commissioned and should not need to be changed for a considerable time. However, as the system load grows over the years, the tapping may need to be changed to maintain the secondary working voltage. This is normally done on all phases together by means of a switch on the transformer tank and **must only be carried out off-load and isolated** - that is, with the transformer dead on both sides. Changes of tap settings may be carried out only by Authorised Persons, and then only on the instructions of the Engineering Department. All tap changers on offshore and onshore oil installations are of the off-load type.

In the larger shore networks **on-load** tap changers may be used to maintain system voltage; they are usually remotely controlled from a Control Centre and are described in para. 6.9.3. On-load tap changers are not used on offshore or onshore oil installations but may be employed on the networks supplying onshore plants.

6.9.2 Off Load Tap Changers

It is usual to provide four additional tappings with off-load tap changers, making a total of five, at 2½% intervals, so that the turns ratio varies by ±2½% ±5%. Tappings are always placed on the high-voltage side; this allows the lowest possible current rating for the tapping switch itself. Thus an 11 000/415V transformer with four such extra tappings would be shown on a drawing as '11 000 ±2½% ±5%/415V' and would actually give 11 000/394, 405, 415, 425 or 436V on load. In order to raise the secondary voltage it is necessary to go to a **lower** (i.e. negative) HV tap.

The tap-changer switch handle can be seen in Figure 6.2. It must always be kept padlocked against unauthorised or accidental operation.

6.9.3 On Load Tap Changers

Large network transformers which are provided with on-load tap changing normally have a much larger number of taps in smaller steps. The principle used is 'make-before-break': this means that the new tap must be connected before the old tap is broken, otherwise there would be a break in supply and an interruption of full-load current by the tapping switch.

The difficulty with this simple idea is that, during the transition period while both taps are made, a small number of turns of the transformer's HV winding are short-circuited by the two taps, and a heavy current will flow through them. Arrangements are therefore made to insert resistance temporarily into this short-circuited loop to limit the current until the tap change is complete and the short-circuit removed. Figure 6.5 shows in principle how this is done.

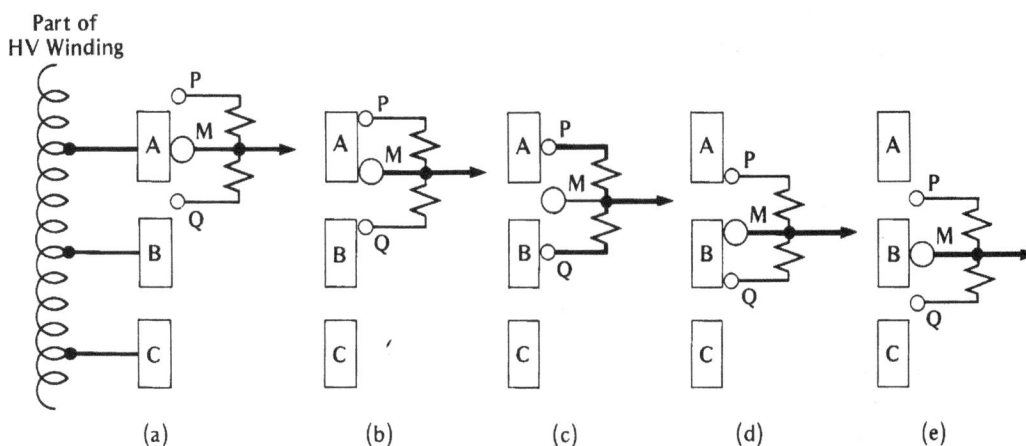

FIGURE 6.5
ON LOAD TAP CHANGING

A, B and C are adjacent taps on an HV winding. In (a) the tapping is on A, and it is desired to move it, on load, to B.

The moving member consists of a main contact M and two 'transition' contacts P and Q which are connected to M each through a resistance. In position (a) M carries the full load, and P and Q are not in contact.

In the first part (b) of the transition the main contact M is still on tap A. Contact Q moves to B and contact P is still on A. Q and M now short-circuit the HV turns between A and B, but the short-circuit current is limited by the lower half of the resistance. Meanwhile M is still carrying the load current from tap A.

At the next stage (c) the moving member has travelled on, and the main contact M leaves tap A. P and Q now share the load current which passes through both halves of the resistance. These two halves also limit the current in the shorted turns between A and B.

At the next stage (d) the main contact M has moved to tap B, so that it is once again carrying the load current, but now from the new tap. P however is still on tap A, so that the current from the shorted turns is limited by the upper half of the resistance.

Finally the moving member is at position (e), where the main contact M is on B and carrying the load, while P and Q are out of contact, as they were in position (a), but now on the new tap.

During these transition stages the load current has never been interrupted, nor has the main contact ever been called upon to break any large current. Moreover the current in the short-circuited turns is always limited by one or both halves of the resistance.

In some designs of tap changer the transition resistors are replaced by reactors. These have a similar limiting effect but are not a source of heat. They also cancel each other out magnetically in stage (c) when both are sharing the load.

During stage (c) the full-load current passes momentarily through both halves of the resistance. To keep them to a reasonable size, they must be short-rated. This poses the problem that, if the driving motor power should be lost at the moment the mechanism reached stage (c), it would stick there and a rapid burnout of the resistors would follow, with inevitable damage to the short-circuited turns. Steps must therefore be taken to prevent this happening.

The philosophy is that the power to operate the tap-changer mechanism must never do so directly but should be used only to store energy. When a tap change is called for, that energy is released and is sufficient to complete the change on its own, even if the external power supply fails.

The stored-energy tap-changer mechanism is usually of one of two types - spring-operated or flywheel-operated. In the former a motor winds and charges a spring. A tap change cannot begin until the spring is fully charged, and, once released, it completes the change on its own.

In the flywheel type a motor runs up a flywheel on receipt of a tap-change signal. When the wheel is up to full speed the motor is disconnected and a clutch engages. The kinetic energy of the flywheel completes the change on its own.

On-load tap changers and their operating mechanisms are usually separate assemblies bolted to the transformer tank, through which the tappings from all three phases are brought out into the changer compartment. This too is usually oil filled but separate from the main tank, so that the tap changer can be drained for maintenance without having to drain the main tank.

Provision is made for manual operation, if that should be necessary, by inserting an operating handle. The speed of the tap change remains the same as with power operation, since the same stored energy is released.

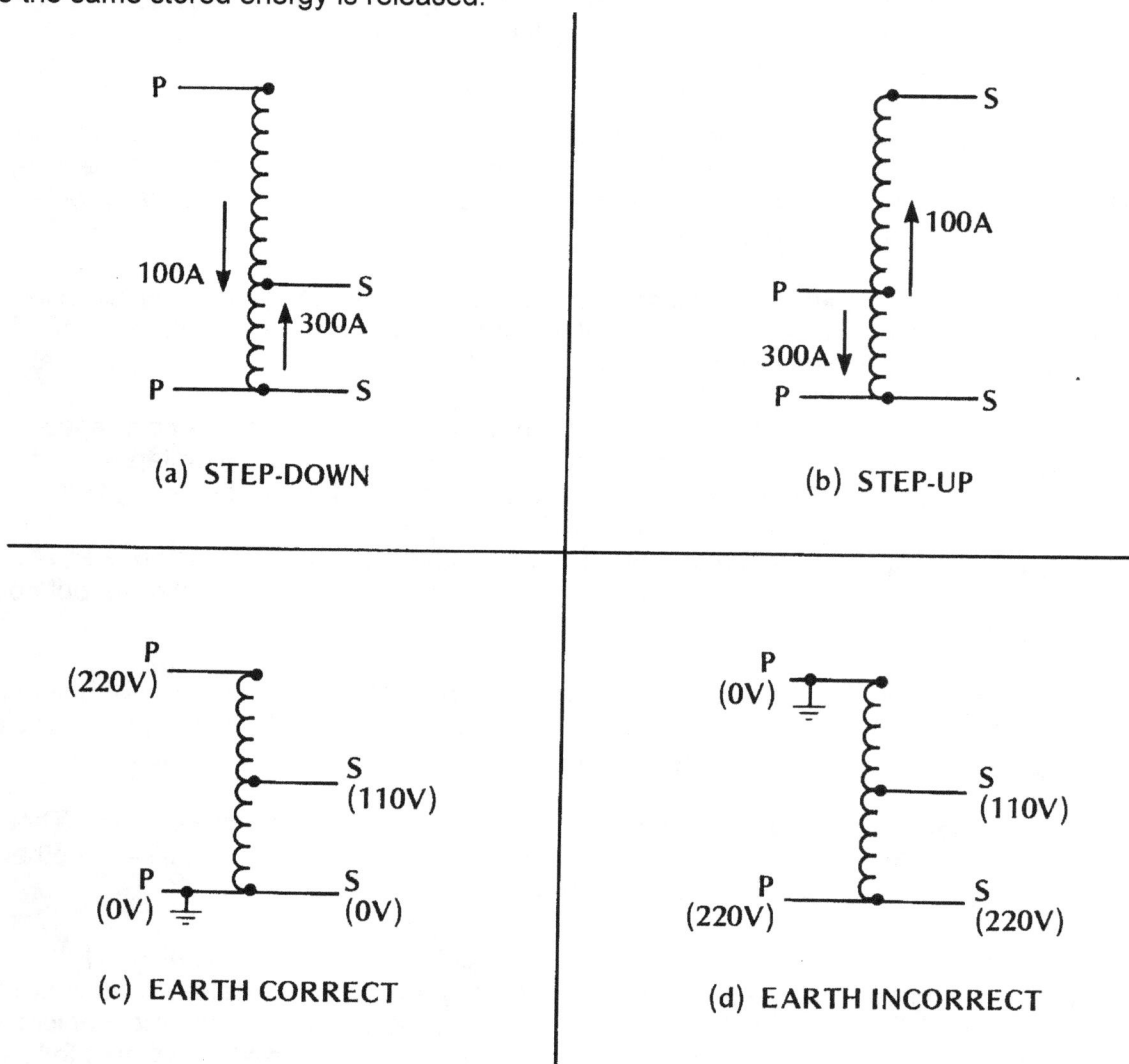

(a) STEP-DOWN

(b) STEP-UP

(c) EARTH CORRECT

(d) EARTH INCORRECT

FIGURE 6.6
AUTO TRANSFORMER CONNECTION

6.10   AUTO TRANSFORMERS

Where a transformer ratio is fairly close - for example 3:1 or less - there is much advantage in both cost and weight in combining the primary and secondary windings, as in Figure 6.6(a). Such an arrangement is called an 'auto-transformer'.

In Figure 6.6(a) the secondary winding is combined with the primary, one terminal of each being common; the other secondary terminal is effectively a tap on the primary winding. This arrangement gives a step-down effect, like a potentiometer, depending on the primary/ secondary turns ratio. Since the primary and secondary currents are in opposition, the net current in the common part is **less** than the secondary current alone. For example, if the primary current were 100A and the ratio 3:1, the secondary current would be 300A, and the net current in the common part would then be the **difference,** namely 200A. This part of the winding could therefore be of lighter construction than would be needed if the transformer had been of the normal double-wound type. Also, because of the closer linkage between the primary and secondary windings, there is less leakage reactance, and the reactance of an auto-transformer is in general less than that of its double-wound counterpart.

Although Figure 6.6(a) shows a voltage step-down arrangement, an auto-transformer can equally be used for stepping up (unlike a potentiometer), as in Figure 6.6(b). This is possible because the primary flux still links the whole of the secondary winding, so developing in it the full emf determined by the secondary turns. Use of an auto-transformer is a very economical way of converting, for example, control supplies from 110V to 220V or vice versa.

Because a double-wound transformer provides complete electrical isolation between the two sides, an earth fault on one side is not carried over into the other. This is not the case however with an auto-transformer. Both sides are electrically connected through the common terminal and the 'tap'.

It is important for reasons of safety that, if one line is earthed on one side, that earth should be applied to the common terminal so that it is also applied to both sides, as shown in Figure 6.6(c). In that case, if the primary voltage were 220V and the secondary 110V, the common earth would ensure that the 'live' secondary terminal would never be more than 110V to earth.

A safety hazard would exist if an auto-transformer were wrongly connected, as shown in Figure 6.6(d). Here the earthed line is not the common one, with the result that there is now no direct earth on the 110V system, one line being at 110V and the other at 220V to earth - a possibly dangerous situation when the secondary circuit is switched in one pole only.

This error can easily arise when domestic equipment which has been designed for the USA 110V system is adapted to operate from the UK 240V supply. Any such adaptations should always be carefully checked for polarity.

6.11   TRANSFORMER TESTING

6.11.1 Manufacturers' Tests

All power transformers are subjected to extensive tests by the manufacturers before delivery to the customer. While operators and maintenance staff are not responsible for carrying out these tests, it is obviously an advantage to have some knowledge of them. They are summarised overleaf. If more details are required, reference should be made to BS 171: - Specifications for Power Transformers.

Tests by the manufacturer are of three kinds:

Routine   A test to which each individual transformer is subjected.

Type   A test made on a transformer which is representative of other transformers, to demonstrate that they comply with specified requirements not covered by routine tests.

Special   A test other than a type test or a routine test, agreed by the manufacturer and the purchaser, and applicable only to one or more transformers of a particular contract.

Routine tests comprise:

(a)   Measurement of winding resistance, using a d.c. source and taking account of temperature.

(b)   Ratio, polarity and phase relationships, in which the voltage ratio is measured on each tapping. The polarity of single-phase transformers and the vector group symbol of 3-phase transformers are checked.

(c)   Impedance voltage, using an a.c. source at the rated frequency, and carried out between pairs of windings. Impedance voltage is defined in para. 6.4.

(d)   Load losses, at rated frequency and carried out between pairs of windings.

(e)   No-load losses and no-load current, measured at rated voltage and frequency.

(f)   Induced overvoltage withstand: a test of dielectric insulation using a source of higher-than-rated frequency to avoid excessive excitation current.

(g)   Separate source voltage withstand: similar to (f) but using a source not less than 80% of the rated frequency.

(h)   The insulation resistance of each winding in turn to all other windings, core and frame or tank, all connected together and to earth. (Note: Where windings are star-connected or delta-connected inside the transformer, phase-to-phase insulation tests cannot be carried out.)

Type tests and special tests are made only if specified by the purchaser. They include:

(j)   Temperature rise test.

(k)   Impulse-voltage withstand tests (with and without chopped waves).

(l)   Measurement of zero-phase sequence impedance.

6.11.2 Users' Tests

Operators and maintenance staff, while not responsible for the manufacturers' tests referred to above, are required to apply certain routine checks and tests to power transformers at the intervals laid down in the appropriate maintenance schedules.

These routine tests include:

(a)   Visual examination of the transformer and its earthing resistor (if any), cable connections and earthing arrangements for tightness, mechanical damage, corrosion and signs of overheating.

(b)   Checking the oil or Askarel levels and inspecting for leaks and clear drains.

(c)   High-voltage insulation resistance test on the HV windings.

(d)   Low-voltage insulation resistance test on the LV windings.

(e)   Simulation of overtemperature and overpressure by manual operation of the protection devices, and checking that the alarm indications appear and the circuit-breaker trips.

6.12   TRANSFORMER PROTECTION

The protection of electrical installations, including transformers, against damage caused by overload or fault conditions is described in Part 10 'Electrical Protection'. To summarise, the protection provided for transformers may consist of one or more of the following:

HV Side   Overcurrent
              Earth fault

LV Side   Restricted earth fault

General   Overpressure ('Qualitrol')
              Overtemperature
              Buchholz (oil-filled only)
              Differential

6.13   INSTRUMENT TRANSFORMERS

There is a range of small transformers, other than power transformers, which are used to operate measuring instruments, meters and protective relays. They comprise voltage transformers (VT) and current transformers (CT) and are covered in Chapter 9.

CHAPTER 7 CABLES

7.1   GENERAL

Cables form an important part of any installation but, because they are static, and in normal service are very reliable, they do not always receive the attention that they deserve.

There are three categories of cables associated with industrial installations - power cables, control cables, and special cables for, for example, communications and data transmission circuits. It is the first two categories which are described in this chapter. A power cable contains one, two, three or four cores each consisting of a copper conductor surrounded by insulating material; a control cable usually has many cores and is known as a 'multicore' cable. Aluminium is sometimes used as a conductor material; although its conductivity is less than that of copper, it is somewhat cheaper. Corrosion problems, however, preclude its use on some installations, particularly offshore.

7.2   POWER CABLES

Cables are designed for both high-voltage and low-voltage transmission of power. Though the general construction is similar in both cases, high-voltage cables have thicker insulation and usually have smaller conductors, since low-voltage cables carrying bulk power handle the heavier currents.

7.2.1 General Construction

A power cable is made up of one, two, three or four insulated conductors enclosed in a bedding. For mechanical protection, wire armouring is wrapped around the bedding, and a coloured outer protective sheath, usually of PVC, is extruded over the armouring, as shown in Figure 7.1. Each insulated conductor is known as a 'core'.

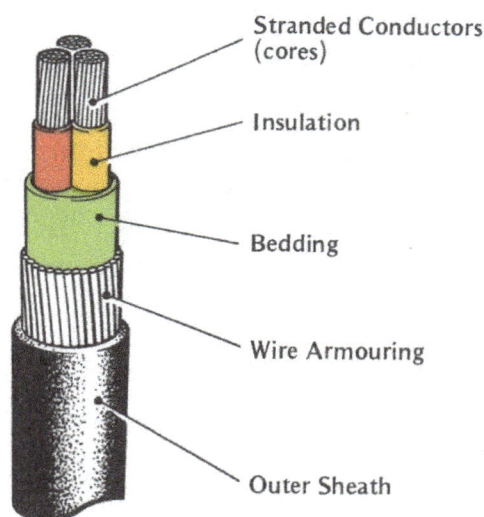

FIGURE 7.1
THREE CORE POWER CABLE

7.2.2 Conductors

The size of the copper conductor forming one of the cores of a cable is expressed in square millimetres (mm^2), and the current rating of the cable is dependent upon the cross-sectional area of each core. The very smallest cables have conductors consisting of only one strand of copper; larger cables however have stranded conductors consisting of many individual strands or wires laid up together; this gives flexibility, allowing the cable to be bent more readily during installation. To achieve a circular conductor, the number of strands follows a particular progression: 3, 7, 19, 37, 61, 127 etc, the diameter of each strand being chosen to achieve the desired cross-sectional area of the whole conductor.

As seen in Figure 7.2, 3-core and 4-core cables in the larger sizes have conductors with the strands laid up in a segmental formation; this achieves a better space factor and reduces the overall diameter of the cable. It also reduces the inductance of the cable due to decreased spacing between phases.

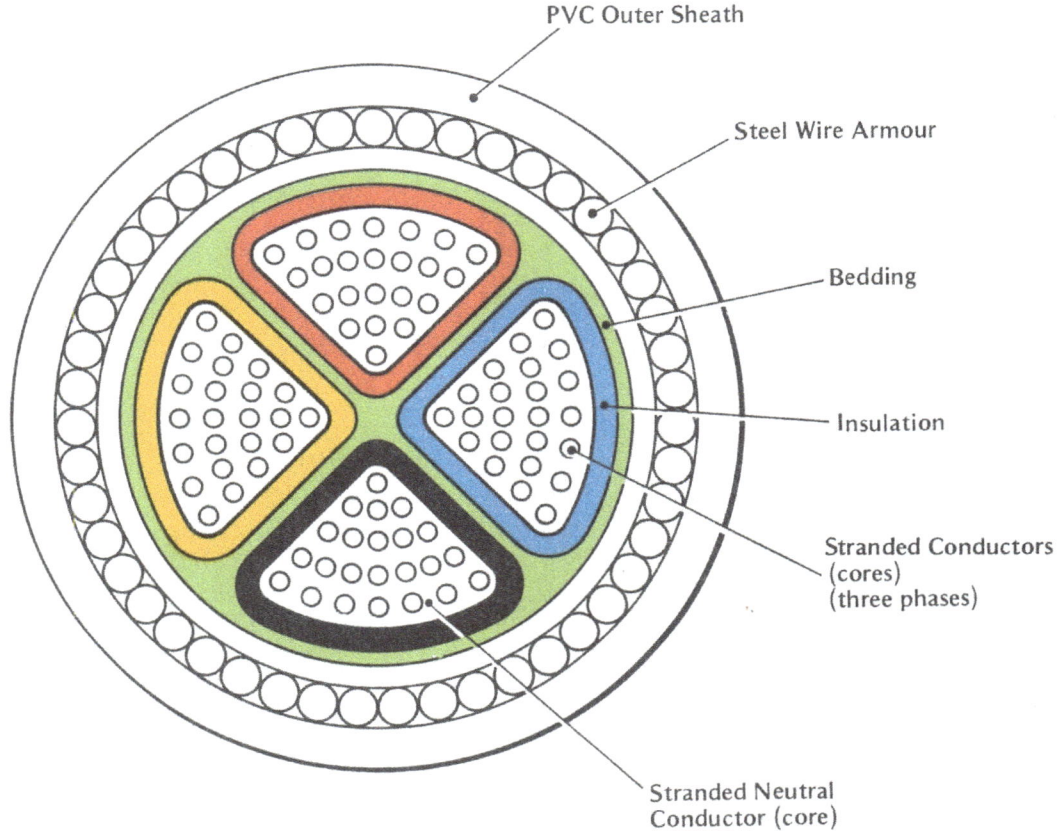

FIGURE 7.2
SEGMENTAL CORES

Standard conductor sizes range from 1.5mm^2 to 400mm^2 for 2-core, 3-core and 4-core cables, and from 50mm^2 to 1 000mm^2 for single-core cables.

7.2.3 Insulation, Covering and Stress Relief

Natural rubber or oil-impregnated paper is no longer used for the insulation of cables up to 3 810/6 600V; synthetic materials are now used. For high-voltage cables the insulation is ethylene propylene rubber (EPR) and for low-voltage cables it is polyvinyl chloride (PVC). EPR has good electrical properties and is resistant to heat and chemicals; it is suitable for a conductor temperature up to 85°C. PVC is a thermoplastic material, therefore care must be taken not to overheat it; it is suitable for conductor temperatures up to 70°C. PVC insulated cables should not be laid when the temperature is less than 0°C because it becomes brittle and is liable to crack.

High-voltage cables have an earthed metallic screen over the insulation of each core. This screen consists of a lapped copper tape or metallic foil, and its purpose is to control the electric field within the insulation and thus the voltage gradient across it, as shown in

Figure 7.3. Also, it avoids any interaction of the electric stresses due to the voltages on different phase conductors within the same cable.

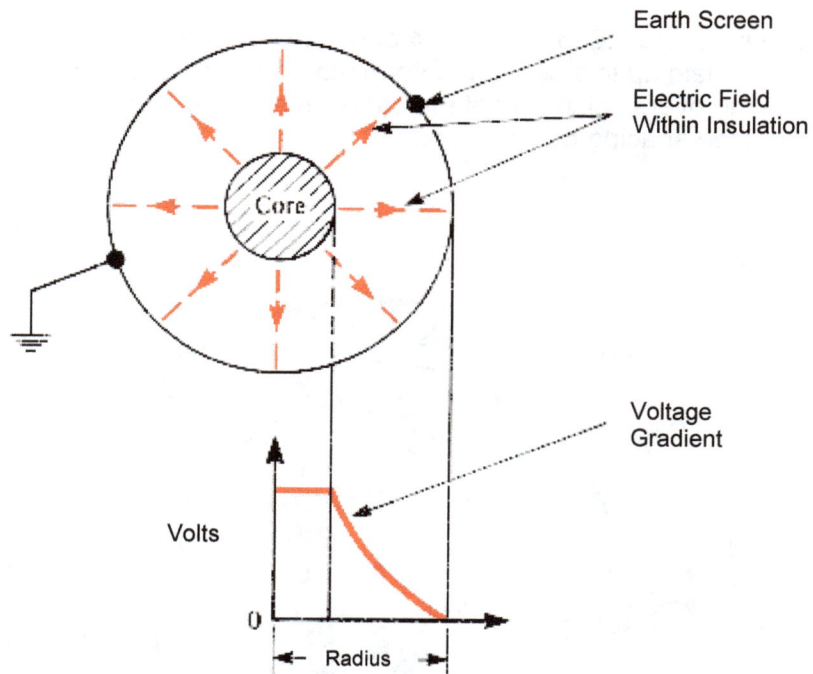

FIGURE 7.3
VOLTAGE GRADIENT ACROSS HIGH VOLTAGE CABLE INSULATION

Core insulation may be coloured red, yellow, blue and black to identify the three phases and neutral. Twin cores are coloured red and black. Single-core cables are identified by coloured PVC tape applied to the outer sheath.

7.2.4 Cable Stress Relief

The copper screen is often terminated in a 'stress cone', which may be seen in Figure 7.7. This is to spread the electric stress which would otherwise tend to concentrate where the screen is cut off at a cable end and could lead to breakdown. This is further discussed in para. 7.6.4.

7.2.5 Bedding

The bedding consists of a layer of PVC extruded over the core insulation as a base for the armouring.

7.2.6 Armouring

Mechanical protection of the cable is provided by a single layer of wire strands laid over the bedding. Steel wire is used for 3-core or 4-core cables, but single-core cables have aluminium wire armouring. With 3-core or 4-core cables the vector sum of the currents in the conductors is zero, and there is virtually no resultant magnetic flux. This is not so however for a single-core cable, where eddy-current heating would occur if a magnetic material were used for the armouring. Armouring is described as Steel Wire Armoured (SWA) or Aluminium Wire Armoured (AWA).

7.2.7 Outer Sheath

The outer sheath of extruded PVC protects the armouring and the cable against moisture and generally provides an overall protective covering.

High-voltage cables are identified by outer sheaths coloured red; a black sheath indicates a low-voltage cable (see also para. 7.7).

7.2.8 Selection of Power Cables

The following considerations are taken into account when selecting a power cable for a particular application:

(a) **The System Voltage and Method of Earthing**

A low-voltage system usually has a solidly earthed neutral so that the line-to-earth voltage cannot rise higher than (line volts) $\div \sqrt{3}$. However, cables for low-voltage use are insulated for 600V rms core to earth and 1 000V rms core to core

High-voltage cables used in some installations are rated 1 900/3 300V or 3 810/6 600V or 6 600/11 000V, phase/line. In selecting the voltage grade of cable, the highest voltage to earth must be allowed for. For example, on a nominal 6.6kV unearthed system, a line conductor can achieve almost 6.6kV to earth under earth-fault conditions. To withstand this, a cable insulated for 6 600/11 000V must be used.

(b) **The Normal Current of the Cable**

The conductors within a cable have resistance, and therefore I^2R heating occurs when currents pass through them. The maximum permissible temperature of the cable depends upon the material of the insulation, and a conductor size must be chosen so that this temperature is not exceeded. Tables giving the continuous current-carrying capacities of different cables are given in manufacturers' literature and in the Regulations for the Electrical Equipment of Buildings published by the Institution of Electrical Engineers.

The temperature of a cable depends not only on the rate of heat input due to the passage of load current but also on the rate at which the heat can be carried away. When using the tables of current ratings it is important to note whether they refer to cables laid in the ground, laid in ducts or laid in air. De-rating may be necessary if a number of cables are run in close proximity to each other.

Another consideration in selecting a cable is the voltage (IR) drop from the source of supply to the load. A drop of 1V in a 440V circuit is of little consequence, but it is a significant percentage when the circuit operates at 24V.

(c)   Abnormal Currents in the Cable

One abnormal condition is a sustained overload; a cable must be protected so that an overload cannot persist long enough to cause damage to the insulation by overheating. For example, for PVC cables laid in air, the overload must not be greater than 1.5 times the continuous maximum rated current and must not persist for longer than four hours.

Another abnormal condition is when a cable has to carry a through short-circuit current. In this case the temperature of the conductor may be allowed to rise to a higher value, say 150°C, for the short interval between the onset of the fault and its disconnection. The short-circuit current that a given cable can withstand depends upon the speed with which the protection operates. For example, a PVC cable having conductors of 185mm^2 has the following short-circuit ratings:

46kA for 0.2s
20.3kA for 1.0s
11.7kA for 3.0s

The 0.2s rating would be suitable for use with fuse protection, but, where relay-operated circuit-breakers are concerned, a longer time rating would be necessary. Again, tables of short-circuit ratings are available in manufacturers' literature.

7.3   CONTROL CABLES

Control cables usually have conductors either 1.50mm^2 or 2.50mm^2 in cross-section. The insulation, bedding and outer sheath are of PVC, and they are steel wire armoured. Multicore cables are available having 2, 3, 4, 7, 12, 19 and 27 cores, each core being identified by a number on the insulation. The outer sheath of control cables is coloured green.

7.4   MINERAL INSULATED CABLES

Mineral-insulated (MI) cables are used where the integrity of a circuit is of great importance. They are particularly resistant to fire and are used in circuits, such as communications or

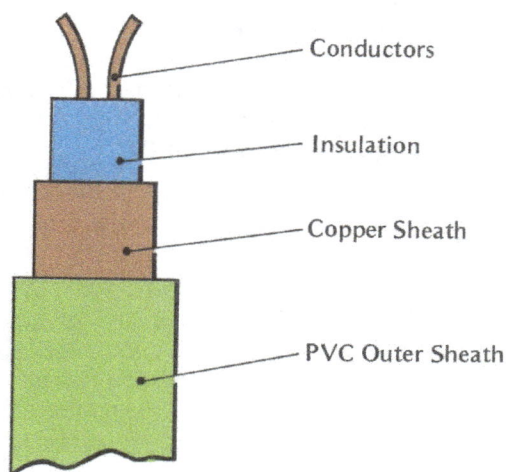

FIGURE 7.4
MINERAL INSULATED TWO CORE CABLE

emergency lighting, which must continue operational as long as possible after fire has broken out. They are also very robust and resistant to mechanical damage.

MI cables are constructed by assembling the single-strand conductor or conductors inside a seamless copper tube. After threading a number of 'tablets' of magnesium oxide insulating material onto the conductors, the whole assembly - conductors, insulation and copper tube - is drawn down through a series of dies until the magnesium oxide is crushed to a powder and the whole cable is solid. The final appearance is as in Figure 7.4.

After annealing to make the cable more flexible, an outer sheath of PVC is applied.

MI cables are available in single-core from $1mm^2$ to $150mm^2$, in 2-core, 3-core and 4-core from $1mm^2$ to $25mm^2$, and in 7-core from $1mm^2$ to $4mm^2$.

Special jointing techniques and materials must be used for terminating MI cables, and great care must be taken to seal the cable ends against the entry of moisture.

7.5   METHOD OF SPECIFYING CABLES

There is a 'shorthand' method used to describe the construction of any cable, using abbreviations to indicate the nature of the various materials. For example, a low-voltage cable might be described as:

(Reference)	(1)	(2)	(3)	(4)	(5)	(6)	(7)	(8)	(9)	(10)
Abbreviation	0.6/1kV	STR 3-core, 150mm^2	CU/	PVC/	PVC/	SWA/	PVC/	HO$_2$/	HCL	

Interpreted this means:

1	0.6kV line to earth
2	1 kV line to line
3	Stranded conductor
4	Copper conductor
5	PVC conductor insulation
6	PVC bedding
7	Steel wire armoured
8	PVC outer sheath
9	see below
10	see below

Another example is:

6.6/11kV   STR   CU/ EPR/ SCR/ PVC/ AWA/ PVC/ HO$_2$/ HCL
1-core, 630mm^2

where   EPR indicates ethylene propylene conductor insulation
SCR indicates screened
AWA indicates aluminium wire armoured.

The last two items (9 and 10) indicate the flammability and the toxicity of the synthetic materials used in the cable. HO$_2$ indicates that a high level of oxygen is required to sustain combustion: in the case of the specification this means more than 30% oxygen in the atmosphere. HCL denotes 'Hydrochloric Level' showing that, when the synthetic materials burn, they produce hydrochloric acid gas (HCl) which is highly poisonous and very corrosive.

In particular PVC, when burnt, releases large quantities of HCl and also produces dense black smoke; for example, a 1m length of cable containing, say, 6kg of PVC can completely black out a room 1 000m^3 in size within five minutes of the fire starting.

7.6   INSTALLATION

7.6.1 Cable Runs

Cables may be laid discretely in the ground, run in ducts or clamped to cable trays; the third method is the most common in offshore installations. Each cable must be identified at each end, using a marker bearing the cable number.

There is a practical limit to the conductor size which can be run as a 3-core or 4-core cable it becomes too stiff and heavy to handle. A 3-phase circuit is then run as three (or four) single-core cables. To minimise the electromechanical forces between the cables under short-circuit conditions, and to avoid eddy-current heating in nearby steelwork due to magnetic fields set up by load currents, the three single-core cables comprising the three phases of a 3-phase circuit are always run clamped in 'Trefoil' formation, as shown in Figure 7.5.

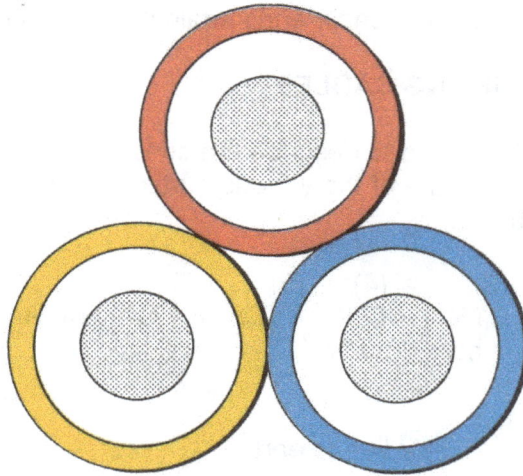

FIGURE 7.5
SINGLE CORE CABLES LAID IN TREFOIL

At any instant in time the net magnetic flux outside the group of cables due to the three line currents in them approximates to zero because of the symmetrical cable layout.

Heavy current cable runs, such as the low-voltage connections from a transformer, may consist of up to four single-core cables in parallel per phase; all 12 cables are run bunched into four 3-phase sets, each set laid in trefoil. In the case of 4-wire systems the neutral conductor needs a smaller cross-sectional area than that of the phase conductors and may be met by one or more smaller single-core cables in parallel.

7.6.2 Cable Terminations

A power cable is terminated in an air-insulated cable box in offshore installations; it enters the box through a compression gland which grips the wire armouring and seals the entry of the cable. The outer sheath, armouring and bedding of the cable are stripped back, enabling the cores to be spread to match up with the fixed bushing terminals, and the insulation is removed to expose the conductors. Such a cable box is often referred to as a 'trifurcating box'.

In some high-voltage onshore installations, especially outdoor ones, the cable box may be filled with compound, a tar-like substance which is poured in hot and then sets hard to exclude moisture. It can only be removed by heating.

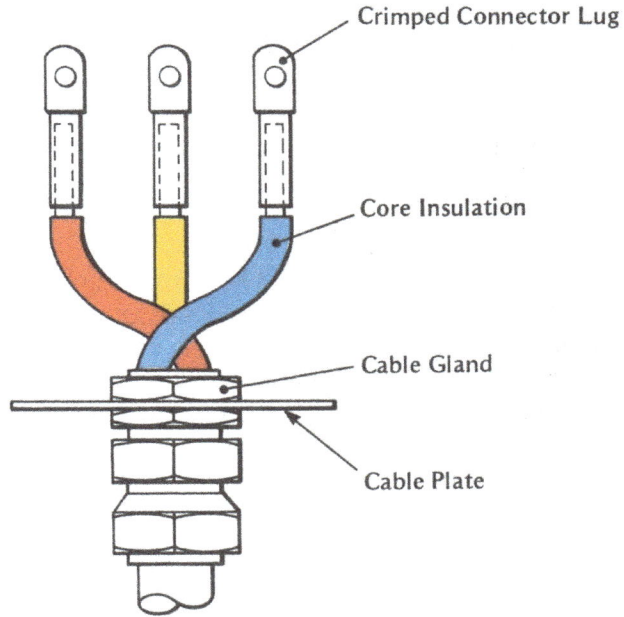

FIGURE 7.6
LOW VOLTAGE CABLE TERMINATION

Conductors are terminated either with lugs bolted to the fixed bushing stems as shown in Figure 7.6 or, for heavier currents, with cylindrical ferrules which are clamped into terminal blocks. In either case the terminations are crimped onto the conductors using either hand or hydraulic crimping tools. To make a good connection it is vital that the lug or ferrule is the correct size for the particular conductor and that the correct die is used in the crimping tool.

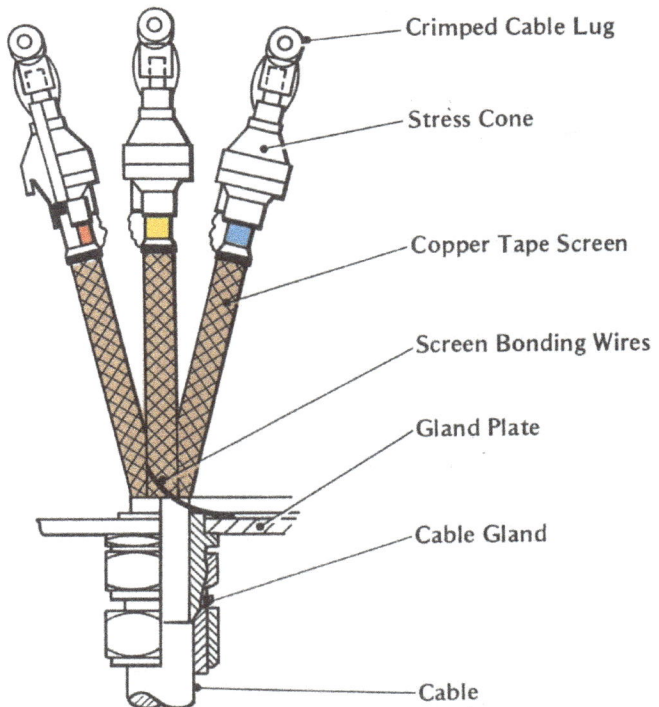

FIGURE 7.7
HIGH VOLTAGE CABLE TERMINATION

Special measures must be adopted, when screened high-voltage cables are terminated, to prevent a concentrated electric field being developed where the copper screening tape is cut back; this strong electric field could lead to the insulation at that point being so overstressed that a breakdown occurs. Special stress cones are fitted which are bonded to the screening tape; they control the electric stress and reduce the resulting voltage gradients to a safe value. This arrangement is shown in Figure 7.7.

7.6.3 Single-core Cables

The conductor of a single-core cable and its surrounding metallic armouring act as a current transformer having a 1:1 turns ratio; load current passing through the conductor produces a magnetic flux which, linking with the wires of the armouring, induces an emf in them. If a circuit is provided between the armouring at one end of the cable and at the other, a current flows in the armouring which, if sufficiently large, causes heating. This is shown in Figure 7.8(a).

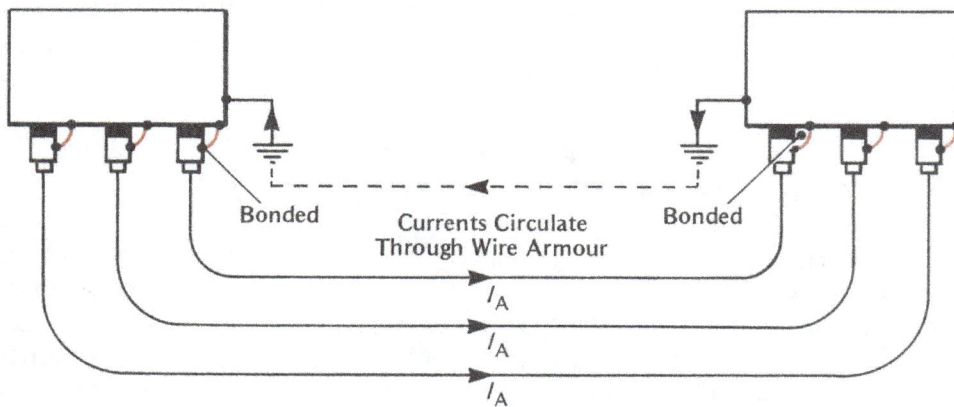

(a) CABLE GLANDS BONDED AT BOTH ENDS

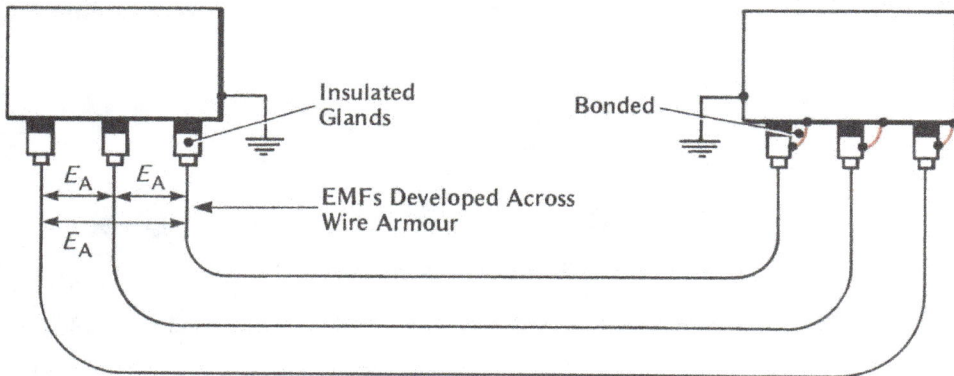

(b) CABLE GLANDS BONDED AT ONE END

FIGURE 7.8
INSULATED CABLE GLANDS

To control these circulating currents insulated cable gland adaptors are used whereby the body of the gland, and consequently the wire armouring of the cable, is electrically isolated from the earthed gland plate of the cable box by a layer of insulation. Figure 7.8(b) shows a 3-phase circuit run with single-core cables using insulated cable glands. To control the

308

voltage of the armouring it must be bonded to earth; this is done by deliberately bridging the gland insulation using bonding links. The armouring can be bonded in one of two ways. In Figure 7.8(a) it is bonded at both ends of the cable run (shown in red); the emf induced in the armouring causes currents (I_A) to circulate in the armouring which in heavy current circuits may lead to an undesirable temperature rise in the armouring. Alternatively, the armouring may be bonded at one end only as in Figure 7.8(b); there is no circuit for current to flow, but a voltage (E_A) is developed across the gland insulation at the unbonded end. Where one end of the circuit is in a hazardous area, it is customary to bond this end so that any arcing that may occur due to emfs induced in the armouring can only take place in the non-hazardous area.

There is one other magnetic problem associated with single-core cables: where such cables enter a cable box or pass through partitions the conductors must pass through holes in the gland plate. If these plates are made of a magnetic material such as steel, the magnetic fields due to the load currents in the conductors induce eddy currents in the gland plate which may cause it to become very hot. For terminating or passing a.c. circuits using single-core cables, gland plates of non-magnetic material must be used.

7.6.4 Plug-in Connectors

One make of high-voltage, plug-in elbow connector used in some installations is the 'Elastimold' type illustrated in Figure 7.9, where the live conducting parts are shown in red.

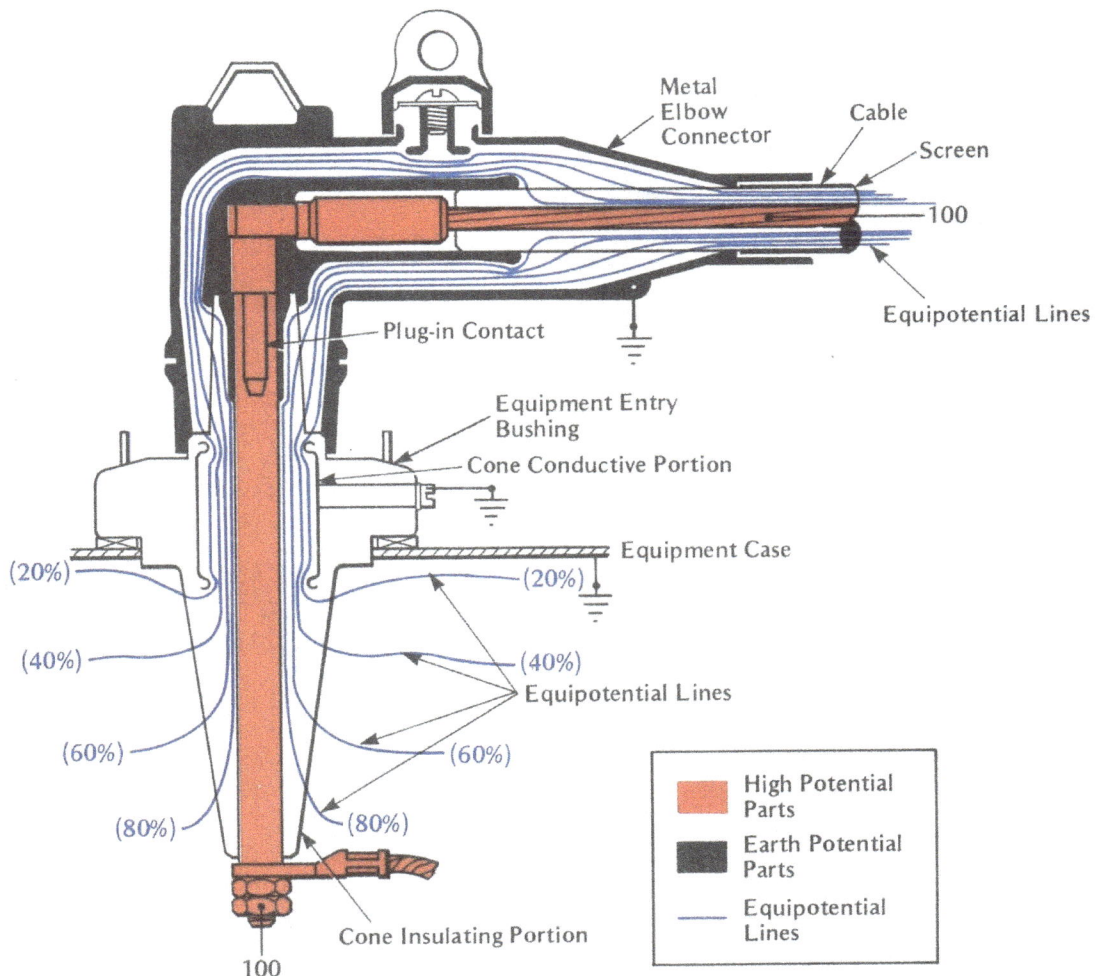

FIGURE 7.9
ELASTIMOLD HIGH VOLTAGE ELBOW CONNECTOR

If a cable core screen is cut or terminated abruptly, the electric field distribution within the cable changes radically outside it. Both the surrounding air and the dielectric material immediately in the vicinity of the terminated screen then become overstressed electrically. The continuity of the earthed cable screen is carried on by the metal elbow past the plug-in connector to the entry bushing on the equipment. To prevent rapid breakdown of the cable, a stress cone is applied at the end of the screen (bottom of Figure 7.9). The cone has an insulating portion to reinforce the primary cable insulation and also an earthed conductive portion to mate with the cable core screen. This is to control the distribution of the equipotential lines (that is, lines of equal voltage) shown in blue, so that, when they finally emerge into the air, they are sufficiently far apart not to cause too great a potential gradient and so not to give rise to ionisation and possible electrical breakdown.

In Figure 7.9 the dispersal of the blue equipotential lines at the cone area (shown as percentages of the core voltage) is clearly seen.

7.6.5 Control Cables

Control cables are also terminated using compression glands. The sheathing and armouring are stripped back to leave tails of the required length. Each core is identified using plastic ferrules bearing the wire number, and terminated using a crimped connector. The cores are either laced up into suitable runs using plastic cable ties, or secured to cable racks within a control panel.

7.7   OUTER SHEATH COLOURS

Standard colours are used in some installations to identify the system to which the various cables belong; they are:

Red	:	High-voltage system.
Black	:	Low-voltage system.
Green	:	Control and instrumentation system.
Orange	:	Fire and gas detection and telecommunications systems
Blue	:	Intrinsically safe systems.
Yellow	:	Thermocouple circuits.

CHAPTER 8 BATTERY SUPPORTED D.C. AND A.C. POWER UNITS

8.1 GENERAL

There are a number of systems, both offshore and onshore, whose functioning is so important that they must not be allowed to cease, even momentarily, even if there were a total power failure. Such systems include:

- Emergency lighting
- Communications
- Certain process instrumentation and control
- Operation of electrical switchgear
- Navigational aids
- Fire and Gas detection.

In order that these systems shall not fail when main power is lost, they must be supplied with power from a source of stored energy _which means, in practice, a battery. And since a battery can store only d.c. energy, these systems must in general operate from d.c.

The d.c. voltages are normally 110V and 24V, but others are used. Separate systems are provided for each, sometimes more than one. Where several similar services are supplied from a common d.c. unit, it is referred to as a 'central' d.c. system.

There are also a number of d.c. power units that supply individual equipments such as navigation lights, foghorns or diesel engine starters; these are referred to as 'dedicated' systems (as distinct from the 'central' systems).

Most d.c. systems are unearthed, but those supplying telecommunications equipments usually have one pole earthed.

Because the main power system of an offshore or onshore installation is a.c., this power must be converted to d.c. before being fed to these systems. The modern method is to convert the a.c. electronically into d.c. through a solid-state 'rectifier'. It is entirely static and has no moving parts; it is also very easily controlled. The d.c. voltage output depends on the level of the a.c. voltage input, so an in-built transformer on the a.c. side causes it to produce any d.c. voltage desired from the 415V or 440V a.c. system. The principle of rectification is discussed more fully in Part 1 Electrical Theory.

8.2 D.C. SUPPLY SYSTEMS

Figure 8.1 shows the basic circuit of a d.c. supply system. An a.c. supply is led to a transformer, which converts the a.c. to whatever voltage is necessary for producing the desired level of d.c. and thence to a 6-element rectifier bridge. The d.c. is taken from the bridge to a distribution fuseboard from which it is fed to all the d.c. services.

Figure 8.1 shows the unit as 3-phase, but d.c. supply systems are also used supplied from a single-phase source. In that case the rectifier bridge is a 4-element one.

In order to regulate the outgoing d.c. voltage level, the rectifiers on one side of the bridge are replaced by 'thyristors' (controlled rectifiers) by which the d.c. output level can be adjusted. The firing of the thyristors is automatically controlled so as to maintain the d.c. voltage at the correct level for the load or for battery charging.

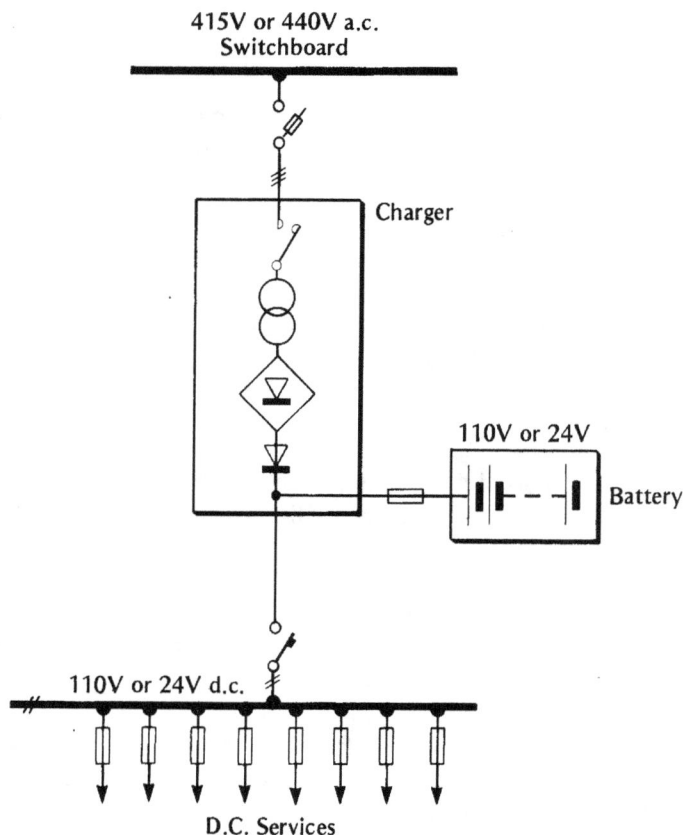

FIGURE 8.1
BATTERY SUPPORTED D.C. SYSTEM

This system will produce d.c. from a.c., but, if the a.c. itself fails, the whole thing stops, and no d.c. is produced either. Such a system therefore would not give the continuity needed. But if now a battery is connected in parallel with the d.c. side, as shown in Figure 8.1, the d.c. current from the rectifier will not only go to the system load but will also keep the battery charged. Suppose now the a.c. system fails. No d.c. current comes out of the rectifier, but the battery remains connected to the load and continues to supply it with d.c. current **without operator action and without interruption.** Indeed, the d.c. loads would not even know that there had been a failure.

This state of affairs would continue so long as the battery held its charge. It would of course begin running down and there would be a small but progressive fall of voltage. How long it continues to supply the load depends on the capacity, or size, of the battery, which may be regarded as an electrical 'ready-use' tank. The designer, knowing the current load on the battery (in amperes) and the time during which it is desired that it continue to operate (in hours), decides the capacity (in ampere-hours, or 'Ah') of the battery to be installed. Thus if the d.c. load is 80A and if it must continue for a minimum of four hours after an a.c. failure, then the battery must have a capacity of at least 80 x 4 = 320 ampere-hours.

When a.c. voltage is restored after the mains failure, the rectifier takes over its original function and starts to convert it to d.c. again. It relieves the battery of its emergency duty and supplies d.c. directly to the load once more. In addition it starts to recharge the battery. It is important to note that the prime duty of the rectifier is to supply d.c. **to the load**. This is not done by the battery normally, which occurs solely on failure of the a.c. It only charges the battery **after** such a failure. The rectifier unit is usually called a 'Charger', but this is not its principal duty. The rectifier must be rated to carry out both functions (d.c. load and battery recharging) together.

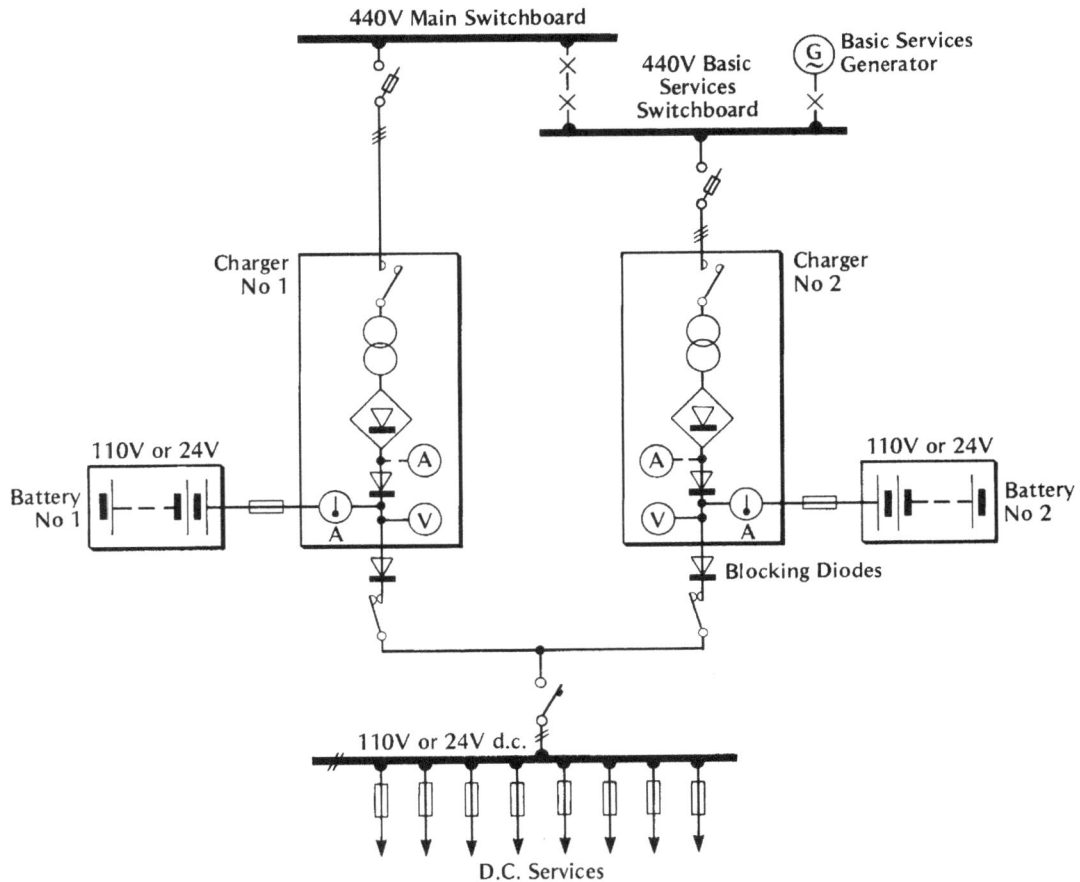

FIGURE 8.2
D.C. DUAL SUPPLY SYSTEM

In Figure 8.1 the transformer-rectifier has been drawn in a straight line with the d.c. output, and the battery has been drawn to one side. This is deliberate and is to emphasise the role of the rectifier. Under normal conditions it is the rectifier which supplies the d.c. load; the battery supplies nothing (indeed it receives a small maintenance charge) and is said to float on the d.c. system. It is only in the abnormal condition when the a.c. supply fails that the battery takes over as the supplier of power.

With 'central' battery systems it is usual to provide two chargers and two batteries, both feeding a common d.c. distribution board as shown in Figure 8.2, which is for an offshore installation.

Each charger is supplied from a different a.c. source, one of which is always a Basic Services or Emergency switchboard to which the emergency generator may be connected. Thus, after a prolonged blackout period which leaves both batteries discharged, at least one can be recharged as soon as the emergency generator can be started.

The blocking diodes seen in the figure are to prevent feedback from the batteries. The lower ones prevent one battery feeding into the other if it is in a discharged state, and the upper ones prevent a battery feeding back into a faulty rectifier.

Both batteries are provided with heavy fuses, and a centre-zero ammeter on the unit switchboard indicates whether the battery is charging or discharging.

Both battery/rectifier d.c. sources can be isolated from the distribution board by normally closed contactors (as shown in the figure) or by manual switches. The purpose of these is explained in para. 8.3.

The battery, which is usually of the nickel-cadmium type delivering approximately 1.4V per cell, is permanently connected through the battery fuses to the d.c. side of the rectifier. There are about 88 cells for each 110V battery and 18 or 19 cells for each 24V.

Where d.c. power units are continuously loaded, two batteries are provided, one associated with each charger. In some power units each battery is capable of supplying the load on its own for the required length of time (referred to as '100% capacity each'); in others both batteries are needed to achieve this (referred to as '50% capacity each').

There are other ways in which the charger, battery and load can be interconnected, but the one described above is by far the most common in offshore and onshore installations for bulk d.c. supplies.

8.3   FLOAT OR BOOST CHARGING

If the batteries become partially discharged, as may happen after an a.c. supply failure, a manual, or 'boost', charge is desirable to recharge them quickly. At the same time the d.c. supply to the load must be maintained without subjecting the load to the higher boost voltage. To do this the equipment is arranged so that only one of the two battery banks can be boost-charged at a time and that this battery, and its associated charger, is disconnected from the load while the boost-charge is in progress. The load is meanwhile supplied from the other charger.

In one typical system the operating mode is selected by a control switch which may be on the panel-front or inside the charger cubicle. It is normal to run both chargers in the 'Float' mode (sometimes also referred to as 'Auto'). Under this condition each charger produces a constant voltage suitable for the load, while the battery, when fully charged, receives a small 'floating' or maintenance charge. In the 'Boost' (or 'Manual') mode a higher voltage is applied to the battery, which is then charged at a higher rate than normal.

Suppose that Charger No 1 in Figure 8.2 is to be used to boost-charge its associated battery. The Float/Boost switch is moved to BOOST, but, to ensure continuity of supply, an interlock is provided to ensure that Charger No 2 is ON and switched to FLOAT before the control of Charger No 1 can be changed to the manual-boost mode of operation. To protect the load from the higher boost voltage, the output contactor of Charger No 1 is opened when the changeover takes place, leaving the load to be supplied by Charger No 2 and its battery. If the a.c. supply to Charger No 2 fails, then Charger No 1 automatically reverts to the Float mode, its output contactor closing to support Battery No 2 which picked up the load on the failure of Charger No 2.

Various methods are employed to achieve these ends. The equipment described uses electrical interlocks and switching but relies on the operator to terminate the boost charge. Other equipments may have key interlocks and manual switching with hand-set or electrical timing devices.

When a discharged battery is first put on charge, or under d.c. fault conditions, the load on the charger will try to exceed its current limit setting. The current is sensed by a current limit sensing shunt which holds the rectifier output current at its rated maximum value by reducing its d.c. output voltage.

Provision is sometimes made to disconnect the load from certain batteries in an emergency by tripping the input moulded-case circuit-breaker to the d.c. distribution board, but this is not common. This would only be done when an offshore installation was abandoned and the d.c. supply was no longer necessary.

Certain power units are fitted with equipment to monitor earth leakage and identify the circuit where this occurs. Six such earth-leakage switches are seen on the right centre panel of Figure 8.4.

It must be emphasised that this description is typical only. Switching and control facilities vary from one make of equipment to another.

8.4   CONTROL, ALARMS AND INDICATION

There is a wide variation in the control circuitry of d.c. power units, which are provided by many different manufacturers. However, certain principles are common to all. A typical control scheme is illustrated in Figure 8.3; only Charger No 1 is shown (the scheme for Charger No 2 is similar).

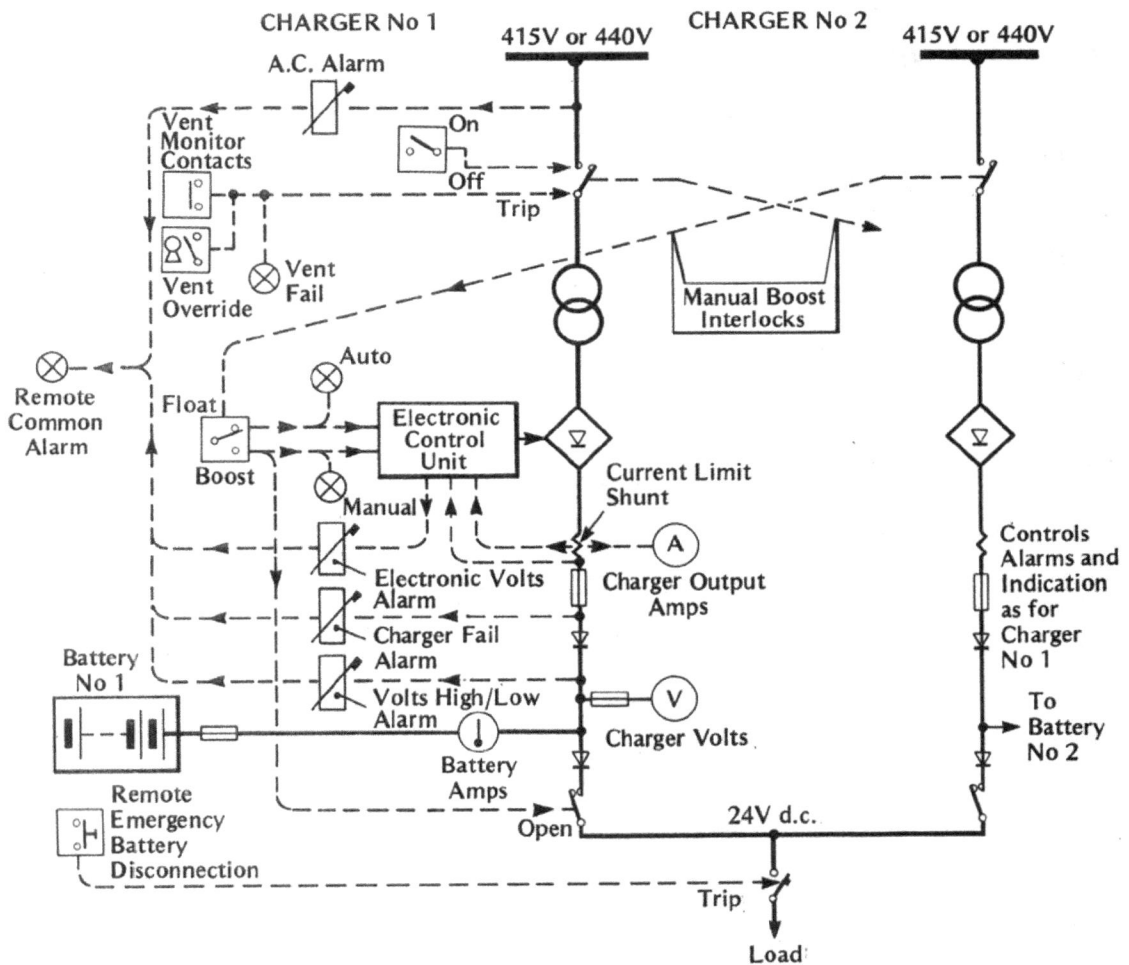

FIGURE 8.3
TYPICAL CHARGER CONTROL SYSTEM

The incoming a.c. supply to each charger is controlled by a contactor operated from an On/Off switch on the front of the board. The supply is armed by a switch in the Ventilation Monitor (see para. 8.8) which in this case only allows charging while ventilation is on.

After transforming, the a.c. supply goes to the diode/thyristor bridge whose d.c. output level is regulated by an Electronic Control Unit. When switched to BOOST this unit raises the d.c. output voltage of the bridge. It also ensures, by sensing the d.c. current and regulating the voltage, that the current output never exceeds a preset level. The d.c. current then passes through the blocking diodes and output contactor to the load distribution fuseboard.

Instruments consist of a d.c. voltmeter and ammeter: also a centre-zero battery ammeter. Alarms are given by flag relays which sense some or all of control unit failure, loss of a.c., charger failure or high or low d.c. voltage. These may also give a common alarm at some remote control point.

Operation of the Float/Boost selector switch to BOOST also causes the corresponding normally closed d.c. output contactor to open. On some systems this function is carried out by extra contacts on the selector switch itself, interlocked by key with the selector switch on the other unit to ensure that it has first been set to FLOAT.

8.5 D.C. POWER SUPPLY UNIT

A typical d.c. power supply unit, incorporating a dual charger and two batteries, is shown in Figure 8.4.

On the centre two panels are the two charger controls, instruments and flag relays. The chargers themselves, the transformers and the various control fuses are inside the cubicles behind doors. The common d.c. distribution fuseboard is also inside.

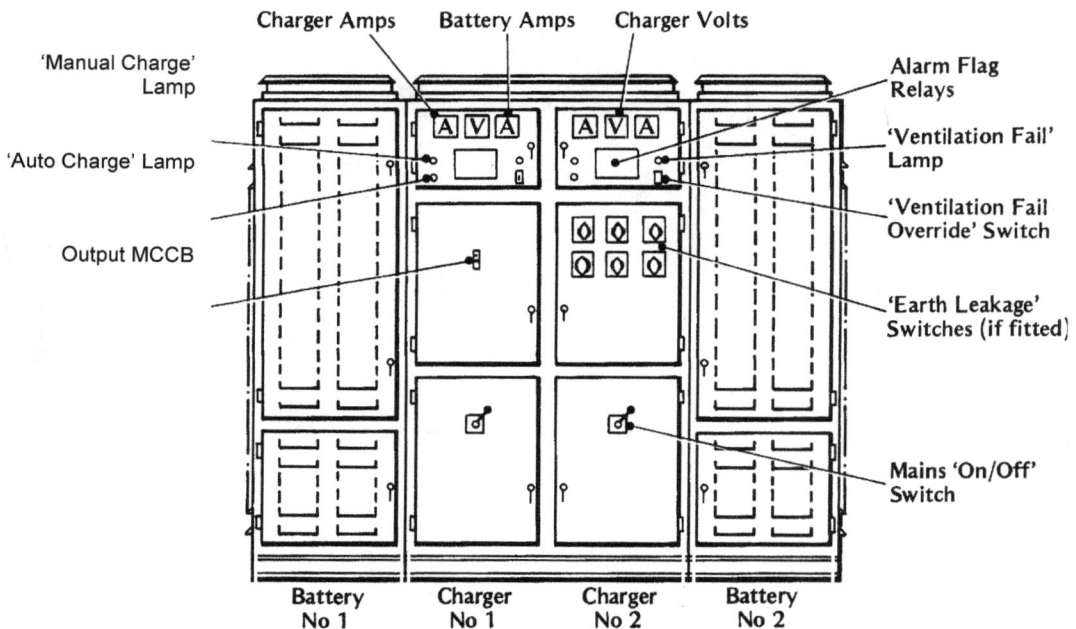

FIGURE 8.4
TYPICAL D.C. POWER SUPPLY UNIT

At either end are cubicles containing the batteries, the cells being arranged in shelves. The 110V battery cells are small and numerous, whereas 24V cells are fewer and generally much larger. Sometimes batteries of high capacity occupy several cubicles on either side.

8.6   D.C. SUPPLIES FOR OTHER PURPOSES

The d.c. power supply unit applies particularly to the 'central' or bulk d.c. supplies to many consumers throughout the installation. In addition there are separate 'dedicated' d.c. supplies used solely for particular equipments. Among them are the supplies for navigational aids, for turbine and diesel engine starting, etc. These are not individually described here, but in principle they are similar - that is, they comprise a transformer-rectifier (charger), floating battery and d.c. output. Dedicated systems usually consist of only a single charger and battery, similar to the arrangement of Figure 8.1.

A special note, however, must be made regarding the d.c. starting of diesel engines and of some gas-turbines. Starting requires a heavy-duty battery (usually 24V) and a large starter motor which consumes power very rapidly while actually starting. The recharge after a start is normally given by a d.c. generator ('dynamo') on the engine itself, as on a car. If however the engine has been standing idle for long periods or has made only short runs after test starts, the battery may gradually lose charge, and there is a danger that it may not be able to start the engine when an emergency arises.

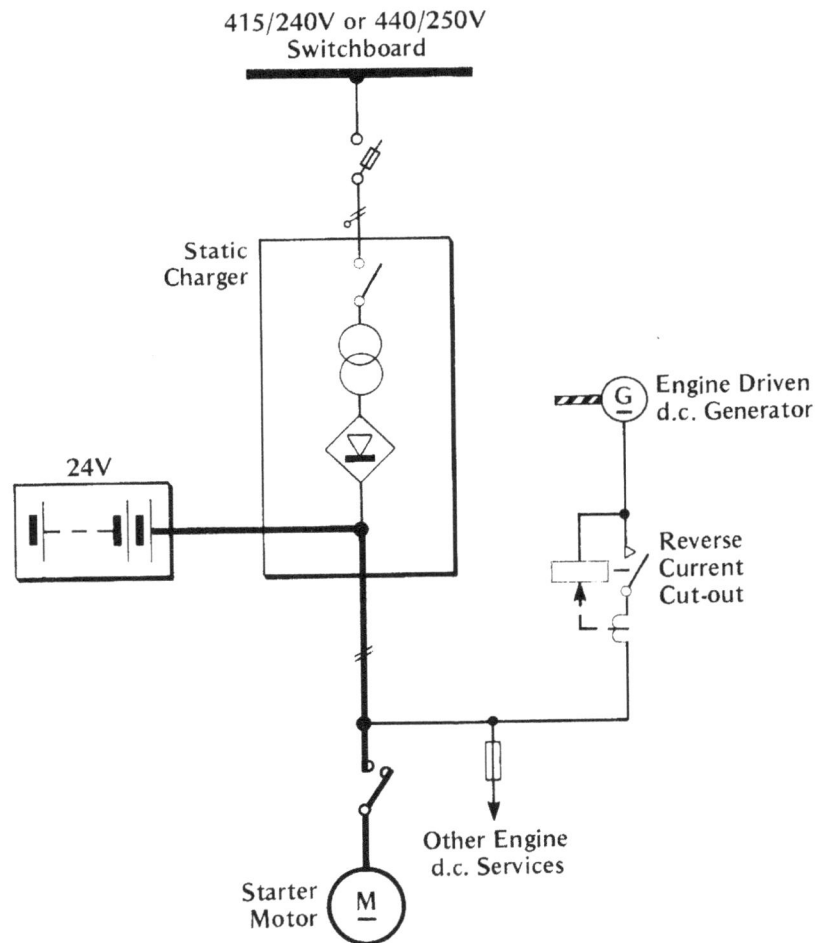

FIGURE 8.5
TYPICAL DEDICATED D.C. SUPPLY – DIESEL ENGINE STARTING

To prevent this, diesel starting batteries are provided also with a static charger powered from the offshore a.c. mains, as shown in Figure 8.5. It is of the same transformer-rectifier and floating battery type as already described, but it is smaller. It is located close to the engine, and in most installations it is left on permanently to give the battery a continuous maintenance charge.

It should be noted that in this case the rectifier is not the normal source of d.c. supply; it is the battery which powers the motor, as shown in heavy line, and the rectifier is purely a charger.

8.7 OTHER D.C. SOURCES

The d.c. sources described so far all employ solid-state rectifiers, but these are not the only methods available, especially when large d.c. powers are required.

8.7.1 Rotating Machines

It is also possible to obtain d.c. supplies from rotating machines. In former times a 'rotary converter' was used, having a common a.c./d.c. armature and both commutator and sliprings. Later these came to be replaced by motor-generator sets, having a standard a.c. motor driving a completely separate d.c. generator. Such rotating equipment however posed a maintenance problem, which has now been largely overcome by the present static rectifiers.

One advantage of the motor-generator type of conversion is that the d.c. system is completely independent electrically of the a.c., and transients in the one are not carried over into the other. This can be important in communications and other electronic systems.

8.7.2 Mercury Arc Rectifiers

An alternative, and static, method uses a mercury arc. This allows electron current to flow only in one direction, as in a thermionic valve, from cathode to anode. A ring of anodes is sealed into an evacuated glass bulb in which a pool of mercury acts as the cathode. Each anode in turn carries the peak voltage in a multi-phase a.c. system, and the arc rotates from each anode to the next, so providing a continuous d.c. current, always in one direction. Some mercury arc rectifiers have six, twelve or even twenty-four anodes supplied from special 6-, 12- or 24-phase transformers. They provide a d.c. supply with very low ripple content, and they also cause lower harmonics in the a.c. system.

An alternative design of mercury arc rectifier uses a steel tank instead of the glass bulb.

8.8 BATTERY CHARGING

The rate of charge which is put into a battery by a charger depends on the d.c. voltage applied to its terminals and to the back-emf developed within the battery, which rises with its state of charge. Under normal running conditions the charger, while supplying the d.c. load at its nominal voltage, maintains the voltage applied to the battery at just over its charged voltage. This results in a minimal charge current, or maintenance charge, going continuously into the battery to maintain its state. The battery contributes nothing to and takes virtually nothing out of the system. It is said to be 'floating'.

After a period of discharge following an a.c. failure and consequent use of the battery as a back-up source, power will eventually return. It is essential that the battery be recharged as quickly as possible, but not at a rate that would damage it. This high charging rate is

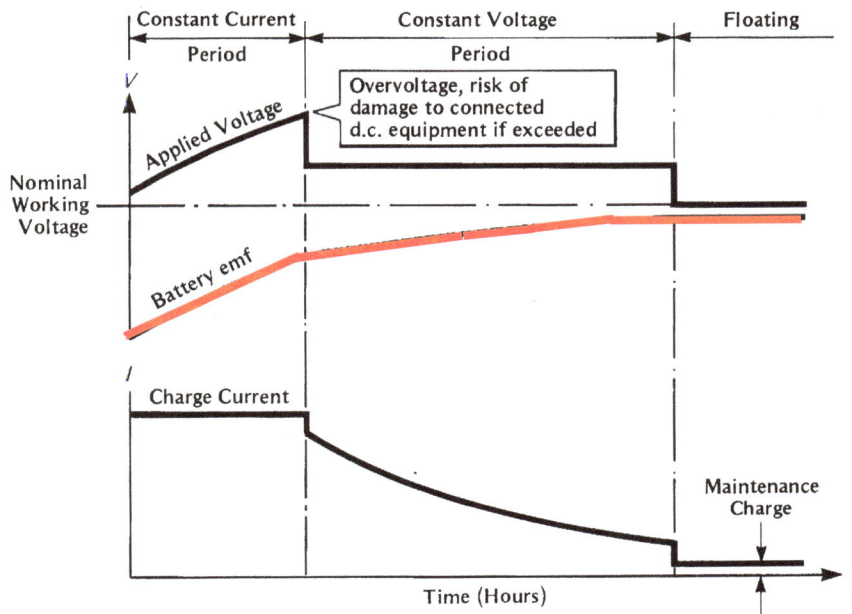

**FIGURE 8.6
RECHARGE CYCLE FOR DISCHARGED BATTERY**

called 'boosting'. Most chargers are provided with electronic circuits which control the rate of charge. Initially, on switching to BOOST, the charger voltage is controlled so that the charge current is limited (the 'constant current' period). After that a constant voltage is applied to the battery so that, as its emf rises, the charge current tapers off. On completion of the charge, the charger reverts to its 'float' mode either by manual switching to FLOAT or automatically, and the battery thereafter receives a maintenance charge only. These stages can be monitored on the charger's d.c. ammeter and are shown graphically in Figure 8.6.

On most chargers the change from float to boost and back again must be done manually by the operator (some systems have a timed return to float). Panel lamps indicate whether the battery is on boost or floating. In some cases a dial may be provided on which the operator can set the hours of boost required. At the end of the set period the dial has worked back to zero and switched the charger automatically back to the 'float' mode. This can be checked from the FLOAT or BOOST lamps. ('HI RATE' is sometimes used instead of BOOST.) With manual control the operator must estimate from his experience how many hours of boost are needed to replace the discharge.

It should be appreciated that while these varying voltages are being applied to the battery to recharge It, they are at the same time being applied to the d.c. loads. They are all higher than nominal, especially during the initial constant-current stage, and they could do damage to some of the d.c. equipments. For this reason only one battery at a time may be boost-charged, and it must be isolated from the d.c. bars while doing so, using the other half of the system to provide the d.c. loads, as described in para. 8.3. On certain makes of equipment this is done automatically, as already described. Most systems have d.c. overvoltage protection to disconnect the load if the voltage exceeds a certain limit. A flag relay or lamp gives an indication if this has happened.

8.9   VENTILATION

When a battery is being charged, especially when near the top of charge, it starts to 'gas'. This is because the charging current, having no more charge to give, electrolyses the water in the cells and breaks it down into hydrogen and oxygen gas. **This gas is a very explosive mixture indeed** over a wide range of hydrogen concentrations (between 4% and 96%). It is therefore essential that batteries - or at least battery rooms where batteries are concentrated - are well ventilated.

The ventilation of the battery rooms is continuously monitored; if the ventilation fails, any charge that is in progress is automatically stopped. This is done by the ventilation flow monitor tripping the a.c. supply into the charger - see Figure 8.3. This condition is indicated by a VENTILATION FAIL lamp. Similarly, if there is no ventilation, a battery boost charge cannot be started.

An uncharged battery however may create a difficult situation, and on many offshore installations facilities are provided to override the Ventilation Fail trip. This is usually a key-operated switch (seen in Figure 8.4), but it may only be used by an Authorised Person after he has satisfied himself that it is safe to do so. This might entail installing temporary fans in the battery room. After use, care must be taken to reset the switch.

8.10   BATTERY CARE

Nickel-cadmium batteries are now used on most offshore and onshore installations. They are robust and will withstand relatively high rates of charge and discharge. The electrolyte is an alkaline solution requiring great care in handling as it attacks the skin and destroys clothing; always wear rubber gloves and wash hands in 10% solution of boric acid or under running water after contact. The electrolyte takes no chemical part in the charge/discharge cycle and has a constant normal specific gravity of 1.210. The state of charge of an alkaline battery is indicated by its cell voltage, not by its gravity. (See Part 1 Electrical Theory.)

The capacity of a battery is expressed in ampere-hours (Ah); a fully charged 200Ah battery will produce 40A for 5 hours before it is discharged; this battery is then discharging at its so-called '5-hour rate'. If the same battery were discharged at 100A it would not last 2 hours but would be serviceable for appreciably less time - say 1.5 hours - and its '200Ah capacity at a 5-hour rate' would be reduced. It is usual to quote the capacities of nickel-cadmium batteries at their 5-hour rate and to make allowance for heavier discharges.

Because a battery is not 100% efficient, it requires more electrical energy to recharge it than was taken out on discharge; a typical value of this 'charge coefficient' is 1.4 for a nickel-cadmium cell.

$$\text{Ah (charge)} = 1.4 \times \text{Ah (discharge)}$$

The only way to determine the state of charge of a nickel-cadmium battery is by measuring the overall voltage and so determining the average cell voltage. This should not be allowed to fall below 1.1V while discharging at the 5-hour rate. A fully charged battery will receive a satisfactory floating maintenance charge when the output of its charger is maintained at 1.4 to 1.45V per cell. These figures vary slightly between different makes.

On boost a battery may be charged at its 5-hour rate, which will bring a discharged battery to full charge in 7 hours. These figures are given as a guide only; they may be varied to suit individual circumstances.

When on charge, a fully charged battery loses electrolyte by gassing. The degree of gassing depends on the charging current; loss of fluid will also take place by evaporation. In both cases only water is lost, not the chemical salts. The level of the electrolyte in each cell must be checked periodically and must not be allowed to fall below the top of the plates; loss of fluid is made up by adding distilled water.

When a battery is on continuous floating charge, it should be discharged periodically and then given a boost-charge, say every six or twelve months, to keep it in good condition; it should also be boost-charged after a mains failure. The tops of battery cells should be kept clean and dry and the terminals lightly coated with a suitable grease to resist corrosion. The terminal connections and inter-cell links must be kept tight.

Lead-acid batteries are not now much used offshore. Where they are used, the above general description still applies, except that the cell voltage at full charge is about 2.1V. The electrolyte is diluted sulphuric acid which, at full charge, has a specific gravity of between 1.200 and 1.300. This falls during discharge, and it must not be allowed to fall below 1.150. The battery must then be recharged at once and never be left in a discharged state.

When sulphuric acid is being diluted for use in a battery, ALWAYS ADD THE ACID SLOWLY INTO THE WATER - never the other way round, as adding water to acid can cause a violent reaction and result in serious danger to the operator.

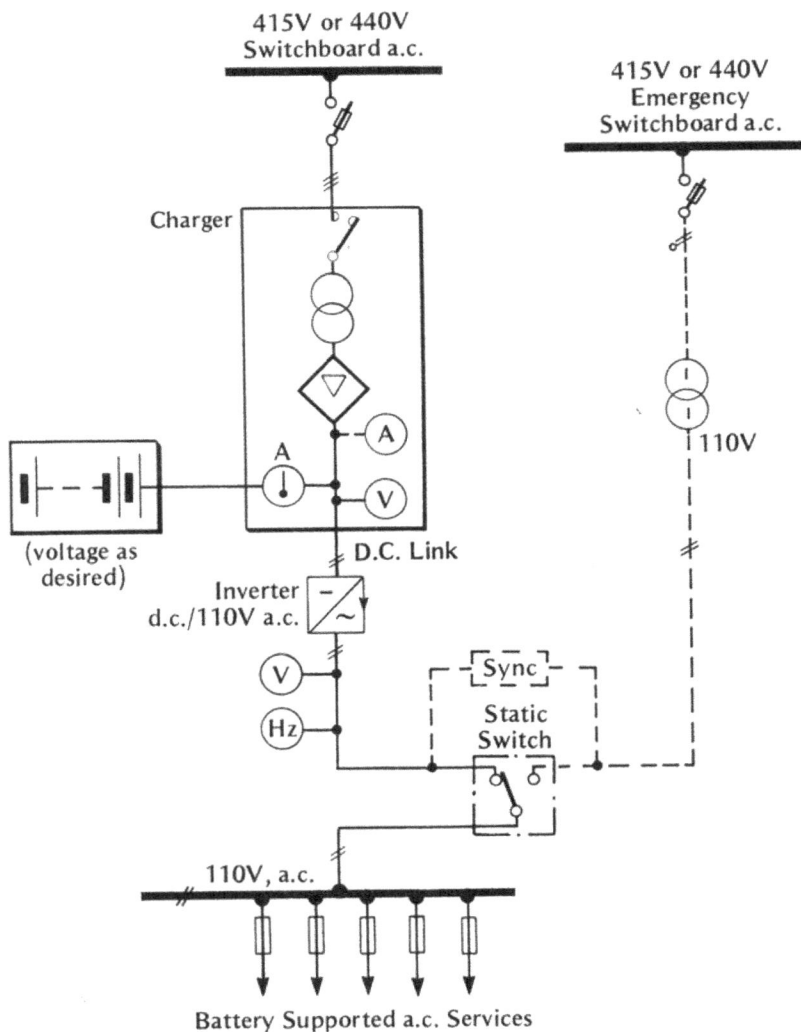

FIGURE 8.7
BATTERY SUPPORTED A.C. SYSTEM

8.11   BATTERY SUPPORTED A.C. SYSTEMS

So far only d.c. systems having battery support have been described. In some cases, such as instrumentation, equally assured a.c. supplies are needed, and clearly they cannot come direct from a battery. What is done, as shown in Figure 8.7, is to provide a battery-supported d.c. system exactly as was done in Figure 8.1, but to take its d.c. output and pass it through an inverter. This is a solid-state static device that converts d.c. into a.c., and moreover at any voltage and frequency desired, so as to distribute the output to the vital a.c. loads. Thus, if the mains a.c. fails, the battery will continue to provide d.c. without operator action or interruption. This unbroken d.c. is inverted to unbroken a.c. and distributed to the various loads. The voltage level of the 'd.c. link' and battery is not important, and any d.c. voltage may be used; in some cases this may even be 220V d.c. Normally only one charger and one battery are needed.

It is usual, with such an arrangement, to provide a standby a.c. supply from the 415V or 440V mains direct to the inverter-fed distribution board, as shown dotted in Figure 8.7. If this distribution were, for example, at 110V a.c. single-phase, a single-phase connection would be taken from a separate main board, transformed to 110V and passed through a 'Static Switch'. This is an electronic switch with no moving parts which normally connects the distribution board to the inverter. If the inverter itself should fail, there would be loss of voltage at the static switch, and it would change over automatically to the direct transformer supply, so re-energising the 110V distribution board. To effect a smooth changeover there is usually an electronic synchronising circuit. This back-up feature is shown dotted in Figure 8.7.

Thus if the charger's a.c. supply or the rectifier should fail, the battery takes over without interruption and the static switch stays on the inverter. But if the inverter itself should fail, the static switch changes over to the alternative supply (normally from an emergency switchboard). Note that this back-up supply is not itself battery-supported.

The actual power unit would look like half the dual d.c. system shown in Figure 8.4, together with an extra cubicle housing the inverter and static switch.

Those systems which are described here, both the battery-supported d.c. and the battery-supported a.c. systems, are sometimes referred to as 'Uninterruptible Power Supplies', or 'UPS' for short.

CHAPTER 9 A.C. MEASUREMENTS

9.1 GENERAL

In a.c. power systems it is necessary continually to monitor the voltage, currents, power and similar quantities in the various parts of the system. This is done by the use of instruments - that is by indicating voltmeters, ammeters, wattmeters etc. The same measured quantities are also used to protect the system by means of relays, which are devices to detect when any of the quantities is going outside the predetermined limit. They initiate whatever automatic action is necessary to restore the situation or disconnect faulty or overloaded apparatus.

Almost all electrical instruments and relays depend for their action on measurements of voltage or current or combinations of the two. Measurements of frequency are obtained from analysing a voltage measurement.

The manner in which the various types of a.c. measuring instruments work is described in Part 1, Chapter 35. These include moving-iron, dynamometer and eddy-current types and also transducer-operated instruments. In the following paragraphs it will be assumed that the appropriate type of instrument is used.

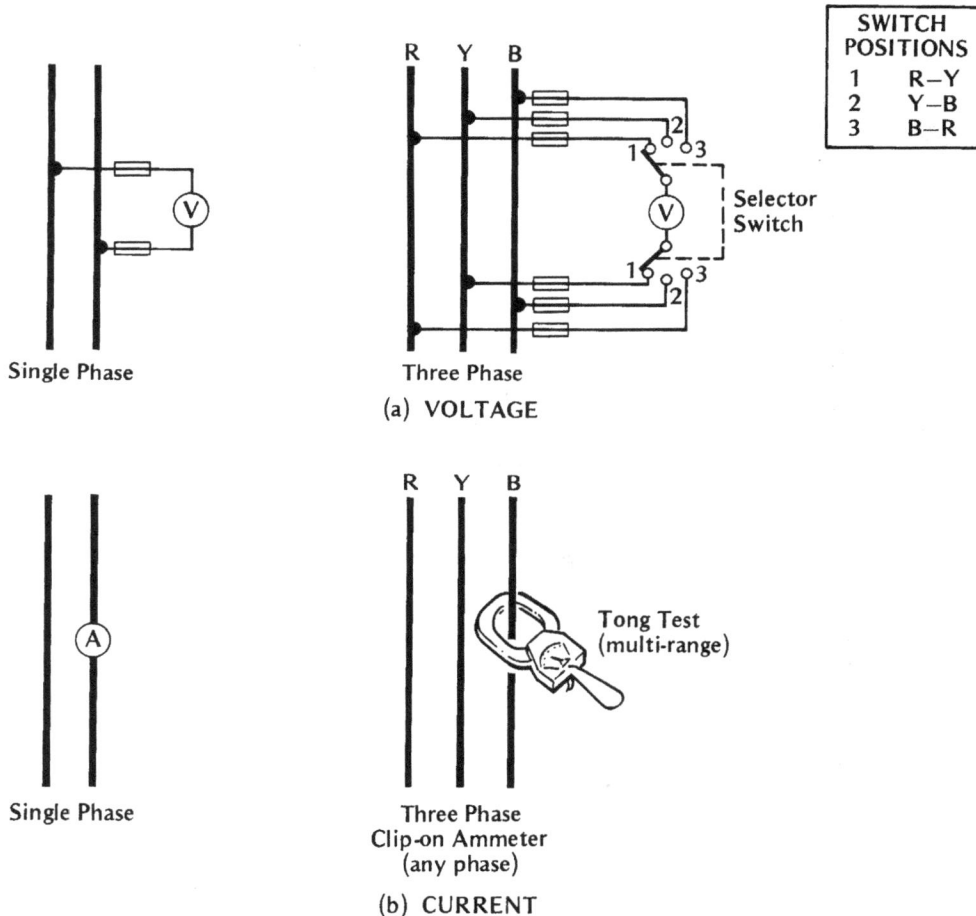

SWITCH POSITIONS	
1	R–Y
2	Y–B
3	B–R

Single Phase   Three Phase

(a) VOLTAGE

Single Phase

Three Phase
Clip-on Ammeter
(any phase)

(b) CURRENT

FIGURE 9.1
DIRECT MEASUREMENT

9.2   DIRECT MEASUREMENT

Voltage and current samples are taken either directly or indirectly from the conductors of the circuit to be monitored. In the simplest case (direct measurement) the voltage is taken by tapping the main conductors. The tappings must always be protected by fuses which, for a voltage-operated instrument or relay, are quite lightly rated, though still able to deal with the full fault capacity of the system. In the 3-phase case a selector switch may be used to measure voltages between any desired phases, as shown in Figure 9.1 (a).

Direct measurement of current in a single-phase circuit is obtained by placing the instrument's current-operated coil in series with a main conductor, shown in Figure 9.1 (b). In the 3-phase case it is not possible to select phases for current measurement unless current transformers are used. It would otherwise be necessary to break each phase to connect the ammeter, and this would not be acceptable. Selection with the use of current transformers is shown under 'Indirect Measurement' in Figure 9.2. Alternatively three separate ammeters may be used.

The currents in the separate phases can, however, be measured independently by use of a clip-on type ammeter (also known by the trade name 'Tong Test'). Different ammeter instruments can be plugged into the tongs to give current ranges from 10A to 1 000A. On some types the range is altered by a switch on the tester.

Direct measurement has serious disadvantages. In high-voltage systems the instrument or relay would have to be insulated up to the full system voltage, which for a normal sized switchboard instrument is not practical. Current-operated instruments would not only have to be insulated up to the full system voltage, they would also have to carry the full normal current of the circuit and to withstand the extreme fault currents. This, too, is not practical except for the lightest circuits.

9.3   INDIRECT MEASUREMENT

To overcome these objections indirect measurement is employed. Transformers are used not only to scale down the quantities actually measured, but also to isolate the instrument or relay from the main system voltage. Such transformers, which are designed specifically for this purpose, are known as instrument transformers.

Instrument transformers are of two types - 'voltage transformers' (VT) and 'current transformers' (CT). They are shown diagrammatically in Figure 9.2 for both single-phase and 3-phase systems. For 3-phase there may be either three separate single-phase VTs (with their ratios adjusted for the star connection) as shown in the inset to the figure, or else a 3-phase unit, which is more usual. Current transformers are always provided as separate single-phase units.

The secondary voltages and currents may be chosen as desired, but in practice the VT secondary voltage is usually 110V line-to-line, and the CT secondary current 5A or 1A (see para. 9.7 for special caution when dealing with CT secondaries).

To select the phases between which voltages are measured, a 3-position selector switch is used, as in Figure 9.1, but connected to the VT secondaries. Further positions may be provided to measure voltages between each phase and neutral.

To select the phases in which currents are measured, a special selector switch is used which inserts the ammeter into the CT secondary of the desired phase and at the same time allows the secondary currents of the other two phases to pass. To avoid open-circuiting the CT secondaries, all contacts are of the make-before-break type. This is shown in Figure 9.2(b), bottom right.

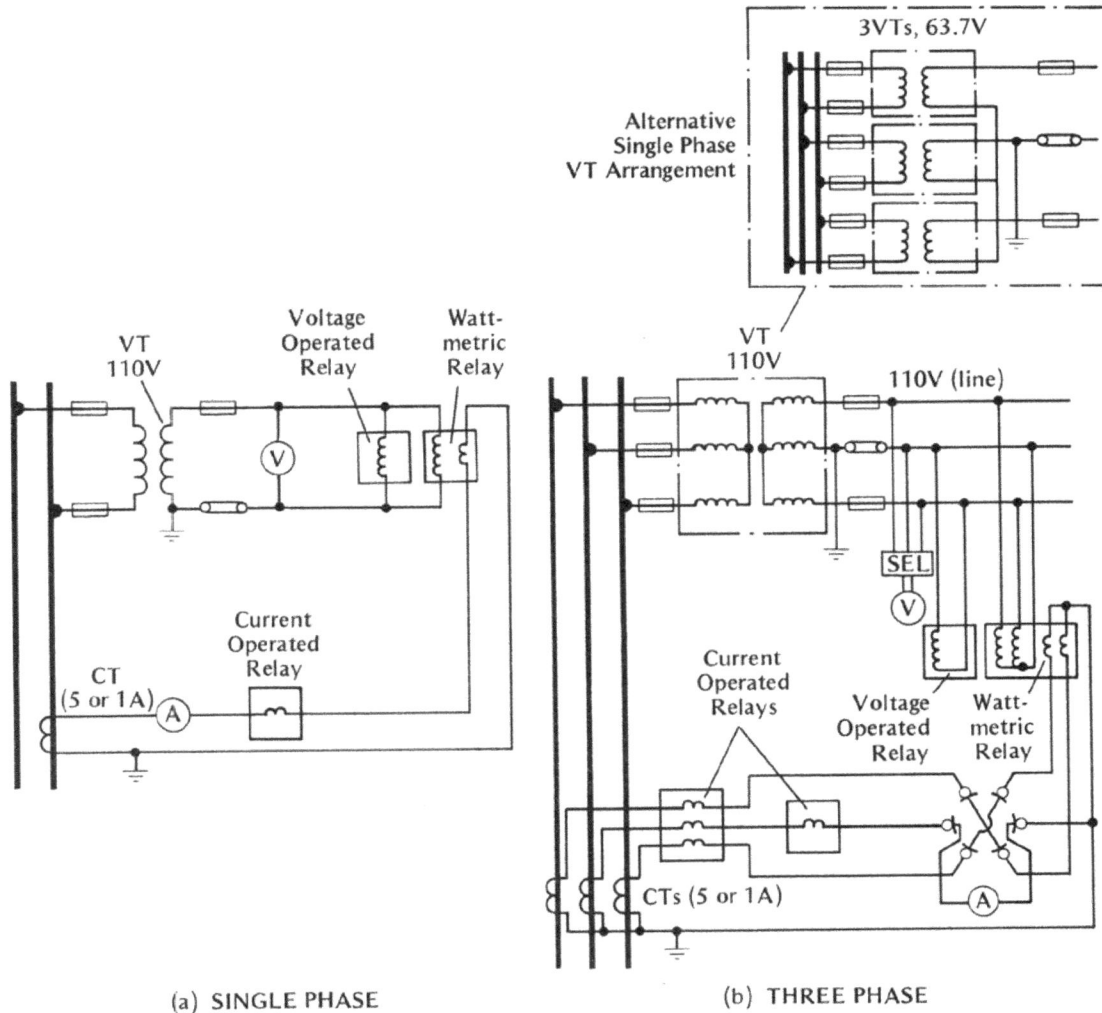

FIGURE 9.2
INDIRECT MEASUREMENT WITH INSTRUMENT TRANSFORMERS

A VT feeds, through secondary fuses (except in the earthed line), all voltage-operated instruments and relays in parallel, single- or 3-phase as required. Current-operated instruments and relays are connected in series with the CT secondary whose phase is being used. Fuses must never be used in a CT secondary circuit, for the reason stated in para. 9.7.

Instrument transformer secondaries must always be earthed. With star-connected VT secondaries it is normal practice to earth one phase (usually the yellow) and not the star-point. CT secondaries are normally commoned at some point, and it is usual to earth this common line, as shown in Figure 9.2(b).

9.4 INSTRUMENT ACCURACY

Since the purpose of instruments and relays is to monitor the actual conditions in the main power line, it is necessary that VTs and CTs reproduce those conditions, to a stepped-down scale, as accurately as possible. That is to say their voltage ratio or current ratio must be correct and constant over their whole range of operation; they must not introduce undue phase shift while doing so (important for wattmeters); and they must reproduce unbalance conditions exactly.

The extent to which these conditions are met determines the accuracy class of the instrument transformer. A distinction is drawn between 'measuring' and 'protective' types. For measurements, the accuracy within, and a little above, the normal working range is important, but accuracy in the overcurrent and fault ranges of current does not matter. On the other hand, a protective CT must deliver accurate currents in the fault range, whereas accuracy in the working range is unimportant. This gives rise to two different design concepts.

The classes of accuracy are laid down by British Standards (BS 3941 for VTs and BS 3938 for CTs). For each type different ranges of accuracy are specified for measurement and for protective transformers according to the purpose for which they are to be used. The ranges are as follows:

VTs (BS 3941)			CTs (BS 3938)		
Class	**Voltage Ratio Error**	**Phase Displ**	**Class**	**Current Ratio Error**	**Phase Displ**
Measurement 0.1 0.2 0.5 1 3	±0.1% ±0.2% ±0.5% ±1.0% ±3.0%	±5' (angle) ±10' ±20' ±40' not spec	0.1 0.2 0.5 1 3 5	±0.25 – 0.1% ±0.5 – 0.2% ±1.0 – 0.5% ±2.0 – 1.0% ±3% ±5%	±10' – 5' ±20' – 10' ±60' – 30' ±120' – 60' not spec not spec
Protective 3P 6P	±3% ±6%	±120' ±240'	5P 10P	±1% ±3%	±60' ±60'
Special			X	as specified	

(Note: These classifications replace the former A-B-C series, which is, however, still found on equipment installed before the change.)

Most indicating instruments on onshore and offshore switchboards are fed from VTs and CTs of Class 0.5, and most protective relays from VTs Class 3P and CTs Class 5P. There are, however, exceptions (for example differential relays are fed from Class X CTs), and it is necessary to refer to drawings for particular cases.

If it is ever necessary to check or recalibrate a switchboard instrument or relay, it must always be done with instrument transformers of a class higher than those with which it normally runs.

9.5 VOLTAGE TRANSFORMER DESIGN

A voltage transformer is made basically like an ordinary open-type power transformer, with separate HV and LV windings. It is, of course, much smaller, having ratings in the range 15 to 200VA per phase. The loading on a VT (or CT) is termed 'burden', not 'load'; an instrument transformer burden is always measured in volt-amperes, never in watts. At voltages up to those found in offshore installations, VTs are always dry-type, often embedded in synthetic resin. They are usually located inside the switchboards. On shore equipments, especially when associated with high-voltage oil circuit-breakers, VTs are often in oil-filled tanks (see Figure 2.1 of Chapter 2).

The high-voltage VT primary fuses are of the HRC type. They have a low current rating but are capable of breaking the full busbar fault current of the HV system. They are located in the VT compartment and with some types are embodied in the VT itself.

Access to the high-voltage VT and its fuses is through the VT compartment door. This cannot be opened until the VT has been isolated. The manner of isolation varies with different manufacturers.

9.6 CURRENT TRANSFORMER DESIGN

A current transformer can take one of two forms. One type is wound like an ordinary transformer, with primary and secondary windings round a common core. As a CT steps current down, it steps voltage up. The primary winding, though connected in the system's high-voltage system, is in fact the LV (high current) winding as far as the transformer is concerned, and the secondary is the HV (low current) winding. Wound-primary CTs are used where the primary current is low and where it is necessary to have several primary turns to achieve enough ampere-turns in the CT. The examples shown in Figure 9.3(a) and (b) are typical; burdens are in the range 5 to 30VA per phase. Wound-primary CTs must be able to withstand the full voltage and fault current of the main system on their primary windings.

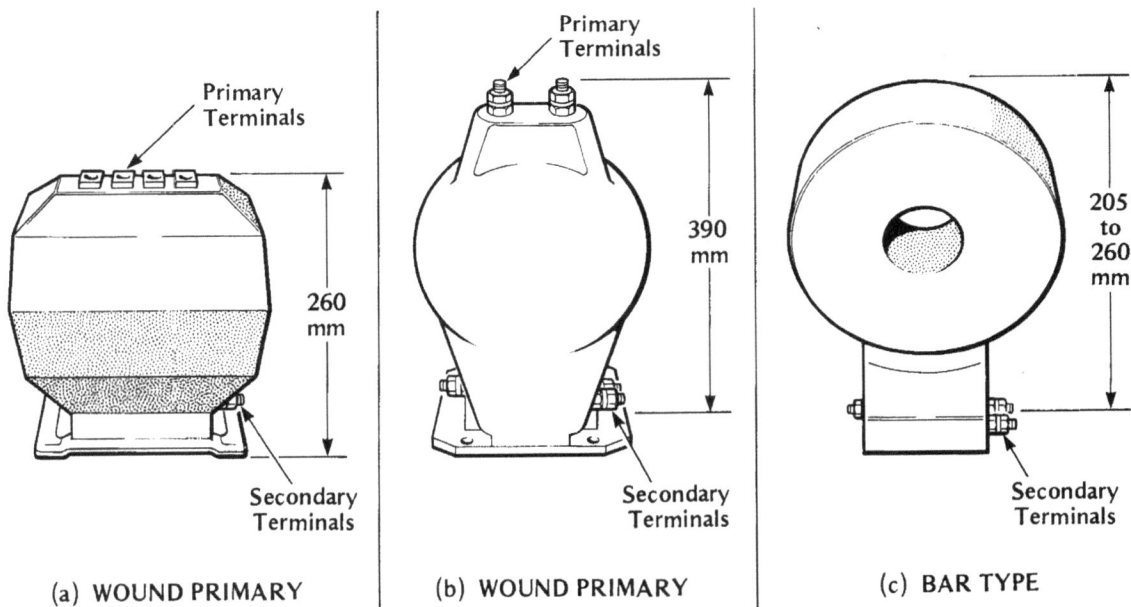

FIGURE 9.3
TYPICAL CURRENT TRANSFORMERS

An alternative form of CT is known as the 'bar' or 'ring' type. It has no primary 'winding' as such but uses the main conductor itself as a 'one-turn' primary. The flux surrounding the conductor, due to the current it is carrying, links the closed iron core of the CT and induces voltage in the secondary winding, which is wound as a toroid around the circular core. The secondary circuit is closed through its burden, and the current which flows in it is an exact scaled-down replica of the primary current in the conductor.

Bar-type CTs are generally used whenever the current ratio (e.g. 1 500/1A) is large enough. They are also convenient in that several can easily be stacked over a single existing conductor. It is very important that they be placed the right way up, otherwise the secondary terminal voltages and current flow will be reversed. By convention the secondary terminal S1 always has the same polarity as primary terminal P1, or as that of the end of the bar emerging from the face marked P1. This type of CT is shown in Figure 9.3(c). Its construction is not limited by the fault current of the main system.

Another important difference between a CT and other types of transformer lies in its magnetisation. The magnetising current, and therefore the flux, of a power transformer or a VT is constant and depends only on the applied voltage. However a CT when it has no burden is effectively short-circuited, and no voltage is present, whatever the primary current; therefore there is no core flux. If the burden is increased, so also is the voltage for a given current, as explained in para. 9.7, and this causes the magnetisation to increase. Thus with a current transformer the magnetisation is variable not only with the current, but it also is increased depending on the burden connected.

In the limit, if the burden is increased beyond the rating of the CT, the core will saturate, and the current ratio of the CT will no longer hold; it will become inaccurate. Moreover the iron losses will rise sharply and may cause severe overheating of the CT and possibly damage to it.

9.7   SPECIAL DANGERS WITH CURRENT TRANSFORMERS

When a CT secondary circuit is closed, a current flows through it which is an exact proportion of the primary current, regardless of the resistance of the burden. In Figure 9.4(a) the secondary of the CT (assumed to have a ratio of 1 000/5A and to have 1 000A flowing in the primary) is carrying exactly 5A, and, since the secondary terminals S1 and S2 are short-circuited, there is no voltage between them.

If now the short-circuit be replaced by a resistance of, say, 0.5 ohm (as in Figure 9.4(b)), the same 5A will flow through, causing a volt-drop of 2.5V and a burden of 5 × 2.5 = 12.5 VA. If the resistance were increased to 5 ohms (as in Figure 9.4(c)), the terminal voltage with 5A flowing would rise to 25V and the burden to 125VA. The greater the resistance, the greater would be the voltage and burden until, as it approached infinity (the open-circuit condition), so also in theory would the voltage (and burden) become infinite. This cannot of course happen in practice because the CT would saturate or the terminals flash over due to the very high secondary voltage between them. But it does show the danger of open-circuiting the secondary of a running CT. Lethal voltages can be produced at the point of opening. **This is why CT secondaries are never fused.**

The danger from an open-circuited CT is twofold. It can produce lethal voltages and so is a very real danger to personnel. The high voltage across the secondary winding could also cause insulation failure in that winding, leading at best to inaccuracy and at worst to burn out or fire.

Before ever an instrument or relay is removed from the secondary loop of a running CT (if such a thing **had** to be done), the wires feeding that instrument must first be securely short-circuited at a suitable terminal box or, better, at the CT itself. Similarly, if a running CT is ever to be taken out of circuit, it must first be firmly shorted. CTs with 1A secondaries are more dangerous than those with 5A, as the induced voltages are higher.

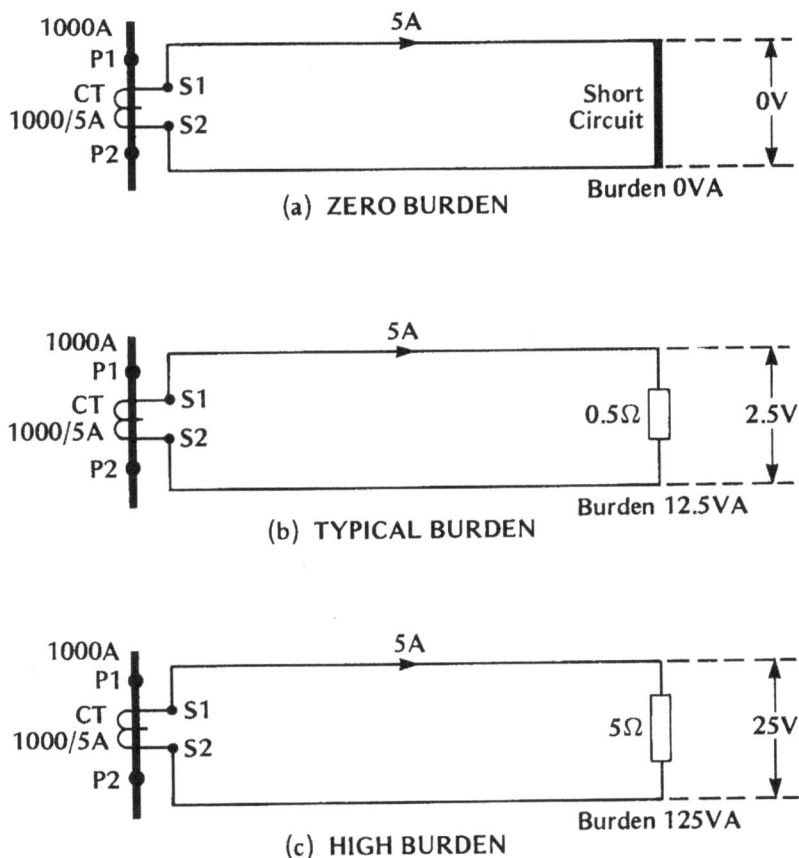

FIGURE 9.4
VOLTAGE AND BURDEN OF A CURRENT TRANSFORMER

To prevent this danger many CT secondaries are permanently short-circuited by a 'metrosil', which is a non-linear element with a high resistance at low voltages but which breaks down to almost a short-circuit at the higher and dangerous voltages. It does, however, somewhat reduce the accuracy of the CT and is not always acceptable for this reason.

There is also a range of CTs designed to saturate if their burden becomes excessive, so that even on open-circuit their secondary voltage will not exceed about 100V. It is not safe, however, to assume that such CTs are fitted in any particular case.

WARNING
WHENEVER POSSIBLE THE MAIN CIRCUIT SHOULD BE MADE DEAD BEFORE INTERFERING WITH CT SECONDARIES OR THEIR INSTRUMENTS OR RELAYS.

9.8   CALCULATION OF AN INSTRUMENT TRANSFORMER BURDEN

Instrument transformers are rated according to the burden that they can carry and still remain within their specified accuracy. The burdens are always given in VA units (i.e. power factor is ignored), and all burdens are simply added together. Manufacturers of instruments and relays similarly state the burdens of these devices in VA. Thus, if a CT operates an ammeter (2VA), a current relay (3VA) and, say, the current coil of a kWh meter (4VA), the total burden on the CT of these three devices will be 9VA.

The burden imposed by long secondary pilot leads, however, cannot be ignored. If, for example, the total resistance of a CT secondary run were 0.5 ohms (go and return) and the CT had a 5A secondary, the total volt-drop across the pilots would be $0.5 \times 5 = 2.5V$. With 5A current flowing in them, the burden of the pilot leads would be $2.5V \times 5A = 12.5VA$, and this would need to be added to that of the instruments (9VA above) to give a total burden on the CT of $12.5 + 9 = 21.5VA$. It must therefore have a rating sufficient to meet this total burden. In general, pilot leads impose far less VA burden on a 1A current transformer than on a 5A.

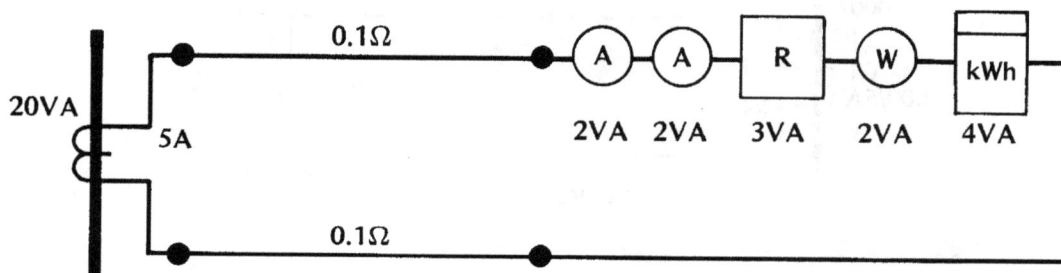

FIGURE 9.5
CALCULATION OF CT BURDEN

In Figure 9.5 a 20VA CT with full-load secondary current of 5A supplies two ammeters, a current relay, a wattmeter and a kWh meter with VA burdens as shown. The pilot leads have a resistance of 0.1 ohm **per core**. Is the 20VA rating of the CT sufficient?

Total instrument burden = $2 + 2 + 3 + 2 + 4 = 13VA$.
Total pilot load resistance = $2 \times 0.1 = 0.2$ ohm.
With 5A secondary current, volt-drop in leads is $5 \times 0.2 = 1V$.
Burden imposed by both leads = $5A \times 1V = 5VA$.
$\therefore$ Total burden on CT = $13 + 5 = 18VA$.

As the CT is rated 20VA, it has sufficient margin.

The reader should work out for himself what would be the total burden if the CT had a 1A secondary.

9.9 LOCATION OF CTS AND VTS

Current and voltage transformers can be located anywhere desired where the primary conductors are available, but in HV switchgear they are usually incorporated in special chambers in the switchgear unit itself. Figures 3.1 and 3.3 in Chapter 3 show views of typical HV circuit-breaker units, where the VT and CT chambers can be clearly seen. The VT can be drawn forward to isolate it from the busbars. Other manufacturers' arrangements differ in detail, especially in the front or back access to the VT chamber.

9.10   INSTRUMENTS

A.C. instruments include voltmeters, ammeters, wattmeters, varmeters, power factor meters, frequency meters and synchroscopes. Voltmeters, ammeters and frequency meters are almost all of the moving-iron or transducer-operated type, with an accuracy of 2% full-scale deflection. Wattmeters and varmeters are of the dynamometer type, and power factor meters and synchroscopes have two sets of fixed coils and a moving-iron armature. All voltage-operated coils (except those for 415V or 440V or less which may be direct-fed) are fed through VTs, and all current-operated coils through CTs at all voltages.

9.11   EXAMPLE - INSTRUMENTATION FOR A GENERATOR

Figure 9.6 shows a typical set of instrumentation for an offshore high-voltage generator. One complete set of indicating instruments is normally located on the electrical control panel in the Electrical Control Room; a second set is mounted on the generator local control panel. A megawatt meter for each generator may also be mounted on the main control panel in the Platform Control Room if this is separate from the Electrical Control Room. The generator circuit-breaker panel usually carries one ammeter and a voltmeter.

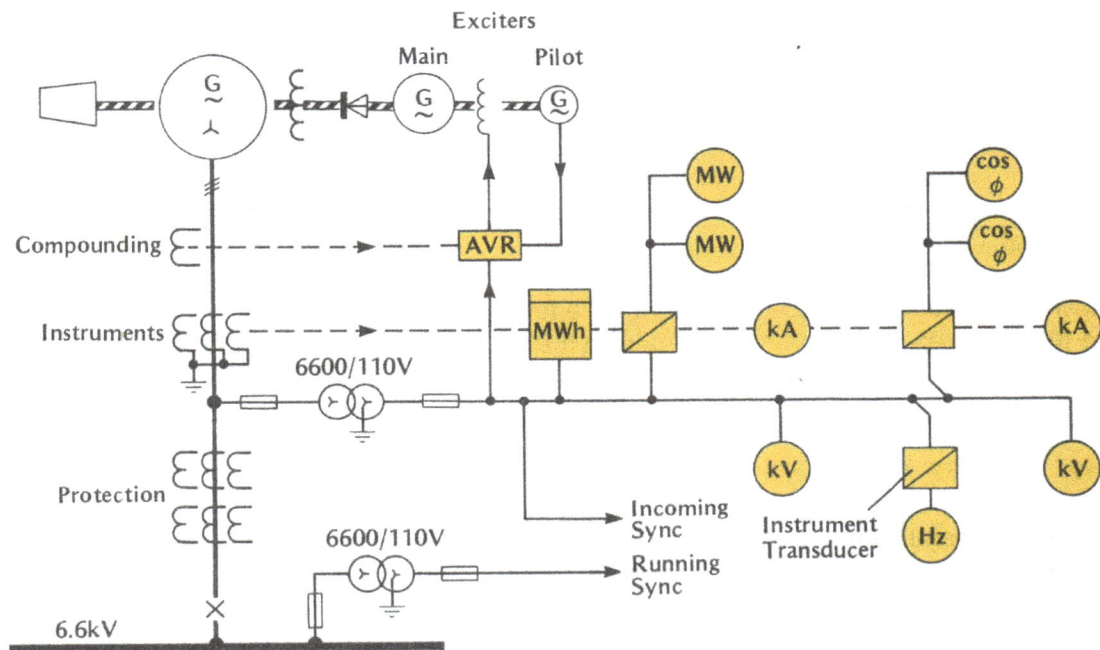

FIGURE 9.6
TYPICAL INSTRUMENTATION FOR A MAIN GENERATOR

Since wattmeter, varmeter, power factor meter and frequency meter movements tend to be expensive, an alternative which is being increasingly used is the transducer-operated instrument. Here the VT and CT signals are fed into static electronic a.c./d.c. transducers, and a d.c. voltage signal is produced from each which faithfully represents the a.c. watts, vars, power factor or frequency. These are led to simple d.c. voltmeter-type moving-coil instruments, but which are scaled in watts, vars, power factor or hertz. Many such instruments can be connected in parallel. Figure 9.7 shows typical connections. They can also be seen in Figure 9.6 where the transducers for wattmeters, power factor meters and frequency meters are indicated by the blocks with diagonal line.

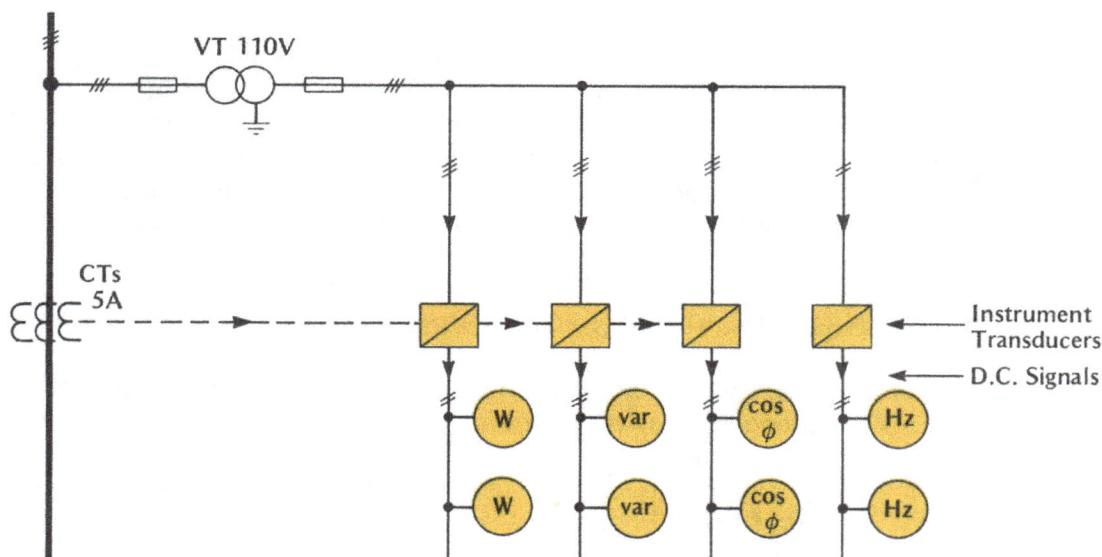

FIGURE 9.7
TRANSDUCER OPERATED INSTRUMENTS

Where two or more such instruments are used from the same transducer, they are connected in parallel. Some instruments have their transducer in the instrument case; others have the transducer in a separate box, especially if it operates more than one instrument.

Kilowatt-hour or megawatt-hour meters are fed through VTs and CTs whose connections are the same as for a wattmeter. As kWh meters are often used onshore as a basis for financial charging, they sometimes operate through VTs and CTs of a higher standard of accuracy.

9.12 TARIFF METERING

In onshore installations, where the supply is taken from the National Grid, the Supply Authority tariff takes account of maximum demand and power factor as well as actual energy consumption in kilowatt-hours. At the main substation therefore, where the supplies are taken, the Supply Authority installs sealed meters to record kilowatt-hours, kilovarhours and maximum demand. In larger installations the maximum demand is measured in kVA; in smaller it is in kW. It is averaged over each successive period of 30 minutes and then it resets.

A fuller description of how tariff metering is carried out will be found in Part 4 Electrical Systems, and methods of power factor control are described in Part 6 System Control.

These meters are the property of the Supply Authority, but they may be mounted on the consumer's main switchboard. Although installed for tariff purposes, they are of use to the consumer by enabling him to reduce load (if possible) to keep the maximum demand below critical level, and to take steps to improve the installation overall power factor.

CHAPTER 10 QUESTIONS AND ANSWERS: CHAPTERS 6 - 9

10.1 QUESTIONS

1. What types of power transformers are likely to be met with onshore and offshore?

2. Why are large oil-filled transformers often fitted with a conservator?

3. What is the purpose of a Buchholz relay?

4. How are large transformers cooled?

5. To which windings is a tapping switch or tap changer connected? Why?

6. What is the advantage of silicone oil over mineral oil?

7. Why is 'Askarel' used in many offshore transformers?

8. What are the disadvantages of Askarel?

9. How is the liquid level checked, and the integrity of the sealing monitored, in a sealed transformer?

10. Why are LV cable boxes in transformers usually much larger than HV boxes?

11. What nameplate voltage ratio would you expect to see on a transformer used to convert from a nominal 11 000V to a nominal 415V system?

12. What do you understand by a transformer's impedance Z? Give a typical value.

13. An Askarel-filled sealed transformer is naturally cooled. What code letters would be used to describe the cooling system?

14. What do you understand by a transformer phase connection 'Dy11'?

15. What precautions would you take before operating an off-load tapping switch?

16. Describe briefly the principle of an on-load tap changer. What types of mechanism are employed to operate it?

17. What site tests would you expect to do on an installed transformer?

18. What is an auto-transformer? Where would it be used? What are its properties as compared with a double-wound transformer of the same rating and voltage ratio?

19. What precautions would you take when connecting an auto-transformer into an earthed system?

20. What material is used for cable conductors on most installations?

21. What insulating materials are used for power cables?

22. Why is steel wire not used for armouring single-core cables?

23. What does the abbreviation HCL mean in a cable description?

24. Name the precautions to be taken when making a crimped cable termination.

25. Why are stress cones used when terminating a high-voltage cable?

26. If a 3-phase circuit is run with single-core cables between a non-hazardous and a hazardous area, at which end must the armouring be bonded to earth?

27. Why are certain vital services designed for operation on d.c.?

28. How are d.c. supplies to such services assured?

29. When should a battery be boost-charged? Why? How is this done?

30. Why must ventilation always be on when a battery is being charged?

31. What do you understand by 'central' and 'dedicated' d.c. supplies? Give an example of both.

32. Why are battery-supported a.c. supplies needed in certain cases? How are they achieved?

33. Why is a battery, when being boost-charged, first given a constant-current charge, then a constant-voltage charge?

34. What are the disadvantages of direct a.c. measurements on high-voltage systems?

35. There are two types of instrument transformers - 'measurement' and 'protective'. What is their main difference?

36. Why must a current transformer secondary never be fused? What are the dangers, and what precaution must be taken when removing an instrument from a live CT circuit?

37. A CT with rated burden of 15VA is feeding a total burden of 5VA through a 200 ft run of pilot leads with resistance of 0.15 ohms per core per 100 ft. It is preferred to use a CT with a 5A secondary; can this be done? If not, what remedy would you propose?

38. What class of accuracy would you expect to find generally used for measurement and protective CTs and VTs in most installations? What class is used with differential protection CTs?

10.2 ANSWERS

(Figures in brackets after each answer refer to the relevant chapter and paragraph in the text.)

1. Oil-filled, sealed or with conservator (not offshore).
   Askarel-filled, sealed.
   Dry type, encapsulated. (6.2)

2. To allow expansion of the oil with rise of temperature, while maintaining static oil pressure in the tank. (6.2.2)

3. A Buchholz relay is inserted in the pipe between the tank and conservator to:

   (a) trap gas bubbles and give a 'gassing' alarm

   (b) to sense any surge of oil due to an internal winding fault and to trip the circuit-breaker. (6.2.2)

4. Liquid-filled transformers (oil or Askarel) have their windings cooled by thermo-syphon action whereby winding heat is transferred to the liquid. The liquid is usually cooled in tubes or radiators by natural convection, sometimes assisted by forced ventilation. (6.2.2, 6.2.3)

   Dry-type encapsulated transformers are cooled by natural air circulation through the encapsulation. This may be assisted by forced fan ventilation at the higher loadings. (6.2.4)

5. Tapping switch or tap changer operates on the HV winding, which has lower currents to switch. (6.9)

6. Silicone oil is non-flammable as compared with mineral oil. (6.2.2)

7. Askarel is used in offshore transformers because it too is non-flammable and has good heat-transfer properties. (6.2.3)

8. Askarel is toxic and risky to handle. If spilt, it must all be carefully recovered and disposed of ashore. If allowed to fall into the sea, it would be destructive of marine life. (6.2.3)

9. By sight-glass on the side of the tank. A pressure/vacuum gauge in the space above the liquid will indicate if the sealing is faulty. (6.2.3)

10. Currents on the LV side are much greater than on the HV side and may require many cables per phase. The LV cable boxes not only carry larger-section conductors but may have to terminate many cables. (6.8)

11. Approximately 11 000/435V (no-load ratio), or 11 000 ±2½ ±5%/435V if tappings are shown. (6.3)

12. Z is the impedance offered to a current passing through a transformer. It is usually expressed as a percentage, being that percentage of the nominal applied voltage which, when applied to the primary windings with the secondary windings short circuited, will give full-load rated current in the secondary. (6.4)

13. LNAN. (6.6)

14. 'Dy' signifies a delta-connected high-voltage winding and a star-connected low-voltage winding. If A, B, C are the high-voltage terminals and a, b, c the corresponding low-voltage terminals, then, taking the vector representing phase 'A' voltage as 12 o'clock, the corresponding vector representing phase 'a' voltage is at 11 o'clock - that is, the LV system leads 30° on the HV. (6.7)

15. Make sure that the transformer is off load and isolated on both the HV and LV sides. (6.9.1)

16. An on-load tap changer changes the tappings without breaking the current by using a 'make-before-break' method. The current in those turns which are temporarily short-circuited during the transition is limited by introducing resistance. To avoid the risk of the changeover becoming stuck during transition, a 'stored energy' mechanism is used which only starts the tap change when there is enough energy stored to complete it without further outside power. The storage of energy may be by spring or flywheel. (6.9.3)

17. (a) Check for leaks, damage, signs of overheating, earthing.

    (b) Insulation resistance testing of HV and LV windings, to earth and, if possible, between phases.

    (c) Checking liquid level and effectiveness of sealing (if applicable).

    (d) Simulate operation of overtemperature or overpressure devices (also of Buchholz relay, if fitted). (6.11.2)

18. In an auto-transformer the secondary and primary sides share part of a common winding in which the secondary and primary current oppose one another. This common part may therefore be of smaller section and usually gives less heating. It may be economically used where the voltage ratio is small - say 3:1 or less.

    Compared with its equivalent double-wound type, it is smaller and gives rise to less heat. Its impedance is usually lower. It does not provide complete electrical isolation between the two sides. (6.10)

19. Where one side is earthed, the earthed line must be the one which is connected to the common primary/secondary terminal in order that the earth may be applied to the other side also. If this is not done, the voltage of one LV line will be the same as that of the HV side. (6.10)

20. Copper. (7.1)

21. Polyvinyl chloride (PVC) or Ethylene Propylene Rubber (EPR). (7.2.3)

22. Because of eddy-current heating. (7.2.5)

23. Hydrochloric Level. (7.5)

24. Use the correct lug or ferrule and correct crimping die. (7.6.2)

25. To control the electric stress where the core screen ends. (7.6.2)

26. In the hazardous area. (7.6.3)

27. Because they must continue in operation after failure of main a.c. power. This means a supply from a battery, which in general requires operation of those services by d.c. (8.1)

28. Power is taken from an a.c. switchboard and is passed through a transformer-rectifier ('charger') unit to provide the d.c. required. A battery floats on the d.c. side ready to take over the supply of d.c. without interruption if the a.c. supply or the rectifier fails. (8.2)

29. After discharge, a battery would take a fairly long time to recharge from the rectifier at the system's constant-voltage rate. This time is shortened by 'boosting' - that is, by increasing the charge rate. Boosting should be done after an appreciable discharge; also at 6- or 12-month intervals to maintain the condition of the battery. (8.3)

30. At top of charge a battery emits hydrogen and oxygen in an explosive mixture. Ventilation ensures that this gas mixture is dissipated. (8.9)

31. Where several d.c. services, usually of a similar type, are grouped to be supplied from a single D.C. Supply System, that system is termed a 'central' one. Where a d.c. supply is provided for a single equipment, that is a 'dedicated' system. Examples of central systems are: main switchgear closing and tripping; fire and gas detection; communications supplies. Examples of dedicated systems are: gas-turbine or diesel engine starting; navigational aids; emergency radio. (8.6)

32. Certain important services such as process instrumentation require unbroken a.c. supplies. This is achieved by providing a battery-supported d.c. system followed by an inverter to convert the assured d.c. power into a.c. (8.11)

33. If the boost-charging voltage were first applied to a discharged battery, the charge current would be so high that the battery might be damaged and the rectifier overloaded. Current-limiting circuits therefore ensure that the charge current cannot exceed a safe value - this is the 'constant-current' mode. After the battery emf has risen to the point where the charge current will not exceed the safe value, the charge automatically becomes constant-voltage, and the charge current tapers off (Fig 3.6). (8.8)

34. Instruments and relays connected directly to the main system must be insulated to withstand the full mains voltage. In HV systems (6.6kV or 11kV) this is not practical. Also current-operated instruments and relays must be able to carry the full fault current of the main system - again not practical. Such devices are therefore operated through instrument transformers (VTs and CTs). (9.2, 9.3)

35. Measurement instrument transformers must maintain their specified accuracy over the normal working range of currents and a little above; accuracy in the fault range is not important. Protective instrument transformers must have their specified accuracy in the range of fault currents; accuracy in the normal working range is not important. (9.4)

36.   A high-resistance burden on a CT gives rise to very high secondary voltages which could be a danger to personnel and could cause insulation breakdown in the CT itself. An open-circuit is an extreme case of this. A blown fuse would cause an open-circuit; therefore CT secondaries must never be fused.

When removing an instrument from a live CT circuit, the CT secondary must first be short-circuited - preferably at the CT secondary terminals - to prevent its becoming open-circuited when the instrument is disconnected.   (9.7)

37.   Instrument burden = 5VA
Pilot leads burden = 15VA
Total = 20VA. This cannot be fed from a 15VA CT.

Either substitute a 20VA CT, or else redesign the instrument system to work on 1A instead of 5A.   (9.8)

38.   Measurement CTs: Class 0.5
Protective CTs: Class 5P

Measurement VTs: Class 0.5
Protective VTs: Class 3P

Differential CTs: Class X   (9.4)

PART 4 ELECTRICAL SYSTEMS

CHAPTER 1 ELECTRICAL DIAGRAMS AND SYMBOLS

1.1 GENERAL

There are many different ways of showing in drawing form how an electrical circuit works; these are all termed 'diagrams' and use simple symbols to represent complete equipments or elements of an equipment.

1.2 WIRING DIAGRAM

The oldest form of diagram is the 'Wiring Diagram', or 'Diagram of Connections', which shows the actual wiring between every terminal in an equipment or group of equipments. In some cases it may show details of the type of wire used or its colour, and it is possible to distinguish between power and control connections by thickness of line. Such a diagram is shown typically in Figure 1.1, which represents a motor control system with local and remote operation, protection and indications.

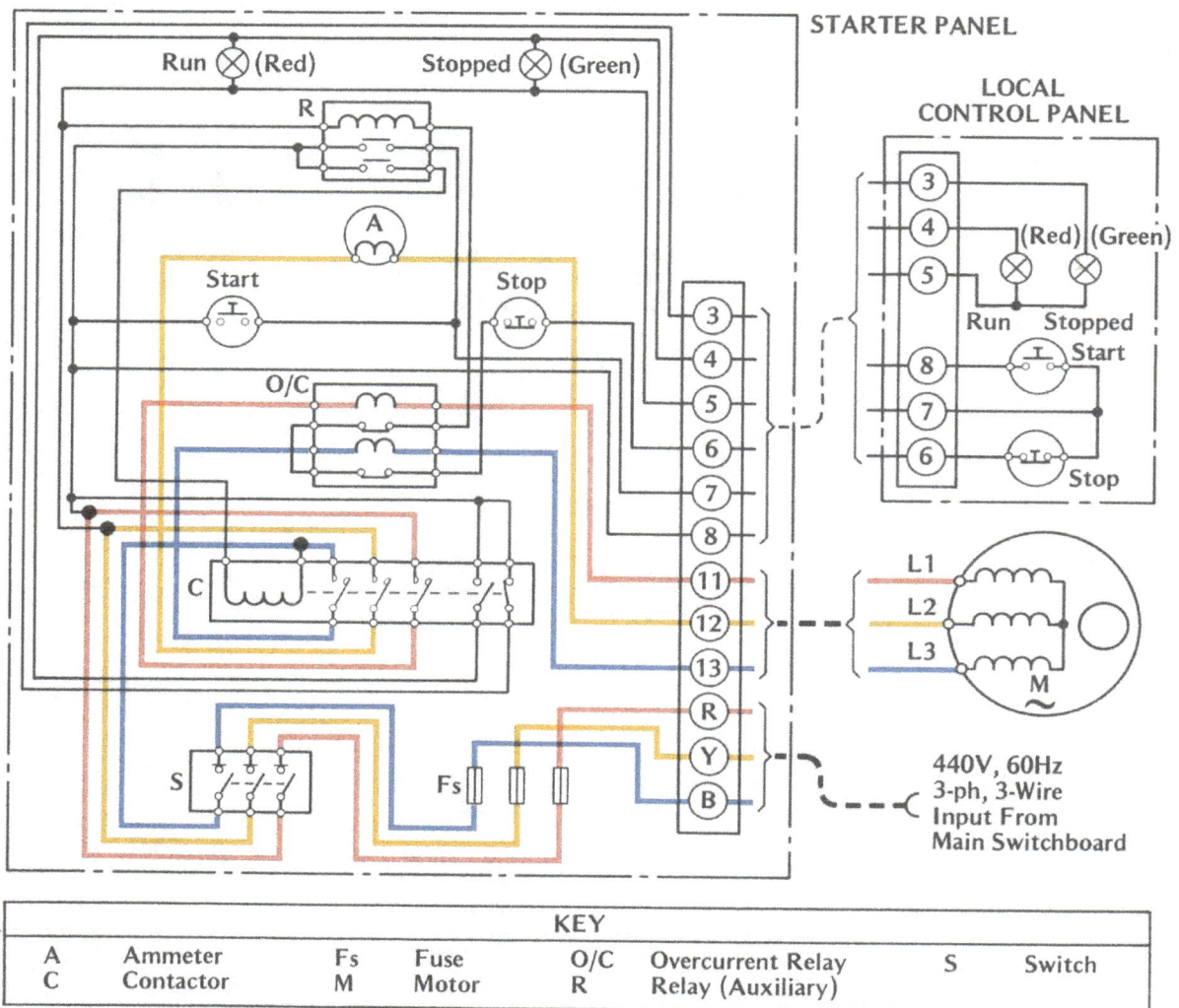

KEY							
A	Ammeter	Fs	Fuse	O/C	Overcurrent Relay	S	Switch
C	Contactor	M	Motor	R	Relay (Auxiliary)		

FIGURE 1.1
TYPICAL WIRING DIAGRAM

This example, which is fairly simple, shows nevertheless how difficult it is to see from a wiring diagram how the system works. With time and patience, however, it can be worked out.

It is obvious that with more complicated systems a wiring diagram would become very difficult to follow, and it would be virtually impossible to see how the system worked. So a new form of diagram was developed, called the 'Schematic' or 'Circuit' diagram.

The wiring diagram is still necessary for the wireman who is building the equipment, but it is of little use to the operator or even the maintainer. Its main use to him is to check an individual connection which has been shown to be suspect from first studying the schematic diagram.

1.3   SCHEMATIC DIAGRAM

Figure 1.2 shows exactly the same motor control system as was shown in Figure 1.1, and the simplification is very obvious. But there are major differences of presentation.

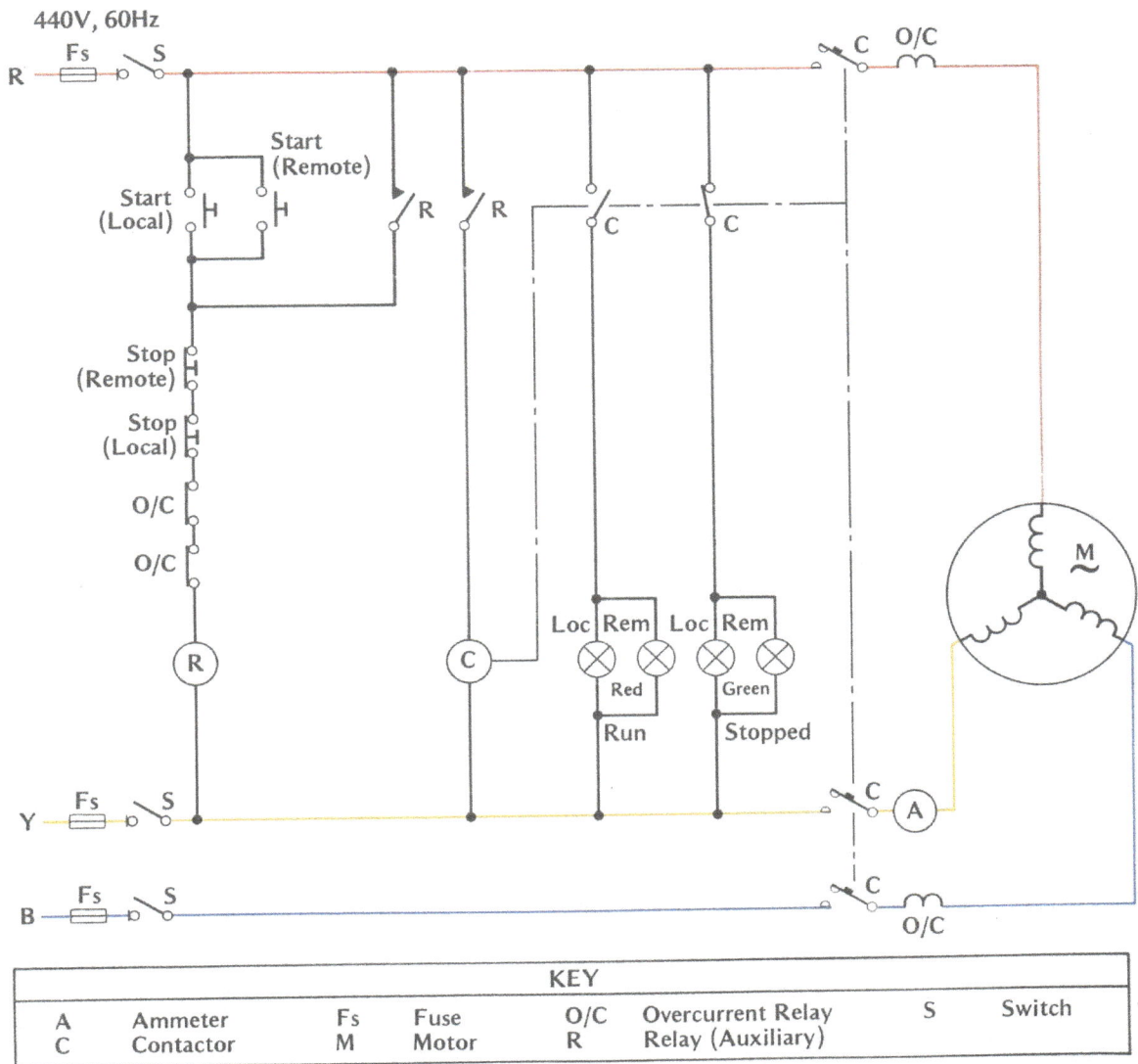

FIGURE 1.2
TYPICAL SCHEMATIC (OR CIRCUIT) DIAGRAM

The most important difference is that no attempt is made to show the contacts of a relay, contactor or circuit-breaker alongside their operating coil (where in fact they are actually located), but rather to show them in the circuit which they electrically control (which may be physically far away). In other words, with this 'detached contact' system no account is taken either of geography or of manufacturing assembly; the contacts are only related to their parent coil by a common identification letter or number. Similarly the different poles of a single switch may be drawn in different places as needed, related only by a common reference number.

This enables the reader to separate out each stage of a succession of operations from the initiating action (perhaps a pushbutton) to the final act (perhaps the running up of a motor). If the system fails to operate correctly, the reader can follow through each stage to the one that has failed. A detailed examination of only those elements (for example other contacts) which may have caused it to fail is not confused by other elements which have no bearing on the problem.

The schematic diagram is an essential tool in trouble-shooting. Any operator who has such a problem and who has only a wiring diagram, or perhaps only the nameplate diagram on the unit's enclosure, should as a first move convert it into a schematic. It greatly simplifies any problem and is itself a far simpler presentation.

1.4   CONVENTIONS FOR SCHEMATIC DIAGRAMS

There are a number of conventions which, if they are observed when preparing a schematic, make the reading of the diagram, and so the location of trouble areas, much easier. They are as follows:

- Every sequence should be drawn from left to right, or top to bottom. If any operation goes in the reverse direction (e.g. feedback), it should be marked with a reverse arrow.
- Each stage should be in strict order of occurrence when read from left to right.
- In complicated circuits where several separate but related functions exist, these can, with advantage, be presented as separate groups of elements - for example, the hoisting, luffing and traverse motions of a crane, which could with greater clarity be displayed on three separate schematic diagrams.
- All contacts and elements which are in series should be drawn, as far as possible, in a straight line with the element which they control.
- All contacts and elements which are in parallel should be drawn, as far as possible, side by side and at the same level so as to emphasise their parallel function.
- All operating coils of relays, contactors, breakers, etc., all indicating lamps and devices across which operating voltage is developed should be drawn in a straight horizontal line between the 'busbars', so that a particular element is quickly found by looking along the line.
- Any urge to align contacts with their operating coils, just because they are physically in the same box, should be firmly resisted. If the common reference number is not sufficient and it is desired to emphasise this association, a dog-legged chain-dotted line should be used.

There are many other conventions, which may be read in the Guiding Principles of BS 3939 and in BS 5070, but if the above are observed when preparing schematic diagrams the use of the diagrams will be simplified and their advantages greatly increased. Figure 1.2 brings out most of these points.

1.5   ONE-LINE DIAGRAM

A variety of diagram much used in both onshore and offshore work is the 'one-line' type as shown in Figure 1.3. It is basically a schematic where electrical rather than physical relationship is the requirement. There is usually no time sequence, but rather a 'hierarchy' running from a bulk supply down to minor distribution. Figure 1.3 actually illustrates a typical offshore installation, but, without the main generators at the top, it could illustrate an onshore network.

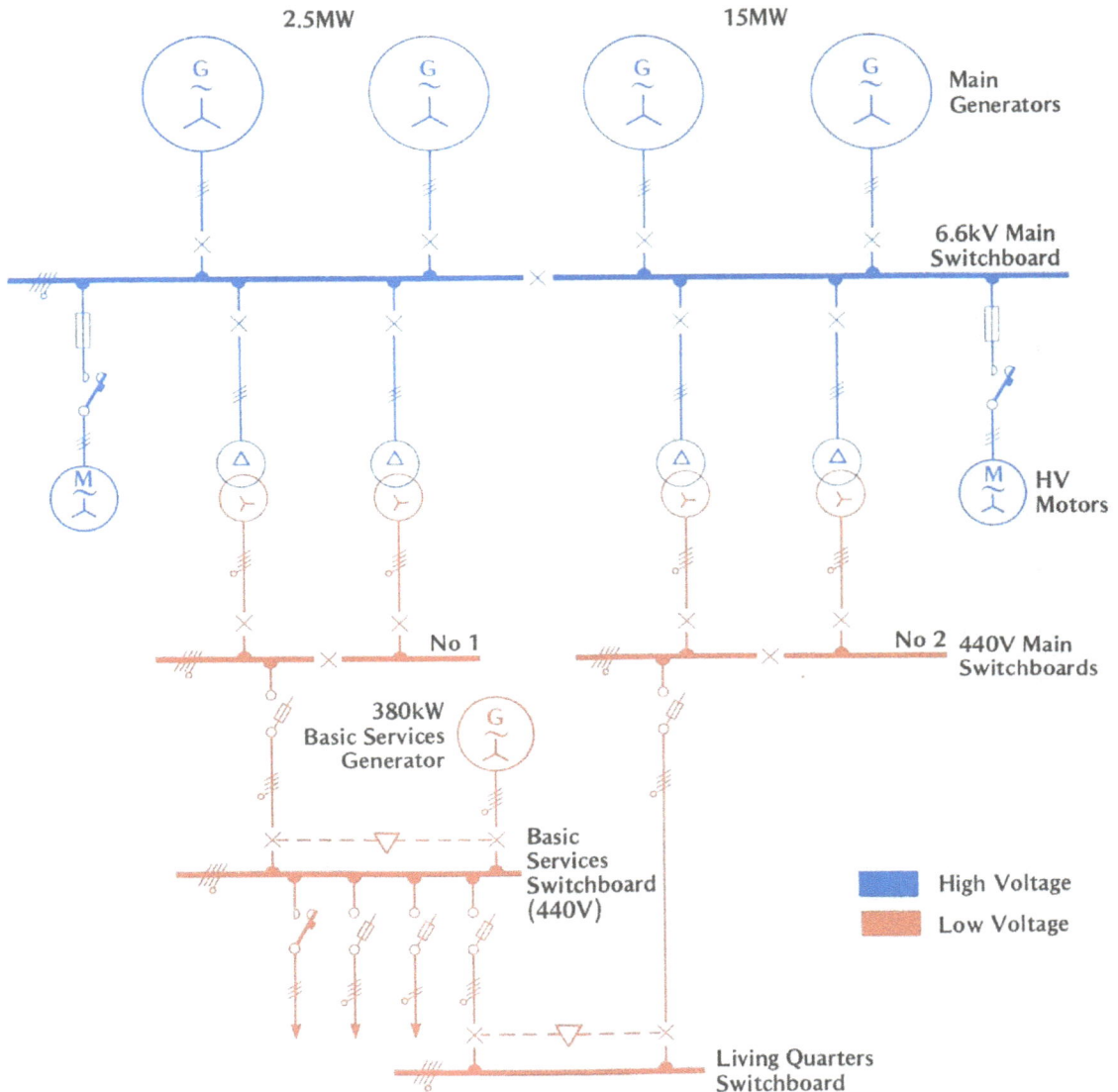

FIGURE 1.3
TYPICAL OFFSHORE NETWORK ONE-LINE DIAGRAM

The connections, which usually represent 3-phase cables, busbars or wiring, are depicted by single lines; 3-part elements such as circuit-breakers, contactors or switch contacts, or sets of fuses, are shown by a single symbol. Cross-marks on the lines may be used to indicate 3-wire or 4-wire connections, with the neutral specially marked with a small 'o'. These can be seen in Figure 1.3.

The one-line diagram is very useful for showing complete networks in simple terms, but giving a great deal of information. Note how much is shown in Figure 1.3, from the fact that there are four 6.6kV main generators to the interlocking of the two inputs to the Living Quarters switchboard.

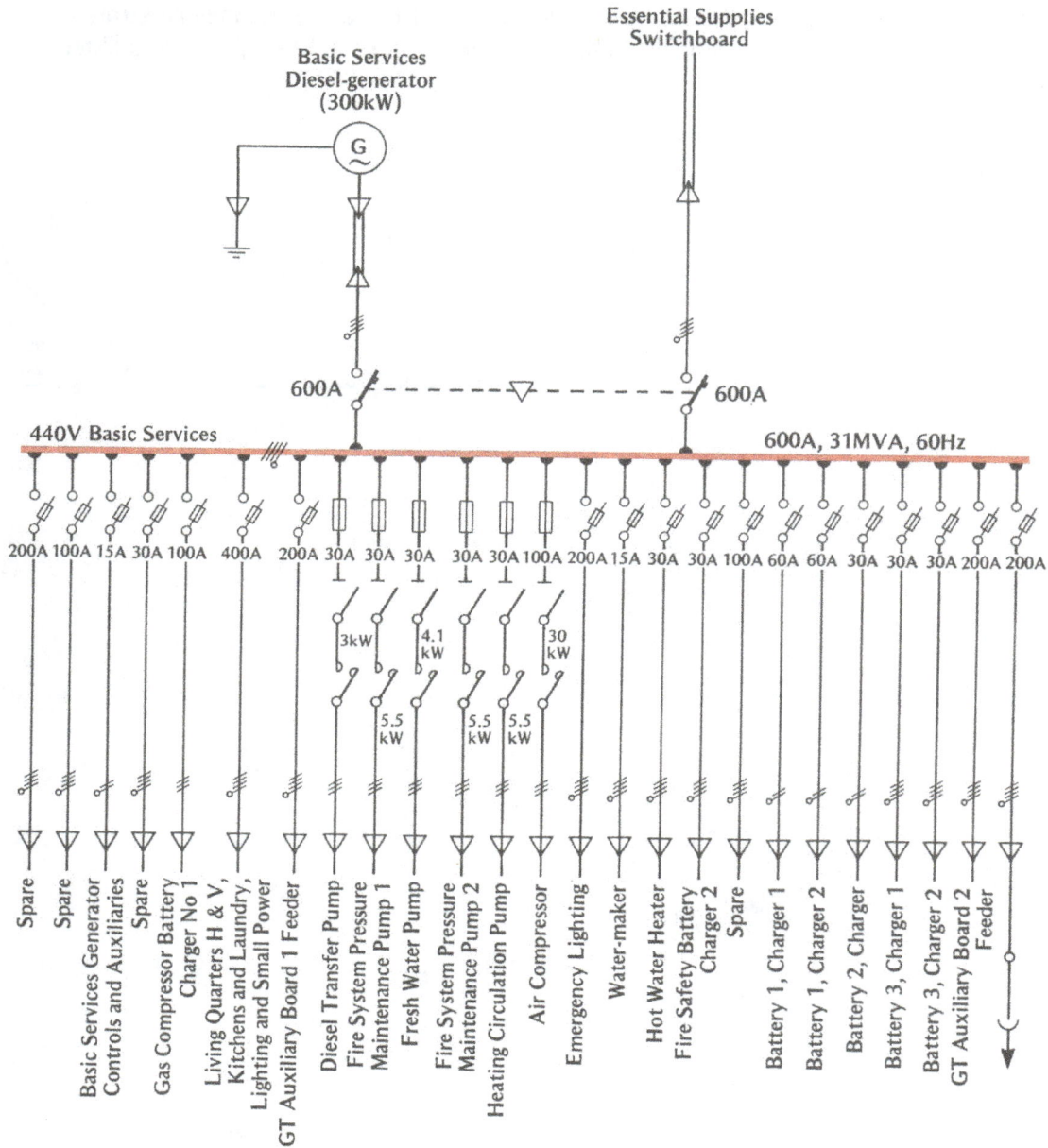

FIGURE 1.4
TYPICAL DISTRIBUTION ONE-LINE DIAGRAM

Figure 1.4 illustrates the one-line technique applied to switchboard distribution. A glance at the services supplied shows that this, too, is part of an offshore installation, but that is of no consequence. Note, again, the wealth of information.

1.6   BLOCK DIAGRAM

A particularly simple form of diagram, used when it is not desired to go into detailed circuits, is the 'block diagram'. Here complete units of equipment, or the separate stages of a piece of equipment, are shown by single blocks. The blocks are connected together with single lines showing the progress of the signals or other function being studied. The blocks are annotated to show their function, or a standard symbol within the block may be used instead. The left-to-right or top-to-bottom progress should be maintained, as shown in the example of Figure 1.5

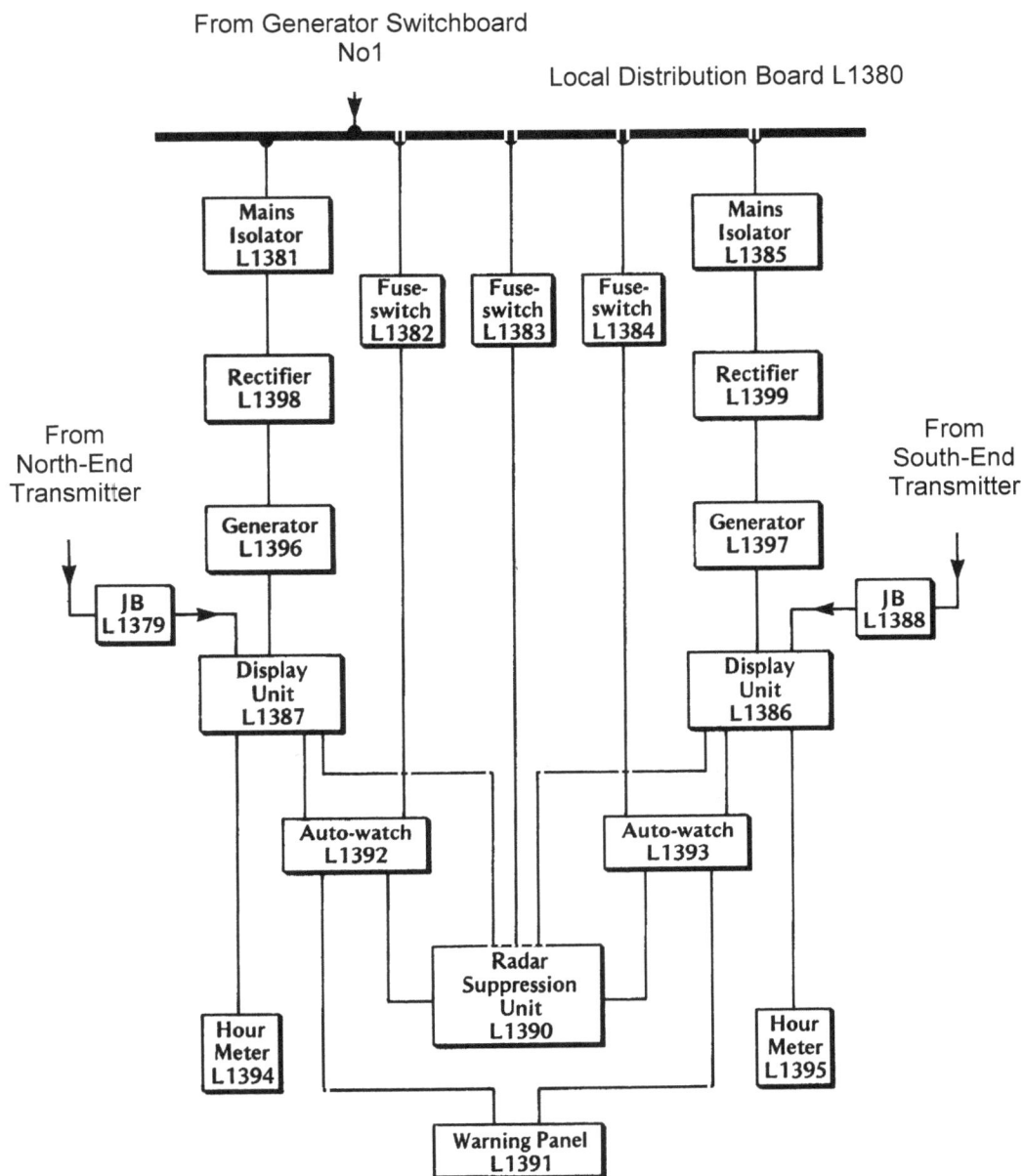

RADIO POWER SUPPLIES

FIGURE 1.5
TYPICAL BLOCK DIAGRAM

1.7   SYMBOLS

The sole authority for electrical symbols used in diagrams prepared in the UK is British Standard (BS) 3939 / IEC60617 now replaced by EN60617

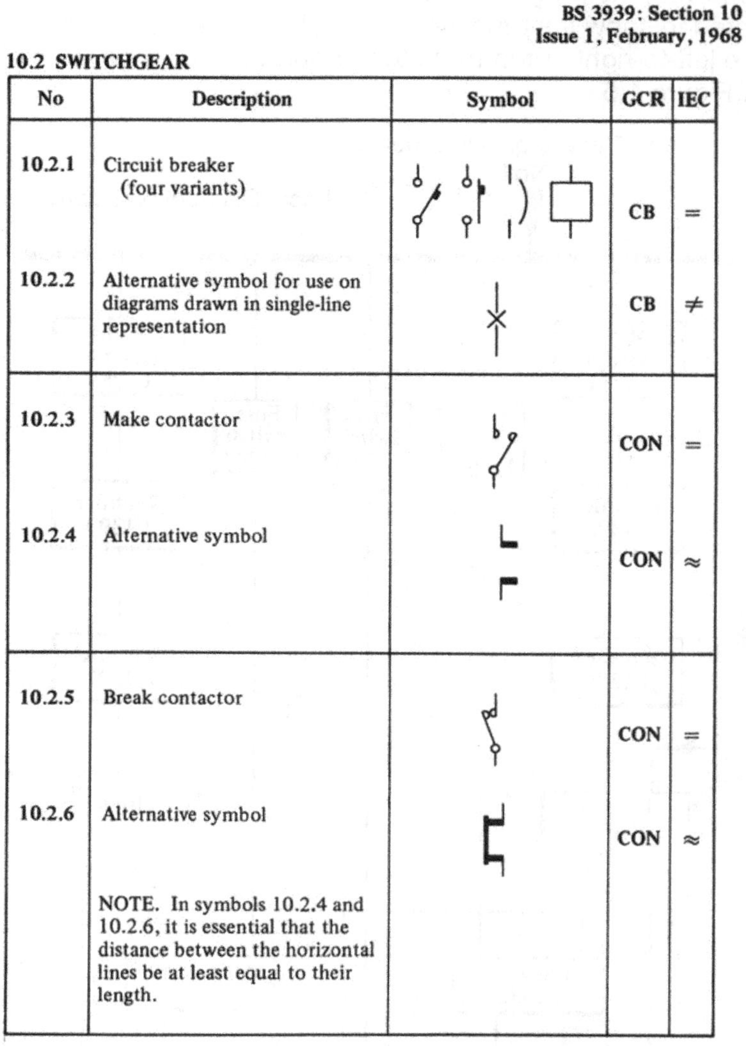

BS 3939: Section 10
Issue 1, February, 1968

10.2 SWITCHGEAR

No	Description	Symbol	GCR	IEC
10.2.1	Circuit breaker (four variants)		CB	=
10.2.2	Alternative symbol for use on diagrams drawn in single-line representation		CB	≠
10.2.3	Make contactor		CON	=
10.2.4	Alternative symbol		CON	≈
10.2.5	Break contactor		CON	=
10.2.6	Alternative symbol		CON	≈
	NOTE. In symbols 10.2.4 and 10.2.6, it is essential that the distance between the horizontal lines be at least equal to their length.			

(By permission of the British Standards Institution)

FIGURE 1.6
SAMPLE PAGE FROM BS 3939

BS 3939, IEC 117 and EN60617 are identical except for a very few symbols. Identicality is indicated by an 'equals' (=) sign in the last column of every page of BS 3939. A few have an 'unequals' (≠) sign, and a few have a 'nearly equals' (≈) sign. All these may be seen in the sample page of Figure 1.6, but throughout most of BS 3939 the 'equals' sign predominates.

Not all the manufacturers and installers of platform and onshore equipment, however, use the British or IEC standard, particularly those following American usage. Diagrams will be found from time to time which use non-standard symbols.

CHAPTER 2 SYSTEM DESIGN

2.1 GENERAL

The purpose of an electrical supply network is to provide a continuous supply of power to meet the requirements of a varying load, and to do it with a high degree of reliability. The detailed design of a network is a complex matter whose features are beyond the scope of this book, but the broad principles are set out below.

To carry out its function as economically as possible, and with a minimum risk of failure, a supply network must have:

- Adequate power, either from the electricity supply authority or from generators, to cope with the highest possible load.

- Provision of surplus power and distributing equipment capacity (i.e. 'redundancy') to take account, for example, of the loss of a running generator or transformer by malfunction and still to supply full load.

- Switchgear, transformers and cables capable of continuous operation in the most arduous conditions to which they could be subjected in normal operation.

- Switchgear, transformers and cables rated to carry the maximum short-circuit fault currents which could possibly occur in their parts of the network. (This is dealt with more fully in Part 10 'Electrical Protection'.)

- A protective system capable of isolating faulty equipment with a minimum of interference to the rest of the network and with a minimum of damage.

- Means of maintaining essential (or basic) services in the event of a total failure of main power.

- Battery-supported supplies, either d.c. or a.c., for those services that cannot tolerate even a short power interruption.

- For offshore installations some means of starting a platform from cold; this is usually known as a 'black start'.

Each item of plant has a continuous maximum rating (CMR), and, although some can tolerate a short period of overload without damage, system design must be based upon the CMR.

So that a system can be designed to carry out its functions as defined above, the following information must be available:

- Total electrical connected load.
- Nature of each item of the load.
- How the installation will be used.
- Diversity of the connected load under various conditions.
- Topography of the installation.

These are dealt with in the following paragraphs.

2.2   LOAD SCHEDULES

Before any system design work can be undertaken, load schedules must be produced. There is no standard method of setting out such schedules, but the following table shows a possible way of doing it. When a complete schedule has been produced the network designer knows the electrical loading of every item of plant and its nature; is it purely resistive (lighting and heating) or is it largely inductive (rotating machinery) or a combination of both?

POSSIBLE LOAD SCHEDULE (Part)					
Ref	Description	KVA	KW	Voltage	Location
P1501	Compressor driver motor	8 600		11 kV	Module XX
P1502	Compressor driver motor	8 600		11 kV	Module XX
P1503	Compressor driver motor	8 600			
SP804	Fire water pump	50		440V	Fire Station
H827	Space heaters – 6		12.0	250V	Fire Station
L280	Lighting – normal		800W	250V	Central Control Room
EL580	Lighting – emergency				

Note that the above table lists the electrical loading of each item of plant, and supplies details of its nature. It also lists the operating voltage, decisions about which must already have been made on technical/economic grounds. The table does not give any idea of how the installation will be used, and local topographical knowledge is needed to interpret the 'Location' heading.

Where an installation may be used under several different operating conditions, a separate load schedule is needed for each condition, or else separate columns must be shown in a composite table. For example, a platform may operate in full production, shut-in, life support only and possibly other states for which the diversity factors greatly differ, and for which the total electrical loadings are also very different.

2.3   DIVERSITY FACTOR

If the total connected (or installed) electrical load of a country were to be switched on at the same time - domestic, industrial and commercial - the total generating capacity of the Central Electricity Generating Board currently in operation would be hopelessly overloaded. All over the country, circuit-breakers would trip to prevent catastrophic damage to equipment, and every activity that uses electricity would stop.

But all the connected load is not switched on at the same time. It never is. There is always diversity in use, and for every installation, from a residential house to the National Grid, there is, at any given moment, a 'diversity factor'. This is defined by BS 4727 as:

'The ratio of the arithmetical sum of the individual maximum demands of a number of consumers, within a specified period, connected to an electrical supply system, to the simultaneous maximum demand of the group within the same period.'

Thus a diversity factor is always greater than unity.

(Note: Although the above is the official definition, sometimes the ratio is used the other way up - namely actual to maximum possible load - in which case the diversity factor is less than unity. Care should be taken not to confuse the two interpretations.)

There are three main causes of diversity of use. One is the way in which an installation is operated: for example, in the table above three large compressor driver motors are listed, but it could be that only two will be used at any time, the third being standby. The second cause is that, even if a motor is running and in use, it will not necessarily be operating at full load and therefore not drawing its full rated power.

The third cause results from the operation of the law of averages - at any given time only a certain number of people are cooking, boiling water, operating lathes, running compressors or driving electric trains. The diversity factor of an installation usually varies in a regular way, according to the time of day, the day of the week, the season of the year, or other cyclic factors, and the larger installation the more reliable load-prediction becomes. It also varies, for a platform, according to the state of operation the platform is in: e.g. full production with injection, normal production, restricted or 'turn-down' production, production shutdown. The first of these gives the lowest diversity factor. The total load for the National Grid, in the absence of disturbing factors, can be predicted with a high degree of confidence. A sudden and unexpected cold spell in winter will upset the regularity of the variation, of course, but, with improved weather forecasting techniques, it is usually possible to bring more generators into use in time to carry the increased load.

On a smaller scale the designer of a distribution network for an installation, either onshore or offshore, should be able to predict the instantaneous diversity factor, with all its cyclic variations, with reasonable accuracy, under the various production states. Naturally, the system must be able to cope with the highest load; switchgear, transformers and cables must all be specified accordingly.

The maximum instantaneous power requirements of the installation can, therefore, be accurately forecast. For an onshore installation this determines the characteristics of the input from the supply authority. For an offshore installation it determines the size and characteristics of the generators.

2.4   LOAD FACTOR

A further factor of interest in system design is the 'load factor', which may be defined as: 'The ratio of the average load over a designated period of time occurring in that period.'
Thus load factor is always less than unity.

Since the consideration of diversity factor already includes the peak load, and the system has been designed to cope with it, load factor is mainly concerned with the prediction of fuel consumption. This has a bearing upon the choice of generator prime-mover characteristics and the bunkering of liquid fuel.

2.5   LOAD FLOW

Sometimes known as 'network analysis', this factor in system design is concerned with the ratings of switchgear, transformers and feeder cables on a topographical basis. In a typical installation, that plant which needs large amounts of power tends to be grouped in particular locations, whereas those services that have comparatively small requirements are usually located elsewhere and fed by different feeders. On a production platform, for example, the accommodation module may be some distance from the pumps, compressors and other plant concerned with production and processing.

The power requirements of the accommodation module, although considerable, are light compared with the large and powerful motors needed to drive pumps. Furthermore the accommodation module is fed at low voltage, while the large motors are fed at high voltage.

Thus, the physical location of equipment within the installation has an effect upon the supply network layout and the ratings of its component parts.

CHAPTER 3 SECURITY OF SUPPLIES

3.1   GENERAL

The nature of the drilling, oil production and processing operations carried out in both offshore and onshore installations demands the highest possible security of supply. Not only are supply interruptions expensive because of production loss; in offshore installations they can also endanger the lives of personnel. Steps are therefore taken, both in system design and in system operation, to raise supply security to the highest possible level. The way in which this is done may vary in detail from one installation to another, but the broad principles are always the same.

They are best described by using an offshore installation as an example, since this incorporates its own power generation. Differences for onshore installations are noted.

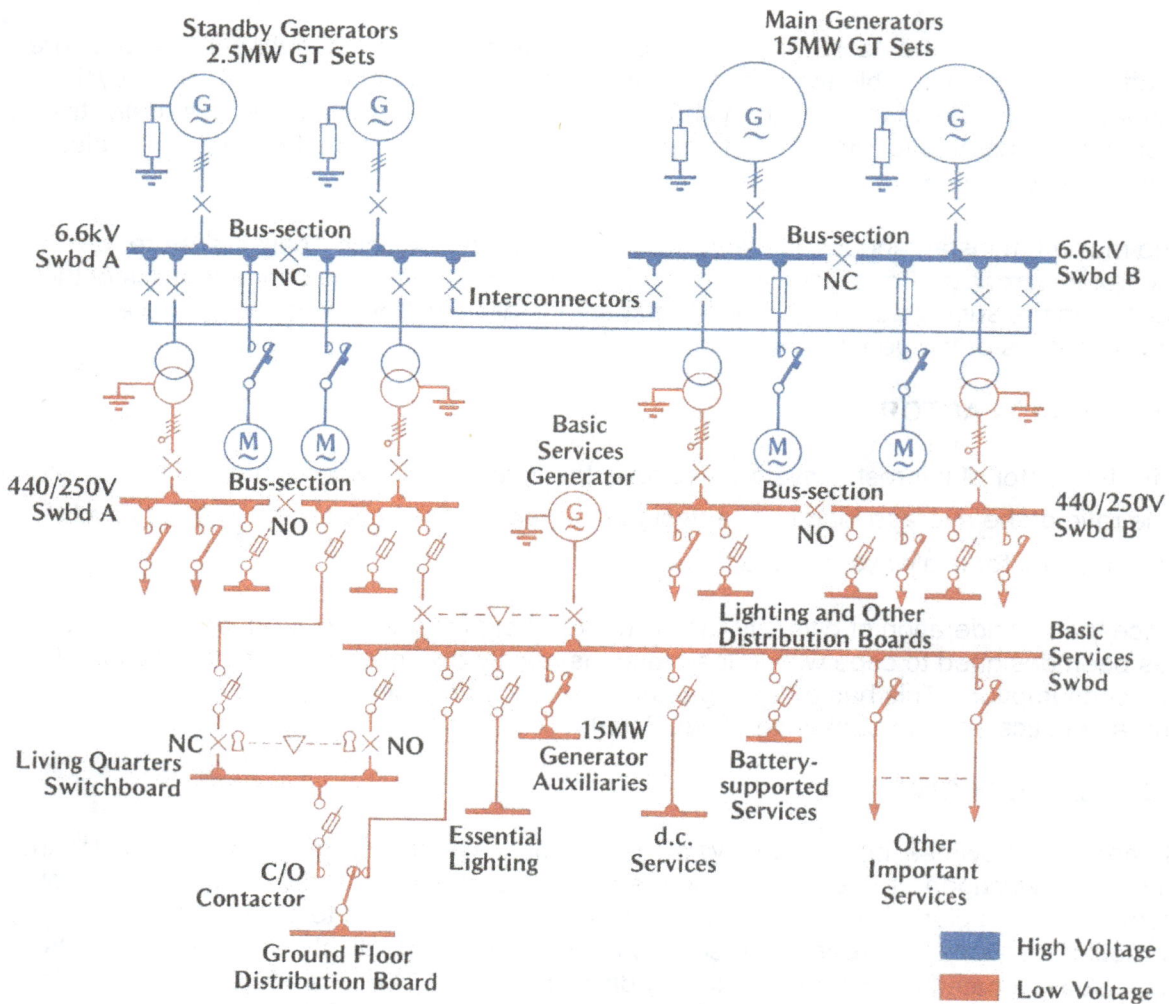

FIGURE 3.1
TYPICAL OFFSHORE NETWORK

Figure 3.1 shows a typical platform power network. It is not an exact reproduction of any particular network but incorporates the more important features that all must have.

3.2 MAIN POWER

Generation is at 6.6kV, 60Hz, by two pairs of gas-turbine-driven generators, one pair being 15MW sets and regarded as the main generators. The other pair are 2.5MW sets and are regarded as the 'sub-main' or standby generators (see para. 3.3).

Each pair of generators feeds its own 6.6kv switchboard through incomer circuit-breakers. Each switchboard is split into two parts by a 'section breaker', and one generator feeds each section. The two switchboards, which may well be in different parts of the platform, are connected to each other by two 'interconnectors', section to section, which have 'interconnector breakers' at each end.

When the platform is in full production, power to the system illustrated in Figure 3.1 is normally supplied to the whole platform from the 15MW sets alone. The 2.5MW sets are only used, apart from platform 'black starting', for 'peak-lopping' if the load on the combined 15MW sets approaches their limit. The 2.5MW sets are then started and paralleled with the 15MW sets to extend their capacity by a maximum of 5MW. When the platform is not in production the load can be supplied by one or both of the smaller sets alone.

When the platform is in production and taking its power from the 15MW sets only, the 6.6kV switchboard associated with the smaller sets takes its power through the interconnectors. It is good practice to use both interconnectors together. It is usually arranged that, if power on the standby 6.6kV board fails (either due to loss of the 15MW sets or of the interconnectors), both 2.5MW sets start automatically, synchronise to each other and connect themselves to their switchboard, at the same time locking open the interconnectors.

To achieve the maximum security of supplies the number of generators running should be such that, if the largest of them should fail, those remaining will not be overloaded, with risk of overcurrent tripping and so of blacking out the whole platform. On some platforms the number of generators installed may not allow this when the platform is in full production, but it is the ideal to be aimed at.

Normal practice is to keep all HV section breakers closed, whether only one or both generators of a pair are running. The corresponding LV section breakers are normally kept open and are only closed when operating with one transformer per board or when transferring load from one transformer to another. The LV section breaker is interlocked with the two LV incomer breakers so that, when all three are closed during load transfer, the section breaker opens automatically after a short delay if one of the incomers has not by that time been opened. Apart from this short period, the two transformers can never be paralleled onto the LV switchboard. This is to reduce the fault level on the LV board to that which would exist with one transformer only (see Part 10 'Electrical Protection'). The short delay should provide enough time to transfer the board's load from one transformer to another without a break, and the additional fault level risk is acceptable for this short period. Synchronising facilities are usually provided across the three LV circuit-breakers to allow them to be closed for this purpose.

The HV and LV breakers on either side of a transformer are also interlocked so that the LV incomer breaker cannot be closed unless the HV feeder breaker has first been closed. There is also an intertrip facility whereby the LV breaker automatically trips in sympathy if the HV breaker opens for any reason, whether manually or by the operation of the protection system.

There is a similar interlock and intertrip arrangement between the breakers at the two ends of an interconnector. This ensures that the heavy loads on the 15MW side cannot be put on the smaller 2.5 MW generators after loss of the 15MW sets.

3.3 STANDBY POWER

The 2.5MW sub-main or standby generators are normally dual-fuel; that is, they can operate on either gas or liquid (diesel) fuel. They can therefore, if circumstances require it, start on a completely dead platform using liquid fuel (provided that their starter batteries are charged), and, once started, can feed the Basic Services switchboard and so provide light, essential services and electric power to start the main 15MW generator sets.

As stated in para. 3.2, they will automatically start on loss of interconnector power on their own switchboard and connect themselves to that board. In this capacity they act as standby to the 15MW main generators.

3.4 MAIN AND STANDBY POWER FOR ONSHORE INSTALLATIONS

In an onshore installation security of main supply is normally ensured by taking two separate supplies from two different sources in the supply authority's network, either capable of supplying on its own the full demand of the installation, and one acting as standby to the other. A typical arrangement is shown in Figure 3.2.

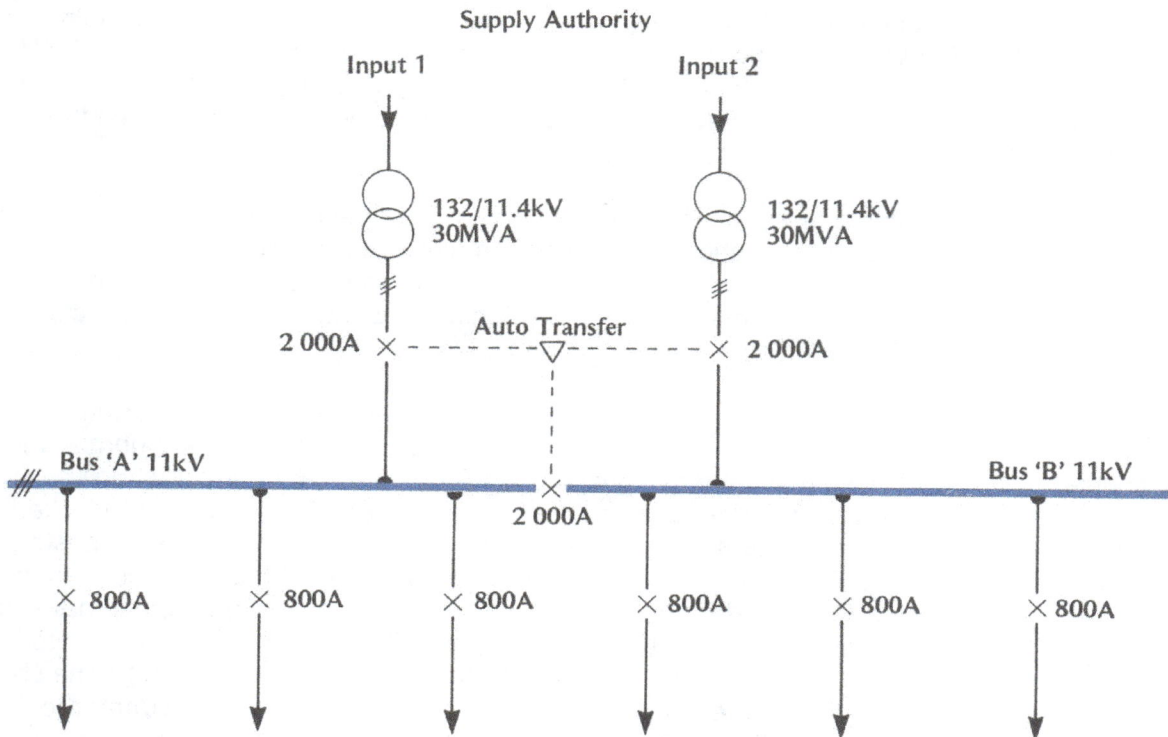

FIGURE 3.2
TYPICAL ONSHORE HV SUPPLY ARRANGEMENT

A 'two-out-of-three' auto-transfer arrangement between the two incomer circuit-breakers and the 11kV bus-section circuit-breaker ensures that the two incoming supplies cannot be run in parallel, and that, if the running supply should fail, there is an automatic transfer to the standby supply.

3.5   BASIC SERVICES POWER

Although the words 'basic services' are used, other terms such as 'essential services' or emergency services' are sometimes used for the same thing.

There are in all installations certain important services, vital for safety, continuity of production and communication, which must continue even if main generation is lost. These are termed 'basic services' and are fed from a separate basic services switchboard (see Figure 3.1). This board is normally supplied from a section of one of the 440V main LV boards and then receives its supply from the main or standby generators. If these should both fail, an alternative supply from a smaller diesel-driven basic services generator is available. In some installations this machine starts automatically on loss of main supply at the basic services switchboard; in others it must be started manually. In either case there is a 'black-out' period between the loss of main and the arrival of the auxiliary supply at the basic services switchboard.

The basic services switchboard feeds, in addition to certain selected vital services direct, such things as the essential lighting distribution boards, supplies to d.c. and battery-supported a.c. services, the Living Quarters switchboard's alternative supply and, most important, the 15MW generator auxiliaries which are further discussed below. The arrangements differ in detail between installations.

Steps must be taken to prevent the basic services generator becoming overloaded by feeding back into the remainder of the 440V main system and thence, through the transformers in reverse, into the 6.6kV system; this would occur if both the normal incomer and the generator breaker of the basic board were closed together. This is prevented by an interlock between the two breakers so that both cannot be closed at the same time.

The Living Quarters have their own 440V switchboard. It is normally supplied from a section of one of the LV 440V switchboards, but there is an alternative supply from the basic services switchboard so that the Living Quarters may continue to receive power from the diesel-generator after main supply has been lost.

There is a similar requirement at the Living Quarters switchboard to prevent the basic services generator feeding back into the 440V system through the two incomer breakers, as was the case with the basic services switchboard. The two incomer breakers are similarly interlocked, but this is usually done by a Castell key.

The basic services switchboard, with its back-up diesel-generator, provides the most secure a.c. supply on the platform. It is, therefore, important that the generator should be tested regularly to ensure its immediate readiness.

In the typical network under consideration the 15MW main generators have their auxiliaries fed from the basic services switchboard, and that switchboard must be energised before the main generators can be started. This arrangement permits the main generators to be started if either one of the standby generators or the basic services generator is running, since the basic services switchboard can be energised from either source.

3.6   BATTERY-SUPPORTED POWER

3.6.1 D.C. Systems

In all installations there are certain important services whose continuation after a complete loss of main power is so vital that only a battery can ensure it. Such services include switchgear operation, navigation lights, foghorns, fire and gas detection, internal communications, alarm systems, some radio communications and certain types of instrumentation which must not only continue but must do so without a break.

In most cases these services are designed to work on direct current. Under normal conditions their d.c. is provided by rectifiers supplied from the a.c. system, but permanently connected to that d.c. system is a battery which has sufficient capacity to carry on alone for a specified time after the rectifier has ceased to provide power.

In a few cases where direct current is unsuitable for the vital service, continuity of a.c. is assured by providing a d.c. system as in the preceding paragraph, but one which feeds a d.c./a.c. inverter to supply the a.c. required. This a.c. is then supported by the battery and so continues as long as the battery lasts. These are 'battery-supported a.c. supplies'.

An installation, either onshore or offshore, may use up to 20 separate d.c. systems, some large and some quite small, each with its own battery-supported power supply unit. (See Part 3 'Electrical Power Distribution'.)

Each d.c. system is usually allocated to a particular purpose which falls into one of the following categories:

- To start up generation.

- To control high- and low-voltage switchgear.

- To operate the fire and gas detection system.

- To start the diesel fire pump.

- To operate process controls, alarms and telecontrol.

- To operate internal and external communications equipment.

- To operate navigational aids.

- To operate emergency lighting in Living Quarters.

In small installations one system may cover several of these categories; in large ones several d.c. systems may be used for a particular purpose.

Those systems which operate several similar units from a single rectifier and battery source are called 'central systems', whereas those which have a single purpose or operate a single unit are referred to as 'dedicated systems'.

Figure 3.3 shows a typical central or 'common' system which is widely used. There is complete duplication: two rectifiers and two batteries, all with duplicated instrumentation and control. Only the d.c. outputs are commoned, with a single d.c. distribution pair of busbars. In some cases each battery alone is capable of providing the full d.c. load for the specified time (referred to as '100% capacity' each); in others both batteries are needed to achieve this (referred to as '50% capacity' each).

The a.c. inputs, which are normally 3-phase, are taken from the 440V system, one of them always from the basic services switchboard, which has the diesel-generator support.

For some applications where the loading is intermittent, such as navigation lights, foghorns and diesel engine starting, a single charger and battery is employed.

In normal use the d.c. load is carried by the rectifier/chargers themselves with the batteries 'floating' on the system. It is only when the a.c. fails that the batteries supply the load, and they then take over without any interruption whatsoever.

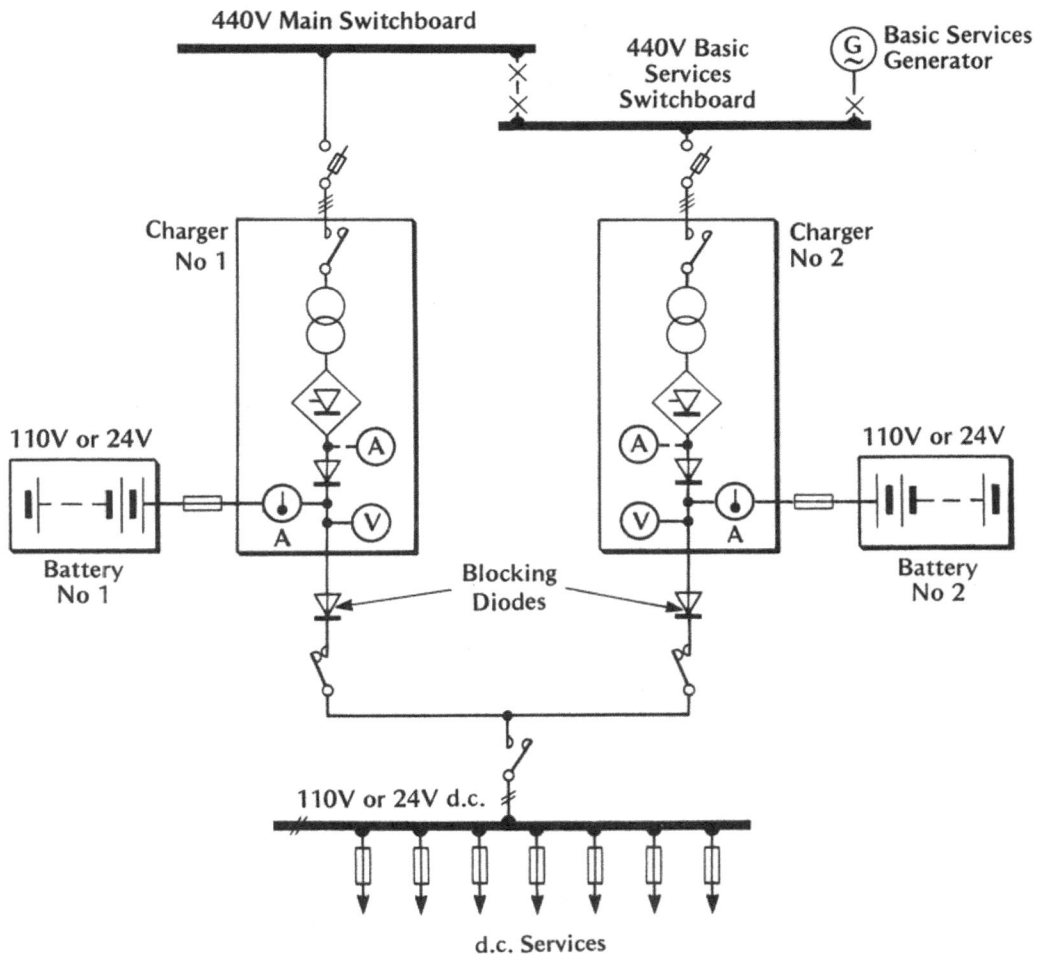

FIGURE 3.3
OFFSHORE D.C. SUPPLY SYSTEM

Direct current supplies for switchgear operation are normally 110V; for other applications 24V is used. Battery capacities range from 30Ah to 1 000Ah. The d.c. systems (except those for communications supplies) are unearthed, and where it is important to detect any earth leakage a circuit, shown in Figure 3.4, is used so that the faulty circuit may be identified without interrupting its supply.

For this purpose a second busbar is provided, in addition to the battery-supported busbar. It is fed from a separate transformer-rectifier. Each outgoing circuit has a selector switch so that, although normally it is fed from the battery busbar, it can, for test purposes, be supplied through the transformer-rectifier. An earth-leakage detection relay is connected to the common battery-supported supply; if leakage occurs from either pole to earth, a remote alarm is initiated and local indicating lamps show whether the positive or negative conductor is down to earth. By switching each outgoing circuit to the rectifier supply in turn until the earth indication disappears, the faulty circuit can be identified while still maintaining a supply to the other loads.

The method of detecting an earth fault or earth leakage in a d.c. system is described fully in Part 10 'Electrical Protection', Chapter 9.

Part 3 'Electrical Power Distribution' describes the various arrangements of battery-charger units and the principles of the charging of batteries; it should be referred to for fuller details.

The above descriptions apply particularly to the 'central' or bulk d.c. supplies to many consumers throughout the installation. In addition there are the separate 'dedicated' d.c. supplies used solely for particular equipments. Among them are the supplies for navigational aids, black start and diesel engine starting, etc. These are not individually described here, but in principle they are similar - that is, they comprise a transformer-rectifier (charger), a floating battery and d.c. output.

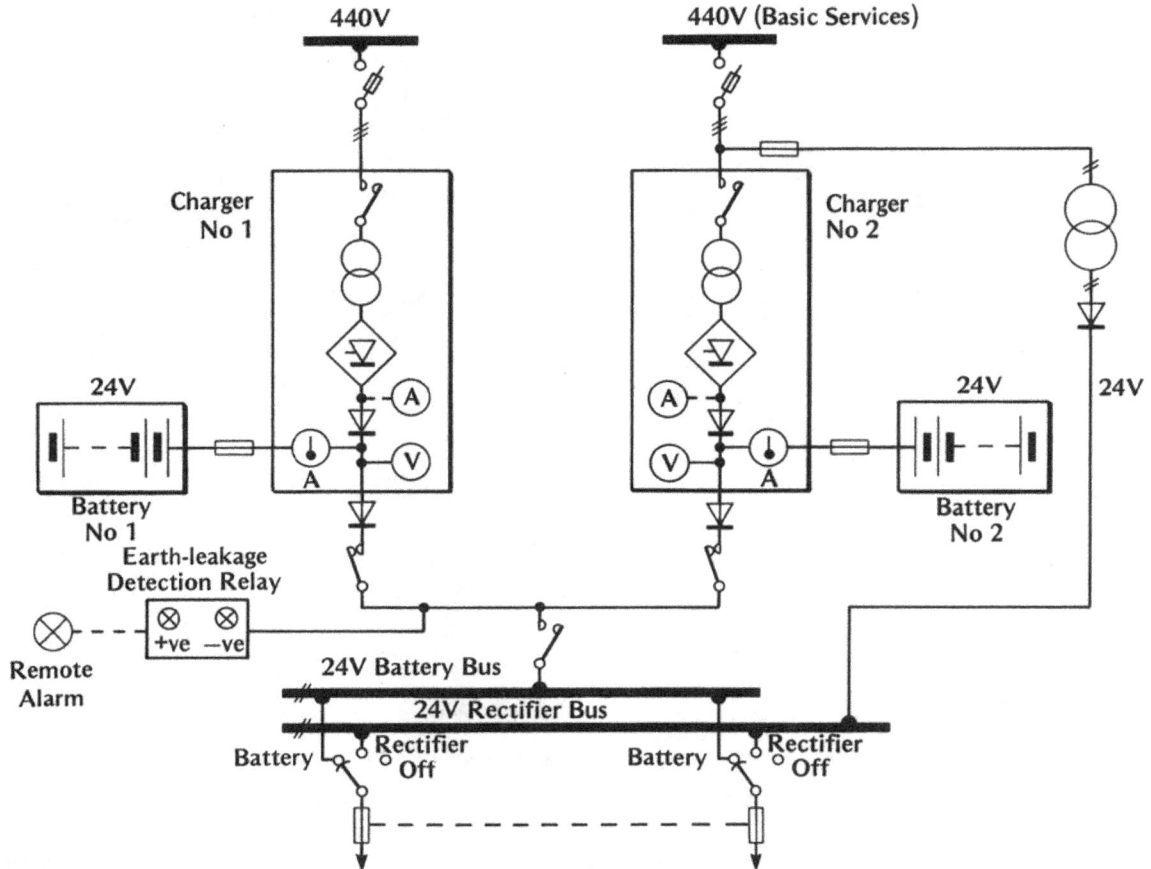

FIGURE 3.4
OFFSHORE D.C. SUPPLY SYSTEM WITH EARTH LEAKAGE DETECTION

3.6.2 A.C. Systems

In some cases it is necessary to have an a.c. supply which has the same unbroken reliability as is provided by a battery d.c. system. This applies particularly to certain vital instrumentation circuits for the process plant.

Such a supply is achieved by first providing a d.c. system exactly like that already described, except that the d.c. is not distributed. It may have duplicate or a single battery and charger. Direct current from the rectifier or battery is passed into a static inverter which converts the d.c. into a.c. at any desired frequency and voltage. The a.c. output may be single- or 3-phase, but single-phase is more usual. Although any frequency is possible, it is normal to use the same as the main system (60Hz offshore or 50Hz onshore). The two systems, main and inverted, are not synchronised unless special steps are taken to do so. The a.c. output from the inverter is distributed via fuses, as was done for the d.c. systems. The arrangement for a single-battery a.c. system is shown in Figure 3.5.

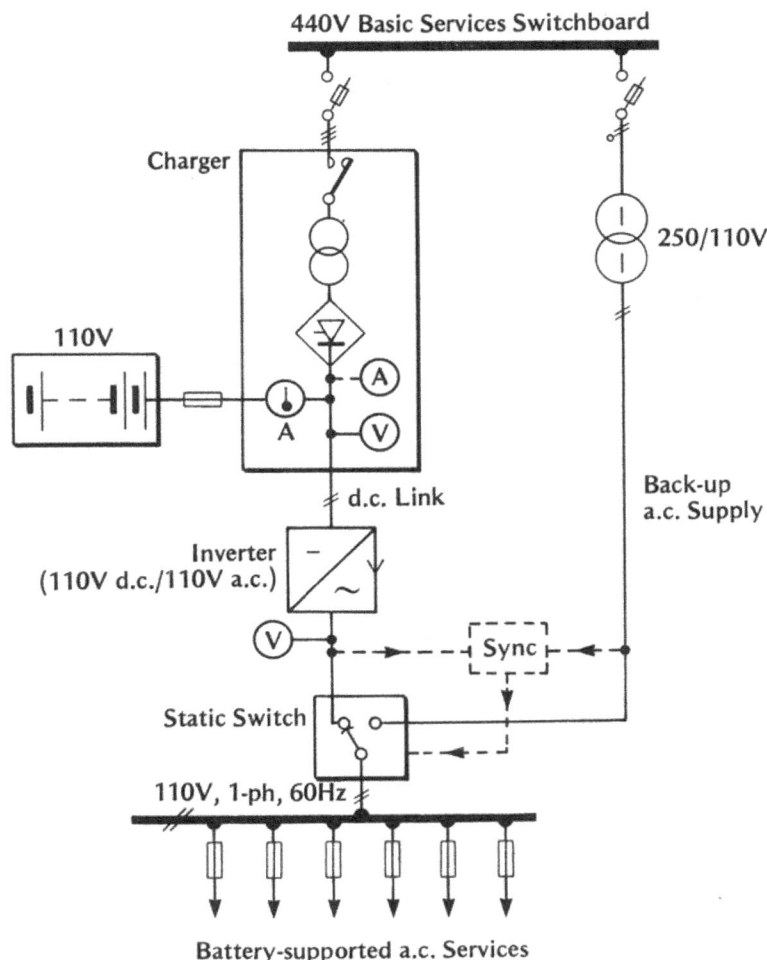

FIGURE 3.5
OFFSHORE BATTERY-SUPPORTED A.C. SUPPLY SYSTEM

The battery floats on the 'd.c. link' between the rectifier and inverter. When either the charger or its a.c. supply fails, the battery takes over the power supply to the inverter without a break, and the a.c. output continues uninterrupted.

It is usual to provide an alternative direct a.c. supply to the battery-supported a.c. output bars; it is taken from one of the more secure 440/250V or 415/240V a.c. switchboards. This is to cover the possibility of a breakdown of the inverter. In such a case the a.c. is not then battery-supported, but in itself it is a form of back-up. This is also shown in Figure 3.5.

Where such a back-up system is provided it is usual to make the changeover automatically through a static (i.e. electronic) switch, and this may include a synchronising circuit to ensure smooth changeover or change-back if both sides are live.

The inverter with its static switches consists wholly of solid-state electronic circuits using thyristor-type rectifying elements. Each of these elements is individually protected by a special quick-acting fuse. The complete assembly, including control switches, indicating instruments and distribution fuses, is housed in a sheet-steel panel. The charger and battery bank, which may be duplicated as described in Part 3 'Electrical Power Distribution', may be housed in a separate panel or suite of panels, or it may be combined into a single suite with the inverter and static switch.

CHAPTER 4 SYSTEM EARTHING

4.1   GENERAL

Offshore electrical power supply systems consist of high-voltage generators, some HV distribution, power step-down transformers and a low-voltage distribution system. Onshore systems have all these except the generators.

If neither the HV nor the LV systems were earthed, conductors could become charged up to any voltage above earth, with risk of breakdown of their insulation to earth. In particular, if there were a failure of insulation between the HV and LV sides of one of the transformers, high voltage could appear on the LV system, whose insulation is not designed to withstand it.

Therefore it is the practice to tie both systems to earth potential, so that neither is free to 'float away', and no part of the system can be at a higher voltage to earth than the nominal voltage of that system. If the earthing point is properly chosen it can be even less.

4.2   UNEARTHED SYSTEMS

Consider a high-voltage generator or transformer with three output terminals R, Y and B and completely unearthed, as shown in Figure 4.1(a). The voltage vector diagram is below, and the three line-to-line voltages V_{RY}, V_{YB} and V_{BR} form a closed triangle. The 'origin' 0, the point of zero potential, does not appear on the diagram because the voltages of the system are not related in any way to earth; they float quite freely. The vector diagram shows only their relationship to each other, not to earth. The above applies whether the generator or transformer is star- or delta-connected.

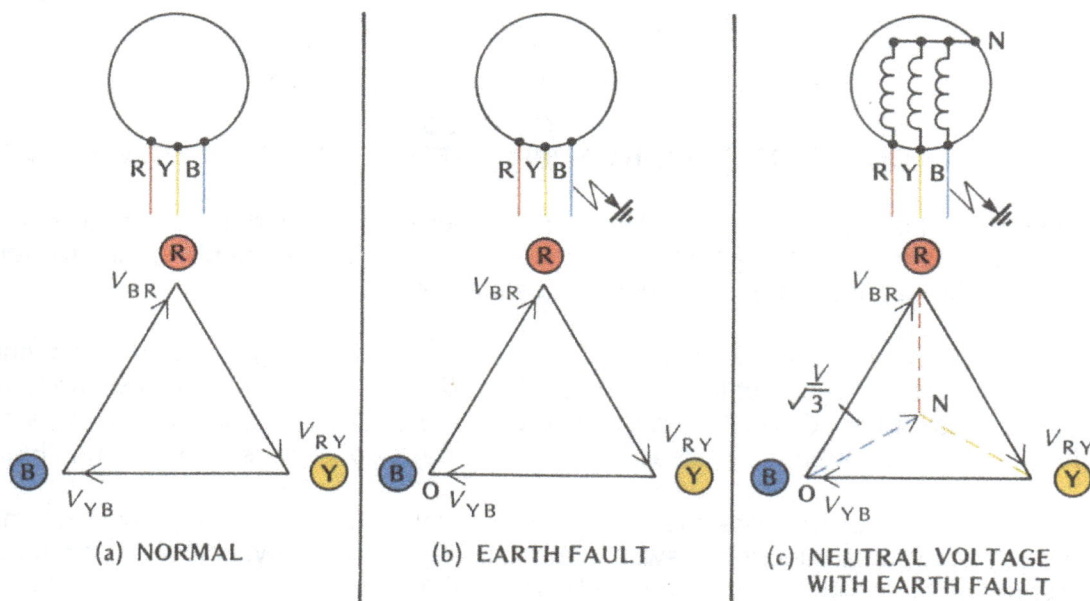

FIGURE 4.1
UNEARTHED SYSTEM

Suppose now a solid earth is applied, say to blue phase, as shown in Figure 4.1(b). Then the point B of the vector diagram becomes the origin 0, which is the point of zero potential (i.e. earth). The shape of the diagram is not altered, and the points R and Y have the same positions relative to B as before. But since B is now at the origin, the actual voltage-to-earth of the red phase is 0R, equal to the line-to-line voltage V_{BR}, and the actual voltage-to-earth of yellow phase is 0Y, equal to the line-to-line voltage V_{YB}.

Hence, in an unearthed system, the accidental earthing of one line will cause both the other lines to take up voltages to earth equal to the line voltage of the system. This applies equally to generator-fed high-voltage and to transformer-fed low-voltage systems.

If the generator or transformer feeding the system is star-connected, the voltage-to-earth '0N' of the neutral point N is then, as seen in Figure 4.1(c), the **phase** voltage (equal to line voltage divided by √3) above earth when one of the lines is accidentally earthed. If the system is 4-wire (three phase and neutral), then the neutral connections of the whole system take up that voltage when a line-earth occurs. This is sometimes referred to as 'neutral shift'.

4.3 EARTHED SYSTEMS

In order to 'anchor' the system voltage to prevent it floating free it is usual (and it is always done at onshore installations and on platforms except the Drilling Packages) to tie one point of the system permanently to earth. For the sake of symmetry the point chosen is the neutral of the supply element, generator or transformer. To provide such a neutral point, the element must be star-connected. Figure 4.2 shows a star-connected earthed system, which may be HV generator-fed or LV transformer-fed.

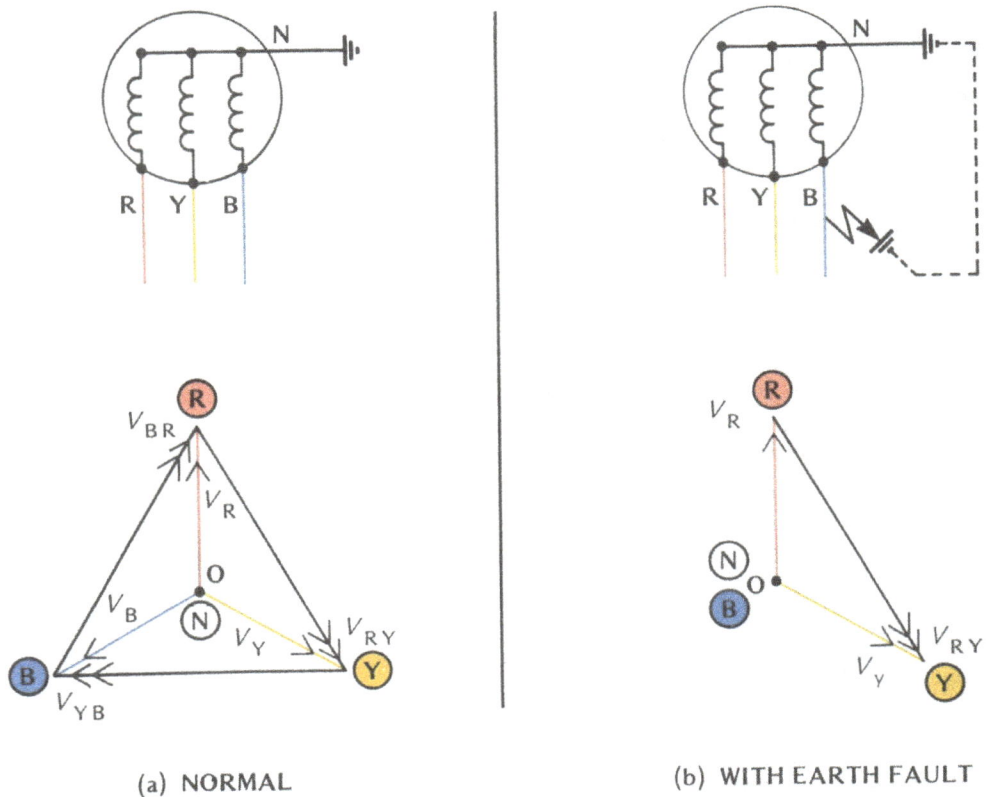

(a) NORMAL

(b) WITH EARTH FAULT

FIGURE 4.2
EARTHED SYSTEM

This arrangement has a further advantage. At the bottom of Figure 4.2(a) is the vector diagram for such a system. The three vectors NR, NY and NB at 120° apart represent the three phase voltages V_R, V_Y and V_B relative to the neutral point N: but as the neutral point N is earthed, it is the same as the origin 0. The three line-to-line voltages V_{RY}, V_{YB} and V_{BR} are represented by the vectors RY, YB and BR.

Since the origin 0 is at the neutral point, the voltage-to-earth of the three terminals R, Y and B can never exceed their **phase** voltage V_R ($= V_Y = V_B$) which is only $1/\sqrt{3}$ of the line-to-line voltage.

Even if one line, say blue, were accidentally earthed, this would still be the case. The situation of Figure 4.2(b) would result. Blue phase would be completely short-circuited, since both ends B and N would be at earth potential; the phase voltage V_B would disappear and B would move to 0. But the other phase voltages V_R and V_Y to earth - that is, the distance of the points R and Y from the origin 0 - would not be affected (unlike with the unearthed system), and they would remain at line-to-line voltage divided by $\sqrt{3}$, just as before the accidental earth.

Thus, whereas in an unearthed system any line can rise to full line voltage-to-earth in the event of an earth fault on a line, in an earthed system the voltage-to-earth of the lines cannot exceed system **phase** voltage - that is, $1/\sqrt{3}$ (or 0.58) times the system line voltage. Therefore an earthed system can use a lower level of insulation and is thus less costly than an unearthed system of the same line voltage. For example, an unearthed 6.6kV system must be insulated throughout for the full 6.6kV to earth, whereas the corresponding earthed system need only be insulated for 3.8kV. This is particularly significant at the higher voltages, especially those used onshore, where insulation becomes increasingly important and costly.

It should be noted that a single line-to-earth fault on an unearthed system will not cause any fault current to flow, since there is no path for it (Figure 4.1(c)). There is nothing to operate an earth-fault relay and so to trip the circuit-breaker. Therefore a **single** earth fault on an unearthed system will not shut down the system. (Note that two simultaneous earth faults become a line-to-line short-circuit, and the ensuing overcurrent will then trip the breaker.)

4.4   EARTH-FAULT CURRENTS

A solidly earthed system is shown in Figure 4.3(a), where an earth fault has appeared on blue phase. It is clear that a short-circuit current will flow between blue phase terminal and the neutral point via the earth link. This short-circuits blue phase and produces a fault current limited only by the impedance of the generator phase winding; such a current could initially be many times the normal designed full-load current of the generator and, if allowed to continue, could permanently damage it by overheating the winding insulation or by mechanical strain. A further hazard is the situation at the fault point itself. The fault is most likely to take the form of an arcing earth, and the fierce short-circuit current could cause intense local heating by the arc at the fault point, with risk to personnel and likelihood of fire.

This might be regarded as a disadvantage of an earthed system. It would not occur on an unearthed machine with a single fault, as there would be no return path for the fault current (see Figure 4.1(c)). A compromise is therefore made, especially in high-voltage systems, whereby the voltage-limiting effect of an earthed system is retained, but the earth-fault currents which result are reduced to a level which is not damaging to the generator, at least for short periods, and which also limits the energy released by the arc at the fault point.

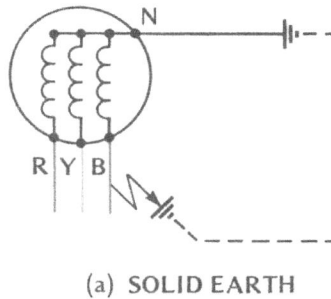

(a) SOLID EARTH

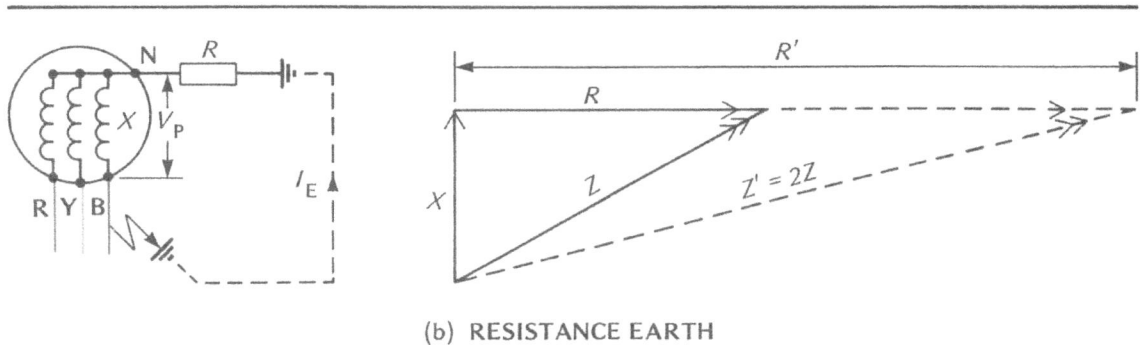

(b) RESISTANCE EARTH

FIGURE 4.3
NEUTRAL EARTHING

4.5   RESISTANCE EARTHING

This is achieved by the method shown in Figure 4.3(b). Instead of earthing the generator neutral point solidly, it is earthed through a heavy-duty, short-time rated resistance of low ohmic value. It can be seen from the figure that, in the event of a line earth, the short-circuit current in the earthed phase is now limited not only by the impedance of the machine's winding, but also by the earthing resistance. Since a generator's impedance is almost wholly reactive (X), it adds vectorially to the earthing resistance (R) to limit the current to produce an impedance (Z) to the fault current, as shown on the right of the figure.

If the value of the resistance is correctly chosen, the earth-fault current can be limited so as not to exceed the normal full-load current of the generator, or indeed it may be chosen to limit it to half, or even a quarter or less, of the full-load current. Clearly it is desirable to limit it as much as possible, but sufficient fault current must be allowed to remain, even with less-serious earth faults, to actuate the protective gear. In the vector diagram of Figure 4.3(b) the reactance vector X is combined with the resistance vector R to give the total impedance Z which limits the current. If the current had to be limited yet further, the resistance would have to be increased as shown dotted in the figure.

With the 15MW generator sets on some platforms, the resistance value chosen was 10 ohms which, with a typical generator reactance value of 0.63 ohm, gives an earth-fault current of about 400A, one-quarter of full load. On other sets the fraction may be a little more. Usually the resistance value is so much greater than the generator's reactance that the latter may be neglected - i.e. $Z = R$

The earthing resistance itself must carry the limited fault current until the generator is tripped, so it must be heavy duty. It is usually arranged to be short-rated for 15 or 30 seconds only. It is customarily used with an isolating link or switch; this must always be opened when megger testing the generator windings for insulation resistance.

Example

An 18MVA, 6.6kV star-connected generator is provided with a neutral earthing resistor. What must be the value of this resistor if it is to limit the earth-fault current of one phase to one-half of the full-load current? (The reactance of the generator winding may be neglected.)

First calculate the full-load current I_F

$$I_F = \frac{kVA}{\sqrt{3}kV} = \frac{18000}{\sqrt{3}\times 6.6} = 1575A$$

Fault current (I_E) is to be limited to half this:

$$I_E = \frac{1575}{2} = 788A \qquad\qquad ....(i)$$

In Figure 4.3(b) the fault current I_E is given from the fault circuit whose emf is the **phase** voltage V_P, where $V_P = \frac{V_L}{\sqrt{3}} = 3.81kV$. Ignoring the generator's phase reactance X, the fault current is given by Ohm's Law:

$$I_E = \frac{V_P}{R}$$

or

$$R = \frac{V_P}{I_E} = \frac{3.81\times 10^3}{788} \quad \text{from (i) above}$$

$$= 4.84 \text{ ohms}$$

4.6 MULTIPLE GENERATOR EARTHING

Ideally a system consisting of several generators in parallel should only be earthed at one point, in order to prevent harmonic currents circulating between the generators through their neutral points and so increasing their loading. This means that, if more than one set is running, only one link should be closed. In practice, however, most generators are now designed to accept such currents, and links are left closed in all machines (see Figure 4.4). There are exceptions; sometimes the links are monitored by a logic system which gives an alarm at the control board if the correct ones have not been closed or opened as necessary.

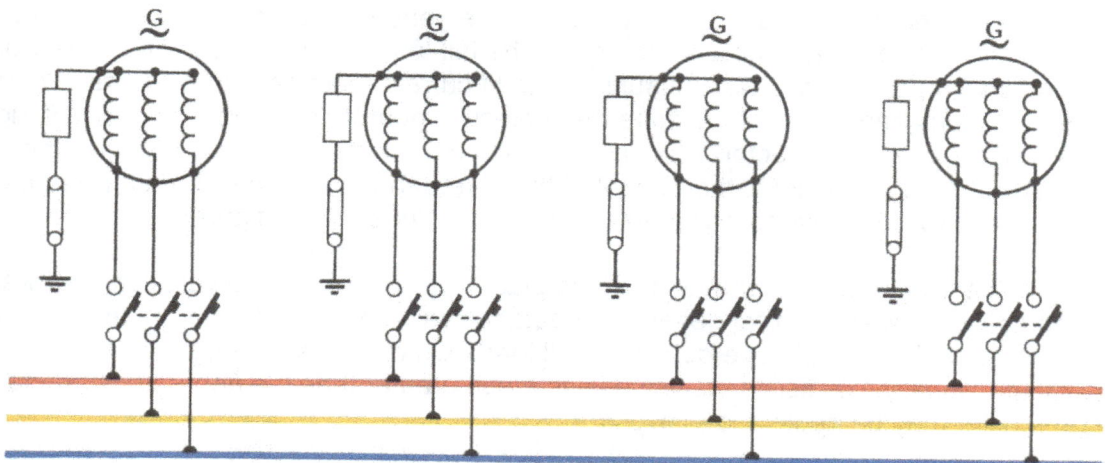

FIGURE 4.4
MULTIPLE EARTHING OF GENERATORS

4.7   TRANSFORMER SECONDARY EARTHING

Whereas resistance-earthing of HV generators is common throughout all platforms, the secondary neutral points of all transformers are solidly earthed, either at the transformer itself or more usually through the neutral busbar of the LV switchboard which it supplies (see Figure 4.5). There is then no resistor to limit the earth-fault current.

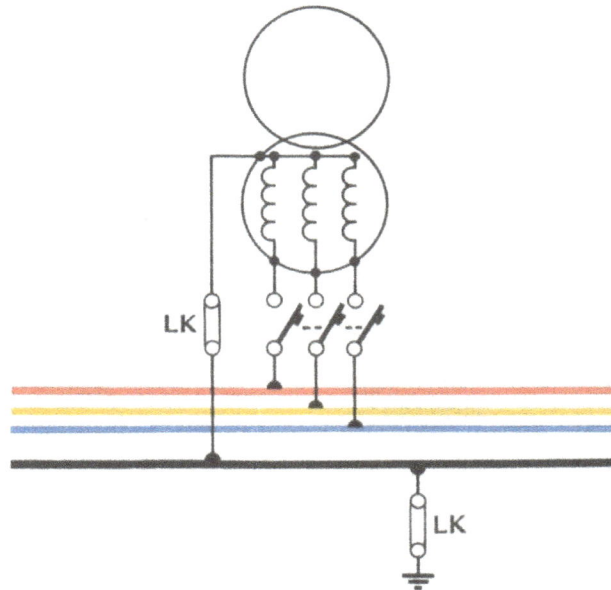

FIGURE 4.5
TRANSFORMER EARTHING

The reason for this is two-fold. First, the secondary (high-current) windings of power transformers are, in any case, robust and can withstand better than an HV generator the full earth-fault currents allowed by a solid earth. Secondly, the amount of energy available to be released at the fault by an arcing earth is much less downstream of a transformer than on the HV side, where it comes straight from the terminals of the generators. This factor is therefore of less importance.

4.8   DRILLING PACKAGES

The power systems of platform drilling packages are usually separate from those of the platform itself; they consist of diesel-driven 600V generators which are delta-connected. As such they cannot be earthed, and drilling systems are consequently run unearthed. At 600V system voltage, any voltage rise on the healthy lines due to an earth fault on one line is not large enough to be significant.

Drillers prefer the use of unearthed systems since, as explained in para. 4.3, a single earth fault will not cause the supply to trip. Loss of power while a drill is in the hole can be serious, leading possibly to the total loss of the equipment.

CHAPTER 5 LIGHTING

5.1 GENERAL

With the advent of modern high-efficiency fluorescent lighting, the requirements for specific standards of illumination in working and living spaces have increased over recent years. Typical standards for lighting levels, some applicable to offshore installations only, are as follows:

Living quarters	100 lux
Control rooms	500 lux
Electrical rooms	500 lux
Radio room	500 lux
Wellhead module	75-150 lux
Working areas	35-75 lux
Walkways	35-75 lux
Stairways and passages	35-75 lux
Other areas	25-50 lux

(**Note:** 'Lux' is the SI unit for illumination per unit area, being the name for 'lumens per square metre'. It is equivalent to 0.093 foot-candles in the old Imperial notation. A normal 4 ft twin-tube fluorescent fitting gives an illuminance of about 250 lux at a point on the ground three metres below it.)

The number, type and positioning of the light fittings in an area are determined before installation, taking into account the light-reflecting properties of surrounding walls and ceilings. After installation the illumination is checked by light-meter. All offshore lighting is supplied at 250V, 60 Hz, single-phase, except for the Living Quarters' emergency system (see para. 5.6); onshore it is at 240V, 50 Hz.

5.2 LAMPS AND LIGHT FITTINGS

The original electric lamp consisted of a carbon filament sealed under vacuum in a glass envelope. Later, filaments of metal - usually tungsten - were used, and inert gas at about atmospheric pressure was introduced into the vacuum to prolong the filament life - the so-called 'gas-filled' lamps.

Filament (or 'incandescent') lamps are still used on platforms and large shore installations in certain special applications, but by far the greatest number of light fittings are now of the 'fluorescent' type using discharge tubes.

The correct name for this type of lamp is 'discharge lamp'. The tube is filled with gas or vapour, normally at a very low pressure, and has electrodes at each end between which a bright glow discharge takes place. The discharge is really an arc, but at these low pressures the arc degenerates into a uniform glow along the tube. The colour of the discharge depends on the gas used: mercury vapour gives a cold blue-white light, sodium vapour a brilliant yellow (as used for street lighting). Different gases are used to give colour to advertising displays.

Discharge tubes operate on most gases, given the right conditions, but pure gases do not give sufficient light output to be useful for general lighting purposes. When lamps are required to give good space illumination, tubes operating on metal vapour are far more satisfactory and are in general use.

In such tubes a minute amount of metal is introduced, together with some inert gas such as argon. The arc first strikes using the argon as medium, and the resulting temperature causes the metal to heat and vaporise. The metal vapour then becomes the arc carrier and gives out much more luminous energy.

The metals used must have a low boiling point, and either mercury or sodium are almost always used. The luminous energy of mercury is mainly in the ultra-violet band, with some of it in the visible spectrum, whereas sodium emits light of almost 'monochromatic' wavelength entirely in the yellow region.

For general lighting purposes neither the yellow of the sodium nor the cold blue-white of the mercury lamp is usually acceptable. Mercury-vapour tubes for this purpose are coated internally with a substance which fluoresces under the excitation of the ultra-violet light from the arc and gives a more acceptable colour. The result is the 'warm white' lamp widely used for domestic and some space lighting. It is this general use of coating material that has caused discharge lamps to be referred to loosely as 'fluorescent' lamps, whereas these are in fact only a special case.

Both the mercury-vapour and the sodium-vapour types may be 'low pressure', operating at about atmospheric pressure, or they may be of the 'high-pressure' type. HP lamps tend to have a different colour spectrum from their LP type, giving a noticeably different colour output. For example, whereas the LP sodium-vapour lamp gives the well-known yellow light, the HP type gives a much whiter light approaching that of a tungsten filament lamp. Most domestic and industrial discharge lamps are of the fluorescent mercury-vapour low-pressure type.

Discharge lamps, as used on platforms and industrial shore installations, are assembled into fittings called 'luminaires' which contain the discharge tubes and their starting and control gear. On platforms and in other hazardous areas the luminaires are usually constructed to be explosion-proof (see Part 9 'Electrical Safety').

5.3 CONTROL OF DISCHARGE LAMPS

5.3.1 General

Discharge lamps can also be classified as 'Cold Cathode' and 'Hot Cathode'. In the former the two electrodes enable the discharge to start without any prior external heating, whereas in the latter the electrodes are independently heated by the external circuit in order that the tube may strike. In both cases, once the tube is running, the electrodes are kept hot by the discharge itself.

Cold cathode tubes have simpler controls and in general have a longer life than the hot cathode type. They are consequently favoured for use where access is difficult - such as with space lighting fittings in large process areas of platforms. On the other hand hot cathode tubes are generally preferred for domestic use. For completeness, both types and their controls are described below.

Discharge lamps are in general much more efficient than incandescent, giving more lumens per watt of electrical power consumed. Of the various kinds of discharge lamp the sodium-vapour type is the most efficient; this is why it is much favoured by Councils responsible for extensive street lighting.

Typical lighting efficiencies, expressed in lumens per watt of electric power consumed, are:

Incandescent:	100W tungsten (coiled-coil)	12 lm/W
Discharge:	250W mercury vapour (low pressure)	45 lm/W
	250W sodium vapour (low pressure)	100 lm/W
	250W sodium vapour (high pressure)	90 lm/W

5.3.2 Cold Cathode Tubes

The single-phase mains supply is stepped up to about 600V in a transformer. On switching on, this voltage is sufficient to strike the tube, even with the electrodes cold, and no special starting arrangements are needed as with the hot cathode type. Cold cathode tubes, therefore, have a quick-start facility. They are particularly suited to long-tube installations, but the longer the tube, the higher the initial striking voltage needed. The arrangement is typically as in Figure 5.1.

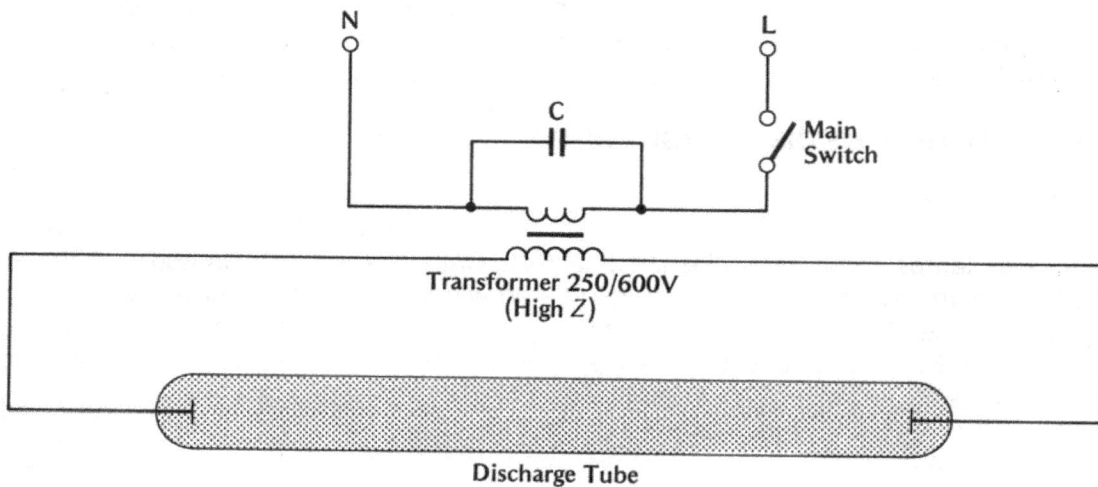

FIGURE 5.1
COLD CATHODE DISCHARGE TUBE

The transformer is a special high-impedance type, and, once the tube has struck and the running current has established itself, the internal drop in the transformer impedance causes the tube voltage to fall to the level necessary to maintain it (as distinct from that required to strike it). This automatic dropping of voltage is known as 'ballasting'; some systems use alternative methods of achieving it.

The high-impedance transformer would cause the power factor of the circuit to be low, so a capacitor 'C' is inserted across the mains input to correct it and bring it near to unity. It can never quite achieve unity, as the harmonics in the tube current will prevent it exceeding about 0.9.

These same harmonics, amounting to some 20%, also tend to cause a poor current waveform where discharge-tube lighting is used. This can cause considerable neutral (3rd harmonic) currents where a lighting system is supplied from a 3-phase network. Discharge tubes may also give rise to some degree of radio interference.

The sodium-vapour lamp, widely used for street lighting and recognised by its yellow light, is usually of the cold cathode type.

5.3.3 Hot Cathode Tubes

In this type of tube the two electrodes have to be pre-heated before the arc can strike. In addition, in the most common types, a very high voltage is applied to the heated electrodes to cause the arc to strike; but once running, only a small voltage is needed to maintain it, and the discharge itself keeps the electrodes heated. These requirements are achieved as shown in the typical arrangement of Figure 5.2.

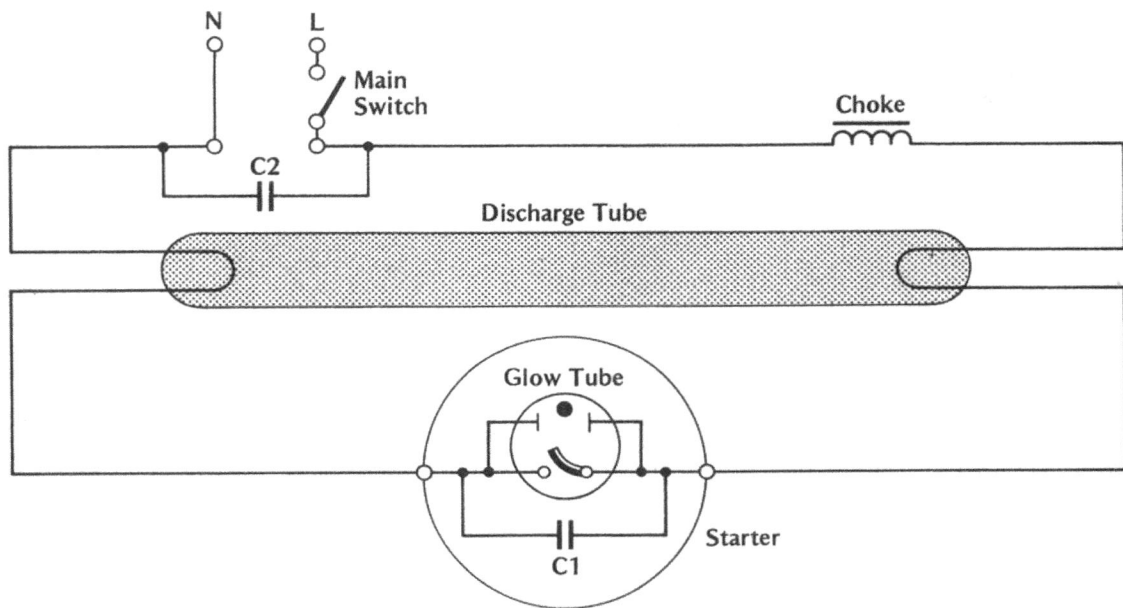

FIGURE 5.2
HOT CATHODE DISCHARGE TUBE - STARTING

The 'glow starter' consists of a separate small tube filled with an inert gas such as argon. Inside is a bimetal contact which closes when heated by the glowing gas. The starter assembly is a separately replaceable unit.

When the main switch is closed a small current flows through the choke, one electrode filament, the glow tube gas and back through the other electrode. The voltage appearing between the two electrodes, which is approximately mains voltage, is at this stage usually not sufficient to strike a discharge arc.

As the glow tube heats, its bimetal contacts close and allow the full current to flow round the loop, so pre-heating both electrodes.

These contacts also short-circuit the glow tube, causing it to cool and the bimetal contacts to reopen. This breaks the loop circuit and causes a sudden collapse of the magnetic field in the choke, resulting in an induced voltage surge of up to 1 000 volts across the tube; it is usually enough to cause the tube to strike and start its discharge.

Once the arc has started, a steady current flows through the tube and choke, which acts as a ballast and causes a steady voltage drop. This ensures that the voltage applied to the running tube is no more than that needed to maintain it (as distinct from that required to strike it). The lower voltage, still applied also to the glow tube, is not enough to cause it to glow and so the starting contacts will not reclose.

Two capacitors are usually provided: C1 is part of the starter unit and is for suppression of radio interference due to switching. C2 is across the mains for power-factor correction. The circuit through the choke and tube is very inductive and would alone have a low power factor. Capacitor C2 corrects it and produces a net power factor approaching unity, apart from the effect of harmonics.

Other starters are made which, instead of a glow tube, use a thermally operated switch, but the principle of operation is similar.

5.4   SPACE LIGHTING FITTINGS

The great majority of lighting fittings for space lighting throughout all installations are of the 1 200mm (4 ft) twin-tube 2 x 40W cold cathode fluorescent type contained in a translucent-covered luminaire, as shown in Figure 5.3.

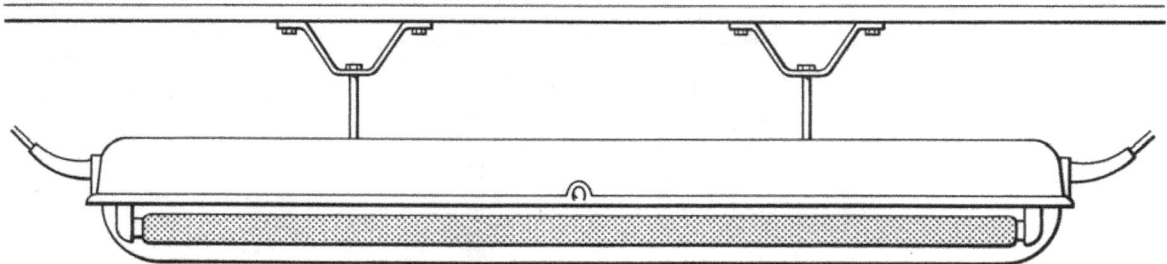

FIGURE 5.3
SPACE LIGHTING LUMINAIRE

Other fittings have single tubes or are of different lengths for special purposes. Except in cabins, offices and similar enclosed spaces the fittings are not individually switched but are supplied in groups from lighting distribution boards where they can be switched by miniature circuit-breakers. The fittings are not individually earthed but are earthed in groups at the distribution board by a separate conductor in the group supply cable.

All space lighting fittings are suitable for use indoors and outside. Except for those in the Quarters areas of offshore installations, they are classed as 'explosion-proof'; they are of the 'increased safety Ex-e' class which can be used in both Zone 2 and Zone 1 hazardous areas. (See Part 9 'Electrical Safety'.)

In the Living Quarters the fittings used are not suitable for hazardous areas. Instead other types of luminaires are used, in particular 1 500mm (5 ft) single- and twin-tube types with 65W tubes for surface mounting; some may have dimmer controls. There is also a small 200mm (8in) 6W single-tube type for mirror lights in cabins.

A limited number of incandescent tungsten filament lamps are used for special purposes, such as a 60W wall-mounted bedhead fitting, a totally enclosed 60W bulkhead-fitted unit for operation at high temperatures in a grease-laden atmosphere (e.g. a kitchen hood) and some 100W ceiling- and wall-mounted fittings.

5.5   ILLUMINATION OF ROTATING MACHINERY

All discharge tubes strike and extinguish twice every cycle, so that the light from them, though it appears to the eye as continuous, consists in fact of 100 or 120 'flashes' per second. Where it illuminates rotating machinery, the moving parts may appear almost stationary - indeed they may seem to be moving very slowly backwards. This is a 'stroboscopic' effect.

Whereas most moving parts of machines are well guarded, there are parts, such as the chuck of a lathe or drill or a grinding wheel, which are accessible to the operator. When they **appear** to be stationary, he might be misled into touching them and endangering himself.

There are no statutory regulations covering the lighting of such machines, other than the illumination levels set down in para. 5.1. If the stroboscopic effect is not acceptable, there are several methods of dealing with it. Amongst them are:

(a)   Use of incandescent lamps in the danger area.

(b)   Use of long-persistence fluorescent tubes in the danger area.

(c)   Use of 3-phase lighting.

(d)   Use of two tubes, one ballasted inductively and the other ballasted capacitively.

Methods (a) and (b) cause the illumination to persist through the current-zero period, whereas with (c), whenever one tube is momentarily extinguished, the other two are providing light. Method (d) produces a similar effect from a single-phase supply by causing the two tubes to flash approximately anti-phase with each other.

5.6   EMERGENCY LIGHTING FITTINGS

For emergency (or 'secure') lighting in all spaces other than the Living Quarters of some platforms, on walkways and on outside staircases, a luminaire similar to that used for normal lighting is fitted. It too is a 1 200mm twin-tube cold cathode fluorescent type similar in appearance to the normal but distinguished from it by having a long tube above the enclosure - see Figure 5.4.

This tube contains a cylindrical 6V rechargeable battery. The fitting also contains an under-voltage relay, a transformer-rectifier and an inverter with its step-up transformer. Under normal conditions both tubes are energised from the 250V a.c. system, and the battery is maintained trickle-charged through the rectifier. The circuit is shown in Figure 5.4.

If main supply is lost, the relay drops off and disconnects one tube. It also transfers the other tube to the inverted supply which provides it with a coarse 250V derived through the step-up transformer from the battery, which is now the sole source of power. Under these conditions the battery has a charge-life of approximately 90 minutes, which should provide ample time for escape. When main or basic supply a.c. returns, the fitting reverts to normal operation, both tubes light and the battery recharges through the rectifier. The charge rate is low, and the battery could take up to 24 hours to recharge from a deep discharge.

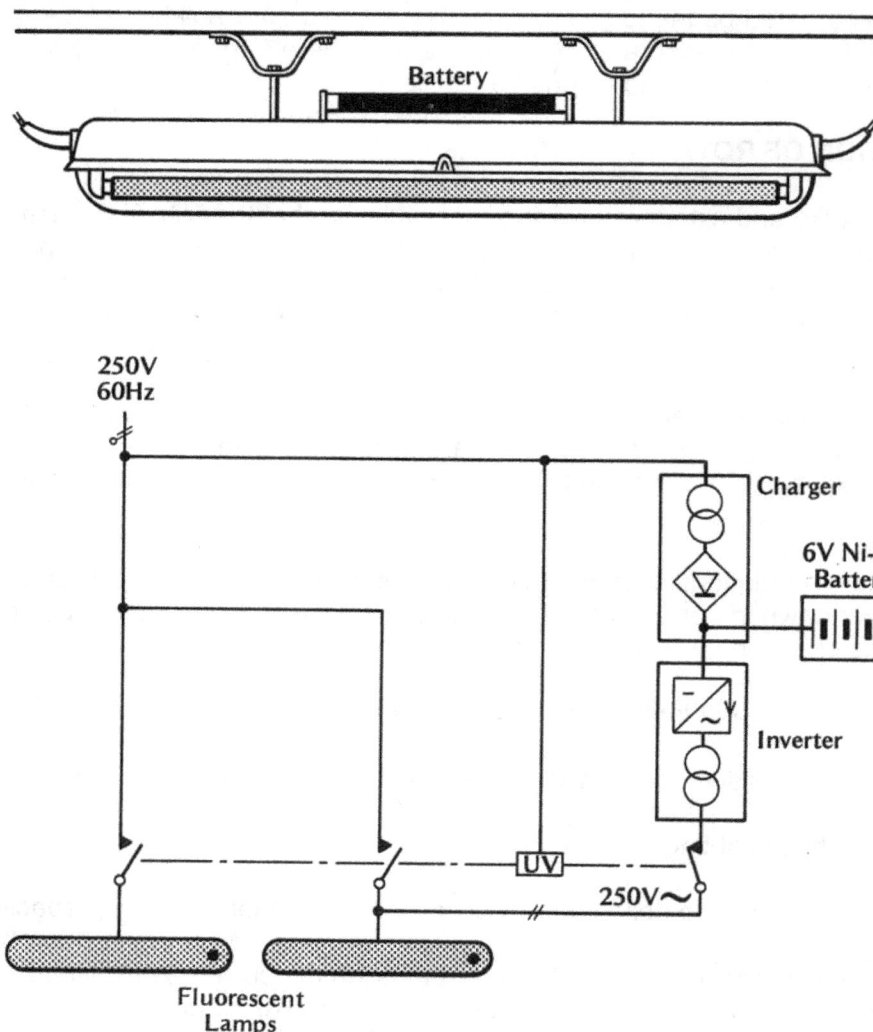

FIGURE 5.4
EMERGENCY SPACE LIGHTING LUMINAIRE

The fitting can be adjusted so that one tube is not disconnected on loss of main power. However, if two tubes are left burning, the battery life is reduced to approximately 45 minutes. Figure 5.4 shows the one-tube emergency arrangement, which is the normal manner of operation.

In large spaces these emergency fittings are installed in a ratio of about one in ten normal fittings. On platform walkways, staircases and escape routes the ratio is much greater - perhaps one in three or even every other one. They receive their power separately from a source different from that for the normal ones - usually from an essential lighting distribution panel. Since this is supplied from the basic services switchboard which has the support of the basic services generator, the time during which the fittings are dependent on their batteries alone should be reasonably short.

To check whether these special fittings are in correct working order, break the circuit to each area in turn at the appropriate essential lighting panel and observe that one tube of the special fittings remains burning.

5.7   EMERGENCY LIGHTING FITTINGS - LIVING QUARTERS

Inside certain Living Quarters the emergency lighting takes a different form.

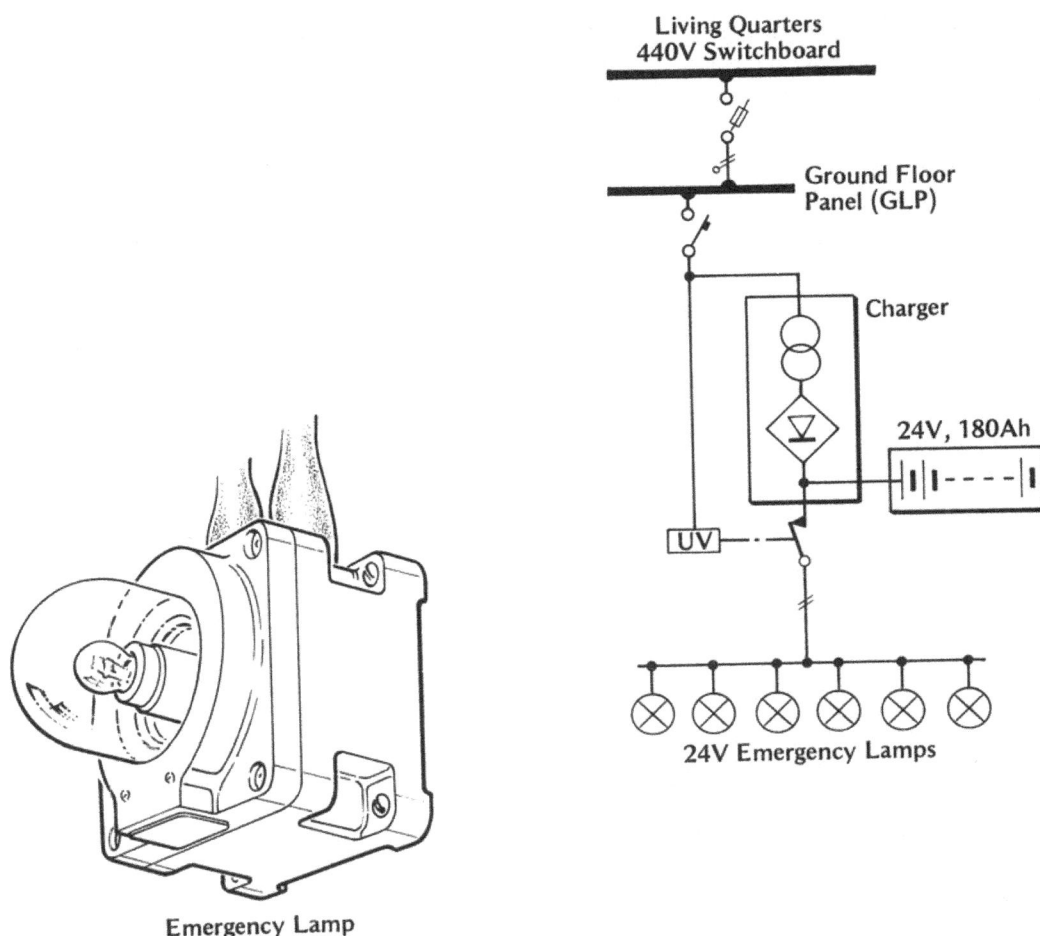

FIGURE 5.5
LIVING QUARTERS D.C. EMERGENCY LIGHTING

All communal areas, passages, etc. are illuminated by wall-mounted, totally enclosed, explosion-proof 24V, 5W d.c. fittings. They are all powered from a common dedicated battery-supported d.c. supply, the battery and charger being housed in the Quarters switch-room. This type of fitting and the power circuit are shown in Figure 5.5.

A.C. power is taken from the Quarters main switchboard through the ground floor lighting panel (GLP) which is in the same switchroom. Alongside it is the 24V, 180Ah emergency lighting battery and charger. There is also an undervoltage relay. The battery, which 'floats' on the system, is maintained trickle-charged through the charger.

Under normal conditions the relay is energised and holds the d.c. supply off the emergency lamps. On failure of the main supply at the ground floor panel, the relay drops off and connects all the emergency lamps to the d.c. system, now supplied from the battery alone. The battery capacity is sufficient for about three hours' lighting. On restoration of the a.c. supply, either from the basic services generator or from the normal mains, the relay re-energises and disconnects all lamps. The battery will also start recharging.

5.8   PORTABLE EMERGENCY HANDLAMPS

In all working areas and rooms portable handlamps are provided, usually near the doors. Each consists of two parts: a fixed part or 'stand' on which the handlamp is stowed, and a removable part which is the handlamp proper. It is shown in Figure 5.6. The stand unit contains a charging transformer and rectifier, and also a time-switch. The handlamp unit contains, as well as the lamp itself, a 2-cell nickel-cadmium battery, an undervoltage relay and a switch on top. The lamp is rated 2.5V, 2W.

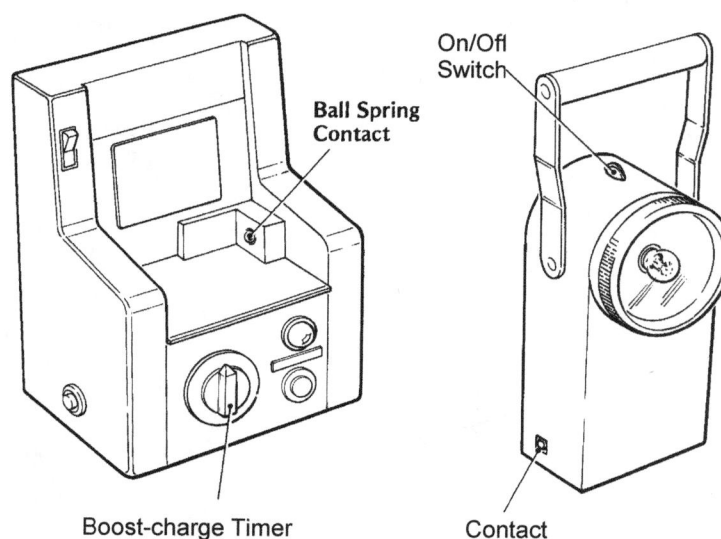

FIGURE 5.6
PORTABLE EMERGENCY HANDLAMP

When the handlamp is 'home' on its stand, its battery is kept trickle-charged through two ball-and-spring contacts. The undervoltage relay is also energised and holds the lamp disconnected, even though the lamp switch is on. When the hand lamp is lifted off, the charge circuit is broken, the relay de-energises and connects the lamp to the battery. When the handlamp is replaced, the lamp goes out and the battery recharges. The lamp will also light while still on its stand if the a.c. mains supply fails, provided that its switch is in the 'on' position.

The battery charge-life will allow adequate light for 2½ hours. After use, when the hand-lamp is replaced on its stand, the battery must be recharged at full rate. This is done by setting the time-switch, and the full-rate charge time depends on this setting. From a deep discharge the maximum time of seven hours will be needed at the full rate. From partial discharge the time-switch must be set correspondingly lower. After completion of the full-rate charge the time-switch automatically causes a return to trickle-charge.

The handlamp stowages are not explosion-proof. In hazardous areas they are sited outside the access to Zone 1 areas and are usually placed near the entrances. The portable hand-lamp itself is explosion-proof and is class 'Ex-s' (see Part 9 'Electrical Safety'). It may be removed at any time and carried to wherever it is temporarily needed. It should be switched off at the lamp switch at all times while it is not actually in use. After use as a portable lantern it must be replaced again without delay to recharge the battery, and the switch must be left on so that the lamp will light automatically in its stowage if a.c. power fails.

To check whether the unit is in working order, simply remove the handlamp from its stand, switch on (if not already switched) and check that it lights at full brilliance. Regular checks should be made on all lamps at not longer than weekly intervals.

5.9   LIGHTING POWER SUPPLIES

The power for normal lighting and floodlights on platforms is derived from a number of lighting distribution panels (LPs), which are supplied from different sections of the principal 440V switchboards. The distribution panels take in power at 440V, 3-phase, 4-wire, through an incomer moulded-case circuit-breaker and distribute it at 250V single-phase through banks of twelve miniature circuit-breakers which provide 2-pole switching. There may be up to three such banks. Alongside each miniature switch is a tally-strip indicating its service. The loads are divided as equally as possible between the phases. Figure 5.7 shows a typical platform lighting distribution panel with three banks of twelve switches; it also shows the sources of power.

All important areas have lighting fittings supplied from at least two boards so that, if one fails, the area is still partially lit.

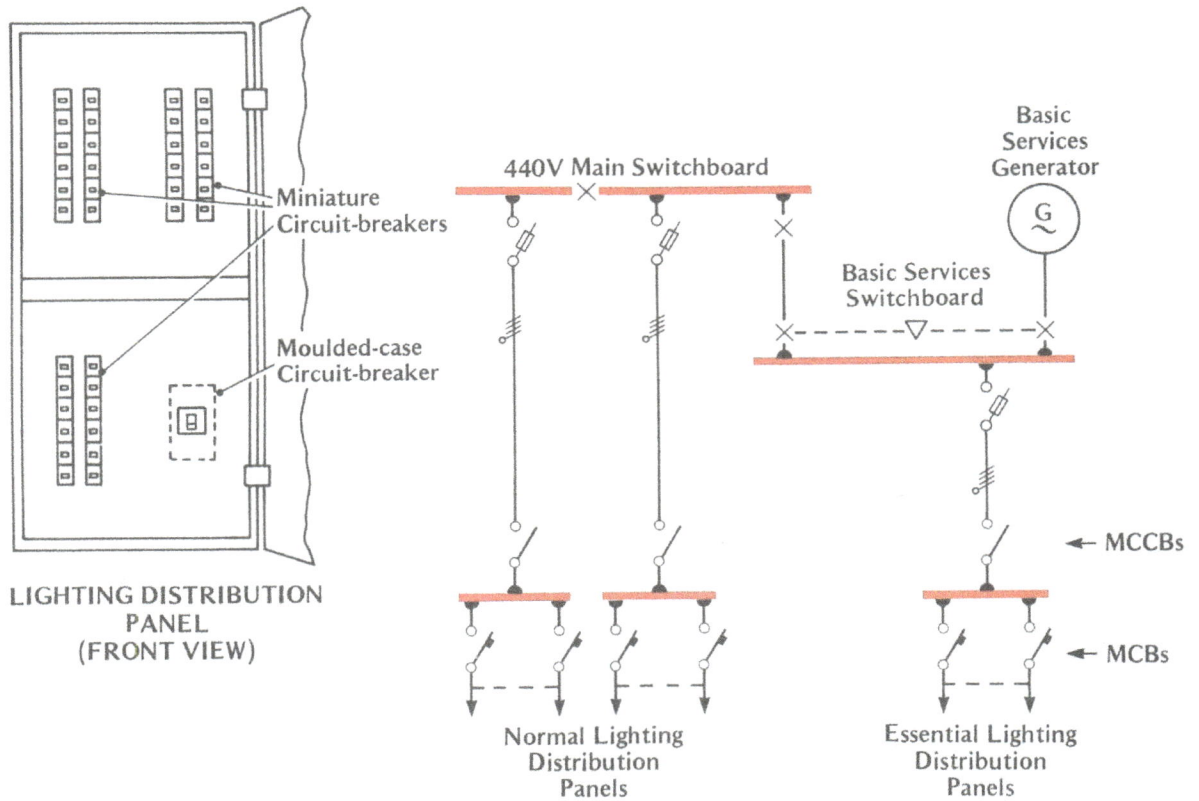

FIGURE 5.7
TYPICAL LIGHTING DISTRIBUTION

The incomer MCCB units, which in other applications are fitted with thermal and electromagnetic overcurrent protection, usually have these devices removed in lighting panels and act merely as incoming isolators. Protection is afforded by the individual distribution MCBs.

Power for the emergency lighting fittings with their individual built-in battery back-up, and also for some of the floodlights, is derived in a similar manner but usually from basic (essential) lighting distribution panels (BLs) supplied from the basic services switchboard, where they have the back-up support of the basic services diesel-generator if normal supply is lost. The BL panels are in appearance exactly like the normal LP panels. Figure 5.7 also shows, on the right, their source of power.

Incomer
2-pole MCBs

22 Distribution
1-pole MCBs

FLOOR LIGHTING PANEL

440V Main Switchboard

Basic
Services
Generator

Basic Services
Switchboard

Living Quarters Switchboard

(R)   (Y)   (B)

MCCBs

MCBs

GLP   1LP   2LP

Floor Lighting Panels

Other
Panels

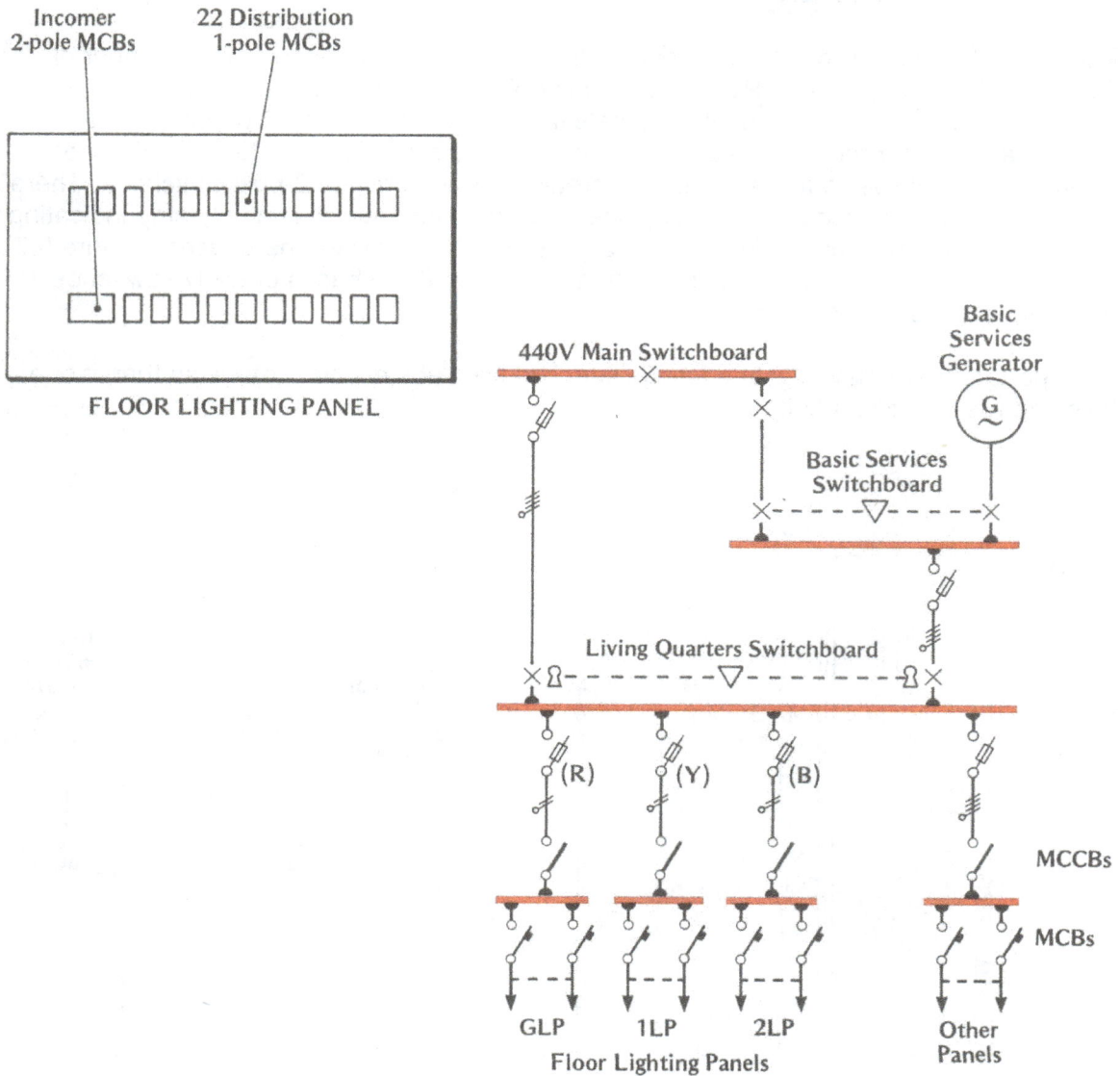

FIGURE 5.8
LIVING QUARTERS LIGHTING DISTRIBUTION

Power for lighting and convenience sockets in certain platform Living Quarters is derived from the Living Quarters switchboard (which has an alternative supply from the diesel-generator supported basic supplies switchboard) through miniature circuit-breakers in small sub-distribution panels on each floor of the Living Quarters. These panels are termed 'GLP', '1LP' and '2LP' for the ground, first and second floors respectively. Each is supplied at 250V single-phase, a different phase being used for each floor. Figure 5.8 shows one of the panels and their source of power.

5.10   FLOODLIGHTS

Floodlights are extensively used for area lighting both on the upper decks of platforms and onshore installations. They are usually in the 200W to 500W range and are provided in a reflector unit which is usually weatherproof.

Floodlights are made for various powers and may be of the tungsten (incandescent) type or discharge tube type. The following are typical of those found on offshore platforms:

- 500W tungsten (incandescent)
- 250W high-pressure mercury vapour (discharge)
- 400W SON/T wide-beam sodium vapour (discharge)
- 2 x 400W SON/T wide-beam sodium vapour (discharge)
- 2 x 400W SON/T narrow-beam sodium vapour (discharge)

All these fittings can be, and offshore usually are, constructed as flame-proof, and all offshore units are powered at 250V single-phase.

The sodium-vapour lamps of the 'SON/T' type are so-called 'high-pressure' discharge lamps. The colour of these lamps is much whiter than the yellow of the normal low-pressure sodium lamps.

5.11   HELIDECK LIGHTING

The helideck on most platforms is situated on the top of the living accommodation module.

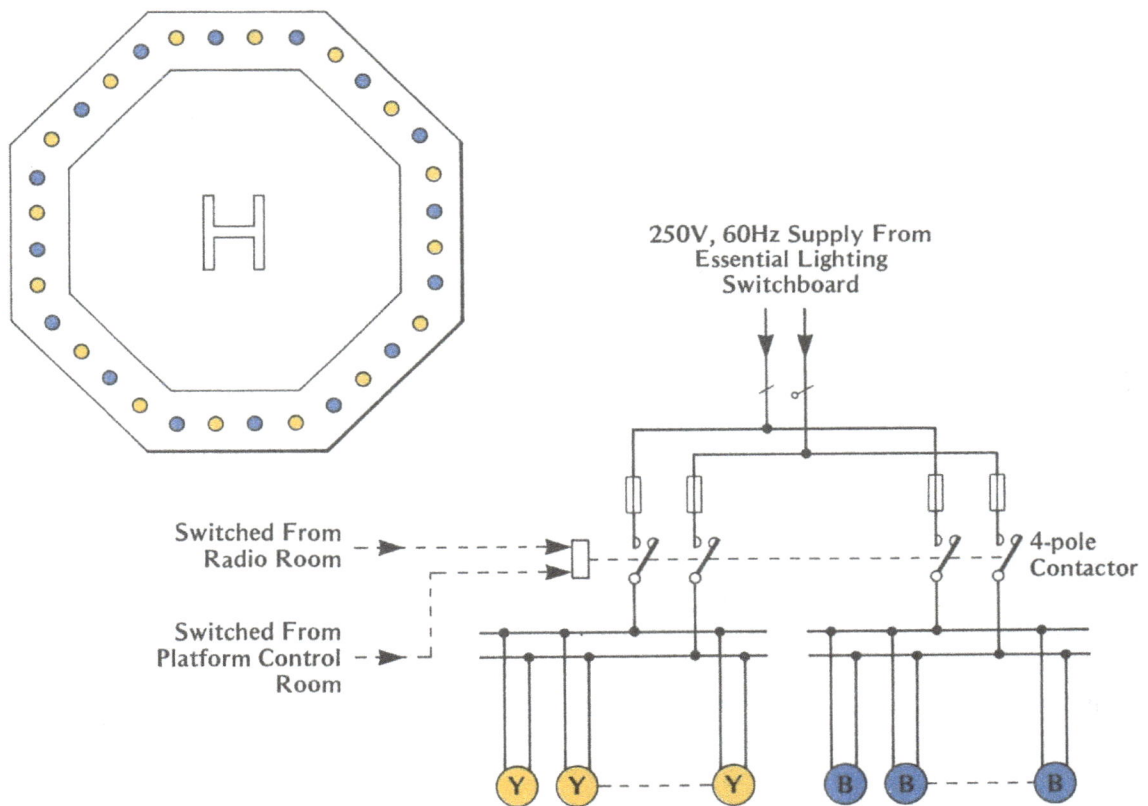

FIGURE 5.9
HELIDECK PERIMETER LIGHTING

Helidecks are marked by a ring of about 30 perimeter lights, alternately blue and yellow, spaced at not more than 3m apart, as shown in Figure 5.9. The fittings are flush-mounted in and just proud of the deck and contain 25W or 40W lamps with coloured lenses.

Power is taken at 250V single-phase usually from an essential lighting distribution switchboard, which in turn takes its supply from the basic services switchboard backed by the basic services diesel-generator. On some platforms it is switched through a 4-pole contactor located in the Plant Room on the second floor of the Living Quarters (just under the helideck itself). This contactor is remotely controlled by switches in the platform Control Room and in the Radio Room.

The yellow and blue lights are on separate circuits, individually fused, so that a fault on one of them which causes the fuse to blow will not leave the helideck unmarked.

The system shown in Figure 5.9 may be varied on some platforms. For instance, on some it is powered from the special emergency switchboard in the Living Quarters Switchroom and is switched directly from there without the intervening contactors.

Other helideck illumination is provided by floodlights which are part of the platform's lighting system. The windsock is also illuminated by floodlight.

5.12   AIRCRAFT OBSTRUCTION LIGHTS

Regulations require that all obstructions exceeding 10m above the upper deck be illuminated during hours of darkness or in poor visibility by red obstruction lights at vertical intervals of not more than 10m, and with all-round and upward visibility.

On offshore platforms such obstructions usually consist of the drilling derrick, flare tower, crane jibs and king-posts. The derrick and flare tower (bottom part) are marked with red lanterns on diametrically opposite corners of the square structure at 10m vertical intervals. However, it is not practical to use lanterns on the upper parts of flare towers, or on flare booms, because of the intense heat of the flare. Those parts are illuminated instead from below by floodlights which are switched with the obstruction lights.

Cranes are marked with one obstruction light at the end of the jib and with others along it as necessary to meet the 10m regulation. The king-post (if there is one) is usually marked by a light at the top, as it projects above the jib when the latter is lowered.

The lanterns contain 60W lamps with red lenses, all powered direct from one of the platform's high-security 440/250V switchboards. The regulations do not require battery support in the event of complete a.c. power failure, and none is provided.

The lights are automatically switched on and off by a sunswitch, but usually there is also a manual override switch in the Radio Room.

CHAPTER 6 CATHODIC PROTECTION

6.1   THE CORROSION PROBLEM

Steels used in platform and pipework construction are not pure materials. Although iron (Fe) forms the major part, steel contains a certain quantity of carbon (C) up to about 1%. The amount of carbon determines the type of steel - for example from mild steels up to hard tool steels. The carbon takes the form of iron carbide (Fe_3C), and, because it is not spread quite evenly through the steel, there are adjacent surface areas of pure iron and iron carbide throughout the steel structure. These individual areas are minute in size and are very close together - a matter of thousandths of an inch.

Iron carbide has a lower potential than pure iron when in the presence of a conducting electrolyte (such as seawater or damp soil). That is to say, a mass of iron/iron-carbide 'cells' are formed in the electrolyte, with the pure iron areas positive and the iron-carbide areas negative.

Each cell, which can be regarded as a primary battery, is short-circuited by the structure metalwork, as shown in Figure 6.1.

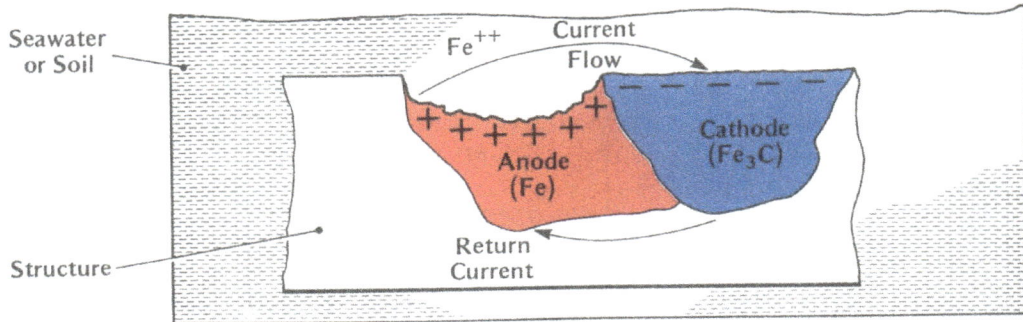

FIGURE 6.1
CORROSION MECHANISM

At the iron surface some of the electrons surrounding the iron atoms break away leaving the atoms of iron positively charged (Fe^{++}). These are called iron 'ions', and they are, in fact, positively charged atoms of the metal. They travel through the sea or soil down the potential gradient from the positive iron areas to the negative iron-carbide areas, forming an electric current through the sea or soil as they do so. This current returns through the main structure metal conducting path.

The positive areas from which current flows into the sea or soil are called 'anodes' (or anodic areas), and the negative areas into which the current flows are 'cathodes' (or cathodic areas).

As the iron metal is removed, atom by atom, from the anode and deposited at the cathode, a corrosion pit is formed as shown in Figure 6.1. Hundreds of thousands of such pits are formed, which appear as general corrosion of the structure member or pipe.

This process can be slowed by painting or by other forms of encapsulation, but it reappears increasingly as the covering wears off. On the other hand marine growth may tend to inhibit the process with time, but no reliance can be placed on it.

6.2   THE CURE

Since corrosion is due to the outflow of iron atoms carried by the sea or soil current, it can be reduced or even prevented if such currents could be stopped or reversed, so causing all the submerged or buried metal parts to receive current instead of giving it out - that is, by making all areas cathodic.

This can be done by placing other, independent electrodes in the sea or soil near the structure and causing them to force current into the structure.

There are two quite distinct methods of achieving this, known as the 'sacrificial anode' and the 'impressed-current' systems. Either method may be used on an offshore installation and on either type of platform, steel-jacket or concrete; in the latter the well risers and their guide-tube structures still have to be protected. The two methods are shown in Figures 6.2(b) and (c).

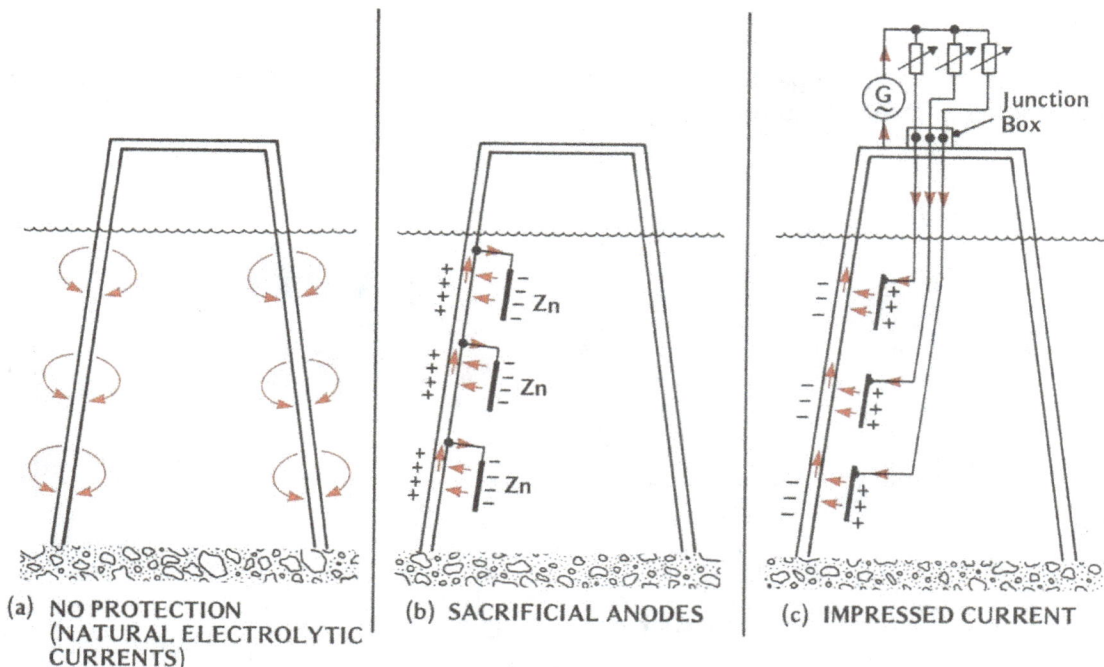

FIGURE 6.2
CATHODIC PROTECTION

Figure 6.2(a) shows an unprotected offshore structure as already described. Myriads of 'seacells' are formed on the surfaces of the structure, causing loops of current to flow into the sea from the positive, anodic areas into the negative, cathodic areas.

The following paragraphs describe the protection systems as applied to an offshore platform. The same methods, however, are used with an onshore installation where, instead of being immersed in seawater, metal is buried in, or stands upon, soil, which always contains water.

6.3   SACRIFICIAL ANODE SYSTEM

The system is shown in Figure 6.2(b). An electrode, called an 'anode', is placed in the sea near the metal to be protected. It is made of a material which is 'base' relative to iron; this means that, in the seawater electrolyte, the anode material becomes negative relative to the iron. Zinc is usually the material chosen, and it is in the form of an array of plates placed near to, and dispersed at points all along, the metal to be protected.

The zincs are each metallically connected or 'bonded' locally to the exposed steelwork at the many dispersed points. The zinc and iron masses, with the seawater between acting as an electrolyte, each form a sea-cell, with the iron positive and the zinc negative. This cell is short-circuited by the bonding, which causes a current to flow from positive to negative through the structure and the bond, as indicated in Figure 6.2(b). It flows from the positive iron, through the low-resistance bonding path, into the negative zinc, and thence back to the iron again through the conducting seawater in between, just like an ordinary short-circuited battery. This sea current is in the opposite direction to the natural electrolytic currents from the anodic areas shown in Figure 6.2(a), and may be considered to swamp the small current loops at the structure.

If the anodes chosen are of sufficient surface area, their current can be made to cancel the natural electrolytic currents exactly; if made larger, they will reverse it. As the current from the zinc anode is accompanied by continuous loss of metal, the anodes eventually become too small to be effective and must be replaced. For this reason they are called 'sacrificial anodes'. Their effectiveness must be periodically monitored to determine when replacement becomes necessary. A life of 10 years is envisaged.

It should be noted that the sacrificial anode system requires energy required to produce the current is derived from the material, which disintegrates when that energy is given up no external power source. The chemical energy of the anode.

Typical figures for a steel-jacket platform are:

Cathode area (structural steel, well risers and foundations) to be protected	84 000m^2
Anode arrays (55 x 0.15m^2)	8.3m^2 (0.01%)
Current per anode array	29A
Total protection current	1 595A

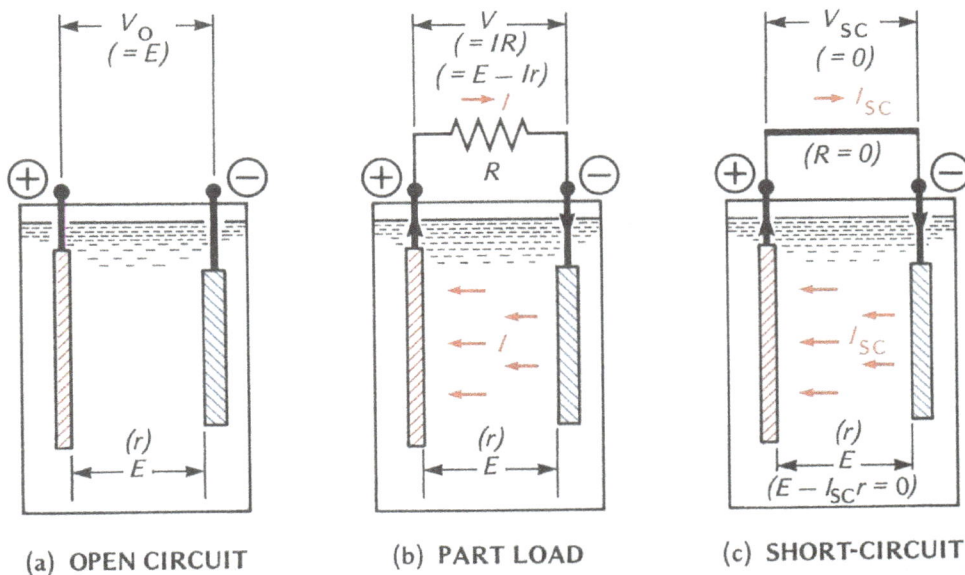

(a) OPEN CIRCUIT   (b) PART LOAD   (c) SHORT-CIRCUIT

FIGURE 6.3
INTERNAL CURRENT IN A BATTERY

To illustrate the effect of anodes of this type, consider first what goes on **inside** an ordinary primary battery cell. Figure 6.3 shows a single two-metal primary battery cell which, in (a), is on open circuit. An emf E is developed between the plates and appears as a voltage between the cell terminals. But no current flows from the positive to the negative terminal, and there is therefore no return current inside the cell from the negative to the positive plate.

Now apply an external load of resistance R as in (b). A current I flows externally through the resistance R from the positive to the negative terminal and returns internally through the electrolyte, path-resistance r, from the negative to the positive plate. The terminal voltage V is reduced from the open-circuit emf E by the internal drop in the cell, so that

$$V = E - I_r$$

If now the ohmic value of the external load is reduced to zero as in (c), $R = 0$ and the cell becomes short-circuited. The load current rises to the short-circuit value I_{sc} limited now only by the internal resistance, so that $E = I_{sc}r$ and $V = 0$. The internal current in the cell from the negative to the positive plate is then also I_{sc}.

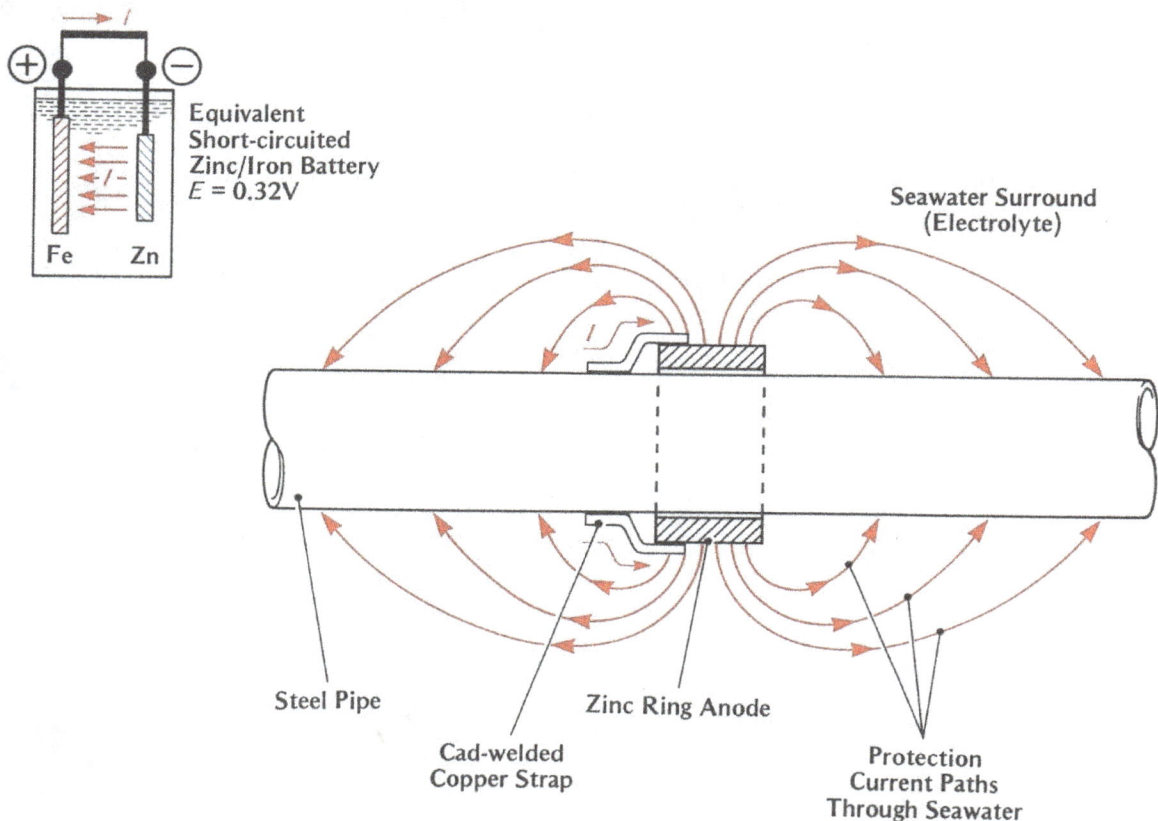

FIGURE 6.4
PRINCIPLE OF SACRIFICIAL ANODE PROTECTION

Consider now Figure 6.4 which shows a steel undersea oil riser pipe with a circular, collar-type zinc anode placed around it and bonded direct to the pipe at several points by cad-welded straps. The steel pipe and the zinc collar form an iron/zinc primary cell with the surrounding seawater acting as electrolyte. The welded straps form the 'external' short-circuit between the zinc and iron.

Normally one thinks of a 'cell' as being a container filled with electrolyte, with two plates immersed in it and brought out to external terminals in air, as inset in Figure 6.4. If these terminals are short-circuited, the situation is exactly as in Figure 6.3(c), the cell having a developed iron/zinc emf $E = 0.32$ volts (see next paragraph).

Different metals immersed in an electrolyte have differing electromotive force potentials, as the examples in the following table show:

	Metal	**emf (volts)**
	Magnesium	-2.37
	Aluminium	-1 .66
BASE ↑	Zinc	-0.76
	Iron	-0.44
	Lead	-0.13
	Copper	+0.34 to +0.52
NOBLE ↓	Silver	+0.80
	Gold	+1.58 to +1.68

These potential voltages reflect the energy used in the refining of the metals. Those needing the greatest energy have the largest negative potentials and corrode most easily, being the least stable and reverting most easily to their original state; these are the 'base' metals. Those needing least energy for refining have positive potentials, are the most stable and have least tendency to corrode; these are the 'noble' metals.

Thus a cell containing zinc and iron, with potentials of -0.76V and -0.44V respectively, has a potential difference, or cell emf, of 0.32V, as stated above, the iron being positive to the zinc.

If the main part of Figure 6.4 is studied, it can be seen that it is, in fact, similar to the inset. The surrounding seawater is the electrolyte and is the 'inside' of the cell, and the welded straps are the 'external' short-circuit. The internal (or return) short-circuit current flow I_{sc} due to the same iron/zinc emf of 0.32V, is from zinc to iron and passes through the surrounding sea, as shown by the red lines. The current from the zinc collar enters into the adjoining iron of the pipe, being densest near the collar. With current thus flowing into the pipe, this is equivalent to **electrons** passing from iron to zinc, and the **ions** (positively charged metallic zinc particles) moving from zinc to iron.

There is thus no loss of iron material - that is to say, the pipe is protected against corrosion - but the zincs are gradually wasted and must eventually be replaced; they are 'sacrificial', as explained in para. 6.3.

As previously stated, no external power source is needed to maintain these protective currents, as in the 'impressed-current' system described below. The energy is derived chemically from the breakdown of the zincs.

Although Figure 6.2(b), in order to simplify the explanation, shows the zincs 'stood off' from the structure but bonded to it, in practice the zincs are permanently fixed to the structure itself (like the collar anodes to the pipes) and are bonded to it by welded straps. The mechanism of current protection is then similar to that described for the collar anodes, the surrounding seawater forming the 'inside' of the cell and the return current path.

6.4   MONITORING OF SACRIFICIAL ANODE SYSTEM

With the sacrificial anode system the protection current is determined solely by the anode area available and, once installed, cannot be controlled (although it may alter slowly with time as paint cover breaks up, or with marine growth, or with the shrinking anodes as they decay). It is therefore necessary to monitor the effectiveness of the protection to decide whether divers are required for maintenance or replacement of anodes.

The normal electrolytic currents which flow out from an unprotected structure cause that structure to be at a potential a little below that of the sea. As protection lessens for any reason, but particularly with loss of anode surface area, the protective current falls off and the structure potential gradually rises from its ideal protected level of approximately -200mV. Therefore measurements of structure or well-riser potentials at various points relative to the sea give a measure of the protection being given at that instant. The results are logged and plotted; these will clearly show any deterioration and will indicate when maintenance or replacement of anodes is necessary.

The effectiveness of the cathodic protection is measured by means of 'reference cells'. These are small silver/silver-chloride half-cells, mounted near the ironwork to be protected but away from any anodes and their current flow. Several reference cells are placed around the platform structure in the case of steel-jacket platforms, or around the well risers and their mounting structures in the case of concrete platforms. The sea-cell formed between the silver of the reference cell and the nearby iron of the structure develops a natural and constant emf of approximately 600mV, the silver end being positive.

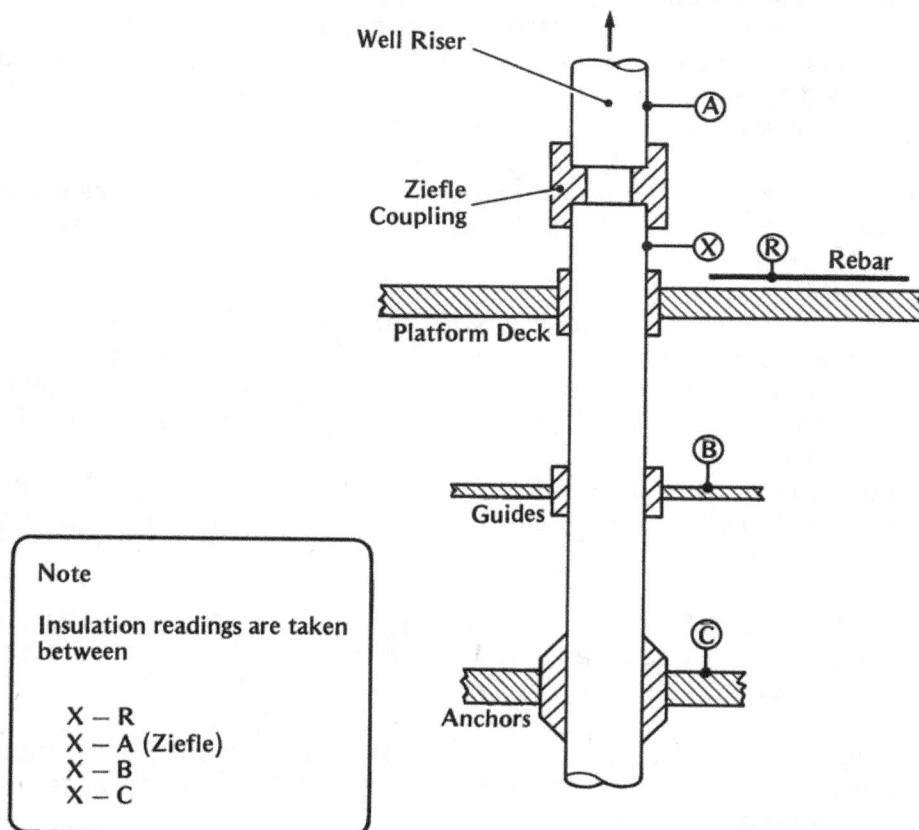

FIGURE 6.5
PIPELINE INSULATION (CONCRETE PLATFORMS)

The reference cell is used to monitor the anode current. When the structure is receiving current from the anode, its potential is depressed below that of the sea, and full protection is assured when that depression is maintained at -200mV. Added to the 600mV of the reference cell, the voltage between the silver electrode and the iron structure will then measure approximately 800mV.

The monitoring equipment measures the potential between the reference cell and the iron structure and notes any departure from the ideal 800mV. With the sacrificial anode system no action can be taken to correct the position other than to inspect and, if necessary, to replace the affected anode by use of divers.

The principle of cathodic protection of platforms standing in a sea environment applies equally to shore pipelines buried in soil; the sea-path for the protective current is simply replaced by the conducting path in the damp soil.

In a concrete platform, while protection current is flowing in the well risers, they are kept at a small potential below that of the sea, and therefore of the platform's deck structure. It is therefore necessary to insulate them as well as their guides from the deck structure - or actually from the platform 'rebar', which is an earth-rail running through the platform. Similarly the underwater part of a riser must be insulated from its top end which goes on into the platform. This is achieved by a special insulating 'Ziefle' coupling in each riser, as shown in Figure 6.5.

Maintenance of this insulation is important, as loss of it could affect the protection given. Also accidental contact with an insulated section, or bad bonding connections, must be avoided because it could give rise to sparking with the large d.c. currents involved, with risk of fire. The latter must be guarded against by regular inspection, and the general insulation by regular routine monitoring.

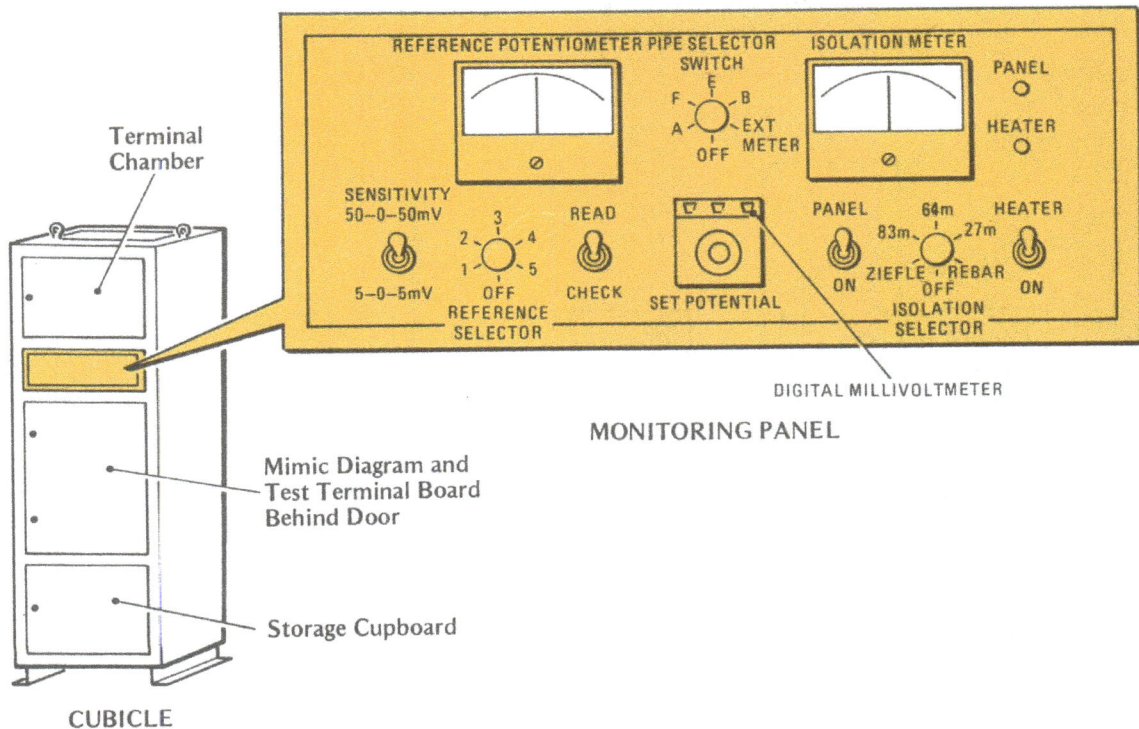

FIGURE 6.6
SACRIFICIAL ANODE MONITORING

Different manufacturers provide different monitoring arrangements, but that shown in Figure 6.6 is typical. Selector switches pick out the well riser or pipeline to be monitored, and further selector switches determine which points on the selected pipe are to be monitored.

Potential readings between the reference cells and each test point are obtained by setting a dial to the desired potential; any departure from it will show as under-protection or over-protection on a centre-zero voltmeter scaled -50, 0, +50mV.

A separate so-called 'Isolation Meter' (actually an insulation meter) is connected by a selector switch to read insulation as a millivolt drop across the Ziefle couplings or other points of insulation for each pipe. The reading should ideally be central. If a large deviation is observed, the area of insulation failure must be inspected.

6.5   IMPRESSED-CURRENT SYSTEM

The impressed-current system is shown in Figure 6.2(c). Unlike the sacrificial anode system, it uses an external power source to provide the protection current. This current is applied in a direction opposite to that of the natural outward electrolytic currents of Figure 6.2(a) and can be regulated at will. Hence the name 'impressed current'.

Anodes are used as in the first system, but the material is not important as they hardly decompose; columbium wound with platinum wire is often used. The anodes are maintained positive relative to the structure, as shown in Figure 6.2(c), by a d.c. supply (in the figure a d.c. generator is shown). In this case the anodes are true 'anodes', being positive and not negative as in the sacrificial method. The structure itself is the 'cathode'.

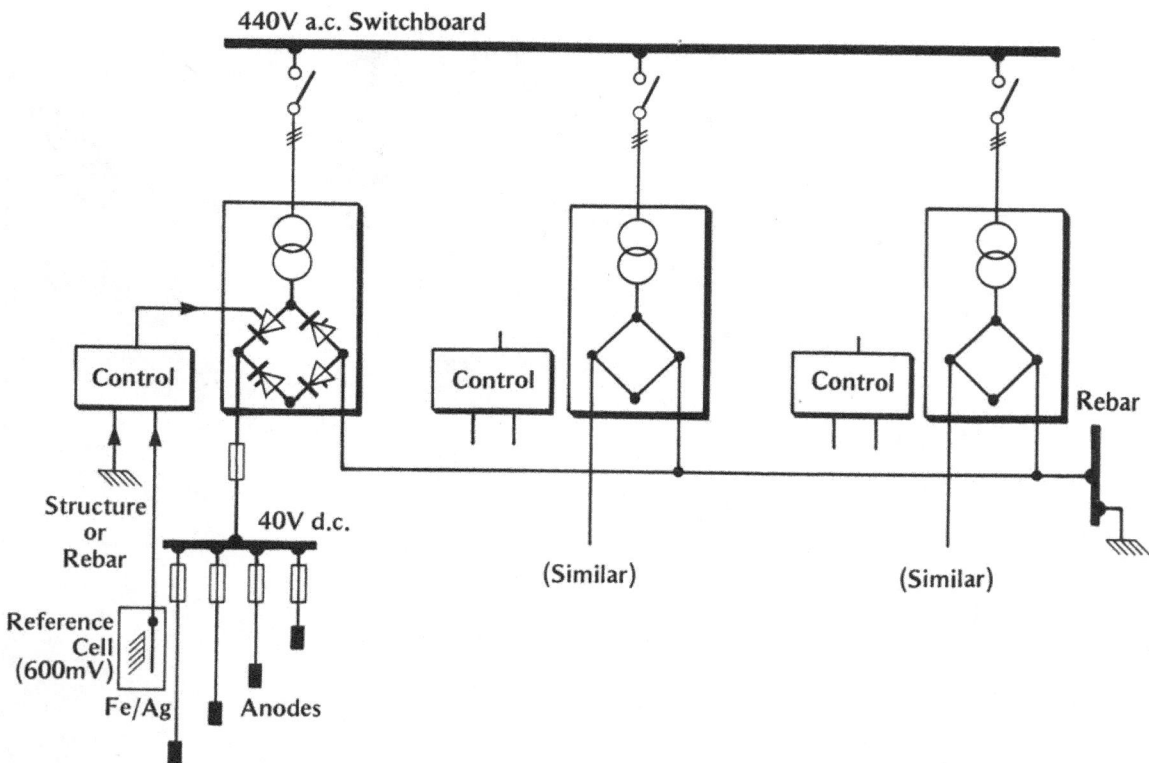

FIGURE 6.7
IMPRESSED CURRENT POWER SYSTEM

6.6   CONTROL OF IMPRESSED CURRENT

The effectiveness of the cathodic protection at any instant is measured by means of reference cells, just as in the case of the sacrificial anode system (see para. 6.4). Ideally the potential between the silver and iron should be approximately 800mV.

The reference cell is used to control the anode current. When the structure is receiving impressed current from the anode, its potential is depressed below that of the sea, and full protection is assured when that depression is maintained at -200mV. Added to the 600mV of the reference cell, the voltage between the silver electrode and the iron structure will then measure approximately 800mV, and the control system automatically regulates the current to keep it at that level.

Figure 6.7 shows a typical impressed-current power circuit. A.C. power from one of the 440V platform switchboards is supplied to a number of thyristor-controlled full-wave rectifiers. Each rectifier feeds d.c. current to a group of anodes. The voltage between the appropriate reference cell and the iron structure is used as sensor, and the control circuit triggers the thyristors, and so the d.c. output current, so that each group maintains the sensing voltage at 800mV. There are a number of adjustments and setting controls, as well as test facilities.

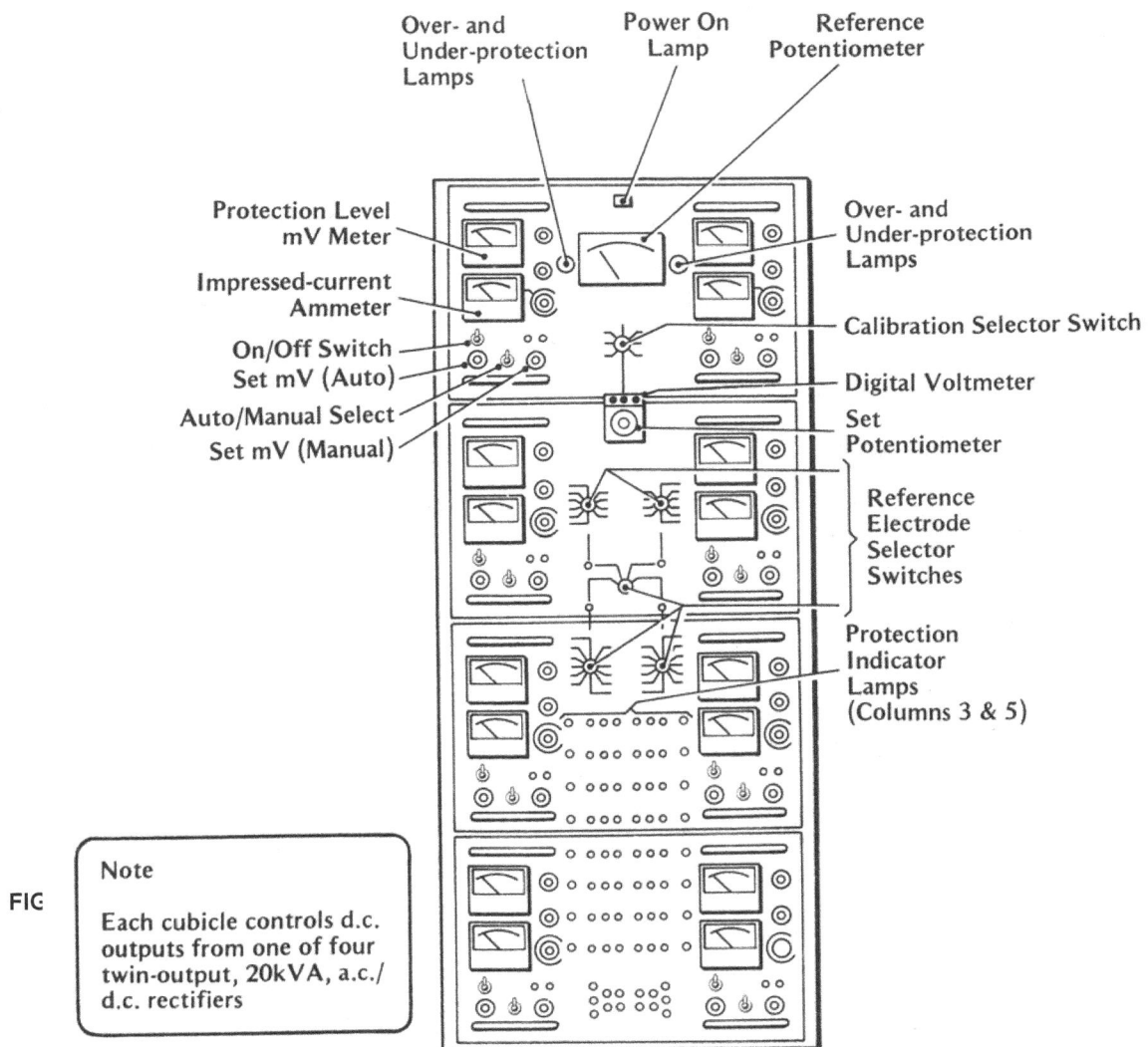

FIGURE 6.8
IMPRESSED-CURRENT SYSTEM - CONTROL PANEL

Where impressed-current protection is used on a concrete platform, the well risers have to be insulated from the platform structure rebar just as with sacrificial anode protection. In this case also the insulation across the Ziefle couplings and at certain other points must be monitored.

Different manufacturers provide different arrangements of control panel; that shown in Figure 6.8 is one of them. It has all the above potential and insulation monitoring controls as well as current-measuring and test facilities.

6.7   MECHANICAL ARRANGEMENT

In a sacrificial anode system each anode plate is bonded locally to the structure which it protects. It can only be replaced by divers. The reference cells however can be withdrawn as described below for the impressed-current anodes.

The impressed-current anodes are long and narrow and can be withdrawn through guide tubes each of which terminates on the platform deck near its allotted position. The connecting cables from all anodes are brought out to junction boxes on the platform, where they are connected in groups and joined to their respective power sources. The reference cells are similarly withdrawn through tubes and connected, through junction boxes, to their control panels.

CHAPTER 7 QUESTIONS AND ANSWERS

7.1   QUESTIONS

1.   What is the disadvantage of a wiring diagram, and in what way is it useful to a maintainer?

2.   What is the most important difference between wiring and schematic diagrams?

3.   In which directions should operational sequence be shown on schematic diagrams?

4.   On Figure 1.4 how is it shown that the basic services switchboard has two alternative inputs (i.e. that they cannot be connected to the switchboard simultaneously)?

5.   What is the standard in the UK for electrical symbols used in diagrams?

6.   What is normally done to safeguard electrical supplies to those services that cannot tolerate even short power interruptions?

7.   What is 'diversity factor'?

8.   Load factor is defined as 'the ratio of the average load over a designated period of time to the peak load occurring in that period'. What factors in system design is it mainly concerned with?

9.   Why is the security of electricity supplies regarded as so important and precautions taken to ensure it?

10.   In the offshore installation shown in Figure 3.1, the LV switchboards are normally run with their section circuit-breakers open. Why?

11.   In an onshore installation, which has no main generators, how is supply security normally ensured?

12.   How is it ensured that the two supplies of Q.11 cannot be connected in parallel?

13.   The 'basic services' generator is intended to supply only a very small load when the main supply fails. How is it normally driven?

14.   A number of important services must be maintained after a complete loss of main power and are therefore battery-supported. Name as many as you can.

15.   In a battery-supported supply with duplicate batteries, what is meant by '100% capacity' batteries?

16.   What d.c. voltage is normally used for switchgear operation and what voltage for other applications?

17.   How is battery support provided for equipment that runs from an a.c. supply?

18.   Why is it normal practice to 'earth' an electricity supply system?

19.   Why is a 3-phase system normally earthed at the neutral?

20.   What are the hazards of an earth fault in an earthed system?

21. How is a compromise made between the voltage-limiting advantage of an earthed system and the disadvantages of the earth-fault currents and high-energy arc?

22. Why are the star-points of transformer secondary windings solidly earthed instead of via resistors?

23. Why are drilling supply systems normally unearthed?

24. What is a 'lux'?

25. What arrangements are made to make a normal twin-tube luininaire suitable for emergency lighting?

26. For about how long will a battery-operated emergency handlamp give adequate light?

27. What switching arrangements are normally made for the incoming supplies to lighting distribution boards?

28. What is the cause of the corrosion of steel structures in seawater or damp soil?

29. Why is painting the steel structure an unsatisfactory method of preventing corrosion?

30. What is the fundamental idea behind the application of cathodic protection as a cure for corrosion?

31. What are the two methods used to provide the protection?

32. What is the principle by which the sacrificial anode system works?

33. What is the advantage of the sacrificial anode system?

34. How can the effectiveness of a sacrificial anode system be monitored?

35. What is the principle by which the impressed-current system works?

36. How is the effectiveness of the impressed-current system monitored and controlled?

37. When sacrificial anodes have deteriorated to the point at which they cease to be effective, what action is necessary?

7.2 ANSWERS

(Figures in brackets after each answer refer to the relevant paragraphs in the text.)

1. Even with a simple system it is difficult to see from a wiring diagram how the system works. With a more complicated system the situation becomes virtually impossible. The main use to a maintainer is checking connections that are already thought to be suspect. (1.2)

2. The 'detached contact' system. (1.3)

3. Left to right, or top to bottom. (1.4)

4. By the inverted triangle and dashed line between the two incomer circuit-breakers. (BS 3939) (Figure 1.3)

5. British Standard (BS) 3939. (1.7)

6. Battery-supported supplies, either d.c. or a.c., are provided. (2.1)

7. The ratio of the arithmetical sum of the individual maximum demands of a number of consumers connected to an electricity supply system to the simultaneous maximum demand of the group. Put simply - the ratio of the total load that **could be** switched on to the total load that **actually is** switched on. (2.3)

8. The prediction of fuel consumption and, thus, the choice of generator driver characteristics. (2.4)

9. In all installations a supply failure is costly because of lost production. In offshore installations it can put lives at risk. (3.1)

10. To reduce the prospective fault level to that which exists with only one transformer. Consult Part 10 'Electrical Protection' for a full explanation of 'fault level'. (3.2)

11. By taking two separate supplies from two different sources in the supply authority's network, either being able to supply the full demand. (3.4)

12. By having a 'two-out-of-three' auto-transfer arrangement between the two incomer circuit-breakers and the bus-section breaker. (3.4)

13. By a diesel engine, often self-starting upon main supply failure. (3.5)

14. Paragraph 3.6 lists: switchgear operation, navigation lights, foghorns, fire and gas detection, internal communications, alarm systems, some radio communications, some instruments. (3.6)

15. Each battery can supply the full load for the specified time. If both batteries are needed for this they are '50% capacity' each. (3.6)

16. Switchgear operation 110V d.c., other applications 24V d.c. (but not all platforms do this). (3.6)

17. By feeding the rectifier/battery output to a d.c./a.c. inverter. The inverter output is normally at mains frequency, but can be at any required frequency. (3.6)

18. If the system is not tied to earth it can become charged to any voltage above (or below) earth with risk of insulation breakdown, particularly if the HV/LV insulation of a transformer should break down. (4.1)

19. This arrangement limits the voltage-to-earth of any phase to the phase voltage, i.e. $1/\sqrt{3}$ of the line voltage. (4.3)

20. Damage to the generator through overheating (if allowed to continue), risk to personnel and possibility of fire by arcing at the point of the fault. (4.4)

21. Resistance earthing. (4.5)

22. There are two reasons: the LV windings of transformers are robust enough to withstand the earth-fault currents; the fault energy is less downstream of a transformer than on the HV side.

23. They are often separate from other platform supplies and fed from 600V diesel-driven generators which cannot be earthed because they are delta-connected. Drillers prefer an unearthed system, as a single earth fault will not trip the supply, leading to possible loss of equipment. (4.8)

24. The SI unit for illumination per unit area, lumens per square metre. (5.1)

25. The luminaire is provided with a battery, charger, inverter, changeover relay and transformer. (5.3)

26. 2½ hours. (5.8)

27. The boards are fed via moulded-case circuit-breakers with the overcurrent trips removed. (5.9)

28. The lack of homogeneity of the steel, resulting in minute adjacent areas of iron and iron carbide. These form primary cells with seawater or soil moisture as the electrolyte. (6.1)

29. Its effect is only temporary: the covering wears off. (6.1)

30. The reversal of the minute electric currents that cause the corrosion. (6.2)

31. 'Sacrificial anode' and 'impressed current'. (6.2)

32. Anodes of zinc ('base' relative to iron) are placed near the structure and bonded to it. The natural cells thus formed reverse the corrosion currents, and the corrosion is transferred to the zinc. (6.3)

33. It needs no external power supply. (6.3)

34. By measuring the potential of various parts of the structure with respect to the sea or soil by means of a reference cell. Its ideal level is about −200mV. (6.4)

35. By applying an externally derived voltage between the structure and an anode. The resultant current is in such a direction as to reverse the corrosion currents. (6.5)

36. By the use of silver/silver chloride 'reference cells'. (6.6)

37. They must be replaced, and this can only be done by divers - a major disadvantage. (6.4)

REFERENCES

Advanced Electrical Technology
H. Cotton, DSc (Pitman)

Alternating Current Electrical Engineering (10th Edition)
Philip Kemp, MSc, CEng, MIEE (Macmillan)

An Introduction to Power Electronics
B. M. Bird & K. G. King (John Wiley)

Applied Electricity (6th Edition)
H. Cotton, DSc (Macmillan)

BS 2769
  'Portable Electric Motor-operated Tools'
  British Standards Institution

BS 3535
  'Safety Isolating Transformers'
  British Standards Institution

BS 7430
  'Earthing'
  British Standards Institution

BS 5345:

  - Selection, Installation and Maintenance
    of Electrical Apparatus for use in
    Potentially Explosive Atmospheres

  - Installation ... of Type of Protection '*d* '
    (Flameproof Enclosures)

  - Installation.., of Type of Protection '*i* '
    (Intrinsically Safe Apparatus and System)

  - Installation ... of Type of Protection '*e* '
    (Increased Safety)

  - Installation ... of Type of Protection '*n* '
    (N-Protection)

  - Installation ... of Type of Protection '*s* '
    (Special)

Cathodic Protection
J.H. Morgan (Leonard Hill)

Electrical Control Engineering, Vols 1 & 2
Poole & Jackson (London Iliffe Books)

Electrical Engineers' Reference Book (13th Edition)
M. G. Say (Butterworth)

Electrical Machines (2nd Edition)
A. Draper, BSc Eng, C Eng, FIFE (Longman)

Electrical Measurements and Measuring Instruments
E. W. Golding and F. C. Widdis (Pitman)

Electrical Technology (5th Edition)
E. Hughes, DSc, PhD, C Eng, MIEE (Longman)

Electric Motor Handbook
E.H. Werninck (McGraw Hill)

Electric Power Systems
B.M. Weedy, PhD (John Wiley)

Electronic Devices and Circuits (2nd Edition)
D.A. Bell (Reston Publishing Co mc)

Electronics: Circuits and Devices
R.J. Smith (John Wiley)

Handbuch fur Explosionsschutz (2nd Edition, 1983)
(Explosion Protection Manual)
Brown Boveri & Cie A/G

Physics and Technology of Semiconductor Devices
Grove (John Wiley)

Principles of Electrical Technology
H. Cotton, DSc (Pitman)

Principles of Inverter Circuits
Bedford & Hoft (John Wiley)

Protective Relays Application Guide
General Electric Co Measurements (GEC)

Standard Handbook for Electrical Engineers (11th Edition)
Fink & Beaty (McGraw Hill)

Storage Batteries (3rd Edition)
G. Smith, C Eng, MIEE (Pitman)

The Power Thyristor and its Applications
Finney (McGraw Hill)

The Lighting of Buildings (2nd Edition)
R.G. Hopkinson, PhD, C Eng & J.D. Kay, AA Dipl, RIBA (Faber & Faber)

Thermocouple and Resistance Thermometry Data
T.C. Limited

Thyristor Physics
Blicher (Springer)

INDEX